HIGHER LEVEL

Physics
for the IB Diploma Programme

Chris Hamper, Emma Mitchell

Published by Pearson Education Limited, 80 Strand, London, WC2R 0RL.

www.pearson.com/international-schools

Text © Pearson Education Limited 2023
Development edited by Martin Payne
Copy edited by Jane Read
Proofread by Kate Blackham and Martin Payne
Indexed by Georgie Bowden
Designed by Pearson Education Limited
Typeset by EMC Design Ltd
Picture research by Integra
Original illustrations © Pearson Education Limited 2023
Cover design © Pearson Education Limited 2023

The right of Chris Hamper and Emma Mitchell to be identified as the authors of
this work has been asserted by them in accordance with the Copyright, Designs
and Patents Act 1988.

First published 2023

25 24 23
10 9 8 7 6 5 4 3

British Library Cataloguing in Publication Data
A catalogue record for this book is available from the British Library

ISBN 978 1 29242 770 6

Printed in Slovakia by Neografia

Acknowledgements

The author and publisher would like to thank the following individuals and
organisations for permission to reproduce photographs, illustrations, and text:

Text extracts relating to the IB syllabus and assessment have been reproduced
from IBO documents. Our thanks go to the International Baccalaureate for
permission to reproduce its copyright.

The "In cooperation with IB" logo signifies the content in this
textbook has been reviewed by the IB to ensure it fully aligns with
current IB curriculum and others high-quality guidance and support
for IB teaching and learning.

KEY (t – top, c – center, b – bottom, l – left, r – right)

Images:

123RF: Dolgachov xxiit, Janos Gaspar 89, Gary Gray 152, Sergiy Kuzmin 240,
Unkreatives 477l; **Alamy Stock Photo:** Photo Researcher/Science History
Images 23, O.Furrer/F1online digitale Bildagentur GmbH 45, Julie Edwards 54,
Bill Grant 91, Photo Researchers/Science History Images 180, Yuri Kevhiev 251,
Noam Armonn 268, Dpa picture alliance archive 281, PhotoSpirit 309c, Jim Kidd
310bl, Aerialarchives 396, James Brittain-VIEW 450, Matteo Omied 540, Iuliia
Bycheva 542; **Algoryx Simulation AB:** Algodoo® xxiic; **Audacity:** Audacity®
software is copyright © 1999-2021 Audacity Team. The name Audacity® is a
registered trademark. 342b, 343; **CERN:** © 2015-2022 CERN 420; **Chris Hamper:**
Chris Hamper xxvi, xxvii, 9; **Getty Images:** SlobodanMiljevic/iStock xxiii, Ray
Bradshaw/Moment 2, Christophe Boisvieux/The Image Bank Unreleased 4, Jody
Amiet/AFP 29, Jeffrey Coolidge/Stone 63, Lionel Bonaventure/AFP 88, Rachel
Husband/Photographer's Choice RF 101, DAJ/Amana images 110, SeongJoon
Cho/Bloomberg 150, Vladimir Vladimirov/E+ 196, Hans Neleman/Stone 216,
Cavan Images 236, Joe McNally/Hulton Archive 264, Antonio Ribeiro/Gamma-

Rapho 266, Tomazl/E+ 289, Westend61 336, Coopder1/iStock 342t, Golf was here/
moment 370, Erik Herrera/EyeEm 372, Mixetto/E+ 433, Sellmore/Moment 470,
BanksPhotos/E+ 488, Bettmann 517, Virtualphoto/E+ 550t; **Giovanni Braghieri:**
Giovanni Braghieri® 572; **Microsoft Corporation:** Microsoft Excel 2021,
Used with permission by Microsoft Corporation. 270, 503; **Pearson Education
Limited:** Naki Kouyioumtzis 7; **NASA:** NASA; ESA; G. Illingworth, D. Magee,
and P. Oesch, University of California, Santa Cruz; R. Bouwens, Leiden University;
and the HUDF09 Team 356, NASA, ESA, CSA, and STScI 528; **PASCO scientific:**
15, 204, 206, 207, 228, 375tl, 375tr; **Paul Falstad:** 314, **Pixabay:** Alexas_Fotos/
Pixabay 308, Image by Olaf from Pixabay 424; **Science Photo Library:** Tek Image
xvi, Edward Kinsman 21, Philippe Plailly 40, Mark Clarke 55, Kaj R. Svensson 71,
Bernhard Edmaier 80, Cern 117, Sheila Terry 124, 551c Ted Kinsman 159,325,
Mark Sykes 166,361, Dr. Arthur tucker 168, John Beatty 169, Charles D. Winters
199, 211, Carlos Munoz Yague/Look At Sciences 250, Martin Bond 290, Alex
Bartel 310b, Andrew Lambert Photography 313c, 313b, 344, 345, 410, 477r,
Erich Schrempp 316, David Parker 317, Science Photo Library 388,425, 551t,
553, Steve Horrell 452, Dept. Of Physics, Imperial College 460b, Giphotostock
478, Goronwy Tudor Jones, University of Birmingham 496, Seymour 530, John
Sanford 534, Physics Dept., Imperial College 537, Noirlab/Nsf/Aura 438, Christian
Darkin 438, B. Mcnamara (University of Waterloo)/NASA/ESA/STSCI 558tl, K. H.
Kjeldsen 559; **Shutterstock:** Daniel M. Silva 48, Peter Gudella 103, DeepSkyTX
189, EpicStockMedia 288, Orxy 309b, Makitalo 310c, XiXinXing 338, Roxana
Gonzalez 339, Yellow Cat 440, FoodAndPhoto 502, Peter Sobolev 521, Travel-Fr
550b, Apiwan Borrikonratchata 551b, Tatiana Popova 552, Nightman1965 555,
Melima and Georgi Fadejev 558bl

Text:

Albert Einstein: Quote by Albert Einstein 550, **Arthur Schopenhauer:** Quote
by Arthur Schopenhauer 550; **British Physics Olympiad:** © 2017 British Physics
Olympiad 62,87,179,287,307,335,369,392,416,487,516; **Carl Sagan:** Quote
by Carl Sagan 559; **Carlo Rovelli:** Carlo Rovelli 555; **Galileo Galilei:** Quote
by Galileo Galilei 558; **Insight Press:** Epstein, Lewis C.. Thinking physics is
Gedanken physics. San Francisco: Insight Press, 1989. 62, 87, 116, 335, 355, 369,
392, 416, 432, 449, 487, 515, 516, 549; **Isaac Asimov:** Quote by Isaac Asimov
559; **Lord Kelvin:** Lord Kelvin 554; **Robert Oppenheimer:** Quote by Robert
Oppenheimer 552; **T. H. Huxley:** Quote by T. H. Huxley 551.

Contents

Syllabus roadmap

The aim of the syllabus is to integrate concepts, topic content and the NOS through inquiry. Students and teachers are encouraged to personalize their approach to the syllabus according to their circumstances and interests.

Skills in the study of physics should be integrated into the syllabus content.

A. Space, time and motion	B. The particulate nature of matter	C. Wave behaviour	D. Fields	E. Nuclear and quantum physics
A.1 Kinematics* A.2 Forces and momentum* A.3 Work, energy and power* A.4 Rigid body mechanics*** A.5 Galilean and special relativity***	B.1 Thermal energy transfers* B.2 Greenhouse effect* B.3 Gas laws* B.4 Thermodynamics*** B.5 Current and circuits*	C.1 Simple harmonic motion** C.2 Wave model* C.3 Wave phenomena** C.4 Standing waves and resonance* C.5 Doppler effect**	D.1 Gravitational fields** D.2 Electric and magnetic fields** D.3 Motion in electromagnetic fields* D.4 Induction***	E.1 Structure of the atom** E.2 Quantum physics*** E.3 Radioactive decay** E.4 Fission* E.5 Fusion and stars*

* Topics with content that should be taught to all students

** Topics with additional HL content

*** Topics with content for HL students only

Authors' introduction to the third edition

Welcome to your study of IB Diploma Programme (DP) Higher Level (HL) physics! This textbook has been written to match the specifications of the new physics curriculum for first assessments in 2025 and gives comprehensive coverage of the course.

Content

The book covers the content that is common to all DP physics students and the additional material for HL students.

HL The additional HL material is labeled as such, and the sequence of the chapters matches the sequence of the subject guide themes, with textbook chapter numbering matching the guide topic numbering.

Each chapter starts with a caption for the opening image, the Guiding Questions, an introduction (which gives the context of the topic and how it relates to your previous knowledge) and the Understandings for the topic. These will give a sense of what is to come, with the Understandings providing the ultimate checklists for when you are preparing for assessments.

Guiding Questions

How can the motion of a body be described quantitatively and qualitatively?

How can the position of a body in space and time be predicted?

The text covers the course content using plain language, with all scientific terms explained. We have been careful to apply the same terminology you will see in IB examinations in worked examples and questions.

Linking Questions that relate topics to one another can be found throughout, with a hint as to where the answer might be located. The purpose of Linking Questions is to connect different areas of the subject to one another – between topics and to the Nature of Science (NOS) more generally. These questions will encourage an open mind about the scope of the course during your first read through and will be superb stimuli for revision.

How does the motion of a mass (A.1) in a gravitational field (D.1) compare to the motion of a charged particle in an electric field (D.2)?

Each chapter concludes with Guiding Questions revisited and a summary of the chapter, in which we describe how we sought to present the material and what you should now know, understand and be able to do. The Guiding Question revisited bulleted lists are available as downloadable PDFs from the eBook to help you with revision.

Guiding Questions revisited

How can we use our knowledge and understanding of the torques acting on a system to predict changes in rotational motion?

If no external torque acts on a system, what physical quantity remains constant for a rotating body?

Aims

Using this textbook as part of your course will help you meet these IB DP physics aims to:

- develop conceptual understanding that allows connections to be made between different areas of the subject, and to other DP sciences subjects
- acquire and apply a body of knowledge, methods, tools and techniques that characterize science
- develop the ability to analyze, evaluate and synthesize scientific information and claims
- develop the ability to approach unfamiliar situations with creativity and resilience
- design and model solutions to local and global problems in a scientific context
- develop an appreciation of the possibilities and limitations of science
- develop technology skills in a scientific context
- develop the ability to communicate and collaborate effectively
- develop awareness of the ethical, environmental, economic, cultural and social impact of science.

Nature of physics

Physicists attempt to understand the nature of the Universe. They seek to expand knowledge through testing hypotheses and explaining observations, and by a commitment to checking and re-checking in a bid to set out basic principles. 'Doing physics' involves collecting evidence to reach partial conclusions, creating models to mediate and enable understanding, and using technology.

Physics flowchart.

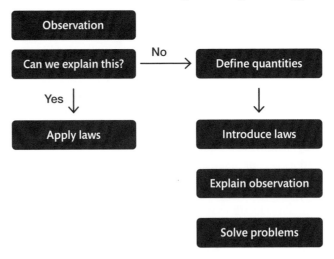

You will find examples of the nature of physics throughout this book, such as the scattering experiments in E.1, the speed of light in A.5, the relationships between pressure, volume and temperature in B.3, and detecting radiation in E.3.

Nature of Science

Throughout the course, you are encouraged to think about the nature of scientific knowledge and the scientific process as it applies to physics. Examples are given of the evolution of physical theories as new information is gained, the use of models to conceptualize our understanding, and the ways in which experimental work is enhanced by modern technologies. Ethical considerations, environmental impacts, the importance of objectivity and the responsibilities regarding scientists' code of conduct are also considered here. The emphasis is not on memorization, but rather on appreciating the broader conceptual themes in context. We have included some examples but hope that you will come up with your own as you keep these ideas at the forefront of your learning.

The following table provides a comprehensive list of the elements of the Nature of Science that you should become familiar with.

Element	Details
Making observations	Using the human senses, or instruments, and identifying new fields for exploration.
Identifying patterns and trends	Using inductive reasoning (from specific cases to more general laws) and classification of bodies (in overlapping ways), and distinguishing between correlation (relationships between two variables) and causation (when one variable has an effect on another).
Suggesting and testing hypotheses	Provisional qualitative and quantitative relationships with explanations before experimentation is carried out, which can then be tested and evaluated.
Experimentation	The process of obtaining data, testing hypotheses, controlling variables, deciding the appropriate quantity of data and developing technology that requires creativity and imagination.
Measuring	Recognizing limitations in precision and accuracy, carrying out repeats for reliability, and accepting the existence of and quantifying the random errors that lead to imprecision and uncertainty and the systematic errors that lead to inaccuracy.
Using models	Artificial representations of natural phenomena that are useful when direct observation is difficult, and simplifications of complex systems in the form of physical representations, abstract diagrams, mathematical equations or algorithms, which have inherent limitations.
Collecting evidence	Used to evaluate scientific claims to support or refute scientific knowledge.
Proposing and using theories	Understanding theories (general explanations with wide applicability), deductive reasoning (from the general to the specific) when testing for corroboration or falsification of the theory, paradigm shifts (new and different ways of thinking) and laws (that allow predictions without explanation).
Falsification	Accepting that evidence can refute a claim but cannot prove truth with certainty.
Perceiving science as a shared endeavor	Making use of agreed conventions, common terminology and peer review in the spirit of global communication and collaboration.
Commitment to global impact	Assessing risk to ensure that no harm is done and the ethical, environmental, political, social, cultural, economic and unintended consequences that work may have through compliance with ethics boards and by communicating honestly and clearly with the public.

Learning physics

Approaches to learning

The IB aspires for all students to become more skilled in thinking, communicating, social activities, research and self-management.

In physics, thinking might include:

- being curious about the natural world
- asking questions and framing hypotheses based upon sensible scientific rationale
- designing procedures and models
- reflecting on the credibility of results
- providing a reasoned argument to support conclusions
- evaluating and defending ethical positions
- combining different ideas in order to create new understandings
- applying key ideas and facts in new contexts
- engaging with, and designing, linking questions
- experimenting with new strategies for learning
- reflecting at all stages of the assessment and learning cycle.

High-quality communication looks like:

- practicing active listening skills
- evaluating extended writing in terms of relevance and structure
- applying interpretive techniques to different forms of media
- reflecting on the needs of the audience when creating engaging presentations
- clearly communicating complex ideas in response to open-ended questions
- using digital media for communicating information
- using terminology, symbols and communication conventions consistently and correctly
- presenting data appropriately
- delivering constructive criticism.

The learning you will do socially could involve:

- working collaboratively to achieve a common goal
- assigning and accepting specific roles during group activities
- appreciating the diverse talents and needs of others
- resolving conflicts during collaborative work
- actively seeking and considering the perspective of others
- reflecting on the impact of personal behavior or comments on others
- constructively assessing the contribution of peers.

You will carry out research, in particular during the Internal Assessment, that includes:

- evaluating information sources for accuracy, bias, credibility and relevance
- explicitly discussing the importance of academic integrity and full acknowledgement of the ideas of others
- using a single, standard method of referencing and citation
- comparing, contrasting and validating information
- using search engines and libraries effectively.

And remember that a significant component of learning comes from you. Maybe you have even reflected on your skills while reading these bullet points! How competent are you at these self-management skills?

- breaking down major tasks into a sequence of stages
- being punctual and meeting deadlines
- taking risks and regarding setbacks as opportunities for growth
- avoiding unnecessary distractions
- drafting, revising and improving academic work
- setting learning goals and adjusting them in response to experience
- seeking and acting on feedback.

Inquiry

Combining the approaches to learning above will facilitate your use of the tools in physics: experimental techniques, technology and mathematics. The next chapter specifically highlights some of these tools; the rest can be found throughout the book.

In turn, these tools will enable you to thrive in the inquiry process, which involves exploring and designing, collecting and processing data, and concluding and evaluating. There are opportunities to practice the inquiry process in this book, and the Internal Assessment chapter includes an eBook link to exemplar work. You are also sure to find the collaborative sciences project to be a highlight, with its:

Tools for physics.

- inclusion of real-world problems
- integration of factual, procedural and conceptual knowledge through study of scientific disciplines
- understanding of interrelated systems, mechanisms and processes
- solution-focused strategies
- critical lens for evaluation and reflection
- global interconnectedness (regional, national and local)
- appreciation of collective action and international cooperation.

Learner profile

There is an abundance of ways in which your physics course will support your all-round growth as an IB learner.

Learning attribute	Advice on how to develop
Inquirer	• Be curious, conduct research and try to become more independent. • Ask questions about the world, search for answers and experiment. • Extend your scientific knowledge and engage with existing research.
Knowledgeable	• Explore concepts, ideas and issues, and seek to deepen your understanding of facts and procedures. • Access a variety of resources. • Apply your knowledge to unfamiliar contexts.
Thinker	• Solve complex problems while reflecting on your strategies. • Analyze methods critically and embrace creativity when seeking solutions. • Practice reasoning and critical thinking (testing assumptions, formulating hypotheses, interpreting data and drawing conclusions from evidence).
Communicator	• Accept opportunities to collaborate. • Step out of your comfort zone during group work, for example, by opening discussions or using scientific language. • Listen to others and share your ideas.
Principled	• Take responsibility for your work, promoting shared values and acting in an ethical manner. • Acknowledge the work of others, cite your sources and reduce waste. • To show integrity during data collection, consider all data, including that which does not match your hypothesis.
Open-minded	• Be aware of the existence of different perspectives and models. • Reject or refine your models due to reasoning, deduction or falsification. • Challenge perspectives and ideas.
Caring	• Protect your environment and aim to improve the lives of others. • Choose sustainable practices. • Connect topics to global challenges (like healthcare, energy supply, food production).
Risk-taker	• Seek opportunities for learning and challenge. • Recognize your freedom to try different techniques or methods of learning. • Collect experimental data in a bid to falsify (not just validate) ideas.
Balanced	• Look holistically at your own development and consider how attentive you are to your tasks. • Have a balanced perspective on scientific issues. • Organize your time to avoid negative impacts on the emotional or social aspects of your life.
Reflective	• Consider why and how success is achieved, and how you might change your approach when learning becomes difficult. • Review your strategies, methods, techniques and approaches, for example, using success criteria. • Reflect on your internal network of knowledge.

How to use this book

The book is written according to the following approach, in which we use electric fields as an example.

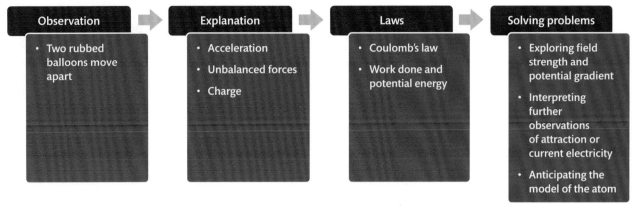

Observation

The aim of the course is to be able to model the physical Universe, so first we must consider a physical process.

A student observes two rubbed balloons moving apart and wonders why they repel. They realize that there must be an unbalanced force. That is the beauty of physical laws; they are always right. The student recognizes a similarity with gravity, which is related to the mass of a body. But gravitational forces are weak and only attractive.

So what is the key property of the body and what is the force? The student does not know, so they have to add something to their model of the Universe.

Explanation

Having studied mechanics and particles, the student has some knowledge of the fundamentals of physics. They know that a body will only accelerate if there is an unbalanced force. We could stop there if this was enough to explain everything, but it is not.

The student reads about a new property, charge. Using what they know about gravitational fields, they expect to learn about field strength (in this case, electric) and wonder if electric forces follow an inverse square law. They carry out an experiment to confirm this.

Laws

Some research reveals that electric forces (like all forces) are vectors, that Coulomb's law applies to point charges, and that moving a charge in an electric field requires a force (so work is done).

They then become curious about the energies involved and read about electric potential energy. They know, using the tool of mathematics, that the area of a graph is the integral of the function and that the reverse of integration is differentiation, so the gradient of a graph of potential energy vs position could be force.

The student is unclear about how field strength can be zero when potential energy is non-zero. They use a simulation and apply the definitions of field strength and potential to a point midway between two equal charges to explore these ideas.

Solving problems

The student makes two further observations. The first is of the attraction between a balloon and a sweater. What might they determine from this? The student then observes their teacher demonstrating a simple electric circuit. What is the connection between the balloons and the circuit?

Based on observations, physicists define quantities and make up a series of rules and laws that fit the observations. They then use these laws to explain further observations, make predictions and solve problems. And it goes on! Having added to their knowledge, the student could now use what they know about mechanics and electricity to develop an understanding of atomic structure.

This example shows how the structure of the book connects factual, procedural and metacognitive knowledge and recognizes the importance of connecting learning with conceptual understanding. Learning physics is a non-linear, ongoing process of adding new knowledge, evolving understanding and identifying misconceptions.

Key to boxes

A popular feature of the book is the different colored boxes interspersed through each chapter. These are used to enhance your learning as explained below.

Nature of Science

This is an overarching theme in the course to promote concept-based learning. Through the book, you should recognize some similar themes emerging across different topics. We hope they help you to develop your own skills in scientific literacy.

Nature of Science

The principle of conservation of momentum is a consequence of Newton's laws of motion applied to the collision between two bodies. If this applies to two isolated bodies, we can generalize that it applies to any number of isolated bodies. Here we will consider colliding balls but it also applies to collisions between microscopic particles such as atoms.

Global context

The impact of the study of physics is global, and includes environmental, political and socio-economic considerations. Examples of this are given here to help you to see the importance of physics in an international context.

Dynamic friction is less than static friction so once a car starts to skid on a corner it will continue. This is also why it is not a good idea to spin the wheels of a car while going round a corner.

Negative time does not mean going back in time – it means the time before you started the clock.

Interesting facts

These give background information that will add to your wider knowledge of the topic and make links with other topics and subjects. Aspects such as historic notes on the life of scientists and origins of names are included here.

Skills

These indicate links to eBook resources that include ideas for experiments, technology and mathematics that will support your learning in the course, and help you prepare for the Internal Assessment. Look out for the grey eBook icons.

To find the decay constant and hence half-life of short-lived isotopes, the change in activity can be measured over a period of time using a GM tube.

Theory of Knowledge

These stimulate thought and consideration of knowledge issues as they arise in context. Each box contains open questions to help trigger critical thinking and discussion.

Color is perceived but wavelength is measured.

Key fact

These key facts are drawn out of the main text and highlighted in bold. This will help you to identify the core learning points within each section. They also act as a quick summary for review.

$$\text{velocity} = \frac{\text{displacement}}{\text{time}}$$

Hint

These give hints on how to approach questions, and suggest approaches that examiners like to see. They also identify common pitfalls in understanding, and omissions made in answering questions.

It is very important to realize that Newton's third law is about two bodies. Avoid statements of this law that do not mention anything about there being two bodies.

Challenge yourself

These boxes contain open questions that encourage you to think about the topic in more depth, or to make detailed connections with other topics. They are designed to be challenging and to make you think. The answers to challenge yourself questions can be found with the exercise and practice question answers.

> ### Challenge yourself
>
> 1. A projectile is launched perpendicular to a 30° slope at $20\,\text{m s}^{-1}$. Calculate the distance between the launching position and landing position.

Toward the end of the book, there are four appendix chapters: Theory of Knowledge as it relates to physics, and advice on the Internal Assessment, External Assessment and Extended Essay.

Questions

In addition to the Guiding Questions and Linking Questions, there are three types of problems in this book.

1. Worked examples with solutions

These appear at intervals in the text and are used to illustrate the concepts covered. They are followed by the solution, which shows the thinking and the steps used in solving the problem.

Worked example

A body with a constant acceleration of $-5\,\text{m s}^{-2}$ is traveling to the right with a velocity of $20\,\text{m s}^{-1}$. What will its displacement be after $20\,\text{s}$?

Solution

$s = ?$

$u = 20\,\text{m s}^{-1}$

$v = ?$

$a = -5\,\text{m s}^{-2}$

$t = 20\,\text{s}$

To calculate s, we can use the equation: $s = ut + \frac{1}{2}at^2$

$$s = 20 \times 20 + \frac{1}{2}(-5) \times 20^2 = 400 - 1000 = -600\,\text{m}$$

This means that the final displacement of the body is to the left of the starting point. It has gone forward, stopped, and then gone backward.

2. Exercises

Exercise questions are found throughout the text. They allow you to apply your knowledge and test your understanding of what you have just been reading. The answers to these (in PDF format) are accessed via icons in the eBook at the start of each chapter. Exercise answers can also be found at the end of the eBook.

Exercise

Q1. Convert the following speeds into m s^{-1}:

    (a) a car traveling at $100\,\text{km h}^{-1}$

    (b) a runner running at $20\,\text{km h}^{-1}$.

3. Practice questions

These questions are found at the end of each chapter. They are mostly taken from previous years' IB examination papers. The mark schemes used by examiners when marking these questions (in PDF format) are accessed via icons in the eBook at the start of each chapter. Practice question answers can also be found at the end of the eBook.

Practice questions

1. Police car P is stationary by the side of a road. Car S passes car P at a constant speed of $18\,\mathrm{m\,s^{-1}}$. Car P sets off to catch car S just as car S passes car P. Car P accelerates at $4.5\,\mathrm{m\,s^{-2}}$ for $6.0\,\mathrm{s}$ and then continues at a constant speed. Car P takes t seconds to draw level with car S.

   (a) State an expression, in terms of t, for the distance car S travels in t seconds. (1)

   (b) Calculate the distance traveled by car P during the first $6.0\,\mathrm{s}$ of its motion. (1)

   (Total 2 marks)

Worked solutions

Full worked solutions to all exercises and practice questions can also be found in the eBook using the grey icons at the start of each chapter.

eBook

In the eBook you will find the following:
- answers and worked solutions to all exercises and practice questions
- links to downloadable lab, activity and recommended simulation worksheets
- interactive quizzes (in the Exercises tab of your eBook account – see screenshot below)
- and links to videos (in the Resources tab of your eBook account – see screenshot below).

Extra eBook resources such as videos and interactive quizzes can be found in the Resources and Exercises tabs of your eBook account.

We hope you enjoy your study of IB physics.

Chris Hamper and Emma Mitchell

Skills in the study of physics

A vernier caliper is a device that relates to all three aspects of the tools in physics: experimental techniques, technology and mathematics.

As discussed in the Introduction, an excellent IB physicist should be aware of the course aims, appreciate the nature of physics (and science more broadly), and know how to learn and how to inquire.

The skills associated with inquiry have already been discussed and will be referred to once again in the Internal Assessment and Extended Essay chapters. In this chapter, you will find out about the three tools that physicists benefit most from: experimental techniques, technology and mathematics.

Read this chapter before embarking on your studies and continue to refer back to the skills addressed, as almost all elements could be required in any of the topics that follow. When preparing for External Assessment (in particular Paper 1B), you may wish to attempt the practice questions that are located in the eBook.

Tool 1: Experimental techniques

Physics is about modeling the physical Universe so that we can predict outcomes, but before we can develop models, we need to define quantities and measure them. To measure a quantity, we first need to invent a measuring device and define a unit. When measuring, we should try to be as accurate as possible but we can never be exact – measurements will always have uncertainties. This could be due to the instrument or the way we use it, or it might be that the quantity we are trying to measure is changing.

Making observations

Before we can try to understand the Universe, we have to observe it. Imagine you are a cave person looking up into the sky at night. You would see lots of bright points scattered about (assuming it is not cloudy). The points are not the same but how can you describe the differences between them? One of the main differences is that you have to move your head to see different examples. This might lead you to define their position. Occasionally, you might notice a star flashing so would realize that there are also differences not associated with position, leading to the concept of time. If you shift your attention to the world around you, you will be able to make further close-range observations. Picking up rocks, you notice some are easy to pick up while others are more difficult, some are hot and some are cold, and different rocks are different colors. These observations are just the start: to be able to understand how these quantities are related, you need to measure them, and before you do that, you need to be able to count.

Figure 1 Making observations came before science.

Standard notation

In this course, we will use some numbers that are very big and some that are very small. 602 000 000 000 000 000 000 000 is a commonly used number, as is 0.000 000 000 000 000 000 160. To make life easier, we write these in standard form. This means that we write the number with only one digit to the left of the decimal place and represent the number of zeros with powers of 10.

It is also acceptable to use a prefix to denote powers of 10.

Prefix	Value
T (tera)	10^{12}
G (giga)	10^{9}
M (mega)	10^{6}
k (kilo)	10^{3}
c (centi)	10^{-2}
m (milli)	10^{-3}
μ (micro)	10^{-6}
n (nano)	10^{-9}
p (pico)	10^{-12}
f (femto)	10^{-15}

If you set up your calculator properly, it will always give your answers in standard form.

Realization that the speed of light in a vacuum is the same no matter who measures it led to the speed of light being the basis of our unit of length.

The meter was originally defined in terms of several pieces of metal positioned around Paris. This was not very accurate so now one meter is defined as the distance traveled by light in a vacuum in $\frac{1}{299\,792\,458}$ of a second.

So:

$$602\,000\,000\,000\,000\,000\,000\,000 = 6.02 \times 10^{23} \text{ (decimal place must be shifted right 23 places)}$$

$$0.000\,000\,000\,000\,000\,000\,160 = 1.60 \times 10^{-19} \text{ (decimal place must be shifted left 19 places).}$$

A number's order of magnitude is the closest whole power of ten. 10^{-2}, 10^{-1}, 10^{0}, 10^{1}, 10^{2} and so on are all orders of magnitude.

Exercise

Q1. Write the following in standard form.

(a) 48 000

(b) 0.000 036

(c) 14 500

(d) 0.000 000 48

Measuring variables

We have seen that there are certain fundamental quantities that define our Universe from which all other quantities can be derived or explained. These include position, time and mass.

Length and distance

Before we take any measurements, we need to define the quantity. The quantities that we use to define the position of different objects are **length** and **distance**. To measure distance, we need to make a scale and to do that we need two fixed points. We take our fixed points to be two points that never change position, for example, the ends of a stick. If everyone used the same stick, we will all end up with the same measurement. However, we cannot all use the same stick so we make copies of the stick and assume that they are all the same. The problem is that sticks are not all the same length, so our unit of length is based on one of the few things we know to be the same for everyone: the speed of light in a vacuum. Once we have defined the unit, in this case, the meter, it is important that we all use it (or at least make it very clear if we are using a different one). There is more than one system of units but the one used in this course is the Système International d'Unités (SI units). Here are some examples of distances measured in meters:

$$\text{distance from the Earth to the Sun} = 1.5 \times 10^{11}\,\text{m}$$
$$\text{diameter of a grain of sand} = 2 \times 10^{-4}\,\text{m}$$
$$\text{the distance to the nearest star} = 4 \times 10^{16}\,\text{m}$$
$$\text{radius of the Earth} = 6.378 \times 10^{6}\,\text{m}$$

Q2. Convert the following into meters (m) and write in standard form:

(a) Distance from London to New York = 5585 km

(b) Height of Einstein = 175 cm

(c) Thickness of a human hair = 25.4 μm

(d) Distance to furthest part of the observable Universe = 100 000 million million million km.

Time

When something happens, we call it an **event**. To distinguish between different events, we use time. The time between two events is measured by comparing to some fixed value, the second. Time is also a fundamental quantity.

Some examples of times:

time between beats of a human heart = 1 s

time for the Moon to go around the Earth = 1 month

time for the Earth to go around the Sun = 1 year

Q3. Convert the following times into seconds (s) and write in standard form:

(a) 85 years, how long Newton lived

(b) 2.5 ms, the time taken for a mosquito's wing to go up and down

(c) 4 days, the time it took Apollo 11 to travel to the Moon

(d) 2 hours 52 min 59 s, the time it took for Concorde to fly from London to New York.

Mass

If we pick up different objects, we find another difference. Some objects are easy to lift up and others are difficult. This seems to be related to how much matter the objects consist of. To quantify this, we define mass measured by comparing different objects to the standard kilogram.

Some examples of mass:

approximate mass of a human = 75 kg

mass of the Earth = 5.97×10^{24} kg

mass of the Sun = 1.98×10^{30} kg

Q4. Convert the following masses to kilograms (kg) and write in standard form:

(a) The mass of an apple = 200 g

(b) The mass of a grain of sand = 0.00001 g

(c) The mass of a family car = 2 tonnes.

---

i The second was originally defined as a fraction of a day but today's definition is 'the duration of 9 192 631 770 periods of the radiation corresponding to the transition between the two hyperfine levels of the ground state of the caesium-133 atom'.

TOK If nothing ever happened, would there be time?

TOK The kilogram was the last fundamental quantity to be based on an object kept in Paris. It is now defined using Planck's constant. What are the benefits of using physical constants instead of physical objects?

Area and volume

The two dimensional space taken up by an object is defined by the area and the three dimensional space is volume. Area is measured in square meters (m^2) and volume is measured in cubic meters (m^3). Area and volume are not fundamental units since they can be split into smaller units (m × m or m × m × m). We call units like these derived units.

A list of useful area and volume equations is located in your data booklet.

Exercise

Q5. Calculate the volume of a room of length 5 m, width 10 m and height 3 m.

Q6. Using the information from pages xviii–xix, calculate:
(a) the volume of a human hair of length 20 cm
(b) the volume of the Earth.

Density

By measuring the mass and volume of many different objects, we find that if the objects are made of the same material, the ratio $\frac{\text{mass}}{\text{volume}}$ is the same. This quantity is called the **density**. The unit of density is $kg\,m^{-3}$. This is another derived unit.

Examples include:

$$\text{density of water} = 1.0 \times 10^3\,kg\,m^{-3}$$
$$\text{density of air} = 1.2\,kg\,m^{-3}$$
$$\text{density of gold} = 1.93 \times 10^4\,kg\,m^{-3}$$

Exercise

Q7. Calculate the mass of air in a room of length 5 m, width 10 m and height 3 m.

Q8. Calculate the mass of a gold bar of length 30 cm, width 15 cm and height 10 cm.

Q9. Calculate the average density of the Earth.

Displacement

So far, all that we have modeled is the position of objects and when events take place, but what if something moves from one place to another? To describe the movement of a body, we define the quantity **displacement**. This is the distance moved in a particular direction.

The unit of displacement is the same as length: the meter.

Referring to the map in Figure 2:
If you move from B to C, your displacement will be 5 km north.
If you move from A to B, your displacement will be 4 km west.

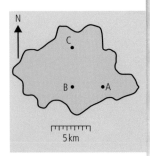

Figure 2 Displacements on a map.

Angle

When two straight lines join, an angle is formed. The size of the angle can be increased by rotating one of the lines about the point where they join (the vertex) as shown in Figure 3. To measure angles, we often use degrees. Taking the full circle to be 360° is very convenient because 360 has many whole number factors it can be easily divided by e.g., 4, 6, and 8. However, it is an arbitrary unit not related to the circle itself.

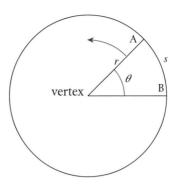

◀ **Figure 3** The angle between two lines.

If the angle is increased by rotating line A, the arc lengths will also increase. So for this circle, we could use the arc length as a measure of angle. The problem is that if we take a bigger circle, then the arc length for the same angle will be greater. We therefore define the angle by using the ratio $\frac{s}{r}$, which will be the same for all circles. This unit is the radian.

Summary – Tool 1: Experimental techniques

So far, you will have become familiar with a range of experimental techniques, including measurements of:

- length
- time
- mass
- volume
- angle.

These tools are prescribed in your subject guide.

There are others still to come throughout the textbook. These include measurements of:

- force (A.2)
- temperature (B.1)
- electric current (B.5)
- electric potential difference (B.5)
- sound intensity (C.2)
- and light intensity (C.2).

You should also be aware of how to recognize and address safety, ethical and environmental issues. Try to spot these throughout the textbook, such as the risks of high-temperature fluids (B.3) or ionizing radiation (E.3), or the environmental impact of using electricity (B.5) or water (C.2) for experimentation.

For one complete circle, the arc length is the circumference = $2\pi r$ so the angle 360° in radians = $\frac{2\pi r}{r} = 2\pi$.

So 360° is equivalent to 2π.

Since the radian is a ratio of two lengths, it has no units.

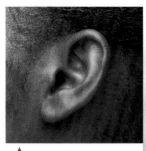

The ear is an example of a sensor. Look out for human-made sensors throughout this book.

Algodoo® is software that enables the simulation of ideas that may or may not be possible in the lab. Gravity can be altered (or removed altogether) and materials or any desired properties can be tested.

TOK

If the system of numbers had been totally different, would our models of the Universe be the same?

In physics experiments, we always quote the uncertainties in our measurements. Shops also have to work within given uncertainties and could be prosecuted if they overestimate the weight of something. An approximation is similar, but not exactly equal, to something else (for example, a rounded number). An estimate is a simplification of a quantity (such as assuming that an apple has a mass of 100 g).

Tool 2: Technology

Technology and physics are closely linked. Technology enables the advancement of physics, and the pursuit of scientific understanding stimulates improvements in technology. The fields impacted are as wide-ranging as communication, medicine and environmental sustainability.

Every measurement requires an instrument, which is itself inherently technological. Technology facilitates collaboration, which is to the benefit of international teams of scientists and IB physicists alike. Technology makes the processes carried out by physicists much faster, for example, when collecting data or performing calculations. Humans can sense light intensity, temperature, sounds, smells, tastes and applied pressure. How might technology replicate or improve upon these senses? What else does technology enable us to measure?

A model is a representation of reality. It can be as concise as a single word (e.g. the brain is like a 'computer') or an equation (e.g. speed is the ratio of distance traveled to time taken). Technology supports physicists in forming new models during exploratory experimental work (e.g. by making it easy to compare the 'fit' of a range of mathematical relationships) and in creating simulations that enable experimentation without need for a lab.

Summary – Tool 2: Technology

Technology can be used to good effect in physics. The Tool 3: Mathematics section of this chapter will reveal that technology can be used to display graphs for representing data. In the remainder of the textbook, you can expect to learn about:

- using sensors (A.2, B.1, B.3, C.1, C.4)
- models and simulations for generation of data (B.2, C.4)
- spreadsheets for manipulation of data (B.5)
- computer modeling for processing data (C.1)
- image analysis of motion (C.5, E.1)
- databases for data extraction (C.5, E.5)
- video analysis of motion (E.3)

Tool 3: Mathematics

When counting apples, we can say there are exactly six apples, but if we measure the length of a piece of paper, we cannot say that it is exactly 21 cm wide. All measurements have an associated uncertainty and it is important that this is also quoted with the value. Uncertainties cannot be avoided, but by carefully using accurate instruments, they can be minimized. Physics is all about relationships between different quantities. If the uncertainties in measurement are too big, then relationships are difficult to identify. Throughout the practical part of this course, you will be trying to find out what causes the uncertainties in your measurements. Sometimes, you will be able to reduce them and at other times not. It is quite alright to have big uncertainties but completely unacceptable to manipulate data so that the numbers appear to fit a predicted relationship.

Summary of SI units

The SI system of units is the set of units that are internationally agreed to be used in science. It is still OK to use other systems in everyday life (miles, pounds, Fahrenheit), but in science, we must always use SI. There are seven fundamental (or base) quantities.

Base quantity	Quantity symbol	Unit	Unit symbol
length	x or l	meter	m
mass	m	kilogram	kg
time	t	second	s
electric current	I	ampere	A
thermodynamic temperature	T	kelvin	K
amount of substance	n	mole	mol
luminous intensity	I	candela	cd

The candela will not be used in this course.

All other SI units are derived units. These are based on the fundamental units and will be introduced and defined where relevant. So far we have come across just three.

Derived quantity	Symbol	Base units
area	m^2	m × m
volume	m^3	m × m × m
density	$kg\,m^{-3}$	$\dfrac{kg}{m \times m \times m}$

By breaking down the units of derived quantities into base quantity units, it is possible to check whether an equation could be correct. This technique is an informal version of dimensional analysis, in which the 'powers of' quantities are compared on either side of an equation. Note, however, that dimensional analysis provides no insights into the constant of proportionality.

Processing uncertainties

The SI system of units is defined so that we all use the same sized units when building our models of the physical world. However, before we can understand the relationship between different quantities, we must measure how big they are. To make measurements, we use a variety of instruments. To measure length, we can use a ruler and to measure time, a clock. If our findings are to be trusted, then our measurements must be accurate, and the accuracy of our measurement depends on the instrument used and how we use it. Consider the following examples.

▲ Even this sophisticated device at CERN has uncertainties.

Measuring length using a ruler

Example 1

A good straight ruler marked in mm is used to measure the length of a rectangular piece of paper as in Figure 4.

The ruler measures to within 0.5 mm (we call this the **uncertainty** in the measurement) so the length in cm is quoted to 2 d.p. This measurement is precise and accurate. This can be written as 6.40 ± 0.05 cm, which tells us that the actual value is somewhere between 6.35 and 6.45 cm.

Figure 4 Length = 6.40 ± 0.05 cm.

Example 2

Figure 5 shows how a ruler with a broken end is used to measure the length of the same piece of paper. When using the ruler, you might fail to notice the end is broken and think that the 0.5 cm mark is the zero mark.

This measurement is precise since the uncertainty is small but is not accurate since the value 6.90 cm is wrong.

Figure 5 Length ≠ 6.90 ± 0.05 cm.

Example 3

A ruler marked only in $\frac{1}{2}$ cm is used to measure the length of the paper as in Figure 6.

These measurements are precise and accurate, but the scale is not very sensitive.

Figure 6 Length = 6.5 ± 0.3 cm.

Example 4

In Figure 7, a ruler is used to measure the maximum height of a bouncing ball. The ruler has more markings, but it is very difficult to measure the height of the bouncing ball. Even though you can use the scale to within 0.5 mm, the results are not precise (the base of the ball may be at about 4.2 cm). However, if you do enough runs of the same experiment, your final answer could be accurate.

When using a scale such as a ruler, the uncertainty in the reading is half of the smallest division. In this case, the smallest division is 1 mm so the uncertainty is 0.5 mm. When using a digital device such as a balance, we take the uncertainty as the smallest digit. So if the measurement is 20.5 g, the uncertainty is ±0.1 g.

In Examples 1 and 2, we are assuming that there is no uncertainty at the 'zero' end of the ruler because it might be possible to line up paper with the long ruler marking. In reality, the uncertainty for Example 1 may be ±0.1 cm, which is the combination of the 0.05 cm uncertainties at each end of the length.

Notice that uncertainties are generally quoted to one significant figure. The uncertainty then dictates the number of decimal places to which the measurement is written.

Figure 7
Height = 4.2 ± 0.2 cm.

Precision and accuracy

To help understand the difference between precision and accuracy, consider the four attempts to hit the center of a target with three arrows shown in Figure 8.

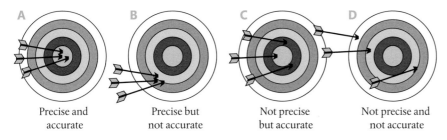

Precise and accuracy   Precise but   Not precise   Not precise and
accurate   not accurate   but accurate   not accurate

◀ **Figure 8** Precision and accuracy

A The arrows were fired accurately at the center with great precision.

B The arrows were fired with great precision as they all landed near one another, but not very accurately since they are not near the center.

C The arrows were not fired very precisely since they were not close to each other. However, they were accurate since they are evenly spread around the center. The average of these would be quite good.

D The arrows were not fired accurately and the aim was not precise since they are far from the center and not evenly spread.

So **precision** is how close to each other a set of measurements are (related to the resolution of the measuring instrument) and the **accuracy** is how close they are to the actual value (often based on an average).

Errors in measurement

There are two types of measurement error – random and systematic.

Random error

If you measure a quantity many times and get lots of slightly different readings, then this called a random error. For example, when measuring the bounce of a ball, it is very difficult to get the same value every time even if the ball is doing the same thing.

Systematic error

A systematic error is when there is something wrong with the measuring device or method. Using a ruler with a broken end can lead to a 'zero error' as in Example 2 on page xxiv. Even with no random error in the results, you would still get the wrong answer.

If you measure the same thing many times and get the same value, then the measurement is precise.
If the measured value is close to the expected value, then the measurement is accurate. If a football player hits the post ten times in a row when trying to score a goal, you could say the shots are precise but not accurate.

TOK
It is not possible to measure anything exactly. This is not because our instruments are not exact enough but because the quantities themselves do not exist as exact quantities. What measurements could you make in the space around you? What might makes these quantities inexact?

Reducing errors

To reduce random errors, you can repeat your measurements. If the uncertainty is truly random, your measurements will lie either side of the true reading and the mean of these values will be close to the actual value. To reduce a systematic error, you need to find out what is causing it and correct your measurements accordingly. A systematic error is not easy to spot by looking at the measurements, but is sometimes apparent when you look at the graph of your results or the final calculated value.

Adding uncertainties

If two values are added together, then the uncertainties also add. For example, if we measure two lengths, $L_1 = 5.0 \pm 0.1$ cm and $L_2 = 6.5 \pm 0.1$ cm, then the maximum value of L_1 is 5.1 cm and the maximum value of L_2 is 6.6 cm, so the maximum value of $L_1 + L_2 = 11.7$ cm. Similarly, the minimum value is 11.3 cm. We can therefore say that $L_1 + L_2 = 11.5 \pm 0.2$ cm.

If $\qquad y = a \pm b \qquad\qquad$ then $\quad \Delta y = \Delta a + \Delta b$

If you multiply a value by a constant, then the uncertainty is also multiplied by the same number.

So $\qquad 2L_1 = 10.0 \pm 0.2$ cm $\quad$ and $\quad \frac{1}{2}L_1 = 2.50 \pm 0.05$ cm.

Example of measurement and uncertainties

Let us consider an experiment to measure the mass and volume of a piece of modeling clay. To measure mass, we can use a top pan balance so we take a lump of clay and weigh it. The result is 24.8 g. We can repeat this measurement many times and get the same answer. There is no variation in the mass so the uncertainty in this measurement is the same as the uncertainty in the scale. The smallest division on the balance used is 0.1 g so the uncertainty is $\pm$ 0.1 g.

So: $\qquad\qquad\qquad\qquad\qquad\qquad$ mass = 24.8 $\pm$ 0.1 g

To measure the volume of the modeling clay, we first need to mold it into a uniform shape: let us roll it into a sphere. To measure the volume of the sphere, we measure its diameter from which we can calculate its radius ($V = \frac{4\pi r^3}{3}$).

Making an exact sphere out of the modeling clay is not easy. If we do it many times, we will get different-shaped balls with different diameters so let us try rolling the ball five times and measuring the diameter each time with a ruler.

Using the ruler, we can only judge the diameter to the nearest mm so we can say that the diameter is 3.5 $\pm$ 0.1 cm. It is actually even worse than this since we also have to line up the zero at the other end, so 3.5 $\pm$ 0.2 cm might be a more reasonable estimate. If we turn the ball round, we get the same value for d. If we squash the ball and make a new one, we might still get a value of 3.5 $\pm$ 0.2 cm. This is not because the ball is a perfect sphere every time but because our method of measurement is not **sensitive** enough to measure the difference.

▲
Ball of modeling clay measured with a ruler.

Let us now try measuring the ball with a vernier caliper.

◀ A vernier caliper has sliding jaws, which are moved so they touch both sides of the ball.

The vernier caliper can measure to the nearest 0.002 cm. Repeating measurements of the diameter of the same lump of modeling clay might give the results in Table 1.

◀ **Table 1**

Diameter/cm								
3.640	3.450	3.472	3.500	3.520	3.520	3.530	3.530	3.432
3.540	3.550	3.550	3.560	3.560	3.570	3.572	3.582	3.582

The reason these measurements are not all the same is because the ball is not perfectly uniform and, if made several times, will not be exactly the same. We can see that there is a spread of data from 3.400 cm to 3.570 cm, with most lying around the middle. This can be shown on a graph but first we need to group the values as in Table 2.

Distribution of measurements

Even with this small sample of measurements, you can see in Figure 9 that there is a spread of data: some measurements are too big and some too small but most are in the middle. With a much larger sample, the shape would be closer to a 'normal distribution' as in Figure 10.

◀ **Table 2**

Range/cm	No. of values within range
3.400–3.449	1
3.450–3.499	2
3.500–3.549	6
3.550–3.599	8
3.600–3.649	1

▲ **Table 2**

◀ **Figure 9** Distribution of measurements of diameter.

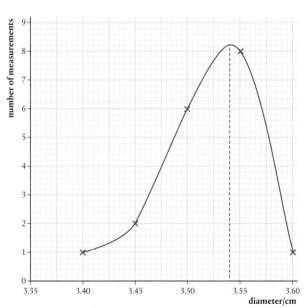

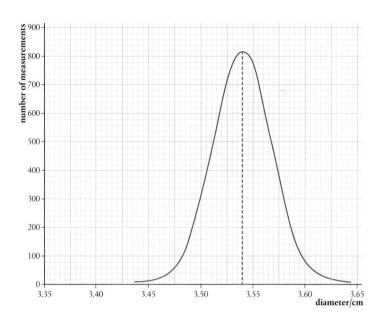

Figure 10 Normal distribution curve.

The mean

At this stage, you may be wondering what the point is of trying to measure something that does not have a definite value. Well, we are trying to find the volume of the modeling clay using the formula $V = \frac{4\pi r^3}{3}$. This is the formula for the volume of a perfect sphere. The problem is we cannot make a perfect sphere. It is probably more like the shape of an egg, so depending on which way we measure it, sometimes the diameter will be too big and sometimes too small. It is, however, just as likely to be too big as too small, so if we take the mean of all our measurements, we should be close to the 'perfect sphere' value which will give us the correct volume of the modeling clay.

The mean or average is found by adding all the values and dividing by the number of values. In this case, the mean = 3.537 cm. This is the same as the peak in the distribution. We can check this by measuring the volume in another way, for example, sinking it in water and measuring the volume displaced. Using this method gives a volume = 23 cm³. Rearranging the formula gives: $r = \sqrt[3]{\frac{3V}{4\pi}}$

Substituting for V gives d = 3.53 cm, which is fairly close to the mean. Calculating the mean reduces the random error in our measurement.

There is a very nice example of this that you might like to try. Fill a jar with jelly beans and get your classmates to guess how many there are. Assuming that they really try to make an estimate rather than randomly saying a number, the guesses are just as likely to be too high as too low. So, if after you collect all the data you find the average value, it should be quite close to the actual number of beans.

Knowing the mean of data enables a calculation of the standard deviation to be performed. Standard deviation gives an idea of the spread of the data.

Smaller samples

You will be collecting a lot of different types of data throughout the course but you will not often have time to repeat your measurements enough to get a normal distribution. With only four values, the uncertainty is not reduced significantly by taking the mean

If the data follows a normal distribution, 68% of the values should be within one standard deviation of the mean.

so *half* the range of values is used instead. This often gives a slightly exaggerated value for the uncertainty – for the example above, it would be ± 0.1 cm – but it is an approach accepted by the IB.

Relationships

In physics, we are very interested in the relationships between two quantities, for example, the distance traveled by a ball and the time taken. To understand how we represent relationships by equations and graphs, let us consider a simple relationship regarding fruit.

Linear relationships

Let us imagine that all apples have the same mass, 100 g. To find the relationship between number of apples and their mass, we would need to measure the mass of different numbers of apples. These results could be put into a table as in Table 3.

In this example, we can clearly see that the mass of the apples increases by the same amount every time we add an apple. We say that the mass of apples is **proportional** to the number. If we draw a graph of mass vs number, we get a straight line passing through the origin as in Figure 11.

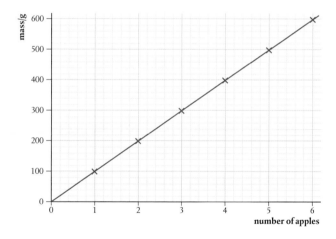

The gradient of this line is given by $\frac{\Delta y}{\Delta x}$ = 100 g/apple. The fact that the line is straight and passing through the origin can be used to test if two quantities are proportional to each other.

The equation of the line is $y = mx$, where m is the gradient, so in this case $y = 100x$ and $m = 100$ g apple^{-1}.

This equation can be used to calculate the mass of any given number of apples. This is a simple example of what we will spend a lot of time doing in this course.

To make things a little more complicated, let us consider apples in a basket with mass 500 g. The table of masses is shown in Table 4.

The slope in Figure 12 is still 100 g/apple, indicating that each apple still has a mass of 100 g, but the intercept is no longer (0, 0). We say that the mass is linearly related to the number of apples but they are *not* directly proportional.

Number (N)	Mass (m)/g
1	100
2	200
3	300
4	400
5	500
6	600

Table 3

◀ **Figure 11** Graph of mass vs number of apples.

Number (N)	Mass (m)/g
1	600
2	700
3	800
4	900
5	1000
6	1100

Table 4

Figure 12 Graph of mass vs number of apples in a basket.

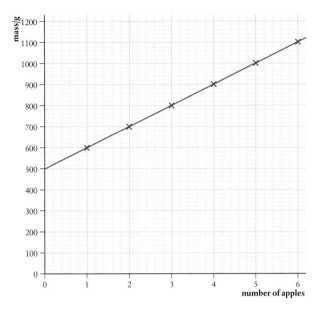

It is much easier to plot data from an experiment without processing it but this will often lead to curves that are very difficult to draw conclusions from. Linear relationships are much easier to interpret so are worth the time spent processing the data.

The equation of this line is $y = mx + c$, where m is the gradient and c the intercept on the y-axis. The equation in this case is therefore $y = 100x + 500$.

Finding the equation that relates two quantities can be useful for interpolation and extrapolation. Both techniques involve inserting a value for one of the quantities into the equation to find the corresponding value of the other. Interpolation can be performed with good confidence as it is done within the range of collected data. Extrapolation is more risky as the values are beyond the range of collected data; you are making a prediction.

Exercise

Q10. The data displayed in the graphs below all show examples of correlation. What other conclusions can you make?

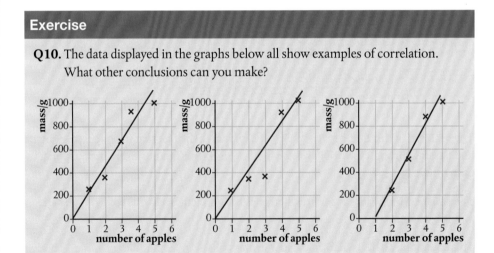

Non-linear relationships

Let us now consider the relationship between radius and the area of circles of paper as shown in Figure 13 on the following page.

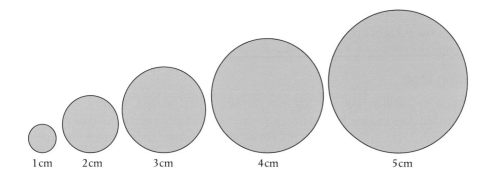

Figure 13 Five circles of green paper.

1 cm   2 cm   3 cm   4 cm   5 cm

The results are recorded in Table 5.

If we now graph the area vs the radius, we get the graph shown in Figure 14.

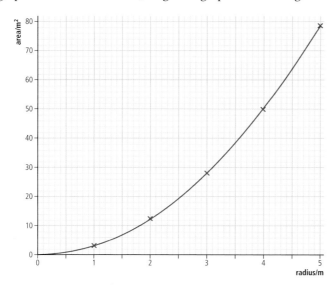

Radius/m	Area/m²
1	3.14
2	12.57
3	28.27
4	50.27
5	78.54

Table 5

Figure 14 Graph of area of green circles vs radius.

This is not a straight line so we cannot deduce that area is linearly related to radius. However, you may know that the area of a circle is given by $A = \pi r^2$, which would mean that A is proportional to r^2. To test this, we can calculate r^2 and plot a graph of area vs r^2. The calculations are shown in Table 6.

Radius/m	r^2/m²	Area/m²
1	1	3.14
2	4	12.57
3	9	28.27
4	16	50.27
5	25	78.54

Table 6

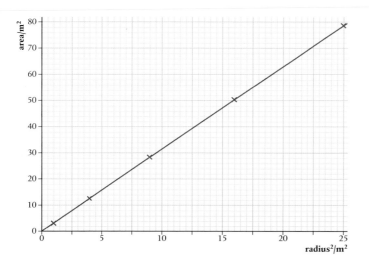

Figure 15 Graph of area of green circles vs radius².

xxix

This time, the graph is linear, confirming that the area is indeed proportional to the radius². The gradient of the line is 3.1, which is π to two significant figures. So the equation of the line is $A = \pi r^2$ as expected.

Using logs

Logs can be useful in your practical work. In the previous exercise, we knew that $A = \pi r^2$, but if we had not known this, we could have found the relationship by plotting a log graph. Let us pretend that we did not know the relationship between A and r, only that they were related. So it could be $A = kr^2$ or $A = kr^3$ or even $A = k\sqrt{r}$.

We can write all of these in the form: $A = kr^n$

Now if we take logs of both sides of this equation, we get: $\log A = \log kr^n = \log k + n\log r$

This is of the form $y = mx + c$, where $\log A$ is y and $\log r$ is x.

So if we plot $\log A$ vs $\log r$, we should get a straight line with gradient n and intercept $\log k$. This is all quite easy to do using a spreadsheet, resulting in Table 7 and the graph in Figure 16.

Table 7 ▶

Radius/m	Area/m²	log (A/m²)	log (r/m)
1	3.14	0.4969	0.0000
2	12.57	1.0993	0.3010
3	28.27	1.4513	0.4771
4	50.27	1.7013	0.6021
5	78.54	1.8951	0.6990

Figure 16 log A vs log r for the green paper disks. ▶

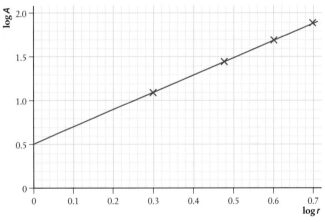

This has gradient = 2 and intercept = 0.5, so if we compare it to the equation of the line:

$$\log A = \log k + n\log r$$

we can deduce that: $\qquad n = 2$ and $\log k = 0.5$

The inverse of $\log k$ is 10^k so $k = 10^{0.5} = 3.16$, which is quite close to π.

Substituting into our original equation $A = kr^n$, we get $A = \pi r^2$.

A	B
1.1	0.524
3.6	0.949
4.2	1.025
5.6	1.183
7.8	1.396
8.6	1.466
9.2	1.517
10.7	1.636

▲
Table 8

Exercise

Q11. Use a log–log graph to find the relationship between A and B in Table 8.

Relationship between the diameter of a modeling clay ball and its mass

So far, we have only measured the diameter and mass of one ball of modeling clay. If we want to know the relationship between the diameter and mass, we should measure many balls of different sizes. This is limited by the amount of modeling clay we have, but should be from the smallest ball we can reasonably measure up to the biggest ball we can make.

◀ Table 9

Mass/g ± 0.1 g	Diameter/cm ± 0.002 cm			
	1	2	3	4
1.4	1.296	1.430	1.370	1.280
2.0	1.570	1.590	1.480	1.550
5.6	2.100	2.130	2.168	2.148
9.4	2.560	2.572	2.520	2.610
12.5	2.690	2.840	2.824	2.720
15.7	3.030	2.980	3.080	2.890
19.1	3.250	3.230	3.190	3.204
21.5	3.490	3.432	3.372	3.360
24.8	3.550	3.560	3.540	3.520

In Table 9, the uncertainty in diameter d is given as 0.002 cm. This is the uncertainty in the vernier caliper: the actual uncertainty in diameter is *more* than this as is revealed by the spread of data which you can see in the first row, which ranges from 1.280 to 1.430, a difference of 0.150 cm. Because there are only four different measurements, we can use the approximate method using $\Delta d = \frac{(d_{max} - d_{min})}{2}$. This gives an uncertainty in the first measurement of ±0.08 cm. Table 10 includes the uncertainties and the mean.

◀ Table 10

Mass/g ± 0.1 g	Diameter/cm ± 0.002 cm				d_{mean}/cm	Uncertainty Δd/cm
	1	2	3	4		
1.4	1.296	1.430	1.370	1.280	1.34	0.08
2.0	1.570	1.590	1.480	1.550	1.55	0.06
5.6	2.100	2.130	2.168	2.148	2.14	0.03
9.4	2.560	2.572	2.520	2.610	2.57	0.04
12.5	2.690	2.840	2.824	2.720	2.77	0.08
15.7	3.030	2.980	3.080	2.890	3.00	0.10
19.1	3.250	3.230	3.190	3.204	3.22	0.03
21.5	3.490	3.432	3.372	3.360	3.41	0.07
24.8	3.550	3.560	3.540	3.520	3.54	0.02

Now, to reveal the relationship between the mass m and diameter d, we can draw a graph of m vs d as shown in Figure 17. However, since the values of m and d have uncertainties, we do not plot them as points but as lines. The length of the lines equals the uncertainty in the measurement. These are called **uncertainty bars**.

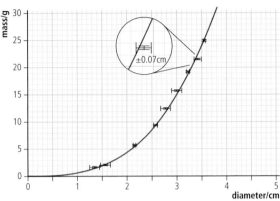

Figure 17 Graph of mass of modeling clay ball vs diameter with error bars.

A worksheet with full details of how to carry out this experiment is available in the eBook.

This curve is the *best fit* for the data collected.

The curve is quite a nice fit but very difficult to analyze. It would be more convenient if we could manipulate the data to get a straight line. This is called **linearizing**. To do this, we must try to deduce the relationship using physical theory and then test the relationship by drawing a graph. In this case, we know that density, $\rho = \dfrac{\text{mass}}{\text{volume}}$ and the volume of a sphere $= \dfrac{4\pi r^3}{3}$, where r = radius.

So:
$$\rho = \frac{3m}{4\pi r^3}$$

Rearranging this equation gives:
$$r^3 = \frac{3m}{4\pi\rho}$$

But:
$$r = \frac{d}{2} \text{ so } \frac{d^3}{8} = \frac{3m}{4\pi\rho}$$

$$d^3 = \frac{6m}{\pi\rho}$$

Since $\dfrac{6}{\pi\rho}$ is a constant, this means that d^3 is proportional to m. So, a graph of d^3 vs m should be a straight line with gradient $= \dfrac{6}{\pi\rho}$. To plot this graph, we need to find d^3 and its uncertainty. The uncertainty can be found by calculating the difference between the maximum and minimum values of d^3 and dividing by 2: $\dfrac{(d_{max}{}^3 - d_{min}{}^3)}{2}$. This has been done in Table 11.

Table 11 ▶

Mass/g ± 0.1 g	Diameter/cm ± 0.002 cm						
	1	2	3	4	d_{mean}/ cm	$d^3{}_{mean}$/ cm³	$d^3{}_{unc.}$/ cm³
1.4	1.296	1.430	1.370	1.280	1.34	2.4	0.4
2.0	1.570	1.590	1.480	1.550	1.55	3.7	0.4
5.6	2.100	2.130	2.168	2.148	2.14	9.8	0.5
9.4	2.560	2.572	2.520	2.610	2.57	17	1
12.5	2.690	2.840	2.824	2.720	2.77	21	2
15.7	3.030	2.980	3.080	2.890	3.00	27	3
19.1	3.250	3.230	3.190	3.204	3.22	33	1
21.5	3.490	3.432	3.372	3.360	3.41	40	2
24.8	3.550	3.560	3.540	3.520	3.54	44	1

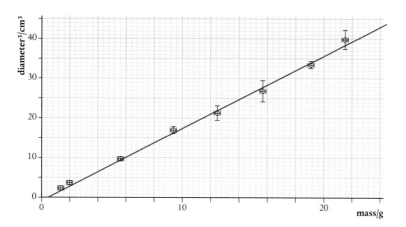

Figure 18 Graph of diameter³ of a modeling clay ball vs mass.

The best fit of these points is now a straight line. Over time, you will learn how to judge whether data is best represented by a linear or non-linear fit, perhaps based on the theory behind an experiment or the positions of the points.

Looking at the line in Figure 18, we can see that due to random errors in the data, the points are not exactly on the line but close enough. What we expect to see is the line touching all of the error bars, which is the case here. The error bars should reflect the random scatter of data. In this case, they are slightly bigger, which is probably due to the approximate way that they have been calculated. Notice how the points furthest from the line have the biggest error bars.

According to the formula, d^3 should be directly proportional to m; the line should therefore pass through the origin. Here we can see that the y-intercept is $-0.3\,cm^3$, which is quite close to the origin and is probably just due to the random errors in d. If the intercept had been more significant, then it might have been due to a **systematic error** in mass. For example, if the balance had not been zeroed properly and instead of displaying zero with no mass on the pan, it read $0.5\,g$, then each mass measurement would be $0.5\,g$ too big. The resulting graph would be as in Figure 19.

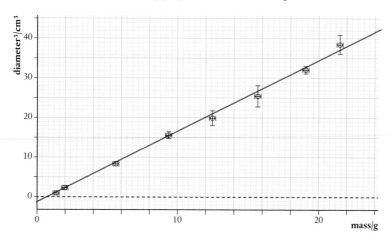

Figure 19 Graph of diameter³ of a modeling clay ball vs mass with a systematic error.

A systematic error in the diameter would not be so easy to see. Since diameter is cubed, adding a constant value to each diameter would cause the line to become curved.

Outliers

Sometimes a mistake is made in one of the measurements. This is quite difficult to spot in a table but will often lead to an outlier on a graph. For example, one of the measurements in Table 12 is incorrect.

Table 12 ▶

Mass/g ± 0.1 g	Diameter/cm ± 0.002 cm			
	1	**2**	**3**	**4**
1.4	1.296	1.430	1.370	1.280
2.0	1.570	1.590	1.480	1.550
5.6	2.100	2.130	2.148	3.148
9.4	2.560	2.572	2.520	2.610
12.5	2.690	2.840	2.824	2.720
15.7	3.030	2.980	3.080	2.890
19.1	3.250	3.230	3.190	3.204
21.5	3.490	3.432	3.372	3.360
24.8	3.550	3.560	3.540	3.520

This is revealed in the graph in Figure 20.

Figure 20 Graph of diameter³ of a modeling clay ball vs. mass with outlier.

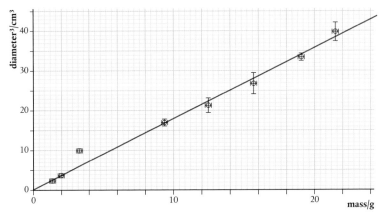

When you find an outlier, you need to do some detective work to try to find out why the point is not closer to the line. Taking a close look at the raw data sometimes reveals that one of the measurements was incorrect. This can then be removed and the line plotted again. However, you cannot simply leave out the point because it does not fit. A sudden decrease in the level of ozone over the Antarctic was originally left out of the data since it was an outlier. Later investigation of this 'outlier' led to a significant discovery.

If asked for a sketch graph, you should consider what shape it will have and where it will cross the axes. Scales are not required.

Uncertainty in the gradient

The general equation for a straight-line graph passing through the origin is $y = mx$. In this case, the equation of the line is $d^3 = \frac{6m}{\pi\rho}$, where d^3 is y and m is x and the gradient is $\frac{6}{\pi\rho}$. You can see that the unit of the gradient is cm³/g. This is consistent with it representing $\frac{6}{\pi\rho}$.

From the graph, we see that gradient = $1.797\,\text{cm}^3\,\text{g}^{-1} = \frac{6}{\pi\rho}$ so $\rho = \frac{6}{1.797\pi}$

$\frac{6}{1.797\pi} = 1.063\,\text{g cm}^{-3}$ but what is the uncertainty in this value?

There are several ways to estimate the uncertainty in a gradient. One of them is to draw the steepest and least steep lines through the error bars as shown in Figure 21.

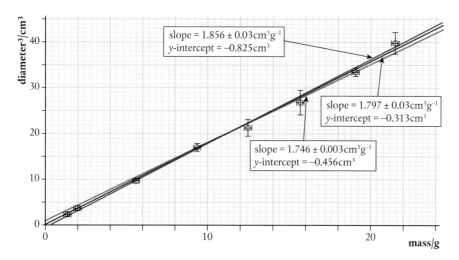

Figure 21 Graph of diameter³ of a modeling clay ball vs mass showing steepest and least steep lines.

This gives a maximum gradient = $1.856 \text{ cm}^3\text{g}^{-1}$ and minimum gradient = $1.746 \text{ cm}^3 \text{ g}^{-1}$.

So: uncertainty in the gradient = $\frac{(1.856 - 1.746)}{2} = 0.06 \text{ cm}^3 \text{g}^{-1}$

Note that the program used to draw the graph (LoggerPro®) gives an uncertainty in the gradient of $\pm 0.03 \text{ cm}^3 \text{g}^{-1}$. This is a more correct value but the steepest and least steep lines method is accepted in IB assessments.

If the y-intercept was of more importance, then constructing steepest and least steep lines would also allow maximum and minimum intercept values to be read off.

The steepest and least steep gradients give maximum and minimum values for the density of:

$$\rho_{max} = \frac{6}{1.746\pi} = 1.094 \text{ g cm}^{-3}$$

$$\rho_{min} = \frac{6}{1.856\pi} = 1.029 \text{ g cm}^{-3}$$

So the uncertainty is: $\frac{(1.094 - 1.029)}{2} = 0.03 \text{ g cm}^{-3}$

The density can now be written as: $1.06 \pm 0.03 \text{ g cm}^{-3}$

Fractional uncertainties

So far, we have dealt with uncertainty as $\pm\Delta x$. This is called the **absolute uncertainty** in the value. Uncertainties can also be expressed as fractions. This has some advantages when processing data.

A value obtained from an experiment can be compared with a 'known' value by seeing if the known value lies within the uncertainty range. Additionally, you could use percentage difference. Find the difference between the experimental and known values and then divide this difference by the known value.

In the previous example, we measured the diameter of modeling clay balls then cubed this value in order to linearize the data. To make the sums simpler, let us consider a slightly bigger ball with a diameter of 10 ± 1 cm.

So the measured value $d = 10$ cm and the absolute uncertainty $\Delta d = 1$ cm.

The fractional uncertainty = $\frac{\Delta d}{d} = \frac{1}{10} = 0.1$ (or, expressed as a percentage, 10%).

During the processing of the data, we found $d^3 = 1000 \text{ cm}^3$.

The uncertainty in this value is not the same as in d. To find the uncertainty in d^3, we need to know the biggest and smallest possible values of d^3. These we can calculate by adding and subtracting the absolute uncertainty:

maximum $d^3 = (10 + 1)^3 = 1331 \text{ cm}^3$

minimum $d^3 = (10 - 1)^3 = 729 \text{ cm}^3$

So the range of values is: $(1331 - 729) = 602 \, \text{cm}^3$

The uncertainty is therefore $\pm 301 \, \text{cm}^3$, which rounded down to one significant figure gives $\pm 300 \, \text{cm}^3$.

This is not the same as $(\Delta d)^3$, which would be $1 \, \text{cm}^3$.

The fractional uncertainty in $d^3 = \frac{300}{1000} = 0.3$. This is the same as $3 \times$ the fractional uncertainty in d. This leads to an alternative way of finding uncertainties when raising data to the power 3.

If $\frac{\Delta x}{x}$ is the fractional uncertainty in x, then the fractional uncertainty in $x^3 = \frac{3\Delta x}{x}$.

More generally, if $\frac{\Delta x}{x}$ is the fractional uncertainty in x, then the fractional uncertainty in $x^n = \frac{n\Delta x}{x}$.

So if you square a value, the fractional uncertainty is $2 \times$ bigger.

Another way of writing this would be that, if $\frac{\Delta x}{x}$ is the fractional uncertainty in x, then the fractional uncertainty in $x^2 = \frac{\Delta x}{x} + \frac{\Delta x}{x}$. This can be extended to any multiplication.

So if $\frac{\Delta x}{x}$ is the fractional uncertainty in x and $\frac{\Delta y}{y}$ is the fractional uncertainty in y, then the fractional uncertainty in $xy = \frac{\Delta x}{x} + \frac{\Delta y}{y}$.

It seems strange but, when dividing, the fractional uncertainties also add. So if $\frac{\Delta x}{x}$ is the fractional uncertainty in x and $\frac{\Delta y}{y}$ is the fractional uncertainty in y, then the fractional uncertainty in $\frac{x}{y} = \frac{\Delta x}{x} + \frac{\Delta y}{y}$.

If you divide a quantity by a constant with no uncertainty, then the fractional uncertainty remains the same.

This is all summarized in the data booklet as:

If $\qquad\qquad\qquad\qquad\qquad\qquad y = \frac{ab}{c}$ then $\frac{\Delta y}{y} = \frac{\Delta a}{a} + \frac{\Delta b}{b} + \frac{\Delta c}{c}$

And if $\qquad\qquad\qquad\qquad\qquad y = a^n$ then $\frac{\Delta y}{y} = n\frac{\Delta a}{a}$

Challenge yourself

1. When a solid ball rolls down a slope of height h, its speed at bottom v is given by the equation:

$$v = \sqrt{\frac{10}{7}gh}$$

where g is the acceleration due to gravity.

In an experiment to determine g, the following results were achieved:

Distance between two markers at the bottom of the slope $d = 5.0 \pm 0.2 \, \text{cm}$

Time taken to travel between markers $t = 0.06 \pm 0.01 \, \text{s}$

Height of slope $h = 6.0 \pm 0.2 \, \text{cm}$.

Given that the speed $v = \frac{d}{t}$, find a value for g and its uncertainty. How might you reduce this uncertainty?

Example

If the length of the side of a cube is quoted as 5.00 ± 0.01 m, what are its volume and the uncertainty in the volume?

$$\text{fractional uncertainty in length} = \frac{0.01}{5} = 0.002$$

$$\text{volume} = 5.00^3 = 125 \text{ m}^3$$

When a quantity is cubed, its fractional uncertainty is 3 × bigger so the fractional uncertainty in volume = 0.002 × 3 = 0.006.

The absolute uncertainty is therefore 0.006 × 125 = 0.75 (approximately 1) so the volume is 125 ±1 m³.

Exercise

Q12. The length of the sides of a cube and its mass are quoted as:
length = 0.050 ± 0.001 m
mass = 1.132 ± 0.002 kg

Calculate the density of the material and its uncertainty.

Q13. The distance around a running track is 400 ± 1 m. If a person runs around the track four times, calculate the distance traveled and its uncertainty.

Q14. The time for 10 swings of a pendulum is 11.2 ± 0.1 s. Calculate the time for one swing of the pendulum and its uncertainty.

Nature of Science

We have seen how we can use numbers to represent physical quantities. By representing those quantities by letters, we can derive mathematical equations to define relationships between them, then use graphs to verify those relationships. Some quantities cannot be represented by a number alone so a whole new area of mathematics needs to be developed to enable us to derive mathematical models relating them.

Vector and scalar quantities

So far we have dealt with six different quantities: length, time, mass, volume, density and displacement.

All of these quantities have a size, but displacement also has a direction. Quantities that have size and direction are **vectors** and those with only size are **scalars**. All quantities are either vectors or scalars. It will be apparent why it is important to make this distinction when we add displacements together.

Example

Consider two displacements one after another as shown in Figure 22.

Starting from A, walk 4 km west to B, then 5 km north to C.

Scalar
A quantity with magnitude only.
Vector
A quantity with magnitude and direction.

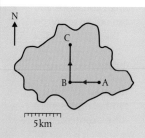

Figure 22 Displacements shown on a map.

The total displacement from the start is not 5 + 4 but can be found by drawing a line from A to C on a scale diagram.

We will find that there are many other vector quantities that can be added in the same way.

Addition of vectors

Vectors can be represented by drawing arrows. The *length* of the arrow is proportional to the magnitude of the quantity and the *direction* of the arrow is the direction of the quantity. The arrow commences at the point of application, the significance of which will become clearer in A.2.

To add vectors, the arrows are simply arranged so that the point of one touches the tail of the other. The resultant vector is found by drawing a line joining the free tail to the free point.

Example

Figure 22 is a map illustrating the different displacements. We can represent the displacements by the vectors in Figure 23.

Calculating the resultant:

If the two vectors are at right angles to each other, then the resultant will be the hypotenuse of a right-angled triangle. This means that we can use simple trigonometry to relate the different sides.

Some simple trigonometry

You will find **cos**, **sin** and **tan** buttons on your calculator. These are used to calculate unknown sides of right-angled triangles.

$$\sin \theta = \frac{\text{opposite}}{\text{hypotenuse}} \rightarrow \text{opposite} = \text{hypotenuse} \times \sin \theta$$

$$\cos \theta = \frac{\text{adjacent}}{\text{hypotenuse}} \rightarrow \text{adjacent} = \text{hypotenuse} \times \cos \theta$$

$$\tan \theta = \frac{\text{opposite}}{\text{adjacent}}$$

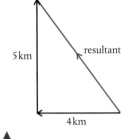

Figure 23 Vector addition.

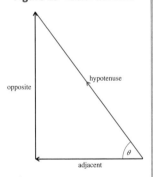

Figure 24 Triangle key terms.

To show that a quantity is a vector, we can write it in a special way. In textbooks, this is often in bold (**A**) but when you write, you can put an arrow on the top. In physics texts, the vector notation is often left out. This is because if we know that the symbol represents a displacement, then we know it is a vector and do not need the vector notation to remind us.

Worked example

Find the side X of the triangle.

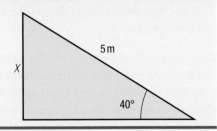

Solution

Side X is the opposite so:

$$X = 5 \times \sin 40°$$

$$\sin 40° = 0.6428 \text{ so } X = 3.2 \text{ m}$$

Exercise

Q15. Use your calculator to find x in each triangle.

(a)

(b)

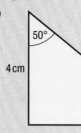

(c)

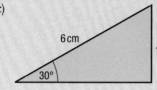

(d)

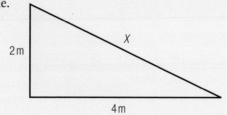

Pythagoras

The most useful mathematical relationship for finding the resultant of two perpendicular vectors is Pythagoras' theorem:

$$\text{hypotenuse}^2 = \text{adjacent}^2 + \text{opposite}^2$$

Worked example

Find the side X on the triangle.

Solution

Applying Pythagoras:

$$X^2 = 2^2 + 4^2$$

So: $X = \sqrt{2^2 + 4^2} = \sqrt{20} = 4.5\text{m}$

Exercise

Q16. Use Pythagoras' theorem to find the hypotenuse in each triangle.

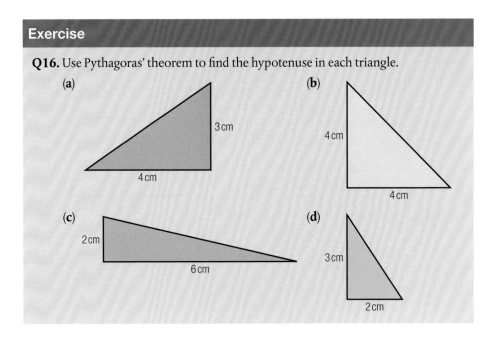

(a) 3 cm / 4 cm

(b) 4 cm / 4 cm

(c) 2 cm / 6 cm

(d) 3 cm / 2 cm

Using trigonometry to solve vector problems

Once the vectors have been arranged point to tail, it is a simple matter of applying the trigonometrical relationships to the triangles that you get.

Exercise

Draw the vectors and solve the following problems using Pythagoras' theorem.

Q17. A boat travels 4 km west followed by 8 km north. What is the resultant displacement?

Q18. A plane flies 100 km north then changes course to fly 50 km east. What is the resultant displacement?

Vectors in one dimension

In this course, we will often consider the simplest examples where the motion is restricted to one dimension, for example, a train traveling along a straight track. In examples like this, there are only two possible directions – forward and backward. To distinguish between the two directions, we give them different signs (forward + and backward –). Adding vectors is now simply a matter of adding the magnitudes, with no need for complicated triangles.

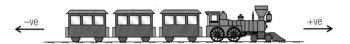

−ve   +ve

▲
Figure 25 The train can only move forward or backward.

You can decide for yourself which you want to be positive but generally we follow the convention below.

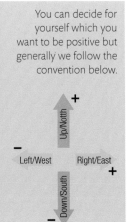

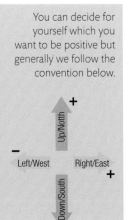

+
Up/North
−
Left/West   Right/East
+
Down/South
−

If a train moves 100 m forward along a straight track then 50 m back, what is its final displacement?

Solution

The vector diagram is as follows.

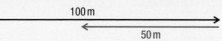

The resultant is 50 m forward.

Subtracting vectors

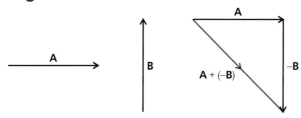

◀ **Figure 27** Subtracting vectors.

🔒 When a vector is multiplied by a scale factor, its alignment is unchanged. If the scale factor is negative, the vector is in the opposite direction. The magnitude is increased by the magnitude of the scale factor.

Now we know that a negative vector is simply the opposite direction to a positive vector, we can subtract vector **B** from vector **A** by changing the direction of vector **B** and adding it to **A**.

$$\mathbf{A} - \mathbf{B} = \mathbf{A} + (-\mathbf{B})$$

Taking components of a vector

Consider someone walking up the hill in Figure 28. They walk 5 km up the slope but want to know how high they have climbed rather than how far they have walked. To calculate this, they can use trigonometry.

$$\text{height} = 5 \times \sin 30°$$

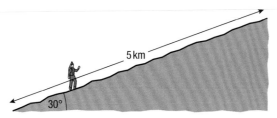

◀ **Figure 28** 5 km up the hill but how high?

The height is called the vertical component of the displacement.

The horizontal displacement can also be calculated.

$$\text{horizontal displacement} = 5 \times \cos 30°$$

This process is called taking components of a vector and is often used in solving physics problems.

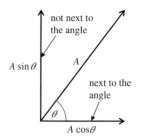

▲ **Figure 29** An easy way to remember which is cos is to say that 'it is becos it is next to the angle'.

Exercise

Q19. If a boat travels 10 km in a direction 30° to the east of north, how far north has it traveled?

Q20. On his way to the South Pole, Amundsen traveled 8 km in a direction that was 20° west of south. What was his displacement south?

Q21. A mountaineer climbs 500 m up a slope that is inclined at an angle of 60° to the horizontal. How high has he climbed?

You will find practice questions and solutions in the eBook.

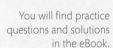

Summary – Tool 3: Mathematics

In terms of mathematics, you should now be aware of:

- scientific notation
- SI prefixes and units
- orders of magnitude
- area and volume
- fundamental units
- derived units in terms of SI units
- approximation and estimation
- dimensional analysis of units for checking expressions
- the significance of uncertainties in raw and processed data
- recording uncertainties in measurements as a range to appropriate precision
- expressing measurement and processed uncertainties to appropriate significant figures or precision
- expressing values to appropriate significant figures or decimal places
- mean and range
- extrapolate and interpolate graphs
- linear and non-linear graphs with appropriate scales and axes
- linearizing graphs
- drawing and interpreting uncertainty bars
- drawing lines or curves of best fit
- constructing maximum and minimum gradient lines by considering all uncertainty bars
- determining uncertainty in gradients and intercepts
- percentage change and percentage difference
- percentage error and percentage uncertainty
- propagation of uncertainties
- scalars and vectors
- scale diagrams
- drawing and labeling vectors
- vector addition and subtraction
- decimals, fractions, percentages, ratios, reciprocals, exponents and trigonometric ratios
- multiplication of vectors by a scalar
- resolving vectors.

There are some mathematical tools that have been introduced here and will be continued, including:

- arithmetic and algebra (see worked example calculations throughout)
- tables and graphs for raw and processed data (see also Sankey diagrams (A.3) and greenhouse gas spectra (B.2))
- direct and inverse proportionality, and positive and negative relationships or correlations (A.1, A.2, B.1, B.2, B.5, C.1, D.1, D.2, D.3)
- interpreting graph features (A.1)
- logarithmic graphs (E.5)
- sketch graphs (labeled but unscaled axes) to qualitatively describe trends (A.1).

The mathematical skills listed in the guide that will be addressed in the textbook content more generally are:

- symbols from the guide and data booklet (throughout)
- selection and manipulation of equations (throughout)
- effect of changes to variables on other variables (throughout)
- use of units (throughout)
- rates of change (A.1, A.3, **HL** D.4)
- neglecting effects and explaining why (A.1)
- free-body diagrams (A.2)
- derivations of equations (B.3, **HL** C.5, **HL** D.1)
- continuous and discrete variables (E.1)
- logarithmic and exponential functions (**HL** E.3).

Video resources

This table lists recommended videos that have been selected to help enhance your learning. The links to the videos can be found in the 'Resources' section of your eBook.

Chapter	Video	Description
Skills in the study of physics	Snooker	A vector quantity, like momentum, has both magnitude and direction.
A.1	Parachute descent	Parachutes are used to reduce the downward acceleration experienced and to enable steering. The parachute used at the end of this video has a larger surface area so that the spacecraft is decelerated.
A.2	Racing car start	The ground exerts a forward force on the tyres and the tyres exert an equal and opposite force on the ground. We can see evidence for this in the movement of the stones.
A.2	Bike riding	The banked wall exerts a force perpendicular to the velocity of the cyclists. They move (temporarily) along the arc of a circle.
A.3	Climbing	This climber, at some point, consumed an energy source and is seeking to convert this into gravitational potential energy. There is also a metaphorical lesson to be learned here about not giving up as you tackle your physics revision!
A.4	Pirouette	Angular momentum and rotational kinetic energy are conserved. When the skater reduces her moment of inertia, by tucking her leg and arms, her angular speed increases.
A.5	Hyperspace	Science fiction writers have long imagined what might happen if travelling at or beyond the speed of light. Physicist Michio Kaku uses the term to refer to higher dimensions in his discussions on the unification of fundamental forces and the fate of the universe.
B.1	Brownian motion	The apparently erratic movement of visible pollen grains can be explained by the presence of many invisible colliding particles in the surrounding fluid.
B.1	Heat conduction in a solid	In high temperature regions, particles vibrate quickly. They transfer energy to neighbouring particles. Eventually, the body as a whole might reach a constant temperature throughout.
B.1	Hot air balloons	These balloons use convection to rise because high temperature gases have a lower density than the surrounding fluid.
B.1	Evaporation	Evaporation of water results in the temperature of the remaining water falling. This is because the particles of highest kinetic energy are more likely to be removed from the surface, meaning that the average kinetic energy of the remaining particles decreases.
B.2	Greenhouse gas emissions	Carbon dioxide release during 2011 and 2012 is represented on this world map. Fossil fuel combustion is not uniformly distributed.
B.2	Albedo	Albedo is the ratio of reflected to incident light. Ice contributes to a higher albedo on Earth.
B.3	Boyle-Mariotte gas law	As the volume of a constant number of moles of gas decreases at constant temperature, pressure increases.
B.3	Gay-Lussac gas law	As the temperature of a constant number of moles of gas increases at constant volume, pressure increases.
B.4	V12 engine	Engines convert heat into work.
B.5	Wall switch	Although the drift speed of electrons in a conductor is roughly 1 mm per second, there is no perceptible delay between pressing a switch and the bulb lighting. This is because the electric force is transmitted at close to the speed of light.

C.1	Seismograph	The recording tip on this seismograph vibrates about a fixed point. The paper moving beneath it shows a wave form.
C.2	Compression wave in a gas	Longitudinal waves are made up of compressions and rarefactions.
C.3	Dispersion	Different wavelengths of light travel at different speeds in a prism and therefore refract by different amounts. White light can be split into its constituent colours.
C.3	Light diffraction through a double slit	Light is diffracted separately at each slit, resulting in an interference pattern. Maxima form where the waves that meet are in phase; minima form where the waves are in antiphase.
C.4	Piano strings	Pianos (as with all stringed and wind instruments) use standing waves. The pitch varies with length and mass per unit length.
C.4	Damped driven pendulums	Displacements are maximized when the driving frequency approximates the natural frequency. Damping reduces the frequency at which the maximum displacement is reached and the amplitude itself.
C.5	Ultrasound imaging	The Doppler effect can be combined with ultrasound imaging to measure the speed of blood flow for diagnosis of cardiovascular conditions.
D.1	Rings of Saturn	The orbital speed of the rocks that make up the rings of Saturn depends on orbital radius but not on the mass of the rocks themselves.
D.2	Magnetic field lines	Iron filings can be used to show the 3D shape of magnetic field lines.
D.3	Aurora	The aurora results from the motion of charged particles from the Sun in the Earth's magnetic field.
D.4	Wind turbines	Inside a wind turbine, a conductor moves relative to a magnetic field. This results in an induced potential difference.
E.1	Rutherford scattering experiment	The (unexpected) outcome of the Geiger–Marsden–Rutherford experiment was that most alpha particles were undeflected but a small proportion were scattered.
E.2	Electron microscope imaging	This image of a mosquito was captured using an electron microscope. The de Broglie wavelength of an electron is smaller than visible light, which means that high resolution is possible.
E.3	Ionizing radiation	Ionizing radiation can damage DNA, which means that precautions must be taken to shield and reduce exposure time by those who work with radioactive isotopes.
E.4	Pressurized water nuclear reactor	This fission reactor contains fuel rods, control rods, a turbine and a generator. Water acts both as the moderator and as the coolant for heat exchange.
E.5	Proton–proton III chain reaction	When nucleons fuse together, energy is released.
E.5	Sunlight	The Suns light forms a spectrum. Shown here are a variety of visible light wavelengths, but the Sun also produces invisible electromagnetic radiation.
E.5	Lifecycle of the Sun	When a nebula collapses under gravity, the temperature might be sufficient for hydrogen nuclei to fuse. The main sequence concludes when the radiation pressure from hydrogen fusion is insufficient to counter gravity, resulting in collapse and then a red giant.
E.5	Milky Way	The Sun is located in the Milky Way galaxy. This time-lapse footage over observatory domes was captured at ESO's La Silla Observatory in the Atacama desert, Chile.

THEME A Space, time and motion

A Space, time and motion

Fireworks displays encapsulate a lot of physics, including thermal energy, light and sound waves, their behavior in the Earth's gravitational field and the effects generated by particular types of atom. As you will see, they also relate to space, time and motion.

There are lots of words that can be used to describe something's motion: distance and displacement, speed and velocity, and acceleration. If you know everything about a body's motion at a specific position and time, then as an IB physicist you will be able to predict its state of motion at another position or time. This is kinematics. When simplified, the equations that govern motion horizontally and vertically can be treated separately. Fireworks are not so simple; they experience air resistance and continue to combust their fuel mid-flight.

The burning of fuels to generate changes in motion relates to forces and momentum. Isaac Newton articulated three laws that describe how a lack of resultant force means there is no change in velocity, a resultant force leads to a change in momentum and the force of one body 'A' on another 'B' means that the same type and size of force must be being exerted by 'B' on 'A' in precisely the opposite direction. There are types of force to contend with, and of course not all forces act in the same direction that the body is already moving in; circular motion results from perpendicular forces and velocities and has its own set of governing equations.

If kinematics is the study of the journey, energy is the study of the start and end. Energy, along with momentum, is a conserved quantity that can be changed only if work is done. Power is another term still; it is the rate at which energy is changed or work is done. Balancing a 100 g apple in your hand requires a force of about 1 N. Lifting it vertically to arm's reach requires you to provide about 1 J of work. Doing so repeatedly every second represents 1 W of power, irrespective of how long in total you do it for.

Some bodies rotate, and you will study how the angles and angular velocities of circular motion can be linked to the kinematics equations in the Rigid Body Mechanics chapter (A.4). Other bodies have velocities similar in magnitude to the speed of light, which leads to relativistic effects like time dilation and length contraction. But fear not. In the first case you will always be solving problems with real-world connections. In the latter, you will get to know about the experimental evidence for these effects as well as how to use the Lorentz transformation and space–time diagrams to your advantage.

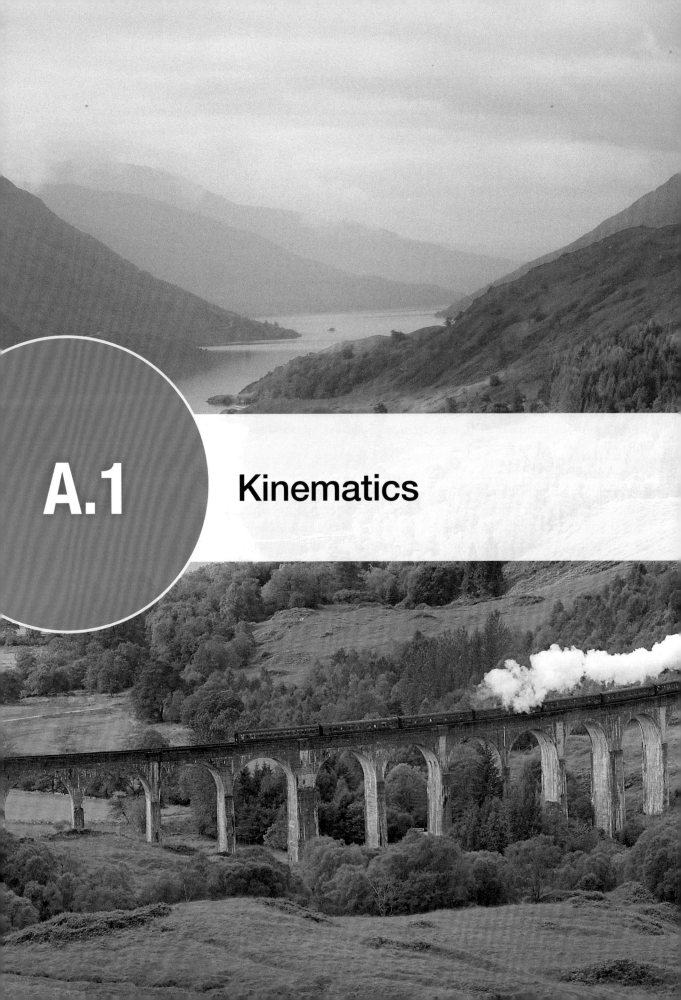

A.1 Kinematics

◀ When can you think of a steam train as a particle?

Guiding Questions

How can the motion of a body be described quantitatively and qualitatively?

How can the position of a body in space and time be predicted?

How can the analysis of motion in one and two dimensions be used to solve real-life problems?

The photograph at the start of this chapter shows a train, but we will not be dealing with complicated systems like trains in their full complexity. In physics, we try to understand everything on the most basic level. Understanding a physical system means being able to predict its final conditions given its initial conditions. To do this for a train, we would have to calculate the position and motion of every part – and there are a lot of parts. In fact, if we considered all the particles that make up all the parts, then we would have a huge number of particles to deal with.

In this course, we will be dealing with one particle of matter at a time. This is because the ability to solve problems with one particle makes us able to solve problems with many particles. We may even pretend a train is one particle.

The initial conditions of a particle describe where it is and what it is doing. These can be defined by a set of numbers, which are the results of measurements. As time passes, some of these quantities might change. What physicists try to do is predict their values at any given time in the future. To do this, they use mathematical models.

Nature of Science

From the definitions of velocity and acceleration, we can use mathematics to derive a set of equations that predict the position and velocity of a particle at any given time. We can show by experiment that these equations give the correct result for some examples, then make the generalization that the equations apply in all cases.

Students should understand:

that the motion of bodies through space and time can be described and analyzed in terms of position, velocity, and acceleration
velocity is the rate of change of position, and acceleration is the rate of change of velocity
the change in position is the displacement
the difference between distance and displacement
the difference between instantaneous and average values of velocity, speed and acceleration, and how to determine them

the equations of motion for solving problems with uniformly accelerated motion as given by:

$$s = \frac{u + v}{2} t$$

$$v = u + at$$

$$s = ut + \frac{1}{2} at^2$$

$$v^2 = u^2 + 2as$$

motion with uniform and non-uniform acceleration

the behavior of projectiles in the absence of fluid resistance, and the application of the equations of motion resolved into vertical and horizontal components

the qualitative effect of fluid resistance on projectiles, including time of flight, trajectory, velocity, acceleration, range and terminal speed.

Further information about the fluid resistance force can be found in A.2.

Nature of Science

In the Tools chapter, we observed that things move and now we are going to mathematically model that movement. Before we do that, we must define some quantities.

Displacement and distance

It is important to understand the difference between distance traveled and displacement. To explain this, consider the route marked out on the map shown in Figure 1.

Displacement is the shortest path moved in a particular direction.

The unit of displacement is the meter (m). Displacement is a vector quantity.

On the map, the displacement is the length of the straight line from A to B, which is a distance of 5 km west.

Distance is how far you have traveled from A to B.

The unit of distance is also the meter (m). Distance is a scalar quantity.

In this example, the distance traveled is the length of the path taken, which is about 10 km.

Sometimes, this difference leads to a surprising result. For example, if you run all the way round a running track, you will have traveled a distance of 400 m but your displacement will be 0 m.

In everyday life, it is often more important to know the distance traveled. For example, if you are going to travel from Paris to Lyon by road, you will want to know that the distance by road is 450 km, not that your final displacement will be 336 km southeast. However, in physics, we break everything down into its simplest parts, so we start by considering motion in a straight line only. In this case, it is more useful to know the displacement, since that also has information about which direction you have traveled in.

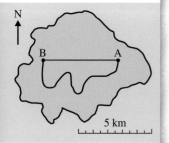

A.1 Figure 1

Note: since displacement is a vector, you should always say what the direction is.

Velocity and speed

Both speed and velocity are a measure of how fast a body is moving.

Velocity is defined as the rate of change of position. Since 'change of position' is displacement and 'rate of change' requires division by time taken:

$$velocity = \frac{displacement}{time}$$

The unit of velocity is m s^{-1}.

Velocity is a vector quantity.

Speed is defined as the distance traveled per unit time:

$$speed = \frac{distance}{time}$$

The unit of speed is also m s^{-1}.

Speed is a scalar quantity.

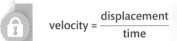

$$velocity = \frac{displacement}{time}$$

$$speed = \frac{distance}{time}$$

Exercise

Q1. Convert the following speeds into m s^{-1}:

    (a) a car traveling at 100 km h^{-1}

    (b) a runner running at 20 km h^{-1}.

Average velocity and instantaneous velocity

Consider traveling by car from the north of Bangkok to the south – a distance of about 16 km. If the journey takes 4 hours, you can calculate your velocity to be $\frac{16}{4}$ = 4 km h^{-1} in a southward direction. This does not tell you anything about the journey, just the difference between the beginning and the end (unless you managed to travel at a constant speed in a straight line). The value calculated is the **average velocity** and in this example it is quite useless. If we broke the trip down into lots of small pieces, each lasting only one second, then for each second the car could be considered to be traveling in a straight line at a constant speed. For these short stages, we could quote the car's **instantaneous velocity** – which is how fast it is going at that moment in time and in which direction.

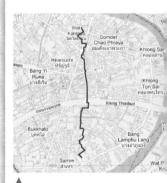

▲
A.1 Figure 2 It is not possible to take this route across Bangkok with a constant velocity.

◀ The bus in the photo has a constant velocity for a very short time.

7

Exercise

Q2. A runner runs once around a circular track of length 400 m with a constant speed in 96 s. Calculate:

(a) the average speed of the runner

(b) the average velocity of the runner

(c) the instantaneous velocity of the runner after 48 s

(d) the displacement after 24 s.

Constant velocity

If the velocity is constant, then the instantaneous velocity is the same all the time so:

$$\text{instantaneous velocity} = \text{average velocity}$$

Since velocity is a vector, this also implies that the direction of motion is constant.

Measuring a constant velocity

From the definition of velocity, we see that:

$$\text{velocity} = \frac{\text{displacement}}{\text{time}}$$

Rearranging this gives:

$$\text{displacement} = \text{velocity} \times \text{time}$$

So, if velocity is constant, displacement is proportional to time. To test this relationship and find the velocity, we can measure the displacement of a body at different times. To do this, you either need a lot of clocks or a stop clock that records many times. This is called a **lap timer**. In this example, a bicycle was ridden at constant speed along a straight road past six students standing 10 m apart, each operating a stop clock as in Figure 3. The clocks were all started when the bike, already moving, passed the start marker and stopped as the bike passed each student.

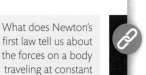

What does Newton's first law tell us about the forces on a body traveling at constant velocity? (A.2)

A.1 Figure 3 Measuring the time for a bike to pass.

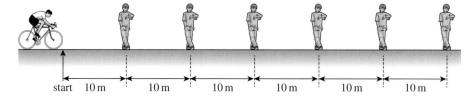

The results achieved are shown in Table 1.

The uncertainty in displacement is given as 0.1 m since it is difficult to decide exactly when the bike passed the marker.

The digital stop clock has a scale with 2 decimal places, so the uncertainty is 0.01 s. However, the uncertainty given is 0.02 s since the clocks all had to be started at the same time.

Since displacement (s) is proportional to time (t), then a graph of s vs t should give a straight line with gradient = velocity as shown in Figure 4.

Displacement/m ±0.1 m	Time/s ±0.02 s
10.0	3.40
20.0	5.62
30.0	8.55
40.0	12.31
50.0	14.17
60.0	17.21

▲ A.1 Table 1

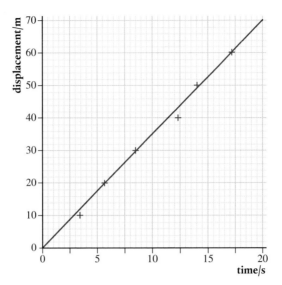

▲ A.1 Figure 4 Graph of displacement vs time for a bike.

Notice that in this graph the line does not pass through all the points. This is because the uncertainty in the measurement in time is almost certainly bigger than the uncertainty in the clock (±0.02 s) due to the reaction time of the students stopping the clock. To get a better estimate of the uncertainty, we would need to have several students standing at each 10 m position. Repeating the experiment is not possible in this example since it is very difficult to ride at the same velocity several times.

The gradient indicates that: velocity = 3.5 m s^{-1}

SKILLS

Most school laboratories are not large enough to ride bikes in so when working indoors, we need to use shorter distances. This means that the times are going to be shorter so hand-operated stop clocks will have too great a percentage uncertainty. One way of timing in the lab is by using photogates. These are connected to a computer via an interface and record the time when a body passes in or out of the gate. So, to replicate the bike experiment in the lab using a ball, we would need seven photogates as in Figure 5, with one extra gate to represent the start.

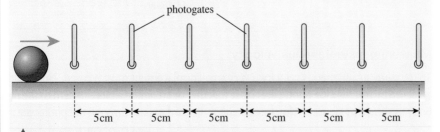

photogates

| 5 cm | 5 cm | 5 cm | 5 cm | 5 cm | 5 cm |

▲ A.1 Figure 5 How to measure the time for a rolling ball if you have seven photogates.

This would be quite expensive so we compromise by using just two photogates and a motion that can be repeated. An example could be a ball moving along a horizontal section of track after it has rolled down an inclined plane. Provided the ball starts from the same point, it should have the same velocity. So, instead of using seven photogates, we can use two – one is at the start of the motion and the other is moved to different positions along the track as in Figure 6.

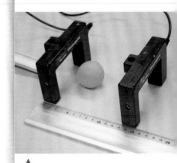

▲ A.1 Figure 6 The ball interrupts the infrared light transmitted across each gate as it passes through them. The times of these interruptions are measured and recorded by a data logger.

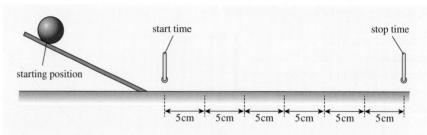

start time   stop time

starting position

5cm 5cm 5cm 5cm 5cm 5cm

Table 2 shows the results obtained using this arrangement.

A.1 Table 2

Displacement/ cm ± 0.1 cm	Time(t)/s ± 0.0001 s					Mean t/s	Δt/s
5.0	0.0997	0.0983	0.0985	0.1035	0.1040	0.101	0.003
10.0	0.1829	0.1969	0.1770	0.1824	0.1825	0.18	0.01
15.0	0.2844	0.2800	0.2810	0.2714	0.2779	0.28	0.01
20.0	0.3681	0.3890	0.3933	0.3952	0.3854	0.39	0.01
25.0	0.4879	0.5108	0.5165	0.4994	0.5403	0.51	0.03
30.0	0.6117	0.6034	0.5978	0.6040	0.5932	0.60	0.01

A.1 Figure 8 Graph of displacement vs time for a rolling ball.

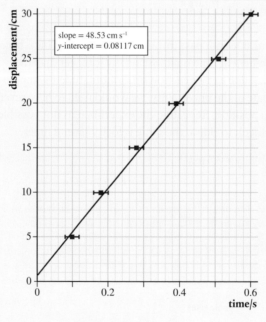

slope = 48.53 cm s⁻¹
y-intercept = 0.08117 cm

displacement/cm vs time/s

Notice that the uncertainty calculated from $\frac{(max - min)}{2}$ is much more than the instrument uncertainty. A graph of displacement vs time gives Figure 8.

From this graph, we can see that within the limits of the experiment's uncertainties the displacement could be proportional to time, so we can conclude that the velocity may have been constant. However, if we look closely at the data, we see that there seems to be a slight curve, indicating that perhaps the ball was slowing down. To verify this, we would have to collect more data.

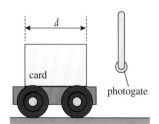

d

card

photogate

A.1 Figure 9 A card and photogate used to measure instantaneous velocity.

Measuring instantaneous velocity

To measure instantaneous velocity, a very small displacement must be used. This could be achieved by placing two photogates close together or attaching a piece of card to the moving body as shown in Figure 9. The time taken for the card to pass through the photogate is recorded and the instantaneous velocity calculated from: $\frac{\text{length of card}}{\text{time taken}} \left(\frac{d}{t} \right)$

Relative velocity

Velocity is a vector so velocities must be added as vectors. Imagine you are running north at $3\,\text{m s}^{-1}$ on a ship that is also traveling north at $4\,\text{m s}^{-1}$ as shown in Figure 10. Your velocity relative to the ship is $3\,\text{m s}^{-1}$ but your velocity relative to the water is $7\,\text{m s}^{-1}$. If you turn around and run due south, your velocity will still be $3\,\text{m s}^{-1}$ relative to the ship but $1\,\text{m s}^{-1}$ relative to the water. Finally, if you run toward the east, the vectors add at right angles to give a resultant velocity of magnitude $5\,\text{m s}^{-1}$ relative to the water. You can see that the velocity vectors have been added.

What is the relative speed of the light from a star measured by a rocket traveling at 0.5 times the speed of light toward the star? (A.5)

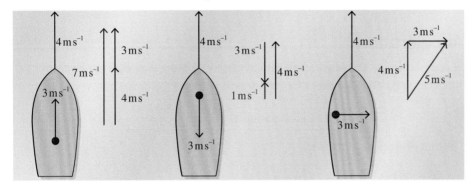

A.1 Figure 10 Running on board a ship.

Imagine that you are floating in the water watching two boats traveling toward each other as in Figure 11.

A.1 Figure 11 Two boats approach each other. The vector addition for the velocity of the green boat from the perspective of the blue boat is shown.

The blue boat is traveling east at $4\,\text{m s}^{-1}$ and the green boat is traveling west at $-3\,\text{m s}^{-1}$. Remember that the sign of a vector in one dimension gives the direction. So, if east is positive, then west is negative. If you were standing on the blue boat, you would see the water going past at $-4\,\text{m s}^{-1}$ so the green boat would approach with the velocity of the water plus its velocity in the water: $-4 + -3 = -7\,\text{m s}^{-1}$. This can also be done in two dimensions as in Figure 12.

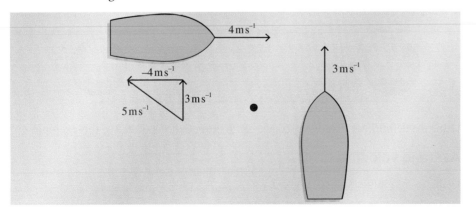

A.1 Figure 12 Two boats traveling perpendicular to each other. The vector addition for the velocity of the green boat from the perspective of the blue boat is shown.

According to the swimmer floating in the water, the green boat travels north and the blue boat travels east, but an observer on the blue boat will see the water traveling toward the west and the green boat traveling due north. Adding these two velocities gives a velocity of $5\,\text{m s}^{-1}$ in an approximately northwest direction.

How effectively do the equations of motion model Newton's laws of dynamics? (A.2)

Q3. An observer standing on a road watches a bird flying east at a velocity of $10\,\text{m}\,\text{s}^{-1}$. A second observer, driving a car along the road northward at $20\,\text{m}\,\text{s}^{-1}$, sees the bird. What is the velocity of the bird relative to the driver?

Q4. A boat travels along a river heading north with a velocity $4\,\text{m}\,\text{s}^{-1}$ as a woman walks across a bridge from east to west with velocity of $1\,\text{m}\,\text{s}^{-1}$. Calculate the velocity of the woman relative to the boat.

Acceleration

In everyday usage, the word **accelerate** means to go faster. However, in physics, acceleration is defined as the rate of change of velocity:

$$\text{acceleration} = \frac{\text{change of velocity}}{\text{time}}$$

The unit of acceleration is $\text{m}\,\text{s}^{-2}$.

Acceleration is a vector quantity.

This means that whenever a body changes its velocity, it accelerates. This could be because it is getting faster, slower, or just changing direction. In the example of the journey across Bangkok, the car would have been slowing down, speeding up and going round corners almost the whole time so it would have had many different accelerations. However, this example is far too complicated for us to consider in this course (and probably any physics course). For most of this chapter, we will only consider the simplest example of accelerated motion, which is constant acceleration.

Constant acceleration in one dimension

In one-dimensional motion, acceleration, velocity and displacement are all in the same direction. This means they can be added without having to draw triangles. Figure 13 shows a body that is starting from an initial velocity u and accelerating at a constant rate a to velocity v in t seconds. The distance traveled in this time is s. Since the motion is in a straight line, this is also the displacement.

A.1 Figure 13 A red ball with constant acceleration.

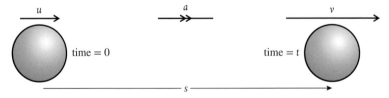

Using the definitions already stated, we can write equations related to this example.

Average velocity

From the definition, $\text{average velocity} = \dfrac{\text{displacement}}{\text{time}}$

$$\text{average velocity} = \frac{s}{t} \qquad (1)$$

Since the velocity changes at a constant rate from the beginning to the end, we can also calculate the average velocity by adding the initial and final velocities and dividing by two:

$$\text{average velocity} = \frac{(u + v)}{2} \qquad (2)$$

Acceleration

Acceleration is defined as the rate of change of velocity:

$$a = \frac{(v - u)}{t} \qquad (3)$$

We can use these equations to solve any problem involving constant acceleration. However, to make problem-solving easier, we can derive two more equations by substituting from one into the other.

Equating equations (1) and (2):

$$\frac{s}{t} = \frac{(u + v)}{2}$$

$$s = \frac{(u + v)\,t}{2} \qquad (4)$$

Rearranging (3) gives: $v = u + at$

If we substitute for v in equation (4), we get: $s = ut + \frac{1}{2}at^2 \qquad (5)$

Rearranging (3) again gives: $t = \frac{(v - u)}{a}$

If t is now substituted in equation (4), we get: $v^2 = u^2 + 2as \qquad (6)$

These equations are sometimes known as the *suvat* equations. If you know any three of s, u, v, a, and t, you can find either of the other two in one step.

These equations are known as the *suvat* equations:

$$a = \frac{(v - u)}{t}$$

$$s = \frac{(v + u)t}{2}$$

$$s = ut + \frac{1}{2}at^2$$

$$v^2 = u^2 + 2as$$

How are the equations for rotational motion related to those for linear motion? (A.4)

A GeoGebra worksheet linked to this topic is available in the eBook.

Worked example

A car traveling at $10\,\mathrm{m\,s^{-1}}$ accelerates at $2\,\mathrm{m\,s^{-2}}$ for 5 s. What is its displacement?

Solution

The first thing to do is draw a simple diagram:

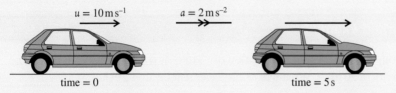

This enables you to see what is happening at a glance rather than reading the text. The next stage is to make a list of *suvat*.

$s = ?$

$u = 10\,\mathrm{m\,s^{-1}}$

$v = ?$

$a = 2\,\mathrm{m\,s^{-2}}$

$t = 5\,\mathrm{s}$

To find s, you need an equation that contains *suat*. The only equation with all four of these quantities is: $s = ut + \frac{1}{2}at^2$

Using this equation gives:

$$s = 10 \times 5 + \frac{1}{2} \times 2 \times 5^2$$

$$s = 75\,\mathrm{m}$$

When the units are consistent, you do not need to include units in all stages of a calculation, just in the answer.

The signs of displacement, velocity and acceleration

We must not forget that displacement, velocity and acceleration are vectors. This means that they have direction. However, since this is a one-dimensional example, there are only two possible directions, forward and backward. We know which direction the vector is in from its sign.

If we take right to be positive:
- A positive displacement means that the body has moved to the right.
- A positive velocity means the body is moving to the right.
- A positive acceleration means that the body is either moving to the right and getting faster or moving to the left and getting slower. This can be confusing so consider the following example.

A.1 Figure 14 A car moves to the left with decreasing speed.

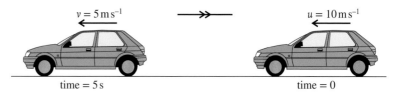

$v = 5\,\mathrm{m\,s^{-1}}$ $u = 10\,\mathrm{m\,s^{-1}}$

time = 5 s   time = 0

The car is traveling in a negative direction so the velocities are negative.

$$u = -10\,\mathrm{m\,s^{-1}}$$

$$v = -5\,\mathrm{m\,s^{-1}}$$

$$t = 5\,\mathrm{s}$$

The acceleration is therefore given by:

$$a = \frac{(v - u)}{t} = \frac{-5 - (-10)}{5} = 1\,\mathrm{m\,s^{-2}}$$

The positive sign tells us that the acceleration is in a positive direction (right) even though the car is traveling in a negative direction (left).

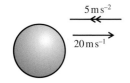

$5\,\mathrm{m\,s^{-2}}$

$20\,\mathrm{m\,s^{-1}}$

A.1 Figure 15 The acceleration is negative so points to the left.

Worked example

A body with a constant acceleration of $-5\,\mathrm{m\,s^{-2}}$ is traveling to the right with a velocity of $20\,\mathrm{m\,s^{-1}}$. What will its displacement be after 20 s?

Solution

$s = ?$

$u = 20\,\mathrm{m\,s^{-1}}$

$v = ?$

$a = -5\,\mathrm{m\,s^{-2}}$

$t = 20\,\mathrm{s}$

To calculate s, we can use the equation: $s = ut + \frac{1}{2}at^2$

$$s = 20 \times 20 + \frac{1}{2}(-5) \times 20^2 = 400 - 1000 = -600\,\mathrm{m}$$

This means that the final displacement of the body is to the left of the starting point. It has gone forward, stopped, and then gone backward.

Q5. Calculate the final velocity of a body that starts from rest and accelerates at $5\,\text{m s}^{-2}$ for a distance of $100\,\text{m}$.

Q6. A body starts with a velocity of $20\,\text{m s}^{-1}$ and accelerates for $200\,\text{m}$ with an acceleration of $5\,\text{m s}^{-2}$. What is the final velocity of the body?

Q7. A body accelerates at $10\,\text{m s}^{-2}$ and reaches a final velocity of $20\,\text{m s}^{-1}$ in $5\,\text{s}$. What is the initial velocity of the body?

Free fall motion

Although a car has been used in the previous examples, the acceleration of a car is not usually constant so we should not use the *suvat* equations. The only example of constant acceleration that we see in everyday life is when a body is dropped. Even then, the acceleration is only constant for a short distance.

Acceleration of free fall

When a body is allowed to fall freely, we say it is in free fall. Bodies falling freely on the Earth fall with an acceleration of about $9.81\,\text{m s}^{-2}$ (depending where you are). The body falls because of gravity. For that reason, we use the letter g to denote this acceleration. Since the acceleration is constant, we can use the *suvat* equations to solve problems.

How does the motion of an object change within a gravitational field? (D.1)

If you jump out of a plane (with a parachute on), you will feel the push of the air as it rushes past you. As you fall faster and faster, the air will push upward more and more until you cannot go any faster. At this point, you have reached terminal velocity. We will come back to this after introducing forces.

Exercise

In these calculations, use $g = 10\,\text{m s}^{-2}$.

Q8. A ball is thrown upward with a velocity of $30\,\text{m s}^{-1}$. What is the displacement of the ball after $2\,\text{s}$?

Q9. A ball is dropped. What will its velocity be after falling $65\,\text{cm}$?

Q10. A ball is thrown upward with a velocity of $20\,\text{m s}^{-1}$. After how many seconds will the ball return to its starting point?

Measuring the acceleration due to gravity

When a body falls freely under the influence of gravity, it accelerates at a constant rate. This means that time to fall t and distance s are related by the equation: $s = ut + \frac{1}{2}at^2$. If the body starts from rest, then $u = 0$ so the equation becomes: $s = \frac{1}{2}at^2$. Since s is directly proportional to t^2, a graph of s vs t^2 would therefore be a straight line with gradient $\frac{1}{2}g$. It is difficult to measure the time for a ball to pass different markers, but if we assume the ball falls with the same acceleration when repeatedly dropped, we can measure the time taken for the ball to fall from different heights. There are many ways of doing this. All involve some way of starting a clock when the ball is released and stopping it when it hits the ground. Table 3 shows a set of results from a 'ball drop' experiment.

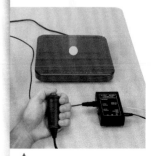

▲
Apparatus for measuring *g*.

A.1 Table 3 ▶

Height(h)/m ± 0.001 m	Time(t)/s ± 0.001 s					Mean t/s	t^2/s²	$\Delta(t^2)$/s²
0.118	0.155	0.153	0.156	0.156	0.152	0.154	0.024	0.001
0.168	0.183	0.182	0.183	0.182	0.184	0.183	0.0334	0.0004
0.218	0.208	0.205	0.210	0.211	0.210	0.209	0.044	0.001
0.268	0.236	0.235	0.237	0.239	0.231	0.236	0.056	0.002
0.318	0.250	0.254	0.255	0.250	0.256	0.253	0.064	0.002
0.368	0.276	0.277	0.276	0.278	0.276	0.277	0.077	0.001
0.418	0.292	0.293	0.294	0.291	0.292	0.292	0.085	0.001
0.468	0.310	0.310	0.303	0.300	0.311	0.307	0.094	0.003
0.518	0.322	0.328	0.330	0.328	0.324	0.326	0.107	0.003
0.568	0.342	0.341	0.343	0.343	0.352	0.344	0.118	0.004

If a parachutist kept accelerating at a constant rate, they would break the sound barrier after about 30 s of flight. By understanding the forces involved, scientists have been able to design wing suits so that base jumpers can achieve forward velocities greater than their rate of falling.

SKILLS

A worksheet with full details of how to carry out this experiment is available in the eBook.

Notice that the uncertainty in t^2 is calculated from: $\dfrac{(t_{max}^2 - t_{min}^2)}{2}$

Notice how the line in Figure 16 is very close to the points and that the uncertainties reflect the actual random variation in the data. The gradient of the line is equal to $\frac{1}{2}g$ so $g = 2 \times$ gradient.

$$g = 2 \times 4.814 = 9.624 \text{ m s}^{-2}$$

The uncertainty in this value can be estimated from the steepest and least steep lines:

$$g_{max} = 2 \times 5.112 = 10.224 \text{ m s}^{-2}$$

$$g_{min} = 2 \times 4.571 = 9.142 \text{ m s}^{-2}$$

$$\Delta g = \frac{(g_{max} - g_{min})}{2} = \frac{(10224 - 9.142)}{2} = 0.541 \text{ m s}^{-2}$$

So, the final value including uncertainty is $9.6 \pm 0.5 \text{ m s}^{-2}$.

This is in agreement with the accepted average value which is 9.81 m s^{-2}.

A.1 Figure 16 Height vs time² for a falling object.

Why would it not be appropriate to apply the *suvat* equations to the motion of a body falling freely from a distance of 2 times the Earth's radius to the surface of the Earth? (D.1)

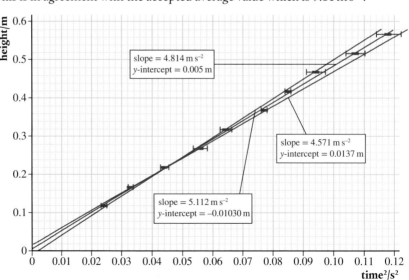

Graphical representation of motion

Graphs are used in physics to give a visual representation of relationships. In kinematics, they can be used to show how displacement, velocity and acceleration change with time. Figure 17 shows the graphs for four different examples of motion.

The best way to sketch graphs is to split the motion into sections then plot where the body is at different times. Joining these points will give the displacement–time graph. Once you have done that, you can work out the v–t and a–t graphs by looking at the s–t graph rather than the motion.

Gradient of displacement–time graph

The gradient of a graph is: $\dfrac{\text{change in } y}{\text{change in } x} = \dfrac{\Delta y}{\Delta x}$

In the case of the displacement–time graph, this will give:

$$\text{gradient} = \frac{\Delta s}{\Delta t}$$

This is the same as velocity.

We can represent the motion of a body on displacement–time graphs, velocity–time graphs and acceleration–time graphs. The three graphs of these types shown in Figure 17 display the motion of four bodies, which are labeled A, B, C and D.

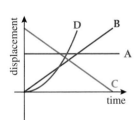 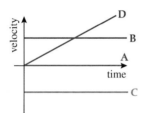 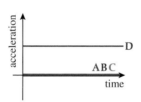

◀ **A.1 Figure 17** Graphical representation of motion.

Body A

A body that is not moving.
Displacement is always the same.
Velocity is zero.
Acceleration is zero.

Body B

A body that is traveling with a constant positive velocity.
Displacement increases linearly with time.
Velocity is a constant positive value.
Acceleration is zero.

Body C

A body that has a constant negative velocity.
Displacement is decreasing linearly with time.
Velocity is a constant negative value.
Acceleration is zero.

Body D

A body that is accelerating with constant acceleration.
Displacement is increasing at a non-linear rate. The shape of this line is a parabola since displacement is proportional to t^2 ($s = ut + \frac{1}{2}at^2$).
Velocity is increasing linearly with time.
Acceleration is a constant positive value.

So, the gradient of the displacement–time graph equals the velocity. Using this information, we can see that line A in Figure 18 represents a body with a greater velocity than line B, and that since the gradient of line C is increasing, this must be the graph for an accelerating body.

You need to be able to:
- work out what kind of motion a body has by looking at the graphs
- sketch graphs for a given motion.

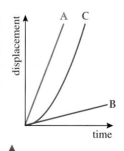

▲ **A.1 Figure 18** Three new bodies to compare.

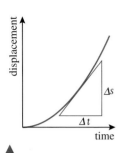

A.1 Figure 19 Finding the gradient of the tangent.

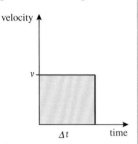

A.1 Figure 20 The area is displacement.

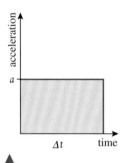

A.1 Figure 21 The area is change in velocity.

How does analyzing graphs allow us to determine other physical quantities? (NOS)

A GeoGebra worksheet linked to this topic is available in the eBook.

SKILLS

Instantaneous velocity

When a body accelerates, its velocity is constantly changing. The displacement–time graph for this motion is therefore a curve. To find the instantaneous velocity from the graph, we can draw a tangent to the curve and find the gradient of the tangent as shown in Figure 19.

Area under velocity–time graph

The area under the velocity–time graph for the body traveling at constant velocity v shown in Figure 20 is given by:

$$\text{area} = v\Delta t$$

But we know from the definition of velocity that: $v = \dfrac{\Delta s}{\Delta t}$

Rearranging gives $\Delta s = v\Delta t$ so the area under a velocity–time graph gives the displacement.

This is true, not only for simple cases such as this, but for all examples.

Gradient of velocity–time graph

The gradient of the velocity–time graph is given by $\dfrac{\Delta v}{\Delta t}$. This is the same as acceleration.

Area under acceleration–time graph

The area under the acceleration–time graph in Figure 21 is given by $a\Delta t$. But we know from the definition of acceleration that: $a = \dfrac{(v - u)}{t}$

Rearranging this gives $v - u = a\Delta t$ so the area under the graph gives the change in velocity.

If you have covered calculus in your mathematics course, you may recognize these equations:

$$v = \frac{ds}{dt}, a = \frac{dv}{dt} = \frac{d^2s}{d^2t} \text{ and } s = \int v dt, v = \int a dt$$

Exercise

Q11. Sketch a velocity–time graph for a body starting from rest and accelerating at a constant rate to a final velocity of $25\,\text{m s}^{-1}$ in 10 seconds. Use the graph to find the distance traveled and the acceleration of the body.

Q12. Describe the motion of the body whose velocity–time graph is shown. What is the final displacement of the body?

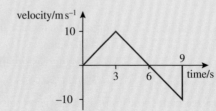

Q13. A ball is released from rest on the hill in the figure below. Sketch the *s–t*, *v–t*, and *a–t* graphs for its horizontal motion.

Q14. A ball rolls along a table then falls off the edge, landing on soft sand. Sketch the *s–t*, *v–t*, and *a–t* graphs for its vertical motion.

Example 1: The *suvat* example

As an example, let us consider the motion we looked at when deriving the *suvat* equations.

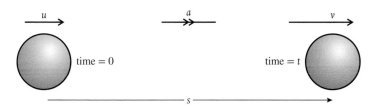

time = 0   time = *t*

s

◀ **A.1 Figure 22** A body with constant acceleration.

Negative time does not mean going back in time – it means the time before you started the clock.

Displacement–time

The body starts with velocity *u* and travels to the right with constant acceleration *a* for a time *t*. If we take the starting point to be zero displacement, then the displacement–time graph starts from zero and rises to *s* in *t* seconds. We can therefore plot the two points shown in Figure 23. The body is accelerating so the line joining these points is a parabola. The whole parabola has been drawn to show what it would look like – the reason it is offset is because the body is not starting from rest. The part of the curve to the left of the origin tells us what the particle was doing before we started the clock.

Velocity–time

Figure 24 is a straight line with a positive gradient showing that the acceleration is constant. The line does not start from the origin since the initial velocity is *u*.

The gradient of this line is $\frac{(v - u)}{t}$, which we know from the *suvat* equations is acceleration.

The area under the line makes the shape of a trapezium. The area of this trapezium is $\frac{1}{2}(v + u)t$. This is the *suvat* equation for *s*.

Acceleration–time

The acceleration is constant so the acceleration–time graph is a horizontal line as shown in Figure 25. The area under this line is $a \times t$, which we know from the *suvat* equations equals $(v - u)$.

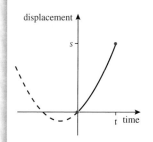

displacement

s

t time

▲ **A.1 Figure 23** Constant acceleration.

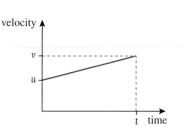

velocity

v

u

t time

◀ **A.1 Figure 24** Constant acceleration.

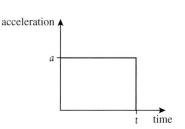

acceleration

a

t time

◀ **A.1 Figure 25** Constant acceleration.

Example 2: The bouncing ball

Consider a rubber ball dropped from position A above the ground onto hard surface B. The ball bounces up and down several times. Figure 26 shows the displacement–time graph for four bounces. From the graph, we see that the ball starts above the ground then falls with increasing velocity (as shown by the increasing negative gradient). When the ball bounces at B, the velocity suddenly changes from negative to positive as the ball begins to travel back up. As the ball goes up, its velocity decreases until it stops at C and begins to fall again.

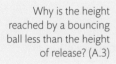

A.1 Figure 26 Vertical displacement vs time.

Why is the height reached by a bouncing ball less than the height of release? (A.3)

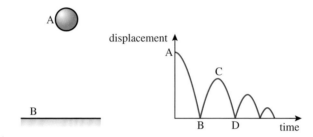

Exercise

Q15. By considering the gradient of the displacement–time graph in Figure 26, plot the velocity–time graph for the motion of the bouncing ball.

Example 3: A ball falling with air resistance

Figure 27 shows the motion of a ball that is dropped several hundred meters through the air. It starts from rest and accelerates for some time. As the ball accelerates, the air resistance increases, which stops the ball from getting any faster. At this point, the ball continues with constant velocity.

A.1 Figure 27 Vertical displacement vs time.

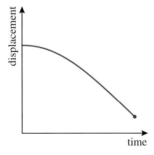

Exercise

Q16. By considering the gradient of the displacement–time graph, plot the velocity–time graph for the motion of the falling ball in Figure 27.

Projectile motion

We all know what happens when a ball is thrown. It follows a curved path like the one in the photo. We can see from this photo that the path is parabolic and later we will show why that is the case.

Modeling projectile motion

All examples of motion up to this point have been in one dimension but projectile motion is two-dimensional. However, if we take components of all the vectors vertically and horizontally, we can simplify this into two simultaneous one-dimensional problems. The important thing to realize is that the vertical and horizontal components are independent of each other. You can test this by dropping an eraser off your desk and flicking one forward at the same time – they both hit the floor together. The downward motion is not changed by the fact that one stone is also moving forward.

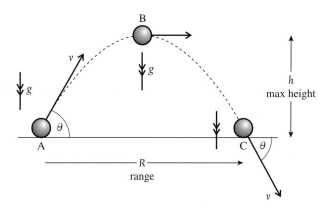

Consider a ball that is projected at an angle θ to the horizontal, as shown in Figure 28. We can split the motion into three parts, beginning, middle and end, and analyze the vectors representing displacement, velocity and time at each stage. Notice that the path is symmetrical, so the motion on the way down is the same as on the way up.

Horizontal components

At A (time = 0)	At B (time = $\frac{t}{2}$)	At C (time = t)
displacement = zero	displacement = $\frac{R}{2}$	displacement = R
velocity = $v \cos \theta$	velocity = $v \cos \theta$	velocity = $v \cos \theta$
acceleration = 0	acceleration = 0	acceleration = 0

Vertical components

At A	At B	At C
displacement = zero	displacement = h	displacement = zero
velocity = $v \sin \theta$	velocity = zero	velocity = $-v \sin \theta$
acceleration = $-g$	acceleration = $-g$	acceleration = $-g$

We can see that the vertical motion is constant acceleration and the horizontal motion is constant velocity. We can therefore use the *suvat* equations.

A stroboscopic photograph of a projected ball.

When can problems on projectile motion be solved by applying conservation of energy instead of kinematic equations? (A.3)

◀ **A.1 Figure 28** A projectile launched at an angle θ.

Note that, at C, we are using the magnitude of θ (which is unchanged from position A). Therefore the negative sign is in place to provide the correct velocity direction; the projectile is moving downward.

Since the horizontal displacement is proportional to t, the path has the same shape as a graph of vertical displacement plotted against time. This is parabolic since the vertical displacement is proportional to t^2.

A.1 Table 4

A GeoGebra worksheet linked to this topic is available in the eBook.

SKILLS

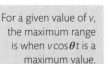

For a given value of v, the maximum range is when $v\cos\theta t$ is a maximum value.

$$t = \frac{2v\sin\theta}{g}$$

If we substitute this for t we get:

$$R = \frac{2v^2\cos\theta\sin\theta}{g}$$

Now, $2\sin\theta\cos\theta = \sin^2\theta$ (a trigonometric identity)

So, $R = \frac{v^2\sin^2\theta}{g}$

This is maximum when $\sin^2\theta$ is a maximum $(\sin^2\theta = 1)$, which is when $\theta = 45°$.

How does the motion of a mass in a gravitational field compare to the motion of a charged particle in an electric field? (D.2)

suvat for horizontal motion

Since acceleration is zero, there is only one equation needed to define the motion.

suvat	A to C
$v = \dfrac{s}{t}$	$R = v\cos\theta t$

suvat for vertical motion

When considering the vertical motion, it is worth splitting the motion into two parts.

suvat	At B	At C
$s = \frac{1}{2}(u+v)t$	$h = \frac{1}{2}(v\sin\theta)\frac{t}{2}$	$0 = \frac{1}{2}(v\sin\theta - v\sin\theta)t$
$v^2 = u^2 + 2as$	$0 = v^2\sin^2\theta - 2gh$	$(-v\sin\theta)^2 = (v\sin\theta)^2 - 0$
$s = ut + \frac{1}{2}at^2$	$h = v\sin\theta t - \frac{1}{2}g\left(\frac{t}{2}\right)^2$	$0 = v\sin\theta t - \frac{1}{2}gt^2$
$a = \dfrac{v-u}{t}$	$g = \dfrac{v\sin\theta - 0}{\frac{t}{2}}$	$g = \dfrac{v\sin\theta - (-v\sin\theta)}{t}$

Some of these equations are not very useful since they simply state that $0 = 0$. However, we do end up with three useful ones (highlighted):

$$R = v\cos\theta t \tag{7}$$

$$0 = v^2\sin^2\theta - 2gh \quad \text{or} \quad h = \frac{v^2\sin^2\theta}{2g} \tag{8}$$

$$0 = v\sin\theta t - \frac{1}{2}gt^2 \quad \text{or} \quad t = \frac{2v\sin\theta}{g} \tag{9}$$

Solving problems

In a typical problem, you will be given the magnitude and direction of the initial velocity and asked to find either the maximum height or range. To calculate h, you can use equation (8), but to calculate R, you need to find the time of flight so must use (9) first. (You could also substitute for t into equation (6) to give another equation but we have enough equations already.)

You do not have to remember a lot of equations to solve a projectile problem. If you understand how to apply the *suvat* equations to the two components of the projectile motion, you only have to remember the *suvat* equations (and they are in the data booklet).

Worked example

A ball is thrown at an angle of 30° to the horizontal at a speed of $20\,\text{m}\,\text{s}^{-1}$. Calculate its range and the maximum height reached.

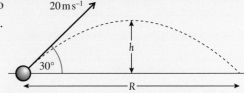

Solution

First, draw a diagram, including labels defining all the quantities known and unknown.

Now we need to find the time of flight. If we apply $s = ut + \frac{1}{2}at^2$ to the whole flight we get:

$$t = \frac{2v\sin\theta}{g} = \frac{(2 \times 20 \times \sin 30°)}{10} = 2\,s$$

We can now apply $s = vt$ to the whole flight to find the range:

$$R = v\cos\theta\, t = 20 \times \cos 30° \times 2 = 34.6\,m$$

Finally, to find the height, we apply $s = ut + \frac{1}{2}at^2$ to the vertical motion, but remember that this is only half the complete flight so the time is $1\,s$.

$$h = v\sin\theta\, t - \frac{1}{2}gt^2 = 20 \times \sin 30° \times 1 - \frac{1}{2} \times 10 \times 1^2 = 10 - 5 = 5\,m$$

When a bullet is fired at a distant target, it will travel in a curved path due to the action of gravity. Precision marksmen adjust their sights to compensate for this. The angle of this adjustment could be based on calculation or experiment (trial and error).

If you have ever played golf, you will know that it is not true that the maximum range is achieved with an angle of 45°. The angle is actually much less. This is because the ball is held up by the air like a plane is. In this photo, Alan Shepard is playing golf on the Moon. Here, the maximum range will be at 45°.

Worked example

A ball is thrown horizontally from a cliff top with a horizontal speed of $10\,m\,s^{-1}$.
If the cliff is 20 m high, what is the range of the ball?

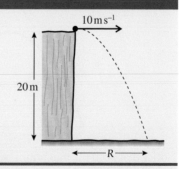

Solution

This is an easy one since there are no angles to deal with. The initial vertical component of the velocity is zero and the horizontal component is $10\,m\,s^{-1}$.
To calculate the time of flight, we apply $s = ut + \frac{1}{2}at^2$ to the vertical component. Knowing that the final displacement is −20 m, this gives:

$$-20\,m = 0 - \frac{1}{2}gt^2 \text{ so } t = \sqrt{\frac{(2 \times 20)}{10}} = 2\,s$$

We can now use this value to find the range by applying the equation $s = vt$ to the horizontal component: $R = 10 \times 2 = 20\,m$

Exercise

Q17. Calculate the range of a projectile thrown at an angle of 60° to the horizontal with a velocity of $30\,\text{m s}^{-1}$.

Q18. You throw a ball at a speed of $20\,\text{m s}^{-1}$.

　(a)　At what angle must you throw the ball so that it will just get over a wall that is 5 m high?

　(b)　How far away from the wall must you be standing?

Q19. A gun is aimed so that it points directly at the center of a target 200 m away. If the bullet travels at $200\,\text{m s}^{-1}$, how far below the center of the target will the bullet hit?

Q20. If you can throw a ball at $20\,\text{m s}^{-1}$, what is the maximum distance you can throw it?

Challenge yourself

1. A projectile is launched perpendicular to a 30° slope at $20\,\text{m s}^{-1}$. Calculate the distance between the launching position and landing position.

How does gravitational force allow for orbital motion? (A.2)

Projectile motion with air resistance

In all the examples above, we have ignored the fact that the air will resist the motion of the ball. Air resistance opposes motion and increases with the speed of the moving object. The actual path of a ball including air resistance is likely to be as shown in Figure 29.

A.1 Figure 29 When air resistance is present, the projectile's motion is asymmetric.

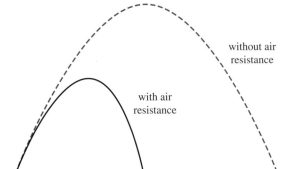

without air resistance

with air resistance

Notice that both the maximum height and the range are less. The path is also no longer a parabola – the way down is steeper than the way up.

The equation for this motion is complex. Horizontally, there is negative acceleration and so the horizontal component of velocity decreases. Vertically, there is increased magnitude of acceleration on the way up and a decreased magnitude of acceleration on the way down. None of these accelerations are constant so the *suvat* equations cannot be used. Luckily, all you need to know is the shape of the trajectory and the qualitative effects on range and time of flight.

Alternative air effects

The air does not always reduce the range of a projectile. A golf ball travels further than a ball projected in a vacuum. This is because the air holds the ball up, in the same way that it holds up a plane, due to the dimples in the ball and its spin.

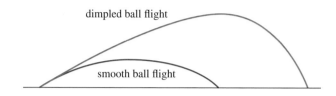

dimpled ball flight

smooth ball flight

A.1 Figure 30 The path of a smooth ball and a dimpled golf ball.

Guiding Questions revisited

How can the motion of a body be described quantitatively and qualitatively?

How can the position of a body in space and time be predicted?

How can the analysis of motion in one and two dimensions be used to solve real-life problems?

In this chapter, we have considered real-life examples to show that:

- Displacement is the straight-line distance between the start and end points of a body's motion and it has a direction.
- Velocity is the rate of change of displacement (and the vector equivalent of speed).
- Acceleration is the rate of change of velocity (and can therefore be treated as a vector).
- Motion graphs of displacement and velocity (or acceleration) against time enable qualitative changes in these quantities to be described and calculations of other quantities to be performed.
- The *suvat* equations of uniformly accelerated motion can be used to predict how position and velocity change with time (or one another) when a body experiences a constant acceleration.
- Vector quantities can be split into perpendicular components that can be treated independently, making it possible to solve problems in two dimensions using the *suvat* equations twice, for example, vertically and then horizontally for a projectile.
- Air resistance changes the acceleration in both perpendicular components, which means that the *suvat* equations cannot be used.

Practice questions

1. Police car P is stationary by the side of a road. Car S passes car P at a constant speed of $18\,\text{m s}^{-1}$. Car P sets off to catch car S just as car S passes car P. Car P accelerates at $4.5\,\text{m s}^{-2}$ for $6.0\,\text{s}$ and then continues at a constant speed. Car P takes t seconds to draw level with car S.

   (a) State an expression, in terms of t, for the distance car S travels in t seconds. (1)

   (b) Calculate the distance traveled by car P during the first $6.0\,\text{s}$ of its motion. (1)

   (c) Calculate the speed of car P after it has completed its acceleration. (1)

   (d) State an expression, in terms of t, for the distance traveled by car P during the time that it is traveling at constant speed. (1)

   (e) Using your answers to (a) to (d), determine the total time t taken by car P to draw level with car S. (2)

   (Total 6 marks)

2. A ball is kicked with a speed of $14\,\text{m s}^{-1}$ at 60° to the horizontal and lands on the roof of a 4 m high building.

   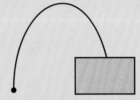

   (a) (i) State the final vertical displacement of the ball. (1)

       (ii) Calculate the time of flight. (3)

       (iii) Calculate the horizontal displacement between the start point and the landing point on the roof. (2)

   (b) The ball is kicked vertically upward. Explain the difference between the time to reach the highest point and the time from the highest point back to the ground. (3)

   (Total 9 marks)

3. Two boys kick a football up and down a hill that is at an angle of 30° to the horizontal. One boy stands at the top of the hill and one boy stands at the bottom of the hill.

   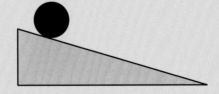

   (a) Assuming that each boy kicks the ball perfectly to the other boy (without spin or bouncing), sketch a single path that the ball could take in either direction. (2)

   (b) Compare the velocities with which each boy must strike the ball to achieve this path. (2)

   (Total 4 marks)

4. The graph shows how the displacement of an object varies with time. At which point (A, B, C or D) does the instantaneous speed of the object equal its average speed over the interval from 0 to 3 s?

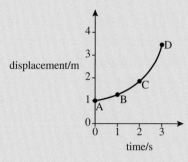

(Total 1 mark)

5. A runner starts from rest and accelerates at a constant rate. Which graph (A, B, C or D) shows the variation of the speed v of the runner with the distance traveled s?

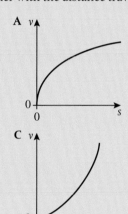
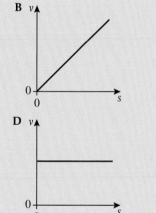

(Total 1 mark)

6. A student hits a tennis ball at point P, which is 2.8 m above the ground. The tennis ball travels at an initial speed of 64 m s^{-1} at an angle of 7.0° to the horizontal. The student is 11.9 m from the net and the net has a height of 0.91 m.

diagram not to scale

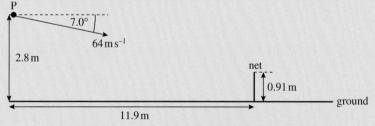

(a) Calculate the time it takes the tennis ball to reach the net. (2)

(b) Show that the tennis ball passes over the net. (3)

(c) Determine the speed of the tennis ball as it hits the ground. (2)

(Total 7 marks)

7. Estimate from what height, under free-fall conditions, a heavy stone would need to be dropped if it were to reach the surface of the Earth at the speed of sound ($330 \, \text{m s}^{-1}$). (2)

    (Total 2 marks)

8. A motorbike is ridden up the left side of a symmetrical ramp. The bike reaches the top of the ramp at speed u, becomes airborne and falls to a point P on the other side of the ramp.

    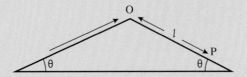

    In terms of u, l and g, obtain expressions for:

    (a) the time t for which the motorbike is in the air (2)

    (b) the distance OP ($= l$) along the right side of the ramp. (3)

    (Total 5 marks)

A.2 Forces and momentum

The launch of the James Webb Space Telescope took place on 25 December 2021. When an object ejects a gas in a downward direction, the gas exerts an equal and upward force on the object. This is an example of a Newton's third law pair.

Guiding Questions

How can we use our knowledge and understanding of the forces acting on a system to predict changes in translational motion?

How can Newton's laws be modeled mathematically?

How can the conservation of momentum be used to predict the behavior of interacting objects?

The motion of a body traveling with constant acceleration can be modeled using the *suvat* equations of uniformly accelerated motion. But what causes the acceleration?

First we must introduce a second body, which interacts with the original body. When two bodies interact, we say that they exert forces on one another in accordance with Newton's three laws of motion. Forces can take many forms, but the presence of a force is required for one body to change the speed or course of another (an acceleration).

A force is a push or a pull.
The unit of force is the newton.

The effect of the force depends upon its direction, so we use an arrow to represent the size and direction of a given force. By adding all the arrows together as vectors, we can calculate the overall size and direction of the resultant force. Using this direction of force and, therefore, acceleration in combination with the *suvat* equations, we can predict the new position and velocity of the original body.

Momentum is the product of mass and velocity, two quantities that we met in the previous chapter. What makes it worth defining in its own right? Momentum is always conserved in any collision provided there are no external forces. This conservation is a direct consequence of Newton's three laws and can be used as a quick way to apply them.

Nature of Science

Newton's three laws of motion are a set of statements, based on observation and experiment, that can be used to predict the motion of a point object from the forces acting on it.

Students should understand:

Newton's three laws of motion
forces as interactions between bodies
forces acting on a body can be represented in a free-body diagram
free-body diagrams can be analyzed to find the resultant force on a system

the nature and use of the following contact forces:

- normal force F_N is the component of the contact force acting perpendicular to the surface that counteracts the body

- surface frictional force F_f acting in a direction parallel to the plane of contact between a body and a surface, on a stationary body as given by $F_f \leq \mu_s F_N$ or a body in motion as given by

  $F_f = \mu_d F_N$ where μ_s and μ_d are the coefficients of static and dynamic friction respectively

- elastic restoring force F_H following Hooke's law as given by $F_H = -kx$ where k is the spring constant

- viscous drag force F_d acting on a small sphere opposing its motion through a fluid as given by $F_d = 6\pi\eta rv$ where η is the fluid viscosity, r is the radius of the sphere, and v is the velocity of the sphere through the fluid

- buoyancy F_b acting on a body due to the displacement of the fluid as given by $F_b = \rho Vg$ where V is the volume of fluid displaced

the nature and use of the following field forces:

- gravitational force F_g as the weight of the body and calculated as given by $F_g = mg$

- electric force F_e

- magnetic force F_m

linear momentum as given by $p = mv$ remains constant unless the system is acted upon by a resultant external force

a resultant external force applied to a system constitutes an impulse J as given by $J = F\Delta t$ where F is the average resultant force and Δt is the time of contact

the applied external impulse equals the change in momentum of the system

Newton's second law in the form $F = ma$ assumes mass is constant whereas $F = \dfrac{\Delta p}{\Delta t}$ allows for situations where mass is changing

the elastic and inelastic collisions of two bodies

explosions

energy considerations in elastic collisions, inelastic collisions, and explosions

bodies moving along a circular trajectory at a constant speed experience an acceleration that is directed radially toward the center of the circle – known as a centripetal acceleration as given by $a = \dfrac{v^2}{r} = \omega^2 r = \dfrac{4\pi^2 r}{T^2}$

circular motion is caused by a centripetal force acting perpendicular to the velocity

a centripetal force causes the body to change direction even if its magnitude of velocity may remain constant

the motion along a circular trajectory can be described in terms of the angular velocity ω which is related to the linear speed v by the equation as given by $v = \dfrac{2\pi r}{T} = \omega r$.

Hooke's law and the elastic restoring force is discussed in A.3. The definitions of elastic and inelastic collisions can also be found in A.3. Information about electric and magnetic forces can be found in D.2.

Force

We can now model the motion of a constantly accelerating body but what makes it accelerate? From experience, we know that to make something move we must push or pull it. We call this **applying a force**. One simple way of applying a force to a body is to attach a string and pull it. Imagine a sphere floating in space with two strings attached. The sphere will not start to move unless one of the astronauts pulls the string as in Figure 1.

A force is a push or a pull.
The unit of force is the newton.

A.2 Figure 1 Two astronauts and a red ball.

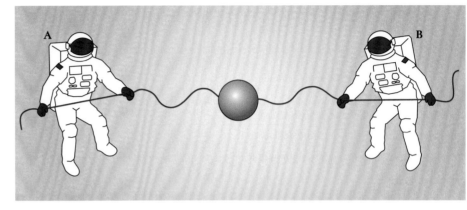

If A pulls the string, then the body will move to the left, and if B pulls the string, it will move to the right. We can see that force is a **vector** quantity since it has **direction**.

If you hold an object of mass 100 g in your hand, then you will be exerting an upward force of about one newton (1 N).

Addition of forces

Since force is a vector, we must add forces vectorially, so if A applies a force of 50 N and B applies a force of 60 N, the resultant force will be 10 N toward B, as can be seen in Figure 2.

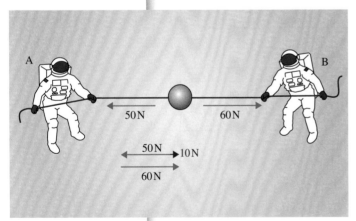

A.2 Figure 2 Astronaut B pulls harder than A.

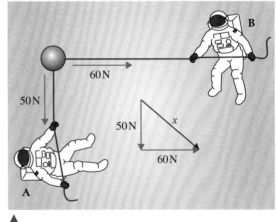

A.2 Figure 3 Astronauts pulling at right angles.

Astronauts in space are considered here so that no other forces (except for very low gravity) are present. This makes things simpler.

Or, in two dimensions, we can use trigonometry as in Figure 3.

In this case, because the addition of forces forms a right-angled triangle, we can use Pythagoras to find x:

$$x = \sqrt{50^2 + 60^2} = 78 \, \text{N}$$

Taking components

As with other vector quantities, we can calculate components of forces. For example, we might want to know the resultant force in a particular direction.

In Figure 4, the component of the force in the x-direction is: $F_x = 60 \times \cos 30° = 52\,\text{N}$

This is particularly useful when we have several forces.

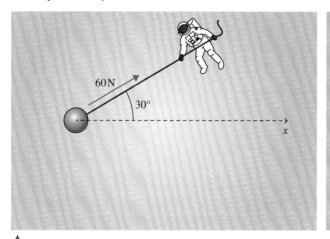

▲ **A.2 Figure 4** Pulling at an angle.

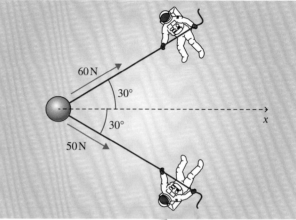

▲ **A.2 Figure 5** Astronauts not pulling in line.

In the example shown in Figure 5, we can use components to calculate the resultant force in the x-direction: $60 \times \cos 30° + 50 \times \cos 30° = 52 + 43 = 95\,\text{N}$

Exercise

Q1. Find the resultant force in the following examples:

(a)

10 N
10 N
10 N
10 N

(b)

3 N
5 N

Equilibrium

If the resultant force on a body is zero, as in Figure 6, then we say the forces are **balanced** or the body is in **equilibrium**.

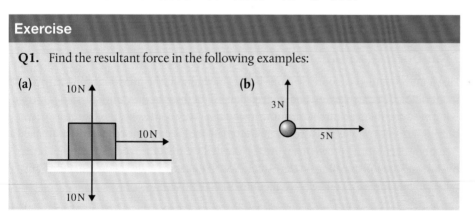

◀ **A.2 Figure 6** Balanced forces.

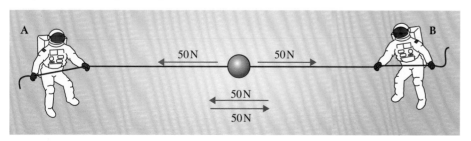

Or with three forces as in Figure 7.

A.2 Figure 7 Three balanced forces.

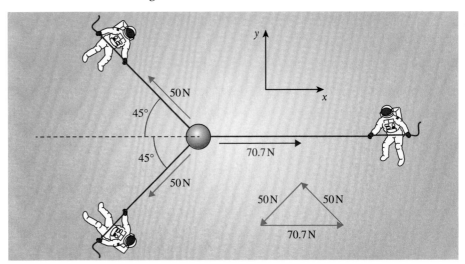

In this example, the two blue forces are perpendicular, making the trigonometry easy. Adding all three forces gives a right-angled triangle. We can also see that if we take components in any direction, then the forces must be balanced.

Taking components in the x-direction:

$-50 \times \cos 45° - 50 \times \cos 45° + 70.7 = -35.35 - 35.35 + 70.7 = 0$

Taking components in the y-direction:

$50 \times \sin 45° - 50 \times \sin 45° = 0$

Free-body diagrams

Problems often involve more than one body. For example, the previous problem involved four bodies, three astronauts, and one red ball. All of these bodies will experience forces, but if we draw them all on the diagram, it would be very confusing. For that reason, we only draw forces on the body we are interested in; in this case, the red ball. This is called a free-body diagram, as shown in Figure 8. Note that we treat the red ball as a point object by drawing the forces acting on the center. Not all forces actually act on the center, but when adding forces, it can be convenient to draw them as if they do.

A.2 Figure 7 Three balanced forces.

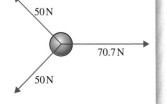

▲
A.2 Figure 8 A free-body diagram of the forces in Figure 7.

Exercise

Q2. In the following examples, calculate the force F required to balance the forces.

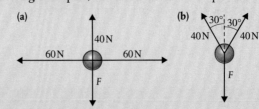

Q3. Calculate the resultant force for the following.

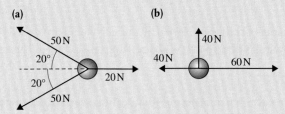

(a)

(b)

Q4. By resolving the vectors into components, calculate if the following bodies are in translational equilibrium or not. If not, calculate the resultant force.

(a)

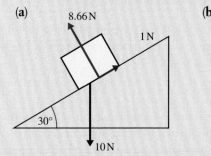

(b)

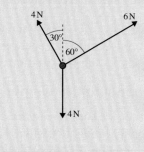

Q5. If the following two examples are in equilibrium, calculate the unknown forces F_1, F_2, and F_3.

(a)

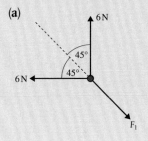

(b)

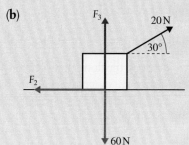

Newton's first law of motion

From observation, we can conclude that to make a body move we need to apply an unbalanced force to it. What is not so obvious is that once moving it will continue to move with a constant velocity unless acted upon by another unbalanced force. Newton's first law of motion is a formal statement of this:

A body will remain at rest or moving with constant velocity unless acted upon by an unbalanced force.

The reason that this is not obvious to us on Earth is that we do not tend to observe bodies traveling with constant velocity with no forces acting on them; in space, it would be more obvious. Newton's first law can be used in two ways. If the forces on a body are balanced, then we can use Newton's first law to predict that it will be at rest or moving with constant velocity. If the forces are unbalanced, then the body will not be at rest or moving with constant velocity. This means its velocity changes – in other words, it accelerates. Using the law the other way round, if a body accelerates, then Newton's first law predicts that the forces acting on the body are unbalanced. To apply this law in real situations, we need to know a bit more about the different types of force.

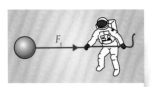

A.2 Figure 9 Exerting tension with a string.

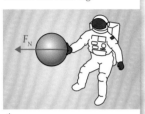

A.2 Figure 10 A normal reaction force is exerted when a hand is in contact with a ball.

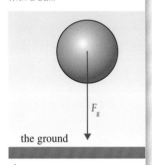

A.2 Figure 11 A ball in free fall.

Types of force

Tension

Tension is the name of the force exerted by the astronauts on the red ball. If you attach a string to a body and pull it, then you are exerting tension, as in Figure 9.

Normal reaction

Whenever two surfaces are in contact with (touching) each other, there will be a force between them. This force is perpendicular to the surface so it is called the **normal reaction force**. If the astronaut pushes the ball with his hand as in Figure 10, then there will be a normal reaction between the hand and the ball.

Note that the force acts on both surfaces so the astronaut will also experience a normal force. However, since we are interested in the ball, not the astronaut, we take the ball as our 'free body' so only draw the forces acting on it.

Gravitational force (weight)

Back on Earth, if a body is released above the ground as in Figure 11, it accelerates downward. According to Newton's first law, there must be an unbalanced force causing this motion. This force is called the **weight**. The weight of a body is directly proportional to its mass: $F_g = mg$ where gravitational field strength, $g = 9.81\,\mathrm{N\,kg^{-1}}$ close to the surface of the Earth. Note that this is the same as the acceleration of free fall. You will find out why later on.

Note that the weight acts at the **center** of the body.

If a block is at rest on the floor, then Newton's first law implies that the forces are balanced. The forces involved are weight (because the block has mass and is on the Earth) and normal force (because the block is in contact with the ground). Figure 12 shows the forces.

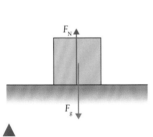

A.2 Figure 12 A free-body diagram of a box resting on the ground.

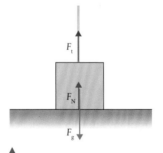

A.2 Figure 13 A string applies an upward force on the box.

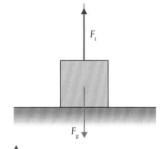

A.2 Figure 14 The block is lifted as the tension is bigger than its weight.

These forces are balanced so: $-F_g + F_N = 0$ or $F_g = F_N$

If the mass of the block is increased, then the normal reaction will also increase.

If a string is added to the block, then we can exert tension on the block as in Figure 13.

The forces are still balanced since $F_t + F_N = F_g$. Notice how F_g has remained the same but F_N has got smaller. If we pull with more force, we can lift the block as in Figure 14. At this point, the normal reaction F_N will be zero. The block is no longer in contact with the ground; now $F_t = F_g$.

The block in Figure 15 is on an inclined plane (slope) so the weight still acts downward. In this case, it might be convenient to split the weight into components, one acting parallel to the slope and one acting perpendicular to the slope.

The component of weight perpendicular to the slope is $F_g \cos \theta$. Since there is no movement in this direction, the force is balanced by F_N. The component of weight parallel to the slope is $F_g \sin \theta$. This force is unbalanced, causing the block to accelerate parallel to the slope. If the angle of the slope is increased, then $\sin \theta$ will also increase, resulting in a greater force down the slope.

Electric and magnetic forces

Electric forces act on charged particles. Magnetic forces act on moving charged particles and magnetic materials.

Like weight (i.e. gravitational forces), electric and magnetic forces act at a distance. Unlike weight, they can be attractive and repulsive.

Friction

There are two types of friction: **static friction**, which is the force that stops the relative motion between two touching surfaces, and **dynamic friction**, which opposes the relative motion between two touching surfaces. In both cases, the force is related to both the normal force and the nature of the surfaces, so pushing two surfaces together increases the friction between them.

$F_f = \mu F_N$ where μ is the coefficient of friction (static or dynamic).

Dynamic friction

In Figure 16, a block is being pulled along a table at a constant velocity.

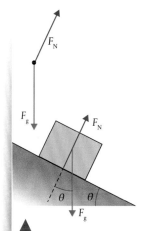

A.2 **Figure 15** Free-body diagram for a block on a slope.

What assumptions (NOS) about the forces between molecules of gas allow for ideal gas behavior? (B.3)

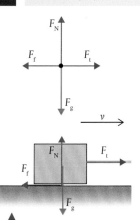

A.2 **Figure 16** The force experienced by a block pulled along a table.

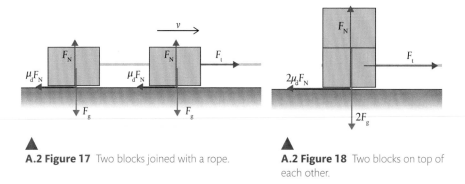

A.2 **Figure 17** Two blocks joined with a rope.

A.2 **Figure 18** Two blocks on top of each other.

Since the velocity is constant, Newton's first law implies that the forces are balanced so $F_t = F_f$ and $F_g = F_N$. Notice that friction does not depend on the area of contact. We can show this by considering two identical blocks sliding at constant velocity across a table top joined together by a rope as in Figure 17. The friction under each cube is $\mu_d F_N$ so the total friction would be $2\mu_d F_N$.

If one cube is now placed on top of the other as in Figure 18, the normal force under the bottom cube will be twice as much so the friction is now $2\mu_d F_N$. It does not matter if the blocks are side by side (large area of contact) or on top of each other (small area of contact); the friction is the same.

If this is the case, then why do racing cars have wide tires with no tread pattern (slicks)? There are several reasons for this but one is to increase the friction between the tires and the road. This is strange because friction is not supposed to depend on area of contact. In practice, friction is not so simple. When one of the surfaces is sticky like the tires of a racing car, the force *does* depend upon the surface area. The type of surfaces we are concerned with here are quite smooth, non-sticky surfaces like wood and metal.

Static friction

If a very small force is applied to a block at rest on the ground, it will not move. This means that the forces on the block are **balanced** (Newton's first law): the applied force is balanced by the static friction.

A.2 Figure 19 μF_N is the maximum size of friction.

In this case, the friction simply equals the applied force: $F_f = F_t$. As the applied force is increased, the friction will also increase. However, there will be a point when the friction cannot be any bigger. If the applied force is increased past that point, the block will start to move; the forces have become **unbalanced** as illustrated in Figure 20. The maximum value that friction can have is $\mu_s F_N$ where μ_s is the coefficient of static friction. The value of static friction is always *greater* than dynamic friction. This can easily be demonstrated with a block on an inclined plane as shown in Figure 20.

A.2 Figure 20 A block rests on a slope until the forces become unbalanced.

In the first example, the friction is balancing the component of weight down the plane, which equals $F_g \sin \theta$, where θ is the angle of the slope. As the angle of the slope is increased, the point is reached where the static friction = $\mu_s F_N$. The forces are still balanced but the friction cannot get any bigger, so if the angle is increased further, the forces become unbalanced and the block will start to move. Once the block moves, the friction becomes dynamic friction. Dynamic friction is less than static friction, so this results in a bigger resultant force down the slope, causing the block to accelerate.

Friction does not just slow things down; it is also the force that makes things move. Consider the tire of a car as it starts to drive away from the traffic lights. The rubber of the tire is trying to move relative to the road. In fact, if there was no friction, the wheel would spin as the tire slipped backward on the road. The force of friction that opposes the motion of the tire slipping backward on the road is therefore in the **forward** direction.

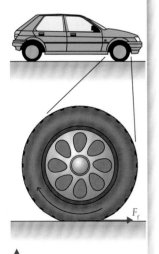

A.2 Figure 21 Friction pushes the car forwards.

If the static friction between the tire and the road is not big enough, the tire will slip. Once this happens, the friction becomes dynamic friction, which is less than static friction, so once tires start to slip, they tend to continue slipping.

Buoyancy

Buoyancy is the name of the force experienced by a body totally or partially immersed in a fluid (a fluid is a liquid or gas). The size of this force is equal to the weight of fluid displaced. It is this force that enables a boat to float and a helium balloon to rise in the air. Let us consider a football and a bucket full of water.

A.2 Figure 22 A football immersed in a bucket of water.

If you take the football and push it under the water, then water will flow out of the bucket (luckily a big bowl was placed there to catch it). The weight of this displaced water is equal to the upward force on the ball. To keep the ball under water, you would therefore have to balance that force by pushing the ball down.

The forces on a floating object are balanced so the weight must equal the buoyant force. This means that the ball must have displaced its own weight of water as in Figure 28.

$$F_g = \rho V_g$$

where V is the volume of fluid displaced and ρ is the density of the fluid.

Air resistance

Air resistance is the force that opposes the motion of a body through the air. More broadly, this is known as fluid resistance or **drag**. The size of this force depends on the speed, size, and shape of the body. At low speeds, the drag force experienced by a sphere is given by Stokes' law:

$$F_d = 6\pi\eta vr$$

where η = viscosity (a constant)

v = velocity

r = radius

When a balloon is dropped, it accelerates downward due to the force of gravity. As it falls through the air, it experiences a drag force opposing its motion. As the balloon's velocity gets bigger so does the drag force, until the drag force balances its weight, at which point its velocity will remain constant (Figure 24). This maximum velocity is called its **terminal velocity**.

The same thing happens when a parachutist jumps out of a plane. The terminal velocity in this case is around 54 m s^{-1} (195 km h^{-1}). Opening the parachute increases the drag force, which slows the parachutist down to a safer 10 m s^{-1} for landing.

A.2 Figure 23 A football floats in a bucket of water.

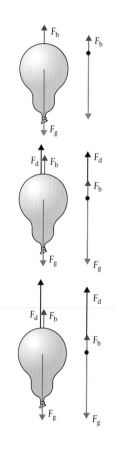

A.2 Figure 24 A balloon reaches terminal velocity as the forces become balanced. Notice the buoyant force is also present.

As it is mainly the air resistance that limits the top speed of a car, a lot of time and money is spent by car designers to try to reduce this force. This is particularly important at high speeds when the drag force is related to the square of the speed.

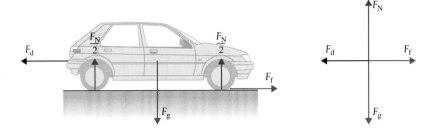

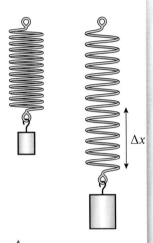

A.2 Figure 25 The forces acting on a car traveling at constant velocity.

A.2 Figure 26 Stretching a spring.

Elastic restoring force

An elastic restoring force, F_H, acts when the shape of an object is changed. An object is stretched when in tension and squashed when in compression. The size of the elastic restoring force increases with the extension (or compression) from the original length.

Speed skiers wear special clothes and squat down like this to reduce air resistance.

How does the application of a restoring force acting on a particle result in simple harmonic motion? (C.1)

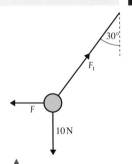

A.2 Figure 27 A ball on a string that is pulled to the side.

Exercise

Q6. A ball of weight 10 N is suspended on a string and pulled to one side by another horizontal string as shown in Figure 27. If the forces are balanced:

(a) write an equation for the horizontal components of the forces acting on the ball

(b) write an equation for the vertical components of the forces acting on the ball

(c) use the second equation to calculate the tension in the upper string, F_t

(d) use your answer to (c) plus the first equation to find the horizontal force F.

Q7. The condition for the forces to be balanced is that the sum of components of the forces in any two perpendicular components is zero. In the 'box on a ramp' example, the vertical and horizontal components were taken. However, it is sometimes more convenient to consider components parallel and perpendicular to the ramp.

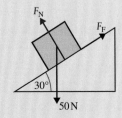

Consider the situation in the figure. If the forces on this box are balanced:

(a) write an equation for the components of the forces parallel to the ramp

(b) write an equation for the forces perpendicular to the ramp

(c) use your answers to find the friction (F_f) and normal force (F_N).

Q8. A rock climber is hanging from a rope attached to the cliff by two bolts as shown in Figure 28. If the forces are balanced:

(a) write an equation for the vertical component of the forces on the knot

(b) write an equation for the horizontal forces exerted on the knot

(c) calculate the tension F_t in the ropes joined to the bolts.

The result of this calculation shows why ropes should not be connected in this way.

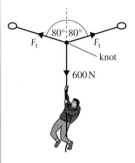

A.2 Figure 28 The rope is attached at two bolts.

The relationship between force and acceleration

Newton's first law states that a body will accelerate if an unbalanced force is applied to it. Newton's second law tells us how big the acceleration will be and in which direction. Before we look in detail at Newton's second law, we should look at the factors that affect the acceleration of a body when an unbalanced force is applied. Let us consider the example of catching a ball. When we catch the ball, we change its velocity, Newton's first law tells us that we must therefore apply an unbalanced force to the ball. The size of that force depends upon two things: the mass and the velocity. A heavy ball is more difficult to stop than a light one traveling at the same speed, and a fast one is harder to stop than a slow one. Rather than having to concern ourselves with two quantities, we will introduce a new quantity that incorporates both mass and velocity: **momentum**.

Nature of Science

The principle of conservation of momentum is a consequence of Newton's laws of motion applied to the collision between two bodies. If this applies to two isolated bodies, we can generalize that it applies to any number of isolated bodies. Here we will consider colliding balls but it also applies to collisions between microscopic particles such as atoms.

Momentum (*p*)

Momentum is defined as the product of mass and velocity: $p = mv$

The unit of momentum is $kg\,m\,s^{-1}$. Momentum is a vector quantity.

Impulse

When you get hit by a ball, the effect it has on you is greater if the ball bounces off you than if you catch it. This is because the change of momentum, Δp, is greater when the ball bounces, as shown in Figure 35.

The unit of impulse is $kg\,m\,s^{-1}$.

Impulse, J, is the change in momentum and is equal to the product of force and the time over which the force is acting. It is a vector.

$$J = \Delta p = F\Delta t$$

41

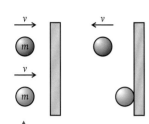

A.2 Figure 29 The change of momentum of the red ball is greater.

Red ball

momentum before $= mv$

momentum after $= -mv$ (remember momentum is a vector)

change in momentum, $J = -mv - mv = -2mv$

Blue ball

momentum before $= mv$

momentum after $= 0$

change in momentum, $J = 0 - mv = -mv$

Exercise

Q9. A ball of mass 200 g traveling at $10\,\mathrm{m\,s^{-1}}$ bounces off a wall as in Figure 29. If after hitting the wall it travels at $5\,\mathrm{m\,s^{-1}}$, what is the impulse?

Q10. Calculate the impulse on a tennis racket that hits a ball of mass 67 g traveling at $10\,\mathrm{m\,s^{-1}}$ so that it comes off the racket at a velocity of $50\,\mathrm{m\,s^{-1}}$.

Newton's second law of motion

The rate of change of momentum of a body is directly proportional to the unbalanced force acting on that body and takes place in same direction.

Let us once again consider a ball with a constant force acting on it as in Figure 30.

A.2 Figure 30 A ball gains momentum.

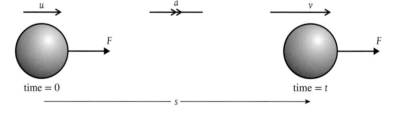

Newton's first law tells us that there must be an unbalanced force acting on the ball since it is accelerating.

Newton's second law tells us that the size of the unbalanced force is directly proportional to the rate of change of momentum. We know that the force is constant so the rate of change of momentum is also constant, which, since the mass is also constant, implies that the acceleration is uniform so the *suvat* equations apply.

If the ball has mass, m we can calculate the change of momentum of the ball.

initial momentum $= mu$

final momentum $= mv$

change in momentum $= mv - mu$

The time taken is t so the rate of change of momentum $= \dfrac{mv - mu}{t}$

If $F =$ $\dfrac{\text{change in momentum}}{\text{time}}$ then change in momentum $=$ force × time. So the unit of momentum is N s. This is the same as $\mathrm{kg\,m\,s^{-1}}$.

This is the same as $\dfrac{m(v - u)}{t} = ma$

Newton's second law states that the rate of change of momentum is proportional to the force, so $F \propto ma$.

To make things simple, the newton is defined so that the constant of proportionality is equal to 1 so:

$$F = ma$$

So when a force is applied to a body in this way, Newton's second law can be simplified to:

The acceleration of a body is proportional to the force applied and inversely proportional to its mass.

Not all examples are so simple. Consider a jet of water hitting a wall as in Figure 31. The water hits the wall and loses its momentum, ending up in a puddle on the floor.

Newton's first law tells us that since the velocity of the water is changing, there must be a force on the water,

Newton's second law tells us that the size of the force is equal to the rate of change of momentum. The rate of change of momentum in this case is equal to the amount of water hitting the wall per second multiplied by the change in velocity. This is not the same as *ma*. For this reason, it is best to use the first, more general, statement of Newton's second law, since this can always be applied.

However, in this course, most of the examples will be of the $F = ma$ type.

▲ **A.2 Figure 31** A jet becomes a puddle.

Example 1: Elevator accelerating upward

An elevator has an upward acceleration of $1\,\mathrm{m\,s^{-2}}$. If the mass of the elevator is 500 kg, what is the tension in the cables pulling it up?

First draw a free-body diagram as in Figure 32. Now we can see what forces are acting. Newton's first law tells us that the forces must be unbalanced. Newton's second law tells us that the unbalanced force must be in the direction of the acceleration (upward). This means that F_t is bigger than *mg*.

Newton's second law also tells us that the size of the unbalanced force equals *ma* so we get the equation:

$$F_t - mg = ma$$

Rearranging gives:

$$\begin{aligned}
F_t &= mg + ma \\
&= 500 \times 10 + 500 \times 1 \\
&= 5500\,\mathrm{N}
\end{aligned}$$

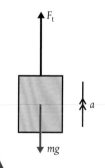

▲ **A.2 Figure 32** An elevator accelerating upward. This could either be going up getting *faster* or going down getting *slower*.

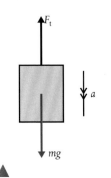

A.2 Figure 33 The elevator with downward acceleration.

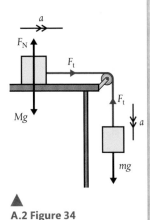

A.2 Figure 34

Example 2: Elevator accelerating downward

The same elevator as in Example 1 now has a downward acceleration of $1\,\text{m s}^{-2}$ as in Figure 33.

This time, Newton's laws tell us that the weight is bigger than the tension so:
$mg - F_t = ma$

Rearranging gives:

$$F_t = mg - ma$$
$$= 500 \times 10 - 500 \times 1$$
$$= 4500\,\text{N}$$

Example 3: Joined masses

Two masses are joined by a rope. One of the masses sits on a frictionless table, while the other hangs off the edge as in Figure 34.

M is being dragged to the edge of the table by m.

Both are connected to the same rope so F_t is the same for both masses. This also means that the acceleration a is the same.

We do not need to consider F_N and mg for the mass on the table because these forces are balanced. However, the horizontally unbalanced force is F_t.

Applying Newton's laws to the mass on the table gives:

$$F_t = ma$$

The hanging mass is accelerating down so mg is bigger than F_t. Newton's second law implies that: $mg - F_t = ma$

Substituting for F_t gives: $mg - ma = ma$ so $a = \dfrac{mg}{M + m}$

Example 4: The free fall parachutist

After falling freely for some time, a free fall parachutist, whose weight is 60 kg, opens his parachute. Suddenly, the force due to air resistance increases to 1200 N. What happens?

Looking at the free-body diagram in Figure 35, we can see that the forces are unbalanced and that, according to Newton's second law, the acceleration, a, will be upward.

The size of the acceleration is given by:

$$ma = 1200 - 600 = 60 \times a$$

So:
$$a = 10\,\text{m s}^{-2}$$

The acceleration is in the opposite direction to the motion. This will cause the parachutist to slow down. As he slows down, the air resistance gets less until the forces are balanced. He will then continue down with a constant velocity.

A.2 Figure 35 The parachutist just after opening the parachute.

Even before opening their parachutes, base jumpers reach terminal velocity.

Exercise

Q11. The helium in a balloon causes an upthrust of 0.1 N. If the mass of the balloon and helium is 6 g, calculate the acceleration of the balloon.

Q12. A rope is used to pull a felled tree (mass 50 kg) along the ground. A tension of 1000 N causes the tree to move from rest to a velocity of 0.1 m s^{-1} in 2 s. Calculate the force due to friction acting on the tree.

Q13. Two masses are arranged on a frictionless table as shown in the figure on the right. Calculate:

(a) the acceleration of the masses

(b) the tension in the string.

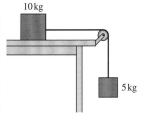

10 kg

5 kg

Q14. A helicopter is lifting a load of mass 1000 kg with a rope. The rope is strong enough to hold a force of 12 kN. What is the maximum upward acceleration of the helicopter?

Q15. A person of mass 65 kg is standing in an elevator that is accelerating upwards at 0.5 m s^{-2}.
What is the normal force between the floor and the person?

Q16. A plastic ball is held under the water by a child in a swimming pool. The volume of the ball is 4000 cm^3.

(a) If the density of water is 1000 kg m^{-3}, calculate the buoyant force on the ball (buoyant force = weight of fluid displaced).

(b) If the mass of the ball is 250 g, calculate the theoretical acceleration of the ball when it is released. Why will the ball not accelerate this quickly in a real situation?

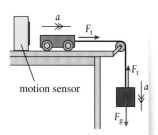

SKILLS

A.2 Figure 36 Apparatus for finding the relationship between force and acceleration.

It is not easy to apply a constant known force to a moving body: just try pulling a cart along the table with a force meter and you will see. One way this is often done in the laboratory is by hanging a mass over the edge of the table as shown in Figure 36.

If we ignore any friction in the pulley or in the wheels of the trolley, then the unbalanced force on the trolley = F_t. Since the mass is accelerating down, then the weight is bigger than F_t so $F_g - F_t = ma$ where m is the mass hanging on the string. The tension is therefore given by:
$F_t = mg - ma = m(g - a)$

There are several ways to measure the acceleration of the trolley; one is to use a motion sensor. This senses the position of the trolley by reflecting an ultrasonic pulse off it. Knowing the speed of the pulse, the software can calculate the distance between the trolley and sensor. As the trolley moves away from the sensor, the time taken for the pulse to return increases; the software calculates the velocity from these changing times. Using this apparatus, the acceleration of the trolley for different masses was measured, and the results are given in the Table 1.

A.2 Table 1

Mass/kg ±0.0001	Acceleration/m s⁻² ±0.03	Tension $(F_t = mg - ma)$/N	Max F_t/N	Min F_t/N	ΔF_t/N
0.0100	0.10	0.097	0.098	0.096	0.001
0.0500	0.74	0.454	0.453	0.451	0.001
0.0600	0.92	0.533	0.532	0.531	0.001
0.1000	1.49	0.832	0.830	0.828	0.001
0.1500	2.12	1.154	1.150	1.148	0.001

A.2 Figure 37 Graph of tension against acceleration.

The uncertainty in mass is given by the last decimal place in the scale, and the uncertainty in acceleration by repeating one run several times. To calculate the uncertainty in tension, the maximum and minimum values have been calculated by adding and subtracting the uncertainties.

These results are shown in Figure 37.

Applying Newton's second law to the trolley, the relationship between F_t and a should be $F_t = ma$ where m is the mass of the trolley. This implies that the gradient of the line should be m. From the graph, we can see that the gradient is 0.52 ± 0.02 kg, which is quite close to the 0.5 kg mass of the trolley.

According to theory, the intercept should be (0, 0) but we can see that there is a positive intercept of 0.05 N. It appears that each value is 0.05 N too big. The reason for this could be friction. If there was friction, then the actual unbalanced force acting on the trolley would be tension – friction. If this is the case, then the results would imply that friction is about 0.05 N.

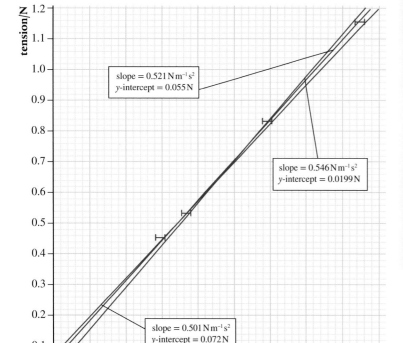

slope = $0.521\,\mathrm{N\,m^{-1}\,s^2}$
y-intercept = $0.055\,\mathrm{N}$

slope = $0.546\,\mathrm{N\,m^{-1}\,s^2}$
y-intercept = $0.0199\,\mathrm{N}$

slope = $0.501\,\mathrm{N\,m^{-1}\,s^2}$
y-intercept = $0.072\,\mathrm{N}$

Newton's third law of motion

When dealing with Newton's first and second laws, we are careful to consider only the body that is *experiencing* the forces, not the body that is *exerting* the forces. Newton's third law relates these forces.

If body A exerts a force on body B, then body B will exert an equal and opposite force on body A.

So if someone is pushing a car with a force F as shown in Figure 38, the car will push back on the person with a force −F. In this case, both of these forces are normal to the car's surface.

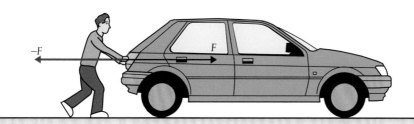

If experimental measurements contain uncertainties, how can laws be developed based on experimental evidence? (NOS)

It is very important to realize that Newton's third law is about two bodies. Avoid statements of this law that do not mention anything about there being two bodies.

A.2 Figure 38 The man pushes the car and the car pushes the man.

You might think that, since these forces are equal and opposite, they will be balanced, and, in that case, how does the person get the car moving? This is wrong. The forces act on different bodies so cannot balance each other.

In summary, a Newton's third law pair is made up of two forces of the same type and magnitude acting in opposite directions on different bodies.

Example 1: A falling body

A body falls freely toward the ground as in Figure 39. If we ignore air resistance, there is only one force acting on the body – the force due to the gravitational attraction of the Earth, which we call weight.

Applying Newton's third law

If the Earth pulls the body down, then the body must pull the Earth up with an equal and opposite force. We have seen that the gravitational force always acts on the center of the body, so Newton's third law implies that there must be a force equal to F_g acting upward on the center of the Earth as in Figure 40.

A.2 Figure 39 A falling body pulled down by gravity.

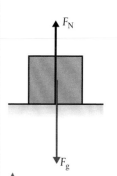

A.2 Figure 41 Forces acting on a box resting on the floor.

A.2 Figure 40 The Earth pulled up by gravity.

Example 2: A box rests on the floor

A box sits on the floor as shown in Figure 41. Let us apply Newton's third law to this situation.

There are two forces acting on the box.

Normal force: The floor is pushing up on the box with a force F_N. According to Newton's third law, the box must therefore push down on the floor with a force of magnitude F_N.

Students often think that Newton's third law implies that the normal force = –weight, but *both* of these forces act *on the box*. If the box is at rest, these forces are indeed equal and opposite but this is due to Newton's *first* law.

Weight: The Earth is pulling the box down with a force F_g. According to Newton's third law, the box must be pulling the Earth up with a force of magnitude F_g as shown in Figure 42.

Example 3: Recoil of a gun

When a gun is fired, the velocity of the bullet changes. Newton's first law implies that there must be an unbalanced force on the bullet. This force must come from the gun. Newton's third law says that if the gun exerts a force on the bullet, the bullet must exert an equal and opposite force on the gun. This is the force that makes the gun recoil or 'kick back'.

Example 4: The water cannon

When water is sprayed at a wall from a hosepipe, it hits the wall and stops. Newton's first law says that if the velocity of the water changes, there must be an unbalanced force on the water. This force comes from the wall. Newton's third law says that if the wall exerts a force on the water, then the water will exert a force on the wall. This is the force that makes a water cannon so effective at dispersing demonstrators.

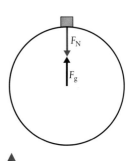

▲ **A.2 Figure 42** Forces acting on the Earth according to Newton's third law.

A boat tests its water cannons.

Exercise

Q17. Use Newton's first and third laws to explain the following:

(a) When burning gas is forced downward out of a rocket motor, the rocket accelerates up.

(b) When the water cannons on the boat in the photo are operating, the boat accelerates forward.

(c) When you step forwards off a skateboard, the skateboard accelerates backward.

(d) A table tennis ball is immersed in a fluid and held down by a string as shown in Figure 43. The container is placed on a balance. What will happen to the reading of the balance if the string breaks?

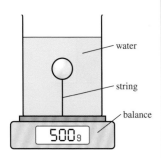

▲ **A.2 Figure 43** A table tennis ball attached to a mass balance

Collisions

In this section, we have been dealing with the interaction between two bodies (gun–bullet, skater–skateboard, hose–water). To develop our understanding of the interaction between bodies, let us consider a simple collision between two balls as illustrated in Figure 44.

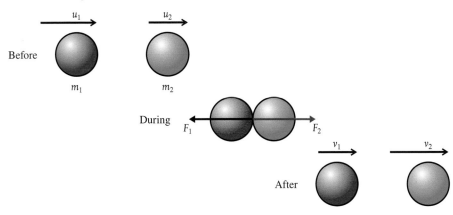

◀ **A.2 Figure 44** Collision between two balls.

Let us apply Newton's three laws to this problem.

Newton's first law

In the collision, the red ball slows down and the blue ball speeds up. Newton's first law tells us that this means there is a force acting to the left on the red ball (F_1) and to the right on the blue ball (F_2).

Newton's second law

This law tells us that the force will be equal to the rate of change of momentum of the balls so if the balls are touching each other for a time Δt:

$$F_1 = \frac{m_1 v_1 - m_1 u_1}{\Delta t}$$

$$F_2 = \frac{m_2 v_2 - m_2 u_2}{\Delta t}$$

Newton's third law

According to the third law, if the red ball exerts a force on the blue ball, then the blue ball will exert an equal and opposite force on the red ball.

$$F_1 = -F_2$$

$$\frac{m_1 v_1 - m_1 u_1}{\Delta t} = \frac{-(m_2 v_2 - m_2 u_2)}{\Delta t}$$

Rearranging gives: $\quad m_1 u_1 + m_2 u_2 = m_1 v_1 + m_2 v_2$

In other words, the momentum at the start equals the momentum at the end. We find that this applies not only to this example but to all interactions.

An isolated system is one in which no external forces are acting. When a ball hits a wall, the momentum of the ball is not conserved because the ball and wall is not an isolated system, since the wall is attached to the ground. If the ball and wall were floating in space, then momentum would be conserved.

The law of the conservation of momentum

For a system of isolated bodies, the total momentum is always the same.

This is not a new law since it is really just a combination of Newton's laws. However, it provides a useful short cut when solving problems.

In many examples, we will have to pretend everything is in space isolated from the rest of the Universe, otherwise they are not isolated and the law of conservation of momentum will not apply.

You may have noticed that some collisions enable the bodies to bounce off one another, while there are other collisions where the bodies stick together. This is discussed further in A.3 Work, energy and power.

> ### Nature of Science
>
> By applying what we know about motion in a straight line, we can develop a model for motion in a circle. This is a common way that models are developed in physics: start simple and add complexity later.

Circular motion

If a car travels around a bend at $30\,\text{km h}^{-1}$, it is obviously traveling at a constant speed, since the speedometer registers $30\,\text{km h}^{-1}$ all the way round. However, it is not traveling at constant velocity. This is because velocity is a vector quantity, and for a vector quantity to be constant, both magnitude and direction must remain the same. Bends in a road can be many different shapes, but, to simplify things, we will only consider circular bends taken at constant speed.

Quantities of circular motion

Consider the body in Figure 45 traveling in a circle radius r, with constant speed v. In time Δt, the body moves from A to B. As it does this, the radius sweeps out an angle $\Delta \theta$.

When describing motion in a circle, we often use quantities referring to the angular motion rather than the linear motion. These quantities are:

Time period (*T*)

The time period is the time taken to complete one circle.

The unit of the time period is the second.

Angular displacement (*θ*)

The angular displacement is the angle swept out by a line joining the body to the center.

The unit of angular displacement is the radian.

How are concepts of equilibrium and conservation applied to understand matter and motion from the smallest atom to the whole Universe? (B.3, D.1)

In which way is conservation of momentum relevant to the workings of a nuclear power station? (E.4)

When dealing with circular motion in physics, we always measure the angle in radians.

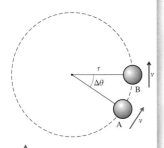

▲ **A.2 Figure 45** Although speed is constant, velocity is changing because direction is changing.

Angular velocity (ω)

The angular velocity is the angle swept out by per unit time.

The unit of angular displacement is the radian s⁻¹.

$$\omega = \frac{\Delta\theta}{\Delta t}$$

The angle swept out when the body completes a circle is 2π and the time taken is by definition the time period T so this equation can also be written:

$$\omega = \frac{2\pi}{T}$$

Frequency (f)

The frequency is the number of complete revolutions per unit time.

$$f = \frac{1}{T}$$

So:

$$\omega = 2\pi f$$

Angular velocity and speed

In a time T, the body in Figure 46 completes one full circle so it travels a distance $2\pi r$, the circumference of the circle. Speed is defined as the $\frac{\text{distance traveled}}{\text{time taken}}$ so $v = \frac{2\pi r}{T}$. In this time, a line joining the body to the center will sweep out an angle of 2π radians so the angular velocity, $\omega = \frac{2\pi}{T}$. Substituting into the equation for v we get:

$$v = \omega r$$

Although the speed is constant, when a body moves in a circle, its direction and velocity are always changing. At any moment in time, the magnitude of the instantaneous velocity is equal to the speed and the direction is perpendicular to the radius of the circle.

Centripetal acceleration

From the definition of acceleration, we know that if the velocity of a body changes, it must be accelerating, and that the direction of acceleration is in the direction of the change in velocity. Let us consider a body moving in a circle with a constant speed v. Figure 46 shows two positions of the body separated by a short time.

To derive the equation for this acceleration, let us consider a very small angular displacement $\delta\theta$ as represented by Figure 47.

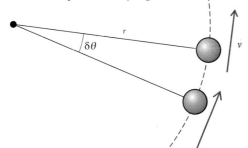

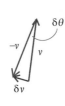

Why is no work done on a body moving along a circular trajectory at constant speed? (A.3)

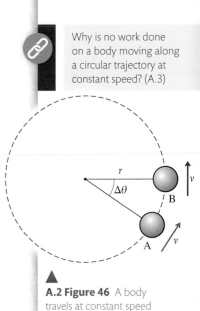

A.2 Figure 46 A body travels at constant speed around a circle of radius r.

A.2 Figure 47 Angular displacement.

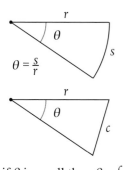

$$\theta = \frac{s}{r}$$

if θ is small then $\theta = \frac{c}{r}$

▲

A.2 Figure 48 A small angle approximation.

You will not be asked to reproduce this derivation in the exam.

If this small angular displacement has taken place in a short time δt, then the angular velocity, $\omega = \frac{\delta\theta}{\delta t}$.

From the definition of acceleration: $a = \dfrac{\text{change of velocity}}{\text{time}}$

If we took only the magnitude of velocity, then the change of velocity would be zero. However, velocity is a vector so change in velocity is found by taking the final velocity vector − initial velocity vector as in the vector addition in Figure 48. This triangle is not a right-angled triangle so cannot be solved using Pythagoras. However, since the angle $\delta\theta$ is small, we can say that the angle $\delta\theta$ in radians is approximately equal to $\frac{\delta v}{v}$.

Rearranging gives:
$$\delta v = v\delta\theta$$

$$\text{acceleration} = \frac{\delta v}{\delta t} = \frac{v\delta\theta}{\delta t}$$

$$a = v\omega$$

But we know that $v = \omega r$ so we can substitute for v and get $a = \omega^2 r$

Or substituting for:
$$\omega = \frac{v}{r}$$

$$a = \frac{v^2}{r}$$

The direction of this acceleration is in the direction of δv. Now, as the angle $\delta\theta$ is small, the angle between δv and v is approximately 90°, which implies that the acceleration is perpendicular to the velocity. This makes it directed *toward* the center of the circle; hence the name **centripetal acceleration**.

Exercise

Q18. A body travels with constant speed of $2\,\text{m s}^{-1}$ around a circle of radius $5\,\text{m}$. Calculate:

(a) the distance moved in one revolution

(b) the displacement in one revolution

(c) the time taken for one revolution

(d) the frequency of the motion

(e) the angular velocity

(f) the centripetal acceleration.

Centripetal force

From Newton's first law, we know that if a body accelerates, there must be an unbalanced force acting on it. The second law tells us that this force is in the direction of the acceleration. This implies that there must be a resultant force acting toward the center. This force is called the **centripetal force**.

From Newton's second law, we can also deduce that $F = ma$ so $F = \dfrac{mv^2}{r} = m\omega^2 r$.

$$F = m\omega^2 r$$

Examples of circular motion

All bodies moving in a circle must be acted upon by a force toward the center of the circle. However, this can be provided by many different forces.

Mass on a string in space

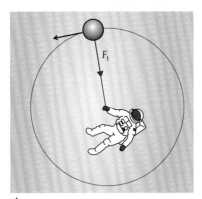

A.2 Figure 49 An astronaut playing with a mass on a string.

If you take a mass on the end of a string, you can easily make it move in a circle, but the presence of gravity makes the motion difficult to analyze. It will be simpler if we start by considering what this would be like if performed by an astronaut in deep space: much more difficult to do but easier to analyze. Figure 49 shows an astronaut making a mass move in a circle on the end of a string. The only force acting on the mass is the tension in the string.

In this case, it is obvious that the centripetal force is provided by the tension so:

$$F_t = \frac{mv^2}{r}$$

From this, we can predict that the force required to keep the mass in its circular motion will increase if the speed increases. This will be felt by the astronaut who, according to Newton's third law, must be experiencing an equal and opposite force on the other end of the string. If the string were to break, the ball would have no forces acting on it so would travel at a constant velocity in the same direction as it was moving when the break occurred. This would be at some tangent to the circle as in Figure 50.

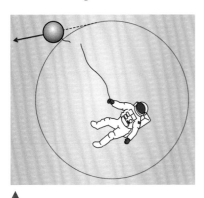

A.2 Figure 50 The string breaks.

How can knowledge of electric and magnetic forces allow the prediction of changes to the motion of charged particles? (D.2)

In this example, the astronaut has a much larger mass than the ball. If this was not the case, the astronaut would be pulled out of position by the equal and opposite force acting on the other end of the string.

TOK

People often think that the mass will fly outward if the string breaks. This is because they feel themselves being forced outward so think that if the string breaks, the mass will move in this direction. Applying Newton's laws, we know that this is not the case. This is an example of a case where intuition gives the wrong answer.

Mass on a string on the Earth (horizontal)

When playing with a mass on a string on the Earth, there will be gravity acting as well as tension. We will first consider how this changes the motion when the mass is made to travel in a horizontal circle as in Figure 51.

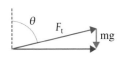
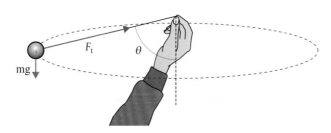

A.2 Figure 51 A mass swung in a horizontal circle.

For the motion to be horizontal, there will be no vertical acceleration so the weight must be balanced by the vertical component of tension ($F_t \cos \theta = mg$). This means that the string cannot be horizontal but will always be at an angle, as shown in Figure 51. The centripetal force is provided by the horizontal component of tension ($F_t \sin \theta = \frac{mv^2}{r}$), which is equal to the vector sum of the two forces.

Mass on a string on the Earth (vertical)

As a mass moves in a vertical circle, the force of gravity sometimes acts in the same direction as its motion and sometimes against it. For this reason, it is not possible to keep it moving at a constant speed, so here we will just consider it when it is at the top and the bottom of the circle as shown in Figure 52.

At the top, the centripetal force $\frac{mv_t^2}{r} = F_{t\,(top)} + mg$ so $F_{t\,(top)} = \frac{mv_t^2}{r} - mg$

At the bottom, $\frac{mv_b^2}{r} = F_{t\,(bottom)} - mg$ so $F_{t\,(bottom)} = \frac{mv_b^2}{r} + mg$

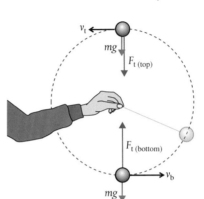

A.2 Figure 52 A mass swung in a vertical circle.

When the mass approaches the top of the circle, its kinetic energy is transferred to potential energy, resulting in a loss of speed. If it were to stop at the top, then it would fall straight down. The minimum speed necessary for a complete circle is when the weight of the ball is enough to provide the centripetal force without any tension. So if $F_{t\,(top)} = 0$, then $\frac{mv_t^2}{r} = mg$.

When you rotate a mass in a vertical circle, you definitely feel the change in the tension as it decreases toward the top and increases toward the bottom.

The wall of death

In the wall of death, motorbikes and cars travel around the inside of a cylinder with vertical walls.

In the wall of death shown in Figure 53, the centripetal force is provided by the normal reaction, F_N. The weight is balanced by the friction between the ball and wall, which is dependent on the normal reaction $F_f = \mu F_N$. If the velocity is too slow, the normal force will be small, which means the friction will not be large enough to support the weight.

Wall of death.

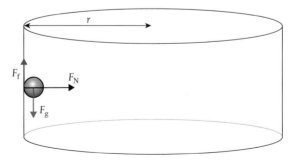

A.2 Figure 53 The wall of death with a ball rather than bike.

Car on a circular track

When a car travels around a circular track, the centripetal force is provided by the friction between the tires and the road. The faster you go, the more friction you need. The problem is that friction has a maximum value given by $F_f = \mu F_N$, so if the centripetal force required is greater than this, the car will not be able to maintain a circular path. Without friction, for example, on an icy road, the car would travel in a straight line. This means that you would hit the curb on the outside of the circle, giving the impression that you have been thrown outward. This is of course not the case since there is no force acting outward.

Dynamic friction is less than static friction so once a car starts to skid on a corner it will continue. This is also why it is not a good idea to spin the wheels of a car while going round a corner.

A.2 Figure 54 A car rounding a bend.

Cyclist on a banked track

A banked track is a track that is angled to make it possible to go faster around the bends. These are used in indoor cycle racing. In the case shown in Figure 55, where the bike is represented by a ball, the centripetal force is provided by the horizontal component of the normal reaction force, so even without friction, the ball can travel around the track. If the track was angled the other way, then it would have the opposite effect. This is called an adverse camber and bends like this should be taken slowly.

A racing cyclist on a banked track in a velodrome.

Remember when solving circular motion problems, centripetal force is not an extra force – it is one of the existing forces. Your task is to find which force (or a component of it) points toward the center.

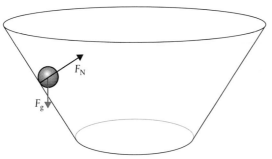

A.2 Figure 55 A ball on a banked track.

Exercise

Q19. Calculate the centripetal force for a 1000 kg car traveling around a circular track of radius 50 m at 30 km h^{-1}.

Q20. A 200 g ball is made to travel in a circle of radius 1 m on the end of a string. If the maximum force that the string can withstand before breaking is 50 N, what is the maximum speed of the ball?

Q21. A rollercoaster is designed with a 5 m radius vertical loop. Calculate the minimum speed necessary to get around the loop without falling down.

Q22. A 200 g ball moves in a vertical circle on the end of a 50 cm long string. If its speed at the bottom is 10 m s^{-1}, calculate:

(**a**) the velocity at the top of the circle

(**b**) the tension at the top of the circle.

Challenge yourself

1. A car of mass 1000 kg is driving around a circular track with radius 50 m. If the coefficient of friction between the tires and road is 0.8, calculate the maximum speed of the car before it starts to slip. What would the maximum speed be if the track was banked at 45°?

Guiding Questions revisited

How can we use our knowledge and understanding of the forces acting on a system to predict changes in translational motion?

How can Newton's laws be modeled mathematically?

How can the conservation of momentum be used to predict the behavior of interacting objects?

In this chapter, we have considered new quantities and accepted laws to describe how:

- Different types of force can be distinguished from one another.
- Forces are vector quantities that can be represented by arrows of appropriate length and direction on free-body diagrams (in which the forces acting on, rather than exerted by, a given body are shown).
- Newton's first law states that an object will remain at constant velocity unless acted upon by a resultant force.
- Newton's second law states that the rate of change of momentum is proportional to and in the same direction as the resultant force.
- A resultant force acting at an angle to a body's velocity leads to a change in direction and, when the force and velocity are perpendicular, circular motion.
- Linear acceleration and centripetal acceleration are both defined as the rate of change of velocity, but are calculated using different equations.
- Newton's third law states that when body A exerts a force on body B, body B exerts an equal and opposite force (of the same type) on body A.

- Momentum, the product of mass and velocity, is conserved in the absence of external forces, which means that the initial momentum and final momentum of a system can be equated.
- When objects interact, they exert forces on one another, which means that momentum is exchanged between them.
- Newton's second law can be rephrased mathematically as impulse, the product of force acting and the time of the interaction, which is equal to change in momentum.

Practice questions

1. **(a)** A car goes round a curve in a road at constant speed. Explain why, although its speed is constant, it is accelerating. (2)

   In the diagram, a marble (small glass sphere) rolls down a track, the bottom part of which has been bent into a loop. The end A of the track, from which the marble is released, is at a height of 0.80 m above the ground. Point B is the lowest point and point C the highest point of the loop. The diameter of the loop is 0.35 m.

   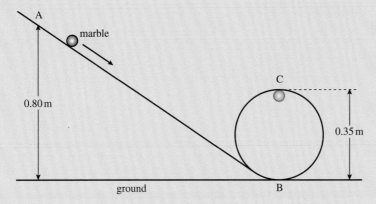

   The mass of the marble is 0.050 kg. Friction forces and any gain in kinetic energy due to the rotating of the marble can be ignored. The acceleration due to gravity, $g = 10\,\mathrm{m\,s^{-2}}$.

   Consider the marble when it is at point C.

   (b) (i) Copy the diagram and on the diagram, draw an arrow to show the direction of the resultant force acting on the marble. (1)

   (ii) State the names of the **two** forces acting on the marble. (2)

   (iii) Deduce that the speed of the marble is $3.0\,\mathrm{m\,s^{-1}}$. (3)

   (iv) Determine the resultant force acting on the marble and hence determine the reaction force of the track on the marble. (4)

   (Total 12 marks)

2. **(a)** Define what is meant by **coefficient of friction**.   (1)

The diagram shows a particular ride at a funfair (sometimes called 'the fly') that involves a spinning circular room. When it is spinning fast enough, a person in the room feels 'stuck' to the wall. The floor is lowered and they remain held in place on the wall. Friction prevents the person from falling.

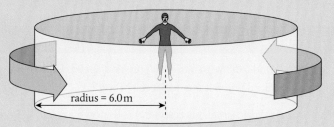

radius = 6.0 m

(b) (i) Explain whether the friction acting on the person is static, dynamic, **or** a combination of both.   (2)

The diagram below shows a cross section of the ride when the floor has been lowered.

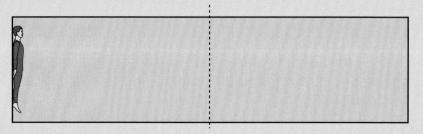

(ii) Copy the diagram and, on your diagram, draw labeled arrows to represent the forces acting on the person.   (3)

(c) Use the data given below:

mass of person = 80 kg

coefficient of friction between the person and the wall = 0.40

radius of circular room = 6.0 m

Calculate each of the following:

(i) the magnitude of the minimum resultant horizontal force on the person   (2)

(ii) the minimum speed of the wall for a person to be 'stuck' to it.   (2)

(Total 10 marks)

3. A 10 000 kg cubic rock (a boulder) rests on the side of a mountain as shown in the diagram below.

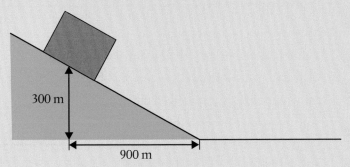

300 m

900 m

(a) Calculate the frictional force acting on the rock. (3)

(b) After a prolonged period of rain, the rock starts to slide down the slope. Show that this will happen when the coefficient of friction between the rock and the slope is equal to 0.33. (3)

(c) Once the block starts to slide, the coefficient of friction is reduced to 0.2. Calculate:

    (i) the acceleration of the rock (2)

    (ii) the speed of the rock when it reaches the bottom of the hill. (2)

(d) The rock then slides along the flat section of ground. Assuming there is no change in the coefficient of friction calculate:

    (i) the frictional force on the rock (1)

    (ii) the distance traveled by the rock. (2)

(Total 13 marks)

4. A student investigates the forces involved in holding a climbing rope by standing on bathroom scales while holding a rope as shown in the diagram on the right.

(a) If both students have mass 60 kg, calculate:

    (i) the tension in the rope (1)

    (ii) the reading on the scales (2)

    (iii) the friction between the rope holder's feet and the scales. (2)

(b) Explain why the rope-holding student should lower the hanging student smoothly. (2)

(Total 7 marks)

30°

scales

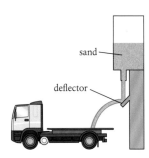

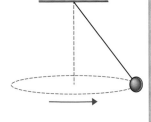

5. Sand flows out of a container at a rate of 5 kg s^{-1} and falls a vertical distance of 1 m, where it is deflected into the back of a truck by a deflector placed at an angle of 45°.

 (a) Calculate:

 (i) the velocity of the sand as it hits the deflector (1)

 (ii) the horizontal velocity of the sand after hitting the deflector (1)

 (iii) the force exerted by the sand on the deflector. (3)

 (b) The sand stops when it hits the back of the truck.

 (i) Explain why there must be a frictional force between the tires and the road. (3)

 (ii) Calculate the force of friction between the truck and the ground. (2)

 (Total 10 marks)

6. Two forces act on an object in different directions. The magnitudes of the forces are 18 N and 27 N. The mass of the object is 9.0 kg. What is a possible value for the acceleration of the object?

 A 0 m s^{-2} **B** 0.5 m s^{-2} **C** 2.0 m s^{-2} **D** 6.0 m s^{-2}

 (Total 1 mark)

7. An object of mass m strikes a vertical wall horizontally at speed U. The object rebounds from the wall horizontally at speed V. What is the magnitude of the change in the momentum of the object?

 A 0   **B** $m(V - U)$ **C** $m(U - V)$ **D** $m(U + V)$

 (Total 1 mark)

8. A sphere is suspended from the end of a string and rotates in a horizontal circle as shown on the left. Which free-body diagram, to the correct scale, shows the forces acting on the sphere?

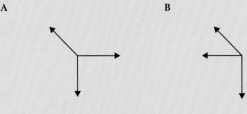

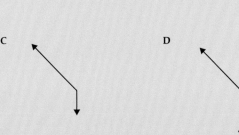

(Total 1 mark)

9. A company delivers packages to customers using a small unmanned aircraft. Rotating horizontal blades exert a force on the surrounding air. The air above the aircraft is initially stationary.

The air is propelled vertically downward with speed v. The aircraft hovers motionless above the ground. A package is suspended from the aircraft on a string. The mass of the aircraft is 0.95 kg and the combined mass of the package and string is 0.45 kg. The mass of air pushed downward by the blades in one second is 1.7 kg.

(a) State the value of the resultant force on the aircraft when hovering. (1)

(b) Outline, by reference to Newton's third law, how the upward lift force on the aircraft is achieved. (2)

(c) Determine v. State your answer to an appropriate number of significant figures. (3)

(d) The package and string are now released and fall to the ground. The lift force on the aircraft remains unchanged. Calculate the initial acceleration of the aircraft. (2)

(Total 8 marks)

10. The Rotor is an amusement park ride that can be modeled as a vertical cylinder of inner radius R rotating about its axis. When the cylinder rotates fast enough, the floor drops out and the passengers stay motionless against the inner surface of the cylinder. The diagram on the right shows a person taking the Rotor ride. The floor of the Rotor has been lowered away from the person.

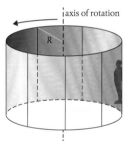

(a) Draw and label the free-body diagram for the person. (2)

(b) The person must not slide down the wall. Show that the minimum angular velocity of the cylinder for this situation is:

$$\omega = \sqrt{\frac{g}{\mu R}}$$

where μ is the coefficient of static friction between the person and the cylinder. (2)

(c) The coefficient of static friction between the person and the cylinder is 0.40. The radius of the cylinder is 3.5 m. The cylinder makes 28 revolutions per minute. Deduce whether the person will slide down the inner surface of the cylinder. (3)

(Total 7 marks)

11. A boulder is many times heavier than a pebble; that is, the gravitational force that acts on a boulder is many times that which acts on the pebble. If you drop a boulder and a pebble at the same time, they will fall together with equal accelerations (neglecting air resistance). The principal reason the heavier boulder does not accelerate more than the pebble has to do with what?

A Energy **B** Weight **C** Mass

D Surface area **E** none of these

(Total 1 mark)

12. The force exerted on a house by a 120 mph hurricane wind is how many times as strong as the force exerted on the same house by a 60 mph gale wind?

A Equally **B** Two times **C** Three times **D** Four times

(Total 1 mark)

13. Two buckets hang from a rope over a frictionless pulley. The bucket on the right has a mass m_2, which is greater than the mass of the bucket on the left m_1. Bucket 2 starts at height h above the ground. If the buckets are released from rest, determine:

(a) the speed with which bucket 2 hits the ground in terms of m_1, m_2, h, and the acceleration due to gravity g (2)

(b) the further increase in height of bucket 1 after bucket 2 hits the ground and stops. (2)

Ignore resistive effects and assume the rope is long compared to the height above the ground.

(Total 4 marks)

A.3 Work, energy and power

◀ A Rube Goldberg machine starts with a small action (like the pressing of a button or the knocking over of a domino tile) that sets off a chain reaction of different events, which continues for an extended time. Objects that are stretched, at a height, or moving, combine – often with entertaining consequences (as seen in films, music videos, advertisements and social media).

Guiding Questions

How are concepts of work, energy and power used to predict changes within a system?

How can a consideration of energetics be used as a method to solve problems in kinematics?

How can transfer of energy be used to do work?

If body A pushes body B, body B may start to move. This movement might cause body B to hit body C, after which, body C moves toward body D (you get the idea). The 'ability to push' seems to be passed on from one object to another. We call this energy, and the fact that it is conserved can lead to a simple way of solving problems that bypasses all the complications of forces and motion.

For example, when calculating the final velocity of a box pushed up a slope by a constant force, we need to find the component of force up the slope, use that to find acceleration, and then apply the *suvat* equations to calculate final velocity. Or, using conservation of energy, we can state that the gain in potential energy and kinetic energy is equal to the work done.

It is the same when we study gases. It is much easier to understand how work done on a gas increases its temperature than to model the motion of each molecule.

Students should understand:

the principle of the conservation of energy
work done by a force is equivalent to a transfer of energy
energy transfers can be represented on a Sankey diagram
work W done on a body by a constant force depends on the component of the force along the line of displacement as given by $W = Fs \cos \theta$
work done by the resultant force on a system is equal to the change in the energy of the system
mechanical energy is the sum of kinetic energy, gravitational potential energy and elastic potential energy
in the absence of frictional, resistive forces, the total mechanical energy of a system is conserved
if mechanical energy is conserved, work is the amount of energy transformed between different forms of mechanical energy in a system, such as: • the kinetic energy of translational motion as given by $E_k = \frac{1}{2}mv^2 = \frac{p^2}{2m}$ • the gravitational potential energy, when close to the surface of the Earth, as given by: $\Delta E_p = mg\Delta h$ • the elastic potential energy as given by $E_H = \frac{1}{2}k\Delta x^2$

| power P is the rate of work done, or the rate of energy transfer, as given by $P = \frac{E}{t} = Fv$ |
| efficiency η in terms of energy transfer or power as given by $\eta = \dfrac{E_{output}}{E_{input}} = \dfrac{P_{output}}{P_{input}}$ |
| energy density of the fuel sources. |

Nature of Science

Scientists should remain skeptical but that does not mean you have to doubt everything you read. The law of conservation of energy is supported by many experiments and is the basis of countless predictions that turn out to be true. If someone now found that energy was not conserved, there would be a lot of explaining to do. Once a law is accepted, it gives us an easy way to make predictions. For example, if you are shown a device that produces energy from nowhere, you know it must be a fake without even finding out how it works because it violates the law of conservation of energy.

In A.1 and A.2, we dealt with the motion of a small red ball and now understand what causes acceleration. We have also investigated the interaction between a red ball and a blue one and have seen that the red one can cause the blue one to move when they collide. But what enables the red one to push the blue one? To answer this question, we need to define some more quantities.

Work

In the introduction to this book, it was stated that by developing models, our aim is to understand the physical world so that we can make predictions. At this point, you should understand certain concepts related to the collision between two balls, but we still cannot predict the outcome. To illustrate this point, let us again consider the red and blue balls. Figure 1 shows three possible outcomes of the collision.

If we apply the law of conservation of momentum, we realize that all three outcomes are possible. The original momentum is 10 N s and the final momentum is 10 N s in all three cases. But which one actually happens? This we cannot say (yet). All we know is that from experience the last option is not possible – but why?

When body A hits body B, body A exerts a force on body B. This force causes B to have an increase in velocity. The amount that the velocity increases depends on how big the force is and over what distance the collision takes place. To make this simpler, consider a constant force acting on two blocks as in Figure 2.

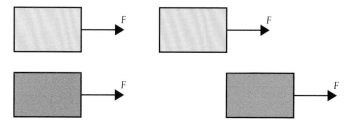

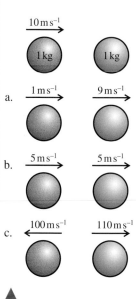

▲ **A.3 Figure 1** The red ball hits the blue ball but what happens?

◀ **A.3 Figure 2** The force acts on the orange block for a greater distance.

65

Both blocks start at rest and are pulled by the same force, but the orange block will gain more velocity because the force acts over a longer distance. To quantify this difference, we say that in the case of the orange block, the force has done more work. Work is done when the point of application of a force moves in the direction of the force.

Work is defined in the following way:

work done = force × distance moved in the direction of the force

The unit of work is the newton meter (N m), which is the same as the joule (J).

Work is a scalar quantity.

How do traveling waves allow for a transfer of energy without a resultant displacement of matter? (C.2, A.1)

Worked example

A tractor pulls a felled tree along the ground for a distance of 200 m. If the tractor exerts a force of 5000 N, how much work will be done?

Solution

work done = force × distance moved in direction of force

work done = 5000 × 200 = 1 MJ

Worked example

A force of 10 N is applied to a block, pulling it 50 m along the ground as shown. How much work is done by the force?

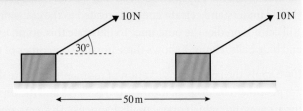

Solution

In this example, the force is not in the same direction as the movement. However, the horizontal component of the force is 10 × cos 30°.

work done = 10 × cos 30° × 50 = 433 N

Worked example

When a car brakes, it slows down due to the friction force between the tires and the road. This force opposes the motion as shown. If the friction force is a constant 500 N and the car comes to rest in 25 m, how much work is done by the friction force?

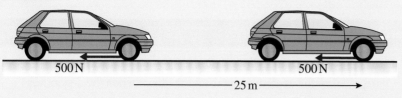

Solution

This time, the force is in the opposite direction to the motion.

$$\text{work done} = -500 \times 25 = -12\,500\,\text{J}$$

The negative sign tells us that the car's kinetic energy is decreasing. The internal energy of the brake disks has increased; positive work has been done on them by friction.

Worked example

The woman in the figure walks along with a constant velocity holding a suitcase.
How much work is done by the force holding the case?

Solution

In this example, the force is acting perpendicular to the direction of motion, so there is no movement in the direction of the force.

$$\text{work done} = \text{zero}$$

General formula

In general:

$$\text{work} = F \cos \theta \times \Delta s$$

where θ is the angle between the displacement, Δs, and force, $F3$.

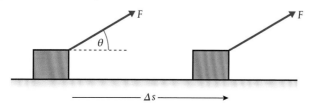

It may seem strange that when you carry a heavy bag you are not doing any work – that is not what it feels like. In reality, lots of work is being done, since to hold the bag you use your muscles. Muscles are made of microscopic fibers, which are continuously contracting and relaxing, so are doing work.

All the previous examples can be solved using this formula.

If $\theta < 90°$, $\cos\theta$ is positive so the work is positive.

If $\theta = 90°$, $\cos\theta = 0$ so the work is zero.

If $\theta > 90°$, $\cos\theta$ is negative so the work is negative.

Exercise

Q1. The figure shows a boy taking a dog for a walk.

    (a) Calculate the work done by the force shown when the dog moves 10 m forward.

    (b) Who is doing the work?

Q2. A bird weighing 200 g sits on a tree branch. How much work does the bird do on the tree?

Q3. As a box slides along the floor, it is slowed down by a constant force due to friction. If this force is 150 N and the box slides for 2 m, how much work is done against the frictional force?

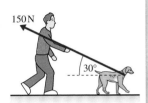

Graphical method for determining work done

Let us consider a constant force acting in the direction of movement pulling a body a distance Δx. The graph of force against distance for this example is as shown in Figure 3. From the definition of work, we know that work done $= F\Delta x$, which in this case is the area under the graph. From this we can deduce that:

work done = area under force vs distance graph

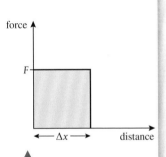

A.3 Figure 3 Force vs distance for a constant force.

Work done by a varying force

Stretching a spring is a common example of a varying force. When you stretch a spring, it gets more and more difficult the longer it gets. Within certain limits, the force needed to stretch the spring is directly proportional to the extension of the spring: $F_H = kx$. This was first recognized by Robert Hooke in 1676, so is named Hooke's law. If we add different weights to a spring, the more weight we add, the longer it gets. If we draw a graph of force against distance as we stretch a spring, it will look like the graph in Figure 4. The gradient of this line, $\frac{F_H}{\Delta x}$ is called the spring constant, k.

The work done as the spring is stretched is found by calculating the area under the graph:

$$\text{area} = \tfrac{1}{2}\text{base} \times \text{height} = \tfrac{1}{2}F_H\Delta x$$

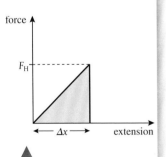

A.3 Figure 4 Force vs extension for a spring.

So: $$\text{work done} = \tfrac{1}{2}F_H\Delta x$$

But if: $$\frac{F_H}{\Delta x} = k \text{ then } F_H = k\Delta x$$

Substituting for F_H gives:

$$\text{work done} = \frac{1}{2}k\Delta x^2$$

This is equal to the elastic potential energy, E_H

Exercise

Q4. A spring of spring constant $2\,\text{N cm}^{-1}$ and length $6\,\text{cm}$ is stretched to a new length of $8\,\text{cm}$.

(a) How far has the spring been stretched?

(b) What force will be needed to hold the spring at this length?

(c) Sketch a graph of force against extension for this spring.

(d) Calculate the work done in stretching the spring.

(e) The spring is now stretched a further $2\,\text{cm}$. Draw a line on your graph to represent this and calculate how much additional work has been done.

Q5. Calculate the work done by the force represented by the figure on the right.

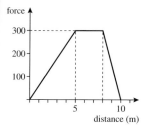

Energy

We have seen that it is sometimes possible for body A to do work on body B, but what does A have that enables it to do work on B? To answer this question, we must define a new quantity: energy.

Energy is the quantity that enables body A to do work on body B.

If body A collides with body B as shown in Figure 5, body A has done work on body B. This means that body B can now do work on body C. Energy has been transferred from A to B.

When body A does work on body B, energy is transferred from body A to body B.

The unit of energy is the joule (J). Energy is a scalar quantity.

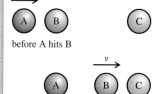

A.3 Figure 5 The red ball gives energy to the blue ball.

Different types of energy

If a body can do work, then it has energy. There are two ways that a simple body such as a red ball can do work. In the example above, body A could do work because it was moving – this is called **kinetic energy**. Figure 6 shows an example where A can do work even though it is not moving. In this example, body A is able to do work on body B because of its position above the Earth. If the hand is removed, body A will be pulled downward by the force of gravity, and the string attached to it will then drag body B along the table. If a body is able to do work because of its position, we say it has **potential energy**.

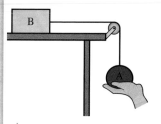

A.3 Figure 6 A has potential energy that could become kinetic energy.

TOK

If we say a body has potential energy, it sounds as though it has the potential to do work. This is true, but a body that is moving has the potential to do work too. This can lead to misunderstanding. It might have been better to call it positional energy.

A.3 Figure 7 A ball gains kinetic energy.

In this section, we only deal with examples of potential energy due to a body's position close to the Earth. However, there are other positions that will enable a body to do work (for example, in an electric field). These will be introduced after the concept of fields has been introduced.

What happens to the energy transferred as we approach the speed of light? (A.5)

Why is the equation for the change in gravitational potential energy only relevant close to the surface of the Earth, and what happens when moving further away from the surface? (D.1)

How are Kirchhoff's and Lenz's laws a consequence of the law of conservation of energy? (B.5, D.4)

Kinetic energy (E_k)

This is the energy a body has due to its movement. To give a body kinetic energy, work must be done on the body. The amount of work done will be equal to the increase in kinetic energy. If a constant force acts on a red ball of mass m as shown in Figure 7, then the work done is Fs.

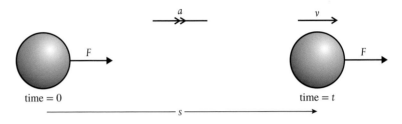

From Newton's second law, we know that $F = ma$, which we can substitute in work = Fs to give work = mas.

We also know that since acceleration is constant, we can use the *suvat* equation $v^2 = u^2 + 2as$, which since $u = 0$ simplifies to $v^2 = 2as$.

Rearranging this gives: $as = \dfrac{v^2}{2}$ so work = $\dfrac{1}{2}mv^2$

This work has increased the kinetic energy of the body so we can deduce that:

$$E_k = \frac{1}{2}mv^2$$

Gravitational potential energy (E_p)

This is the energy a body has due to its position above the Earth.

For a body to have potential energy, it must have at some time been lifted to that position. The amount of work done in lifting it equals the potential energy. Taking the example shown in Figure 9, the work done in lifting the mass, m, to a height h is mgh (this assumes that the body is moving at a constant velocity so the lifting force and weight are balanced).

If work is done on the body then energy is transferred so:

$$\text{gain in } E_p = mgh$$

The law of conservation of energy

We could not have derived the equations for kinetic energy or potential energy without assuming that the work done is the same as the gain in energy. The law of conservation of energy is a formal statement of this fact.

Energy can neither be created nor destroyed; it can only be transferred from one store to another.

This law is one of the most important laws that we use in physics. If it were not true, you could suddenly find yourself at the top of the stairs without having done any work in climbing them, or a car suddenly has a speed of 200 km h^{-1} without anyone touching the accelerator pedal. These things just do not happen, so the laws we use to describe the physical world should reflect that.

Looping the loop

When looping the loop on a rollercoaster, the situation is very similar to the vertical circle example, except that the tension is replaced by the normal reaction force. This also gives a minimum speed at the top of the loop when $\frac{mv_t^2}{r} = F_g$.

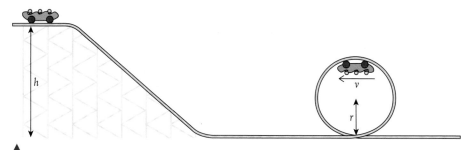

▲
If the ride is propelled by gravity, then the designer must make sure that the car has this minimum speed when it reaches the top.

▲
A.3 Figure 8 Looping the loop.

Applying the law of conservation of energy to the car in Figure 8, if no energy is lost, then the E_p at the top of the hill = $E_p + E_k$ at the top of the loop.

The minimum speed to complete the loop is $mg = \frac{mv^2}{r}$ so at the top of the loop:
$$\tfrac{1}{2}mv^2 = \tfrac{1}{2}mgr$$

The height of the car at the top of the loop is $2r$ so $E_p = mgh = mg2r$.

So E_p at top of slope = $2mgr + \tfrac{1}{2}mgr = 2.5mgr$, which means the height of the slope = $2.5r$. In any real situation, there will be energy lost due to work done against friction and air resistance so the slope will have to be a bit higher.

Worked example

A ball of mass 200 g is thrown vertically upward with a velocity of $2\,\mathrm{m\,s^{-1}}$ as shown in the figure. Use the law of conservation of energy to calculate its maximum height.

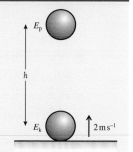

A.3 Figure 9

Solution

At the start of its motion, the body has kinetic energy. This enables the body to do work against gravity as the ball travels upward. When the ball reaches the top, all the kinetic energy has been transferred to potential energy. So applying the law of conservation of energy:

$$\text{loss of } E_k = \text{gain in } E_p$$
$$\tfrac{1}{2}mv^2 = mgh$$

So:
$$h = \frac{v^2}{2g} = \frac{2^2}{2 \times 10} = 0.2\,\mathrm{m}$$

This is exactly the same answer you would get by calculating the acceleration from $F = ma$ and using the *suvat* equations.

Worked example

A block slides down the frictionless ramp shown in the figure. Use the law of conservation of energy to find its speed when it gets to the bottom.

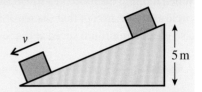

Solution

This time the body loses potential energy and gains kinetic energy so applying the law of conservation of energy:

$$\text{loss of } E_p = \text{gain of } E_k$$

$$mgh = \tfrac{1}{2}mv^2$$

So:
$$v = \sqrt{2gh} = \sqrt{(2 \times 10 \times 5)} = 10\,\text{m s}^{-1}$$

Again, this is a much simpler way of getting the answer than using components of the forces.

Exercise

Use the law of conservation of energy to solve the following:

Q6. A stone of mass 500 g is thrown off the top of a cliff with a speed of 5 m s^{-1}. If the cliff is 50 m high, what is its speed just before it hits the ground?

Q7. A ball of mass 250 g is dropped 5 m onto a spring as shown in the figure on the right.

    (a) How much kinetic energy will the ball have when it hits the spring?

    (b) How much work will be done as the spring is compressed?

    (c) If the spring constant is 250 kN m^{-1}, calculate how far the spring will be compressed.

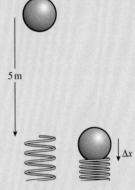

> In this example, the spring is compressed not stretched, but Hooke's law still applies.

Q8. A ball of mass 100 g is hit vertically upward with a bat. The bat exerts a constant force of 15 N on the ball and is in contact with it for a distance of 5 cm.

    (a) How much work does the bat do on the ball?

    (b) How high will the ball go?

Q9. A child pushes a toy car of mass 200 g up a slope. The car has a speed of 2 m s^{-1} at the bottom of the slope.

    (a) How high up the slope will the car go?

    (b) If the speed of the car were doubled, how high would it go now?

Stores of energy

When we are describing the motion of simple red balls, there are only two stores of energy, kinetic energy and potential energy. However, when we start to look at more complicated systems, we discover that we can do work using a variety of different machines, such as petrol engine, electric engine, etc. To do work, these machines must be given energy and this can come from many stores, for example:

- petrol   - solar   - gas   - nuclear   - electricity

All of these (except solar) are related to either the kinetic energy or potential energy of particles.

Fuels

A petrol car gets its energy from petrol, which is mixed with air in the engine and ignited by a spark, causing it to explode. The explosion pushes up a piston that turns a crank, which converts the linear motion of the piston into rotation of the wheels. Petrol is an example of a fuel – a chemical that can be burned to produce heat, which can be used to enable an engine to do work.

Different fuels contain different amounts of energy. Physicists compare the energy contents of fuels in terms of the energy per unit volume (energy density).

Fuel	Energy density/MJ L^{-1}
Coal	72
Crude oil	37
Diesel	36
Sugar	26
Vegetable oil	30
Wood	3

A.3 Table 1

During the process of burning and moving parts of the engine, some energy is lost. The flow of energy can be represented on a Sankey diagram. The width of an arrow represents the amount of that type of energy and, because the total width of all output arrows is equal to the input width, conservation of energy is displayed.

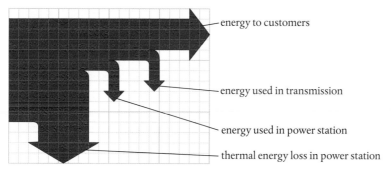

energy to customers

energy used in transmission

energy used in power station

thermal energy loss in power station

Where do the laws of conservation apply in other areas of physics? (NOS)

How is the equilibrium state of a system, such as the Earth's atmosphere or a star, determined? (B.2, E.5)

A.3 Figure 10 A Sankey diagram for a coal-fired power station.

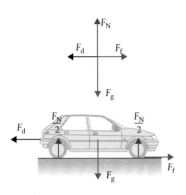

▲
A.3 Figure 11 The forces acting on a car traveling at constant velocity.

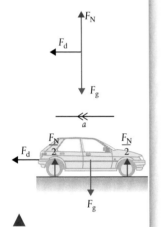

▲
A.3 Figure 12 The forces on a car traveling at high speed with the foot off the accelerator pedal.

If you go over a hump back bridge too quickly, your car might leave the surface of the road. This is because the force needed to keep you moving in a circle is more than the weight of the car.

Energy transfer

Taking the example of a petrol engine, the energy stored in the petrol is transferred to mechanical energy of the car by the engine. 1 liter of petrol contains 36 MJ of energy. Let us calculate how far a car could travel at a constant 36 km h^{-1} on 1 liter of fuel; that is pretty slow but 36 km h^{-1} is 10 m s^{-1} so it will make the calculation easier.

The reason a car needs to use energy when traveling at a constant speed is because of air resistance. If we look at the forces acting on the car, we see that there must be a constant forward force (provided by the friction between tires and road) to balance the air resistance or drag force.

So work is done against the drag force, and the energy to do this work comes from the petrol. The amount of work done = force × distance traveled. So to calculate the work done, we need to know the drag force on a car traveling at 36 km h^{-1}. One way to do this would be to drive along a flat road at a constant 36 km h^{-1} and then take your foot off the accelerator pedal. The car would then slow down because of the unbalanced drag force.

This force will get less as the car slows down but here we will assume it is constant. From Newton's second law, we know that $F = ma$, so if we can measure how fast the car slows down, we can calculate the force. This will depend on the make of car but to reduce the speed by 1 m s^{-1} (about 4 km h^{-1}) would take about 2 s. Now we can do the calculation:

$$\text{acceleration of car} = \frac{(v - u)}{t} = \frac{(9 - 10)}{2} = -0.5 \, \text{m s}^{-2}$$

$$\text{drag force} = ma = 1000 \times -0.5 = -500 \, \text{N}$$

So to keep the car moving at a constant velocity, this force would need to be balanced by an equal and opposite force, $F = 500$ N.

Work done = force × distance so the distance traveled by the car = $\dfrac{\text{work done}}{\text{force}}$.

So if all of the energy in 1 liter of fuel is converted to work, the car will move a distance $= \frac{36 \times 10^6}{500} = 72$ km. Note that if you reduce the drag force on the car, you increase the distance it can travel on 1 liter of fuel.

Circular motion and work

In the previous chapter (A.2), we learned that for an object to move in a circle, a centripetal force (resultant force perpendicular to velocity) was required. An alternative way of deducing that the force acts toward the center is to consider the energy. When a body moves in a circle with constant speed, it will have constant kinetic energy. This means that no work is being done on the mass. But we also know that since the velocity is changing, there must be a force acting on the body. This force cannot be acting in the direction of motion, since if it was, then work would be done and the kinetic energy would increase. We can therefore deduce that the force must be perpendicular to the direction of motion; in other words, toward the center of the circle.

Efficiency

A very efficient road car driven carefully would not be able to drive much further than 20 km on 1 liter of fuel so energy must be lost somewhere. One place where the energy is lost is in doing work against the friction that exists between the moving parts of the engine. Using oil and grease will reduce this but it can never be eliminated. The efficiency of an engine is defined by the equation:

$$\text{efficiency} = \frac{\text{useful work out}}{\text{total energy in}}$$

So if a car travels 20 km at a speed of $10\,\text{m s}^{-1}$, the useful work done by the engine:

$$= \text{force} \times \text{distance} = 500 \times 20000 = 10\,\text{MJ}$$

The total energy put in = 36 MJ so the efficiency $= \dfrac{10}{36} = 0.28$.

Efficiency is often expressed as a percentage so this would be 28%.

Where does all the energy go?

In this example, we calculated the energy required to move a car along a flat road. The car was traveling at a constant speed so there was no increase in kinetic energy and the road was flat so there was no increase in potential energy. We know that energy cannot be created or destroyed so where has the energy gone? The answer is that it has been transferred to the particles that make up the air and car.

There has been a lot of research into making cars more efficient so that they use less fuel. Is this to save energy or money?

Exercise

Q10. A 45% efficient machine lifts 100 kg through 2 m.

    (**a**) How much work is done by the machine?

    (**b**) How much energy is used by the machine?

Q11. A 1000 kg car accelerates uniformly from rest to $100\,\text{km h}^{-1}$.

    (**a**) Ignoring air resistance and friction, calculate how much work was done by the car's engine.

    (**b**) If the car is 60% efficient, how much energy in the form of fuel was given to the engine?

    (**c**) If the fuel contains 36 MJ per liter, how many liters of fuel were used?

Energy and collisions

One of the reasons that we brought up the concept of energy was related to the collision between two balls as shown in Figure 13. We now know that if no energy is lost when the balls collide, then the kinetic energy before the collision = kinetic energy after. This enables us to calculate the velocity afterwards and the only solution in this example is quite a simple one. The red ball transfers all its kinetic energy to the blue one, so the red one stops and the blue one continues, with velocity = $10\,\text{m s}^{-1}$. If the balls become squashed, then some work needs to be done to squash them. In this case, not all the kinetic energy is transferred, and we can only calculate the outcome if we know how much energy is used in squashing the balls.

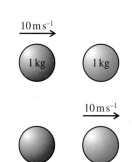

A.3 Figure 13 The red ball gives energy to the blue ball.

Elastic collisions

An elastic collision is a collision in which both momentum and kinetic energy are conserved.

Example: two balls with equal mass

Two balls with equal mass m collide as shown in Figure 14. As you can see, the red ball is traveling faster than the blue one before and slower after. If the collision is perfectly elastic, then we can show that the velocities of the balls simply swap so $u_1 = v_2$ and $u_2 = v_1$.

A.3 Figure 14 Collision between two identical balls.

If the collision is elastic, then momentum and kinetic energy are both conserved. If we consider these one at a time we get:

Conservation of momentum:

$$\text{momentum before} = \text{momentum after}$$

$$mu_1 + mu_2 = mv_1 + mv_2$$

$$u_1 + u_2 = v_1 + v_2$$

Conservation of kinetic energy:

$$E_k \text{ before} = E_k \text{ after}$$

$$\frac{1}{2}mu_1^2 + \frac{1}{2}mu_2^2 = \frac{1}{2}mv_1^2 + \frac{1}{2}mv_2^2$$

$$u_1^2 + u_2^2 = v_1^2 + v_2^2$$

So we can see that the velocities are such that both their sums are equal and the squares of their sums are equal. This is only true if the velocities swap, as in Figure 15.

A.3 Figure 15 A possible elastic collision.

Collision in two-dimensions between two identical balls

Anyone who has ever played pool or snooker will know that balls do not always collide in line; they travel at *angles* to each other. Figure 16 shows a possible collision.

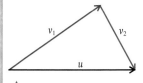
A.3 Figure 16 A two-dimensional collision.

before   after

In this case, the blue ball is stationary so applying the conservation laws we get slightly simpler equations:

$$\vec{u} = \vec{v}_1 + \vec{v}_2$$

$$\vec{u}^2 = \vec{v}_1^2 + \vec{v}_2^2$$

A.3 Figure 17 Adding the velocity vectors.

Note the vector notation to remind us that we are dealing with vectors. The first equation means that the sum of the velocity vectors after the collision gives the velocity before. This can be represented by the triangle of vectors in Figure 17.

The second equation tells us that the sum of the squares of two sides of this triangle = the square of the other side. This is Pythagoras' theorem, which is only true for right-angled triangles. So after an elastic collision between two identical balls, the two balls will always travel away at right angles (unless the collision is perfectly head-on). This of course does not apply to balls rolling on a pool table since they are not isolated.

Pool balls may not collide like perfectly elastic isolated spheres but, if the table is included, their motion can be accurately modeled, enabling scientists to calculate the correct direction and speed for the perfect shot. Taking that shot is another matter entirely.

Inelastic collisions

There are many outcomes of an inelastic collision but here we will only consider the case when the two bodies stick together (coalesce). We call this a **totally inelastic collision**.

Why is the internal energy of an ideal gas equal to the sum of the kinetic energies but not the potential energies? (B.3)

Example

When considering the conservation of momentum in collisions, we used the example shown in Figure 18. How much work was done to squash the balls in this example?

A.3 Figure 18 An inelastic collision.

before   after

According to the law of conservation of energy, the work done squashing the balls is equal to the loss in kinetic energy.

$$E_k \text{ loss} = E_k \text{ before} - E_k \text{ after} = \frac{1}{2} \times 0.1 \times 6^2 - \frac{1}{2} \times 0.6 \times 1^2$$

$$E_k \text{ loss} = 1.8 - 0.3 = 1.5 \, \text{J}$$

So: work done = 1.5 J

How do collisions between charge carriers and the atomic cores of a conductor result in thermal energy transfer? (B.5, B.1)

Explosions

Explosions can never be elastic since, without doing work, the parts that fly off after the explosion would not have any kinetic energy and would therefore not be moving. The energy to initiate an explosion often comes from the chemical energy contained in the explosive.

Example

Consider an exploding ball (shown in Figure 19). How much energy was supplied to the ball by the explosive?

A.3 Figure 19 An explosion. ▶

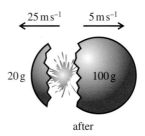

before   after

According to the law of conservation of energy, the energy from the explosive equals the gain in kinetic energy of the ball.

$$E_k \text{ gain} = E_k \text{ after} - E_k \text{ before}$$

$$E_k \text{ gain} = (\tfrac{1}{2} \times 0.02 \times 25^2 + \tfrac{1}{2} \times 0.1 \times 5^2) - 0 = 6.25 + 1.25 = 7.5 \, J$$

> The result of this example is very important; we will use it when dealing with nuclear decay later on. So remember, when a body explodes into two unequal bits, the small bit gets most energy.

Exercise

Q12. Two balls are held together by a spring as shown in the figure. The spring has a spring constant of $10 \, N \, cm^{-1}$ and has been compressed a distance $5 \, cm$.

   (a) How much work was done to compress the spring?

   (b) How much kinetic energy will each gain?

   (c) If each ball has a mass of $10 \, g$, calculate the velocity of each ball.

Q13. Two pieces of modeling clay as shown in the figure collide and stick together.

   (a) Calculate the velocity of the lump after the collision.

   (b) How much kinetic energy is lost during the collision?

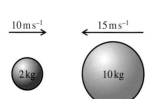

Q14. A red ball traveling at $10 \, m \, s^{-1}$ to the right collides with a blue ball with the same mass traveling at $15 \, m \, s^{-1}$ to the left. If the collision is elastic, what are the velocities of the balls after the collision?

Challenge yourself

1. A $200 \, g$ red ball traveling at $6 \, m \, s^{-1}$ collides with a $500 \, g$ blue ball at rest, such that after the collision the red ball travels at $4 \, m \, s^{-1}$ at an angle of $45°$ to its original direction. Calculate the speed of the blue ball.

Power

We know that to do work requires energy, but work can be done quickly or it can be done slowly. This does not alter the energy transferred but the situations are certainly different. For example, we know that to lift one thousand 1 kg bags of sugar from the floor to the table is not an impossible task – we can simply lift them one by one. It will take a long time but we would manage it in the end. However, if we were asked to do the same task in 5 seconds, we would either have to lift all 1000 kg at the same time or move each bag in 0.005 s; both of which are impossible. Power is the quantity that distinguishes between these two tasks.

Power is defined as:

power = work done per unit time

The unit of power is the $J\,s^{-1}$ which is the same as the watt (W). Power is a scalar quantity.

Example 1: The powerful car

We often use the term power to describe cars. A powerful car is one that can accelerate from 0 to 100 km h^{-1} in a very short time. When a car accelerates, energy is being transferred from the chemical energy in the fuel to kinetic energy. To have a big acceleration, the car must gain kinetic energy in a short time; hence be powerful.

Example 2: Power lifter

A power lifter is someone who can lift heavy weights, so should we not say they are strong people rather than powerful? A power lifter certainly is a strong person (if they are good at it) but they are also powerful. This is because they can lift a big weight in a short time.

Worked example
A car of mass 1000 kg accelerates from rest to 100 km h^{-1} in 5 seconds. What is the average power of the car?

Solution

$$100\,km\,h^{-1} = 28\,m\,s^{-1}$$

gain in kinetic energy of the car $= \frac{1}{2}mv^2 = \frac{1}{2} \times 1000 \times 28^2 = 392\,kJ$

If the car does this in 5 s, then:

$$power = \frac{work\ done}{time} = \frac{392}{5} = 78.4\,kW$$

If power $= \frac{work\ done}{time}$ then we can also write

$$P = \frac{F\Delta s}{t}$$

So $\qquad P = F\frac{\Delta s}{t}$

which is the same as

$$P = Fv$$

where v is the velocity.

This equation is a useful shortcut for calculating the power of a body moving at constant velocity.

Which other quantities in physics involve rates of change? (e.g. A.1, B.5, C.1, E.3)

Horsepower is often used as the unit for power when talking about cars and boats.

746 W = 1 hp

So in the Worked example, the power of the car is 105 hp.

Example 3: Hydroelectric power

It may not be obvious at first, but the energy converted into electrical energy by hydroelectric power stations comes originally from the Sun. Heat from the Sun turns water into water vapor, forming clouds. The clouds are blown over the land and the water vapor turns back into water as rain falls. Rain water falling on high ground has potential energy that can be converted into electricity (see Figure 20). Some countries like Norway have many natural lakes high in the mountains and the energy can be utilized by simply drilling into the bottom of the lake. In other countries rivers have to be dammed.

The Hoover Dam in Colorado can generate 1.5×10^9 watts.

The energy stored in a lake at altitude is gravitational potential energy. This can be calculated from the equation $E_p = mgh$ where h is the height difference between the outlet from the lake and the turbine. Since not all of the water in the lake is the same height, the average height is used (this is assuming the lake is rectangular in cross section).

A.3 Figure 20 The main components in a hydroelectric power station.

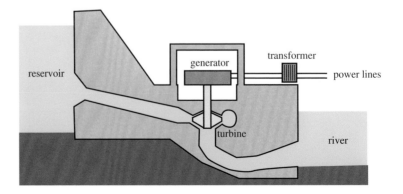

Worked example

Calculate the total energy stored and power generated in the figure if water flows from the lake at a rate of $1\,m^3$ per second.

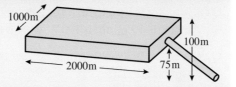

Solution

The average height above the turbine is
$$\frac{(100 + 75)}{2} = 87.5\,m$$

$$\text{Volume of the lake} = 2000 \times 1000 \times 25 = 5 \times 10^7\,m^3$$

$$\text{Mass of the lake} = \text{volume} \times \text{density} = 5 \times 10^7 \times 1000$$
$$= 5 \times 10^{10}\,kg$$

$$E_p = mgh = 5 \times 10^{10} \times 9.8 \times 87.5 = 4.29 \times 10^{13}\,J$$

If the water flows at a rate of $1\,m^3$ per second then $1000\,kg$ falls $87.5\,m$ per second

So the energy lost by the water $= 1000 \times 9.8 \times 87.5 = 875\,000\,J\,s^{-1}$

$$\text{Power} = 875\,kW$$

Exercise

Q15. A weightlifter lifts $200\,kg$ to a height of $2\,m$ in $5\,s$. Calculate the power of the weightlifter in watts.

Q16. In $25\,s$, a trolley of mass $50\,kg$ runs down a hill. If the difference in height between the top and the bottom of the hill is $50\,m$, how much power will have been dissipated?

Q17. A car moves along a road at a constant velocity of $20\,ms^{-1}$. If the resistance force acting against the car is $1000\,N$, what is the power developed by the engine?

Efficiency and power

Efficiency is a quantity that gives a sense of the proportion of input energy that is transferred to useful stores. We define efficiency, η, by the equation:
$$\text{efficiency} = \frac{\text{useful work out}}{\text{total work in}}$$

If the work out is done at the same time as the work in, then we can divide the numerator and the denominator by time to give the equivalent equation:
$$\text{efficiency} = \frac{\text{useful power out}}{\text{total power in}}$$

Exercise

Q18. A motor is used to lift a 10 kg mass 2 m above the ground in 4 s. If the power input to the motor is 100 W, what is the efficiency of the motor?

Q19. A motor is 70% efficient. If 60 kJ of energy is put into the engine, how much work is got out?

Q20. The drag force that resists the motion of a car traveling at 80 km h^{-1} is 300 N.

    (**a**) What power is required to keep the car traveling at that speed?

    (**b**) If the efficiency of the engine is 60%, what is the power of the engine?

Guiding Questions revisited

How are concepts of work, energy and power used to predict changes within a system?

How can a consideration of energetics be used as a method to solve problems in kinematics?

How can transfer of energy be used to do work?

In this chapter, we have provided an alternative model for analyzing the physical changes in a system that requires an understanding of how:

- The work done on a body is the product of the force exerted and the displacement of the body in the direction of the force.
- Gravitational and elastic potential energies are associated with position (relative to positions of zero potential energy that are selected strategically).
- Kinetic energy is associated with momentum (mass and velocity).
- Gravitational potential, elastic potential and kinetic energies are collectively referred to as mechanical energies.
- By considering the types of energy of all bodies at different positions and times, kinematics problems can be solved; unlike the *suvat* equations (which require uniform acceleration), energies can be calculated irrespective of the route taken.
- Energy is a conserved quantity, which means that it can be transferred but not created or destroyed.
- Work is done when energy is transferred and energy is transferred when work is done.
- Power is the rate at which work is done or energy is transferred.
- The upper limit of efficiency, the ratio of the useful energy (or power output) to the total energy (or power input), is 1.
- Sankey diagrams are a visual representation of the input and output energy types and the efficiency of the system.

Practice questions

1. A competition includes an obstacle where the competitor has to jump onto a hanging cylinder, causing it to move along an inclined wire with negligible friction.

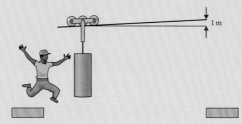

(a) The cylinder has mass 100 kg and the competitor has mass 80 kg. The competitor is moving at 10 m s^{-1} when they catch the cylinder and leave the ground. Using $g = 10\,\text{m s}^{-2}$, calculate:

   (i) the impulse experienced by the cylinder   (2)

   (ii) the impulse experienced by the competitor   (1)

   (iii) the kinetic energy of the cylinder and competitor just after catching the cylinder.   (1)

(b) Determine whether the competitor will get to the far end of the wire.   (2)

(Total 6 marks)

2. A 10 kg block is pulled a distance of 4 m along a frictionless ramp inclined at an angle of 20° by an 8 kg hanging mass as shown in the diagram on the right.

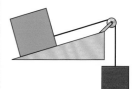

(a) Calculate:

   (i) the work done by the gravitational force   (1)

   (ii) the increase in potential energy of the sliding mass.   (1)

   (iii) Why are the answers to (i) and (ii) different?   (1)

(b) The hanging mass is replaced by an electric motor and winch, which consumes 400 J of energy lifting the block to the same height. If the purpose of the machine is to raise the block, calculate the efficiency of the motor/ramp system.   (2)

(Total 5 marks)

3. A 0.25 kg ball is launched from the ground with initial speed of 20 m s^{-1} and reaches a maximum height of 10 m.

(a) Calculate the speed of the projectile when it reaches maximum height.   (2)

(b) The projectile lands in a bucket with wheels. Calculate the velocity of the bucket plus ball after the ball has landed in the bucket if the mass of the bucket is 1.25 kg.   (2)

(c) The bucket and ball are stopped by a buffer, which is made from a spring that becomes compressed. If the change in length of the spring is 10 cm, calculate the spring constant.   (2)

(Total 6 marks)

4. A stone is falling at a constant velocity vertically down a tube filled with oil. Which of the following statements about the energy transfers of the stone during its motion are correct?

    I.   The gain in kinetic energy is less than the loss in gravitational potential energy.

    II. The sum of kinetic and gravitational potential energy of the stone is constant.

    III. The work done by the force of gravity has the same magnitude as the work done by friction.

    A I and II only   **B** I and III only   **C** II and III only   **D** I, II and III

    (Total 1 mark)

5. The Sankey diagram shows the energy input from fuel that is eventually transferred to useful domestic energy in the form of light in a filament lamp. What is true for this Sankey diagram?

    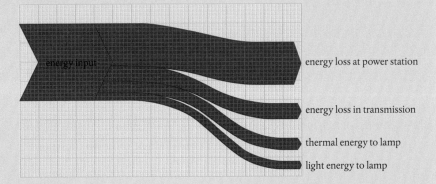

    A The overall efficiency of the process is 10%.

    B Generation and transmission losses account for 55% of the energy input.

    C Useful energy accounts for half of the transmission losses.

    D The energy loss in the power station equals the energy that leaves it.

    (Total 1 mark)

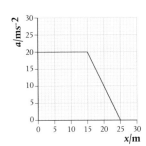

6. The graph on the left shows how the acceleration a of an object varies with distance traveled x. The mass of the object is 3.0 kg. What is the total work done on the object?

    A 300 J   **B** 400 J   **C** 1200 J   **D** 1500 J

    (Total 1 mark)

7. An object of mass m is initially at rest. When an impulse I acts on the object, its final kinetic energy is E_k. What is the final kinetic energy when an impulse of $2I$ acts on an object of mass $2m$ initially at rest?

    A $\dfrac{E_k}{2}$ **B** E_k **C** $2E_k$ **D** $4E_k$

    (Total 1 mark)

8. A train on a straight horizontal track moves from rest at constant acceleration. The horizontal forces on the train are the engine force and a resistive force that increases with speed. Which graph represents the variation with time t of the power P developed by the engine?

A

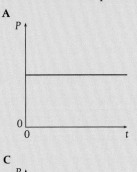

B

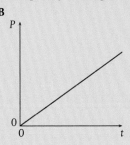

C

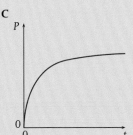

D

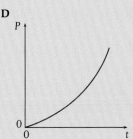

(Total 1 mark)

9. A car traveling at a constant velocity covers a distance of 100 m in 5.0 s. The thrust of the engine is 1.5 kN. What is the power of the car?

A 0.75 kW   **B** 3.0 kW   **C** 7.5 kW   **D** 30 kW

(Total 1 mark)

10. The energy density of a substance can be calculated by multiplying its specific energy (energy per unit mass) with which quantity?

A mass   **B** volume   **C** $\dfrac{\text{mass}}{\text{volume}}$ **D** $\dfrac{\text{volume}}{\text{mass}}$

(Total 1 mark)

11. The diagram shows part of a downhill ski course that starts at point A, 50 m above level ground. Point B is 20 m above level ground. A skier of mass 65 kg starts from rest at point A and, during the ski course, some of the gravitational potential energy is transferred to kinetic energy.

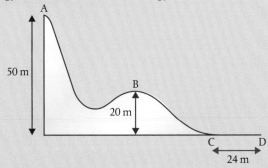

(a) From A to B, 24% of the gravitational potential energy is transferred to kinetic energy. Show that the velocity at B is 12 m s^{-1}.   (2)

(b) The dot on the following diagram represents the skier as she passes point B. Draw and label the vertical forces acting on the skier. (2)

(c) The hill at point B has a circular shape with a radius of 20 m. Determine whether the skier will lose contact with the ground at point B. (3)

(d) The skier reaches point C with a speed of 8.2 m s⁻¹. She stops after a distance of 24 m at point D. Determine the coefficient of dynamic friction between the base of the skis and the snow. Assume that the frictional force is constant and that air resistance can be neglected. (3)

(e) At the side of the course, flexible safety nets are used. Another skier of mass 76 kg falls normally into the safety net with speed 9.6 m s⁻¹. Calculate the impulse required from the net to stop the skier and give an appropriate unit for your answer. (2)

(f) Explain, with reference to change in momentum, why a flexible safety net is less likely to harm the skier than a rigid barrier. (2)

(Total 14 marks)

12. A company designs a spring system for loading ice blocks onto a truck. The ice block is placed in a holder H in front of the spring, and an electric motor compresses the spring by pushing H to the left. When the spring is released, the ice block is accelerated toward a ramp, ABC. When the spring is fully decompressed, the ice block loses contact with the spring at A. The mass of the ice block is 55 kg. Assume that the surface of the ramp is frictionless and that the masses of the spring and the holder are negligible compared to the mass of the ice block.

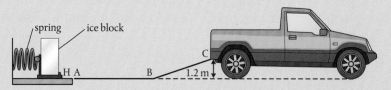

(a) (i) The block arrives at C with a speed of 0.90 m s⁻¹. Show that the elastic energy stored in the spring is 670 J. (2)

(ii) Calculate the speed of the block at A. (2)

(b) Describe the motion of the block:

(i) from A to B with reference to Newton's first law (1)

(ii) from B to C with reference to Newton's second law. (2)

(c) Copy the axes below and sketch a graph to show how the displacement of the block varies with time from A to C. (You do not have to put numbers on the axes.) (2)

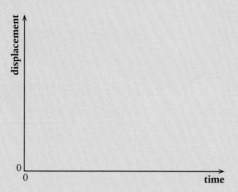

(d) The spring decompression takes 0.42 s. Determine the average force that the spring exerts on the block. (2)

(e) The electric motor is connected to an electrical source of power 816 W. The motor takes 1.5 s to compress the spring. Estimate the efficiency of the motor. (2)

(Total 13 marks)

13. (a) Will hanging a magnet in front of an iron car, as shown in the diagram on the right, make the car go? (1)

    A Yes, it will go.

    B It will move if there is no friction.

    C It will not go.

(b) Explain your answer. (1)

(Total 2 marks)

14. A neutron moving through heavy water strikes an isolated and stationary deuteron (the nucleus of an isotope of hydrogen) head on in an elastic collision.

(a) Assuming the mass of the neutron is equal to half that of the deuteron, find the ratio of the final speed of the deuteron to the initial speed of the neutron. (2)

(b) What percentage of the initial kinetic energy is transferred to the deuteron? (2)

(c) How many collisions would be needed to slow the neutron down from 10 MeV to 0.01 eV? (2)

(Total 6 marks)

 eV is a unit of energy and a conversion into J is not required.

A.4

Rigid body mechanics

◀ Japanese break-dancer, Ami Yuasa, is shown mid-spin in Mumbai in 2019. We learn in A.2 that an object in motion continues with constant velocity unless acted upon by a resultant force. The same is true for rotation. Ami will continue to spin with constant angular speed unless acted upon by a resultant torque.

H L

Guiding Questions

How can we use our knowledge and understanding of the torques acting on a system to predict changes in rotational motion?

If no external torque acts on a system, what physical quantity remains constant for a rotating body?

The balls and boxes we have considered in the previous chapters are rigid bodies. However, for simplicity, we have treated them like points by assuming that all the forces act on the center of the body in question. This is fine if the forces do act on or through the center, but what if they do not?

If you lift one end of a plank of wood, it will rotate. To solve mechanics problems properly, we need to understand the relationship between force and rotation. Fortunately for physicists, the relationships between rotational quantities are very similar to the linear ones. We even use the same words but starting with 'angular': angular displacement, angular speed, etc. The letters used are from the Greek alphabet, but the relationships are the same; the *suvat* equations now become the (not so catchy) $\theta\omega_i\omega_f\alpha t$ equations.

Nature of Science

Treating bodies as if they are points is appropriate up to a point, but insufficient for dealing with real-life examples. However, a rigid body is made up of many points so we can use what we know about point bodies. These models can then be applied to practical problems such as the design of buildings and bridges.

◀ The Millau Viaduct in France took three years to build and is higher than the Eiffel Tower.

Students should understand:

torque τ of a force about an axis as given by $\tau = Fr\sin\theta$
bodies in rotational equilibrium have a resultant torque of zero
an unbalanced torque applied to an extended, rigid body will cause angular acceleration
the rotation of a body can be described in terms of angular displacement, angular velocity and angular acceleration
equations of motion for uniform angular acceleration can be used to predict the body's angular position θ, angular displacement $\Delta\theta$, angular speed ω and angular acceleration α, as given by: $$\Delta\theta = \frac{\omega_f + \omega_i}{2}t$$ $$\omega_f = \omega_i + \alpha t$$ $$\Delta\theta = \omega_i t + \frac{1}{2}\alpha t^2$$ $$\omega_f^2 = \omega_i^2 + 2\alpha\Delta\theta$$
the moment of inertia, I, depends on the distribution of mass of an extended body about an axis of rotation
the moment of inertia for a system of point masses as given by $I = \Sigma mr^2$
Newton's second law for rotation as given by $\tau = I\alpha$ where τ is the average torque
an extended body rotating with an angular speed has an angular momentum L as given by $L = I\omega$
angular momentum remains constant unless the body is acted upon by a resultant torque
the action of a resultant torque constitutes an angular impulse ΔL as given by $\Delta L = \tau\Delta t = \Delta(I\omega)$
the kinetic energy of rotational motion as given by $E_k = \frac{1}{2}I\omega^2 = \frac{L^2}{2I}$

Rotational motion

Up to this point in the course, we have dealt with the motion of a small particle (a red ball), defining quantities related to its motion, deriving relationships relating those quantities, and introducing the concepts of force, momentum and energy to investigate the interaction between bodies. These models were then used to solve problems related to larger bodies, cars, people, etc., by treating them like particles. This works fine provided all the forces act at the center of mass, but what if they do not? Consider the two equal and opposite forces acting on the bar in Figure 1 (notice the bar is floating in space so no gravity is acting on it).

If the bar in Figure 1 was made of rubber, then the problem would be even more complicated as it would also bend. Here we will only consider **rigid** bodies. These are bodies that are made of atoms that do not move relative to one another; in other words, bodies with a fixed shape.

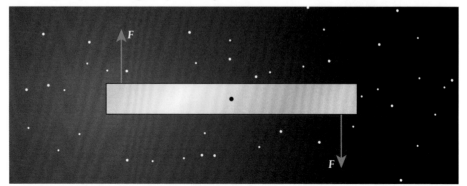

A.4 Figure 1 Forces on a bar.

Let us apply Newton's first law to the body. The forces are balanced so the body will be at rest or moving with a constant velocity. However, if we observe what happens, we find that although the center of mass of the body remains stationary, the body rotates. We need to extend our model to include this type of motion.

Torque (τ)

(a) accelerating

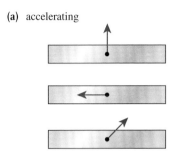

(b) accelerating and rotating

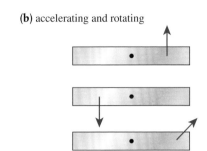

◀ **A.4 Figure 2** Forces do not always cause rotation.

If an unbalanced force acts on the center of mass of a rigid body, then it will have linear acceleration but it will not rotate. All the bodies in Figure 2(a) would have the same magnitude of acceleration. However, if the unbalanced force does not act on the center of mass, as in the examples in Figure 2(b), the bodies will rotate as well as accelerate. We can define the center of mass as *the point on a body through which an unbalanced force can act without causing rotation.*

Describing forces acting on bodies floating in space is rather difficult to imagine since it is not something we deal with every day. To make things more meaningful, let us consider something more down to Earth: a seesaw.

A seesaw is a rigid bar with two movable masses. It only works in a region where the masses are under the influence of gravity, e.g. on the Earth. The forces involved are as shown in Figure 3.

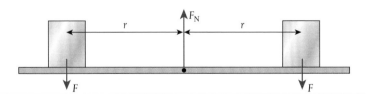

◀ **A.4 Figure 3** Balanced seesaw.

◀ A balanced seesaw only moves when you push with your legs.

Here we can see that the forces up = the forces down so there will be no acceleration. There is also no rotation so the turning effect of the two children must be balanced. The normal reaction that holds the bar up does not turn the bar since it acts at the center of mass. If, however, one child was to get off, then the bar would turn.

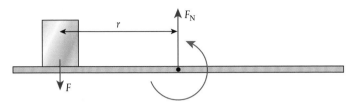

A.4 Figure 4 Seesaw with one child.

The seesaw is held in position by an axle fixed to the center of the bar. This point is called the **pivot**. The axle prevents the bar from accelerating by exerting a force that is equal and opposite to the weight of the children (assuming the bar has negligible weight), but allows it to rotate.

The bar would also turn if one child moved toward the center or was replaced by a child with less weight.

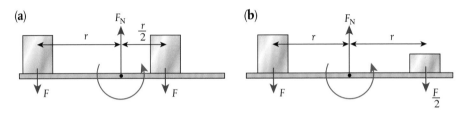

A.4 Figure 5 Unbalanced seesaws.

Balancing the forces when two people lift a heavy object up a flight of stairs, one would expect that each person would exert a force equal to half the weight. But if that is the case, why is it easiest to be at the top? Balancing torques gives the answer.

The turning effect of the force depends upon the force and how far the force is from the pivot. The **torque** gives the turning effect of the force:

torque = force × perpendicular distance from the line of action of the force to a point

So the torque in Figure 4 is $F \times r$. This torque turns the bar in an anticlockwise direction. The torques in Figure 3 are balanced because the clockwise torque = anticlockwise torque, but in Figure 5(a) and (b), the anticlockwise torque ($F \times r$) is greater than the clockwise torque ($F \times \frac{r}{2}$) so the bar will rotate anticlockwise. If we take anticlockwise torques to be positive and clockwise negative, we can say *the bar is balanced when the sum of torques is zero*.

Angular speed and angular acceleration

When the bar rotates, we can define the speed of rotation by the **angular speed**. This is the angle swept out by the bar per unit time. If the torques on the bar are unbalanced, then it will begin to rotate. This means there is change in the angular speed (from zero to something); we can say that the bar has **angular acceleration**:

In this example, the mass of the bar (also called a beam) is negligible but even if it was not, we would not have to consider it since the force at the pivot acts in the same place.

> *angular speed (ω) is the angle swept out per unit time;*
> *angular acceleration (α) is the rate of change of angular speed.*

Equilibrium

When dealing with point masses, we say that a body is in equilibrium when at rest or moving with constant velocity. However, when we define equilibrium for larger, rigid bodies, we should add that there should be no angular acceleration. This means that not only must the forces be balanced but so should the torques.

The sum of all the forces acting on the body is zero

If all the forces acting on a body are added vectorially, the resultant will be zero. With many forces, adding the vectors can lead to some confusing many-sided figures so it is often easier to take components in two perpendicular directions, often vertical and horizontal, then sum these separately. If the total force is zero, then the sum in any two perpendicular directions will also be zero.

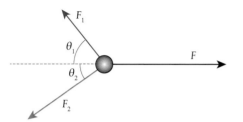
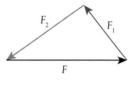

◀ **A.4 Figure 6** Summing vectors or taking components.

If the red ball is in equilibrium, the sum of the forces must be zero so the vector sum has a zero resultant as shown by the triangle. This can be solved but it is not a right-angled triangle so is not simple. An easier approach is to take perpendicular components:

vertical: $F_1 \sin \theta_1 - F_2 \sin \theta_2 = 0$
horizontal: $F - F_1 \cos \theta_1 - F_2 \cos \theta_2 = 0$

In other words:

$$\text{sum of the forces left} = \text{sum of the forces right}$$

and

$$\text{sum of the forces up} = \text{sum of the forces down}$$

The sum of all the torques acting on the body is zero when in equilibrium

In the seesaw example, we obviously considered torques about the pivot but if a body is in equilibrium, then the sum of the torques about *any* point will be zero. Take the example in Figure 7.

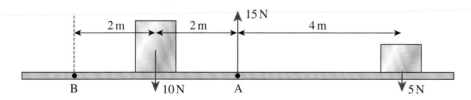

◀ **A.4 Figure 7** Torques can be calculated about A, B or anywhere else.

Taking torques about A:
clockwise = $5 \times 4 = 20$ N m
anticlockwise = $2 \times 10 = 20$ N m

Taking torques about B:
clockwise = $5 \times 8 + 10 \times 2 = 60$ N m (If B was a pivot, both forces would cause a clockwise rotation.)
anticlockwise = $15 \times 4 = 60$ N m (Here we have taken the normal reaction. If this was the only force and B was a pivot, it would cause the bar to rotate in an anticlockwise direction.)

When solving problems, you can choose the *most convenient* place to take torques about. It does not have to be the pivot.

The balanced beam

There are many variations of this problem. In some cases, you can ignore the weight of the beam (as in the seesaw) but in others it must be taken into account.

Worked example

Calculate the weight of the beam balanced as in the figure.

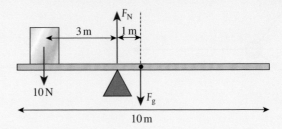

Solution

Taking torques about the pivot we get:

clockwise torques $= F_g \times 1$
anticlockwise torques $= 10 \times 3$
Since balanced: $F_g = 30\,\text{N}$

Worked example

Calculate the length L between the 40 N weight and the pivot needed to balance the beam shown in the figure.

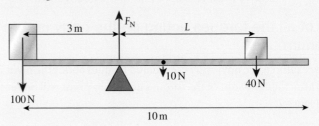

Solution

Taking torques about the pivot:

clockwise torques $= 10 \times 2 + 40 \times L = 20 + 40L$
anticlockwise torques $= 100 \times 3 = 300$
Since balanced: $300 = 20 + 40L$
                       $280 = 40L$
                          $L = 7\,\text{m}$

Exercise

Q1. A 1.0 m ruler is balanced on the 30 cm mark by placing a 300 g mass 10 cm from the end. Calculate the mass of the ruler.

Q2. A 100 g mass is placed at the 10 cm mark on a 20 g ruler. Where must a 350 g mass be placed so that the ruler balances at the 60 cm mark?

Levers

We have seen that the force required to balance the bar depends on how far from the pivot you apply the force. This is the principle of levers and has many applications.

Exercise

Q3. Calculate the unknown force *F* in each of the situations shown below.

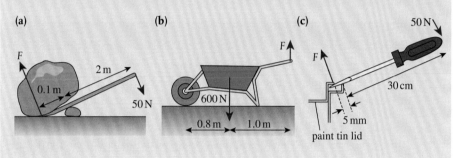

(a)   (b)   (c)

The bridge

A simple bridge consists of a rigid construction spanning the gap between two supports. This may seem nothing to do with rotation, and if built properly, it is not. However, we can use the condition for equilibrium to calculate the forces on the supports.

Advances in engineering have made it possible to construct bridges connecting isolated communities, changing the way people live their lives.

Worked example

A mass of 500 g is placed on the bridge as shown below. If the mass of the bridge is 1.0 kg, calculate the force on each of the supports.

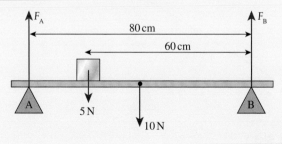

Solution

In this case, if we calculated the torques about the center, we would have two unknowns in the equation so it would be better to find torques about one of the ends. Let us consider end B:

clockwise torques = $F_A \times 0.8$
anticlockwise torques = $5 \times 0.6 + 10 \times 0.4 = 7$ N m

$$F_A = \frac{7}{0.8} = 8.75 \text{ N}$$

To find F_B, we can now use the fact that the vertical forces must also be balanced so:

$$F_A + F_B = 10 + 5$$

$$F_B = 15 - 8.75 = 6.25 \text{ N}$$

Exercise

Q4. A 5.0 m long ladder is held horizontally between two men. A third man with mass 80 kg sits on the ladder 1.0 m from one end. Calculate the force each man exerts if the mass of the ladder is 10 kg.

Q5. A 1.0 m long ruler of mass 200 g is suspended from two vertical strings tied 10 cm from each end. The force required to break the strings is 6.0 N. An 800 g mass is placed in the middle of the ruler and moved toward one end. How far can the mass move before one of the strings breaks?

Non-perpendicular forces

When a force acts at an angle to the bar as in Figure 8, the perpendicular distance from the line of action to the pivot is reduced so $\tau = F \times L \sin\theta$. This is the same component of the force perpendicular to the bar multiplied by the distance to the pivot. The parallel component does not have a turning effect since the line of action passes through the pivot.

A.4 Figure 8 A non-perpendicular force.

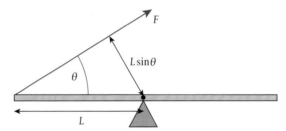

The hanging sign

Signs and lights are often hung on brackets fixed to a wall. This can result in a lot of force on the fixings so they are often supported by a wire as shown in Figure 9. Note that in this case the sign hangs from the center of the bar.

Here we can see that, because the wire is attached to the wall, it makes an angle θ with the supporting bar. This must be balanced by an equal and opposite force from the wall; this is the normal reaction F_N. Calculating torques around the point where the wire joins the bar, we see that the bar and sign cause a clockwise torque. This is balanced by the anticlockwise torque caused by the force F at the wall. This force is provided by the fixing plate or by inserting the bar into a hole in the wall.

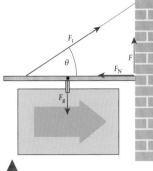

A.4 **Figure 9** A hanging sign.

Exercise

Q6. A sign is hung exactly like the one in Figure 9. The sign has a mass of 50 kg and the bar 10 kg. The bar is 3.0 m in length and the wire is attached 50 cm from the end and makes an angle of 45° with the bar. Calculate:

  (a) the tension F_t in the wire
  (b) the normal force F_N
  (c) the upwards force F.

Q7. Repeat Q6 with the sign hanging from the end of the bar.

How does a torque lead to simple harmonic motion? (C.1)

The leaning ladder

If you have ever used a ladder to paint the wall of a house, you might have wondered what angle the ladder should be: too steep and you might fall backward; not steep enough and it might slip on the ground. By calculating torques, it is possible to find out if the ladder is in equilibrium, but remember the forces change when you start to climb the ladder.

Figure 10 shows a ladder leaning against a frictionless wall in equilibrium. Brick walls are not really frictionless but it makes things easier to assume that this one is. The problem is to find the friction force on the bottom of the ladder.

First we can balance the forces: vertical forces: $F_{N\,(ground)} = F_g$
                                                     horizontal forces $F_{N\,(wall)} = F_f$

Then, calculating torques about the top of the ladder:

sum of clockwise torques = sum of anticlockwise torques

$$F_{N\,(ground)} \times d = F_f \times h + F_g \times \frac{d}{2}$$

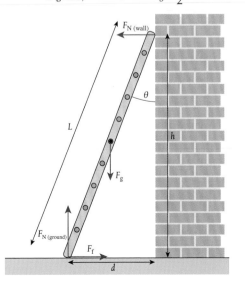

A.4 **Figure 10** A leaning ladder.

When a ladder leans against a wall, the friction at the bottom balances the normal force at the top. As you climb the ladder, the normal force increases so the friction must also increase. However, friction cannot be bigger than $\mu F_{\text{N (ground)}}$. If this is less than the normal force at the top, the ladder will slip. The moral of this tale is that just because the ladder does not slip when you start to climb does not mean it will not slip when you get to the top.

If we were to calculate torques around the bottom of the ladder, we get:

$$F_g \times \frac{d}{2} = F_{\text{N (wall)}} \times h$$

$$F_{\text{N (wall)}} = F_g \times \frac{d}{h} \times \frac{1}{2}$$

$$= F_g \times \frac{\tan \theta}{2}$$

But $F_{\text{N (wall)}} = F_f$:
$$F_f = F_g \times \frac{\tan \theta}{2}$$

So as the angle increases, the friction at the bottom (F_f) increases. This has a maximum value of $\mu F_{\text{N (ground)}}$ (μ is the coefficient of friction) that limits the maximum angle of the ladder.

Exercise

Q8. A ladder of length 5 m leans against a wall such that the bottom of the ladder is 3 m from the wall. If the weight of the ladder is 20 kg, calculate the friction between the ground and the bottom of the ladder.

Q9. If the ladder in Q8. is moved a little bit further out and it begins to slip. Calculate the coefficient of static friction between the ground and the ladder.

Constant angular acceleration

Consider a bar pivoted at one end as in Figure 11. As the bar rotates, it sweeps out an angle $\Delta \theta$. This is the **angular displacement** of the bar and is measured in radians.

A.4 Figure 11 An angle is swept out.

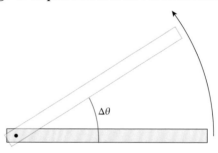

If the time taken for the bar to sweep out angle $\Delta \theta$ is Δt, then the average **angular speed** of the bar ω is given by the equation:
$$\omega = \frac{\Delta \theta}{\Delta t}$$

An unbalanced torque applied to the bar will cause it to rotate faster. The average rate of change of angular speed is the **angular acceleration**, α.
$$\alpha = \frac{\Delta \omega}{\Delta t}$$

These quantities are the rotational equivalents of linear displacement, velocity and acceleration. If the angular acceleration is constant, they are related in the same way, giving angular equivalents of the *suvat* equations (the $\theta \, \omega_i \, \omega_f \, \alpha \, t$ equations!).

Constant angular acceleration equations

A bar rotating at an initial angular speed of ω_i is acted upon by a torque that causes an angular acceleration α, increasing the angular speed to a final value of ω_f in t seconds. During this time, the bar sweeps out an angle θ.

To perform a triple somersault, a gymnast must first initiate the rotation using friction between their feet and the floor. Once the body is rotating, the legs and arms are pulled in to a tucked position, reducing the rotational inertia and resulting in an increase in angular speed. It is also possible to perform a triple somersault with a straight body. In this case, a lot of speed must be built up before take-off to give a high enough angular speed.

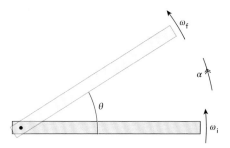

These quantities are related by the equations shown in Table 1.

These angular equations are used to solve problems in exactly the same way as the linear equations.

Angular	Linear
$\omega_f = \omega_i + \alpha t$	$v = u + at$
$\omega_f^2 = \omega_i^2 + 2\alpha\Delta\theta$	$v^2 = u^2 + 2as$
$\Delta\theta = \omega_i t + \frac{1}{2}\alpha t^2$	$s = ut + \frac{1}{2}at^2$
$\Delta\theta = \frac{\omega_i + \omega_f}{2}t$	$s = \frac{u + v}{2}t$

A.4 **Table 1**

Worked example

A body rotating at $10\,\text{rad s}^{-1}$ accelerates at a uniform rate of $2\,\text{rad s}^{-2}$ for 5 seconds. Calculate the final angular speed.

Solution

The data given is:
$\omega_i = 10\,\text{rad s}^{-1}$
$\alpha = 2\,\text{rad s}^{-2}$
$t = 5\,\text{s}$

We wish to find ω_f so the equation to use is $\omega_f = \omega_i + \alpha t$:

$$\omega_f = 10 + 2 \times 5 = 20\,\text{rad s}^{-1}$$

Worked example

Calculate the angle swept out by a body that starts with an angular speed of $2\,\text{rad s}^{-1}$ and accelerates for 10 s at a rate of $5\,\text{rad s}^{-2}$.

Solution

The data given is:
$\omega_i = 2\,\text{rad s}^{-1}$
$\alpha = 5\,\text{rad s}^{-2}$
$t = 10\,\text{s}$

We wish to find $\Delta\theta$ so the equation to use is $\Delta\theta = \omega_i t + \frac{1}{2}\alpha t^2$:

$$\Delta\theta = 2 \times 10 + \frac{1}{2}5 \times 10^2 = 20 + 250 = 270\,\text{rad}$$

This is $\frac{270}{2\pi}$ revolutions.

1 revolution is 2π radians.

Q10. A wheel is pushed so that it has a uniform angular acceleration of $2\,\text{rad}\,\text{s}^{-2}$ for a time of $5\,\text{s}$. If its initial angular speed was $6\,\text{rad}\,\text{s}^{-1}$, calculate:

(**a**) the final angular speed

(**b**) the number of revolutions made.

Q11. The frictional force on a spinning wheel slows it down at a constant acceleration until it stops. Initially, the wheel was spinning at 5 revolutions per second. If the wheel was slowed down to stop in one revolution, calculate:

(**a**) the angular acceleration

(**b**) the time taken.

Graphical representation

As with linear motion, angular motion can be represented graphically. In the example considered previously, a bar rotating at an initial angular speed of ω_i is acted on by a torque that causes an angular acceleration α, increasing the angular speed to a final value of ω_f in t seconds. During this time, the bar sweeps out an angle θ. This can be represented by the three graphs shown in Figure 13.

A.4 Figure 13 Rotational motion graphs.

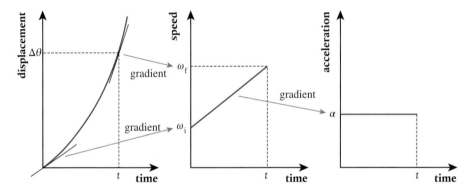

As with the linear equivalents, the gradient of displacement/time $\left(\frac{\Delta\theta}{\Delta t}\right)$ gives speed and the gradient of speed/time $\left(\frac{\Delta\omega}{\Delta t}\right)$ gives acceleration. Working the other way around, the area under acceleration/time gives the change of speed and the area under speed/time gives displacement.

Relationship between angular motion and linear motion

Circular motion can be split into two components: one perpendicular to the circumference and one tangential to it. The perpendicular component is dealt with in A.2 (Circular motion) when we considered only bodies moving with constant speed. In this case, there is acceleration toward the center – the centripetal acceleration – but no tangential acceleration. When an unbalanced torque acts, then there will be an increasing centripetal acceleration plus a tangential acceleration in the directions shown in Figure 14.

We know that if $\Delta\theta$ is measured in radians, $\Delta\theta = \dfrac{\Delta s}{r}$ so $\Delta s = \Delta\theta \times r$.

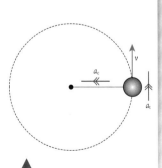

A.4 Figure 14 Centripetal acceleration is along a radius and tangential acceleration is along a tangent.

The average speed of the body is given by: $\dfrac{\Delta s}{\Delta t} = \left(\dfrac{\Delta\theta}{\Delta t}\right)r = \omega r$

If $\Delta\theta$ is a small angle, then we can assume that the velocity does not change significantly so we can say that the instantaneous tangential velocity, $v = \omega r$.

The tangential acceleration of the body, $a_t = \dfrac{\Delta v}{\Delta t} = \dfrac{\Delta \omega r}{\Delta t} = \left(\dfrac{\Delta\omega}{\Delta t}\right)r$

But:
$$\left(\dfrac{\Delta\omega}{\Delta t}\right) = \alpha$$

So:
$$a_t = \alpha r$$

From these equations, we can deduce that if a rigid body is rotating with constant angular acceleration, all points will have the same instantaneous angular speed and angular acceleration, but tangential speed and acceleration will be greater for points furthest away from the axis of rotation as illustrated in Figure 16.

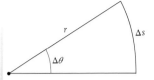

A.4 Figure 15 An angle being swept out relates to a linear displacement.

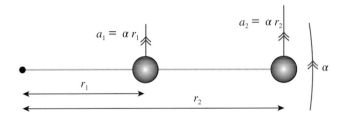

A.4 Figure 16 The greater the radius, the greater the tangential acceleration.

How does rotation apply to the motion of charged particles or satellites in orbit? (D.3, D.1)

Exercise

Q12. A 5 m long ladder is lying on the ground. One end is lifted with constant acceleration $2\,\text{m s}^{-2}$. Calculate:

(a) the angular acceleration of the ladder

(b) the tangential acceleration of the middle of the ladder.

Q13. Two children, each of mass 20 kg, are enjoying a ride on a roundabout as in the photo. One is 0.5 m from the center and the other is 2 m from the center. If the roundabout is rotating at 0.25 revolutions per second, calculate:

(a) the angular speed of the roundabout

(b) the speed of each child

(c) the force required to hold each child onto the roundabout.

Newton's second law applied to angular motion

We have seen that the angular acceleration of a body is related to the torque applied. Here we will derive that relationship by considering a force acting on a particle attached to a rod of negligible mass as shown in Figure 17.

The angular speed is the same for all, but the children on the outside travel faster.

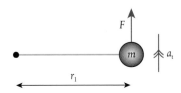

A.4 Figure 17 A mass on a massless rod.

If we apply Newton's second law to this particle, we get:

$$F = ma_t$$

But since the rod is pivoted at the end, the mass will move in a circle of radius r. The angular acceleration of this body will be $\alpha = \frac{a_t}{r}$ so:

$$F = m\alpha r$$

The rotation is caused because the force F provides a torque, Fr about the pivot. Multiplying by r gives:

$$Fr = m\alpha r^2$$

$$\tau = m\alpha r^2$$

A rigid body is made up of lots of particles. When a torque is applied to the body, each particle experiences a small torque turning it in the direction of rotation. Let us consider the body in Figure 18 made of two masses joined with a massless rod rotating about the end with angular acceleration α.

A.4 Figure 18 Two masses on a massless rod.

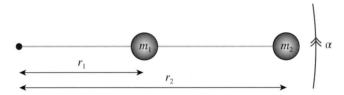

We can apply the formula $\tau = m\alpha r^2$ to find the torque on each mass:

$$\tau_1 = m_1\alpha r_1^2$$

$$\tau_2 = m_2\alpha r_2^2$$

The total torque on the whole body is therefore:

$$\tau = \tau_1 + \tau_2 = m_1\alpha r_1^2 + m_2\alpha r_2^2$$

But the body is rigid so both masses have the same angular acceleration α.

$$\tau = (m_1 r_1^2 + m_2 r_2^2)\alpha$$

So the torque = the sum of $mr^2 \times \alpha$

$$\tau = \left(\Sigma mr^2\right) \times \alpha$$

Moment of inertia

The value Σmr^2 is known as the moment of inertia of the body.

$$I = \Sigma mr^2$$

The unit of moment of inertia is $\mathrm{kg\,m^2}$.

For the body in Figure 18, this is simply $m_1 r_1^2 + m_2 r_2^2$. For more complicated bodies, it can be calculated by performing an integration. This is beyond this course so we will either consider simple bodies or give the equation for the moment of inertia.

The equation for the total torque becomes:

$$\tau = I\alpha$$

This is the rotational equivalent of Newton's second law $F = ma$. We can think of the moment of inertia as being equivalent to the mass in linear motion. If the mass of a body is spread out a long way from the pivot, then I is large, so even though the two objects in Figure 19 have the same mass, the object in Figure 19(a) has the greater moment of inertia and would therefore require a bigger torque to make it rotate with the same angular acceleration.

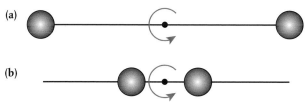

A.4 Figure 19 Same masses but different positions.

The moment of inertia of a body depends on the axis of rotation so the cylinder in Figure 24 will have a greater moment of inertia if rotated about its center than if rotated about its long axis.

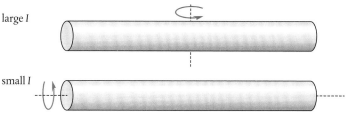

large I

small I

A.4 Figure 20 Same cylinders but different axes.

A bicycle wheel has an easy-to-calculate moment of inertia. If we assume that all of the weight is in the rim and tire and none in the spokes or hub, then all of the mass is the same distance from the center so $I = Mr^2$ where m is the mass of the wheel.

A bicycle wheel.

Worked example

Two 1 kg masses are attached to a rod of negligible mass as shown below. Calculate the moment of inertia of the body if rotated around axes A and B.

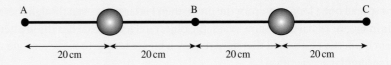

Solution

About A: $I = 1 \times 0.2^2 + 1 \times 0.6^2 = 0.4\,\text{kg m}^2$

About B: $I = 1 \times 0.2^2 + 1 \times 0.2^2 = 0.08\,\text{kg m}^2$

Worked example

If a force of 100 N is applied perpendicular to the rod at point C, calculate the angular acceleration for rotation about A and B.

Solution

About A: torque = $100 \times 0.8 = 80$ N m

$$\tau = I\alpha \quad \text{so} \quad \alpha = \frac{\tau}{I} = \frac{80}{0.4} = 200\,\text{rad s}^{-2}$$

About B: torque = $100 \times 0.4 = 40$ N m

$$\alpha = \frac{\tau}{I} = \frac{40}{0.08} = 500\,\text{rad s}^{-2}$$

Exercise

Q14. Calculate the angular acceleration when a force of 20 N is applied tangentially to the tire of a 2.5 kg bicycle wheel that has a radius of 50 cm.

Q15. A 2.5 kg bicycle wheel with radius 50 cm, rotating at 1 revolution per second, is brought to rest in 1 s by applying the brakes. Calculate the force of the brakes.

Q16. Two forces are applied to a body as shown (the rod has negligible mass). Calculate the angular acceleration of the body.

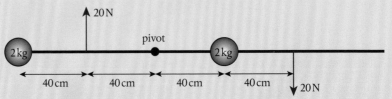

Q17. The forces in Q16. are moved. Calculate the angular acceleration of the body.

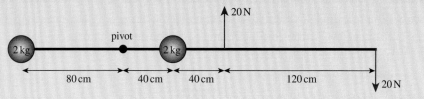

Two identical parallel forces as in Q17 are called a **couple**. The resultant torque of a couple = one force × perpendicular distance between the forces. This is the same about any point.

Some common shapes and their moments of inertia

Although you will not have to derive the formula for the moment of inertia of anything but simple point masses on massless rods, you might come across examples in your practical work where you will need to use the moment of inertia. These are given in Table 2.

◄ **A.4 Table 2** Moments of inertia for different shapes.

Shape		Moment of inertia
A.4 Figure 21 Thin hollow cylinder.		Mr^2
A.4 Figure 22 Solid cylinder.		$\frac{1}{2}Mr^2$
A.4 Figure 23 Sphere.		$\frac{2}{3}Mr^2$ (hollow) $\frac{2}{5}Mr^2$ (solid)
A.4 Figure 24 Rod (length L).		$\frac{1}{12}ML^2$ (center) $\frac{1}{3}ML^2$ (end)

Nature of Science

It is tricky to test out the two types of cylinder because of the need to obtain the same mass with equal radii while ensuring that one is thin and the other is solid. If you are able to construct them, you could experiment with moment of inertia by rolling both down the same ramp.

Exercise

Q18. A metal cylinder is allowed to rotate along the center of its long axis as in the figure below. A string of length 1 m is wrapped around a metal cylinder and pulled with a constant force of 10 N. If the mass of the cylinder is 2 kg and its radius 2 cm, calculate:

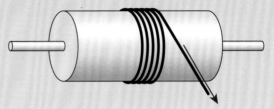

(a) the angular acceleration of the cylinder

(b) the total number of revolutions completed when the string comes to an end

(c) its angular speed after the string is pulled free.

Q19. The cylinder in Q18. spins about its central axis at 100 revolutions per second. It is slowed down by the friction of a rope that is put over the cylinder and pulled down as in the figure below. If the tension in one end of the rope is 10 N and the other 15 N, calculate:

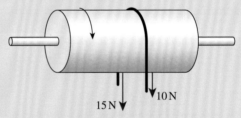

(a) the resultant torque acting on the cylinder

(b) the angular acceleration of the cylinder

(c) the time taken for the cylinder to stop.

Q20. A 5 m long, 20 kg wooden pole lies on the ground. One end is lifted with a vertical force of 200 N. Calculate:

(a) the sum of the torques acting on the pole

(b) the instantaneous angular acceleration of the pole.

Rotational kinetic energy

> For a rigid body, all parts will have the same angular speed no matter how far from the axis they are.

When a rigid body rotates, each particle of the body is moving in a circle, so although the body is not moving forward, each particle has kinetic energy. We call this **rotational kinetic energy**. This energy was transferred to the body by the tangential force that caused the rotation. To calculate the rotational kinetic energy of a body, we can again consider a point mass on a massless rod but this time it has a constant speed v as in Figure 25.

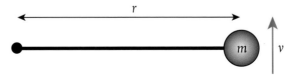

A.4 Figure 25 A mass on a massless rod.

For mass m:
$$E_k = \tfrac{1}{2}mv^2$$

But:
$$v = \omega r$$

So:
$$E_k = \tfrac{1}{2}m\omega^2 r^2 = \tfrac{1}{2}mr^2\omega^2$$

For a body made of many particles:
$$E_k = \tfrac{1}{2}\Sigma mr^2 \times \omega^2$$
$$E_k = \tfrac{1}{2}I\omega^2$$

Again we can see that the moment of inertia is equivalent to mass in linear motion.

Work done

When a force moves in the direction of the force, work is done. When a tangential force causes a body to have angular acceleration, the direction of the force is always changing. However, if we were to consider small movements, the direction is almost constant. The total work done along an arc s will be the sum of all the work done in all of these small movements.

$$\text{work done} = Fs$$

But $\theta = \dfrac{s}{r}$ so work $= Fr\theta = \tau\theta$.

How is the speed of a satellite related to orbital radius? (D.1)

Worked example

Calculate the kinetic energy of a solid sphere of radius 10 cm and mass 2 kg rotating at 10 revolutions per second.

Solution

First find the angular speed $= 10 \times 2\pi = 20\pi\,\text{rad s}^{-1}$

moment of inertia of a solid sphere $= \tfrac{2}{5}Mr^2 = \tfrac{2}{5} \times 2 \times 0.1^2 = 0.008\,\text{kg m}^2$

$E_k = \tfrac{1}{2}I\omega^2 = \tfrac{1}{2} \times 0.008 \times (20\pi)^2 = 16\,\text{J}$

Exercise

Q21. Calculate the rotational kinetic energy of a 3 kg metal cylinder of length 4 m and radius 2 cm rotated at 1 revolution per second:

(a) about its center

(b) about one end

(c) about its long axis.

Q22. A bicycle wheel of radius 45 cm and mass 500 g rotating at 2 revolutions per second is stopped by applying the brakes. Estimate the thermal energy transferred to the brakes and wheel.

Rolling ball

When a ball is released on an inclined plane, the forces acting are as shown in Figure 26. The weight and normal reaction both act through the center of mass. However, the friction does not so will cause rotation about the center causing the ball to roll.

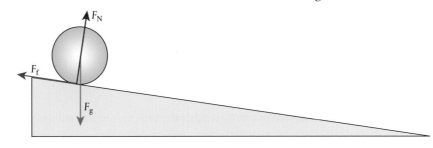

A.4 Figure 26 Forces on a rolling ball.

As the ball rolls down the hill, it loses gravitational potential energy and gains both rotational and translational kinetic energy. If the ball has a vertical displacement h, then we can say;

$$mgh = \tfrac{1}{2}mv^2 + \tfrac{1}{2}I\omega^2$$

If a ball slips down the slope, then there will be no rotation so $mgh = \tfrac{1}{2}mv^2$. Comparing these two equations, we can deduce that the rolling ball will travel down the hill slower than the sliding one. If the ball rolls without slipping, then the point of the ball in contact with the surface of the slope has an instantaneous speed of zero, and the point furthest from the slope has an instantaneous speed of $2v$. The angular velocity of the ball and translational speed of the ball are related through the equation: $\omega = \dfrac{v}{r}$ where r is the radius of the ball. If the ball is solid, then $I = \tfrac{2}{5}mr^2$.

So: $$mgh = \tfrac{1}{2}mv^2 + \frac{\tfrac{1}{2}(\tfrac{2}{5}mr^2)v^2}{r^2} = \tfrac{1}{2}mv^2 + \tfrac{2}{10}mv^2$$

$$gh = \tfrac{7}{10}v^2$$

$$v = \sqrt{\frac{10gh}{7}}$$

It is interesting to note that the velocity does not depend on either the mass or the radius.

How can rotation lead to the generation of an electric current? (D.4)

Exercise

Q23. Following a similar procedure as with the solid ball, derive an equation for the velocity of a rolling hollow ball after it has moved a vertical displacement h. If a hollow ball and a solid ball are rolled down the same slope, which would take the less time?

Q24. A solid sphere of mass 500 g rolls down the slope as shown.

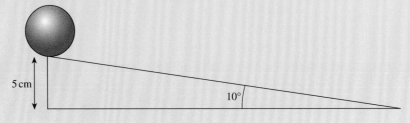

Calculate:

(a) the total kinetic energy at the bottom of the slope

(b) the velocity at the bottom of the slope

(c) the distance traveled down the slope

(d) the time taken to roll down the slope (assume constant acceleration).

Angular momentum (*L*)

Newton's law gives us the relationship $F = ma = (m \times \frac{\Delta v}{\Delta t})$ but it can also be written $F = \frac{\Delta mv}{\Delta t}$ where mv is the momentum.

In rotational motion, we have a similar relationship: $\tau = I\alpha = I \times \frac{\Delta \omega}{\Delta t}$

This can be written $\tau = \frac{\Delta I\omega}{\Delta t}$ where $I\omega$ is the **angular momentum**.

If we consider the particle of mass rotating on a massless bar in Figure 27, we can see that it has instantaneous linear momentum = mv. The angular momentum of this particle would be $I\omega$. In this simple case, $I = mr^2$ so angular momentum = $mr^2\omega$.

But $\omega = \frac{v}{r}$ so angular momentum = mvr, which is linear momentum $\times r$.

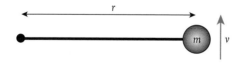

▶ **A.4 Figure 27** A mass on a massless rod.

Exercise

Q25. Calculate the angular momentum of a 5 cm radius solid cylinder of mass 400 g rotating about its center at 10 revolutions per second.

Q26. Calculate the angular momentum of a 10 cm radius solid sphere of mass 750 g rotating about its center at 5 revolutions per second.

Conservation of angular momentum

If no external torques act, then the angular momentum of a system of isolated bodies is conserved. This has some fun applications used to great effect by ballet dancers, ice skaters, and gymnasts. A pirouette is when a ballet dancer spins around very fast. At the start of the spin, the dancer holds her arms outstretched. Her arms are then bought closer to her body, reducing her moment of inertia. Since there are no external torques acting, her angular momentum is conserved, resulting in an increased angular speed. The same principle is used when gymnasts do a triple somersault. The rotation is started with the body stretched. To speed up the rotation, gymnasts pull in their arms and legs to make a tight ball, enabling them to make three rotations before landing.

How does conservation of angular momentum lead to the determination of the Bohr radius? (E.2)

How are the laws of conservation and equations of motion in the context of rotational motion analogous to those governing linear motion? (A.1)

A ballet dancer performs a pirouette. ▶

Worked example

A horizontal disk of radius 20 cm and mass 2 kg is rotating at 3 rotations per second. A second disk of radius 10 cm and mass 1 kg is dropped onto the first one so that the centers are coincident. Calculate the new rotational frequency.

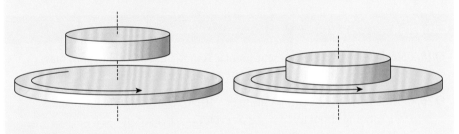

Solution

In this example, the moment of inertia of the turntable has increased so the angular speed will decrease.

moment of inertia of turntable $= \frac{1}{2}mr^2 = 0.5 \times 2 \times (0.2)^2 = 0.04\,\text{kg}\,\text{m}^2$

moment of inertia of disk $= \frac{1}{2}mr^2 = 0.5 \times 1 \times (0.1)^2 = 0.005\,\text{kg}\,\text{m}^2$

combined moment of inertia $= 0.045\,\text{kg}\,\text{m}^2$

initial angular speed $= 2\,\pi \times 3 = 6\pi\,\text{rad}\,\text{s}^{-1}$

Since no external torques act, angular momentum is conserved:

$$I_i \omega_i = I_f \omega_f$$

$$0.04 \times 6\pi = 0.045 \times \omega_f$$

$$\omega_f = 5.3\,\pi\,\text{rad s}^{-1}$$

This is 2.7 revolutions per second.

Exercise

Q27. The body of an ice skater can be considered to be a vertical cylinder of radius 15 cm and mass 60 kg as in the below figure. The outstretched arms act like two 2 kg masses at the end of 1 m long massless rods. The ice skater is spinning with arms stretched at 1 revolution per second. The skater then pulls her arms in so she rotates with radius 25 cm. Calculate:

(a) the moment of inertia of the skater with outstretched arms

(b) the moment of inertia with her arms pulled in

(c) her frequency of rotation after pulling in her arms

(d) her rotational kinetic energy before and after.

(e) Suggest where her extra rotational kinetic energy has come from.

Q28. (a) A turntable is a horizontal rotating disk of mass 1 kg and radius 15 cm. If the turntable rotates at 0.5 revolutions per second, calculate the moment of inertia of the turntable.

A 100 g mass is dropped onto the turntable 10 cm from the center. Calculate:

(b) the new moment of inertia of the turntable + mass

(c) the angular frequency of the turntable.

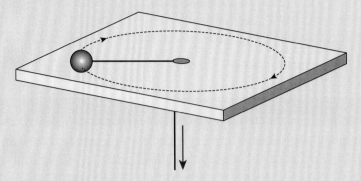

Q29. This one is rather contrived but imagine a frictionless table with a hole in the middle. A string is threaded through the hole and a 500 g ball attached to the end. The ball is now made to travel in a circle while you hold the other end of the string (under the table).

The speed of the ball is $2\,\text{m s}^{-1}$ and the radius is 50 cm.

(a) Calculate the angular momentum of the ball.

(b) If the string is pulled so that it is shortened to 20 cm, calculate the new speed of the ball.

Guiding Questions revisited

How can we use our knowledge and understanding of the torques acting on a system to predict changes in rotational motion?

If no external torque acts on a system, what physical quantity remains constant for a rotating body?

In this chapter, we have considered the analogies between translational and rotational motion to introduce:

- Torque, the product of force and its perpendicular distance from a pivot, as the rotational equivalent of force.
- Angular displacement, the angle subtended between the final and starting positions of a rotating body around a pivot, as the rotational equivalent of displacement.
- Angular speed, the rate of change of angular displacement, as the rotational equivalent of velocity.
- Angular acceleration, the rate of change of angular speed, as the rotational equivalent of acceleration.
- Newton's second law as applying to the proportionality of torque and angular acceleration (just as it does to force and acceleration), which implies that the absence of an external torque will result in a constant angular speed.
- Moment of inertia, which depends on the mass of all particles in a body and their squared distances from a pivot, as the rotational equivalent of mass.
- Angular momentum, angular impulse and rotational kinetic energy as the rotational equivalents of momentum, impulse and kinetic energy.
- The conservation of angular momentum in the absence of external torques.
- The sharing of kinetic energy and rotational kinetic energy in a rolling body (that can be reduced to just kinetic energy for a sliding body).

Practice questions

1. A children's roundabout consists of a rotating disk of radius 2 m and mass 100 kg, and a handrail of radius 1.5 m and mass 20 kg.

   (a) Ignoring the structure supporting the handrail, calculate the moment of inertia of the roundabout. (3)

   (b) The roundabout is set in motion by pushing the handrail with a force of 200 N for a distance of 1 m. This is repeated 10 times. Calculate:

      (i) the torque applied to the roundabout (1)

      (ii) the work done each time the roundabout is pushed (1)

      (iii) the kinetic energy gained by the roundabout (1)

      (iv) the final angular speed of the roundabout (1)

      (v) the final angular momentum of the roundabout. (1)

   (c) A child of mass 40 kg jumps on the roundabout and holds on, such that their center of mass is 2 m from the center.

      (i) Explain why the roundabout slows down. (3)

      (ii) Calculate the new angular speed. (2)

      (iii) Calculate the centripetal force on the child. (2)

   (d) The air resistance of the child is causing the roundabout to slow down.

      (i) If the air resistance is 20 N, calculate the angular acceleration of the roundabout. Ignore friction. (2)

      (ii) The time taken for the roundabout to slow down could be calculated using the equation:

   $$\alpha = \frac{\omega_f - \omega_i}{t}$$

   Explain why this method is not valid and state whether the answer obtained would give a higher or lower value for time than in reality. (2)

   (Total 19 marks)

2. (a) A ramp is constructed by hanging a 2 kg piece of wood from a rope as in the diagram. Calculate the tension in the upper rope. (2)

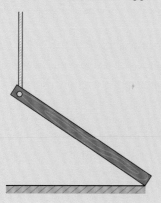

(b) The diagram below shows an alternative construction.

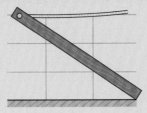

    (i)   State what additional force must be present and where it acts.   (2)

    (ii) Calculate the magnitude of this force.   (3)

(Total 7 marks)

3. A bar rotates horizontally about its center, reaching a maximum angular speed in six complete rotations from rest. The bar has a constant angular acceleration of 0.110 rad s^{-1} . The moment of inertia of the bar about the axis of rotation is 0.0216 kg m^2.

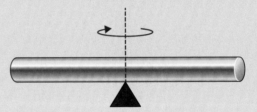

(a) Show that the final angular speed of the bar is about 3 rad s^{-1}.   (2)

(b) Copy the axes below and draw the variation with time t of the angular displacement θ of the bar during the acceleration.   (1)

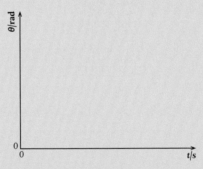

(c) Calculate the torque acting on the bar while it is accelerating.   (1)

(d) The torque is removed. The bar comes to rest in 30 complete rotations with constant angular deceleration. Determine the time taken for the bar to come to rest.   (2)

(Total 6 marks)

4. The first diagram shows a person standing on a turntable that can rotate freely. The person is stationary and is holding a bicycle wheel. The wheel rotates anticlockwise when seen from above.

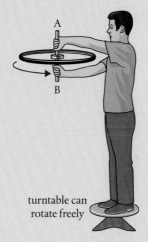

The wheel is flipped, as shown in the second diagram, so that it rotates clockwise when seen from above.

(a) Explain the direction in which the person–turntable system starts to rotate. (3)

(b) Explain the changes to the rotational kinetic energy in the person–turntable system. (2)

(Total 5 marks)

5. A solid sphere of radius r and mass m is released from rest and rolls down a slope without slipping. The vertical height of the slope is h. The moment of inertia I of this sphere about an axis through its center is $\frac{2}{5}mr^2$. Show that the linear velocity v of the sphere as it leaves the slope is $\sqrt{\frac{10gh}{7}}$. (3)

(Total 3 marks)

6. A solid cylinder of mass M and radius R rolls without slipping down a uniform slope. The slope makes an angle θ to the horizontal. The diagram shows the three forces acting on the cylinder. F_N is the normal reaction force and F_f is the frictional force between the cylinder and the slope.

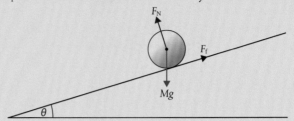

(a) State why F_f is the only force providing a torque about the axis of the cylinder. (1)

(b) (i) The moment of inertia of a cylinder about its axis is $I = \frac{1}{2}MR^2$. Show that, by applying Newton's laws of motion, the linear acceleration of the cylinder is $a = \frac{2}{3}g\sin\theta$. (4)

(ii) Calculate, for $\theta = 30°$, the time it takes for the solid cylinder to travel 1.5 m along the slope. The cylinder starts from rest. (2)

(c) A block of ice is placed on the slope beside the solid cylinder and both are released at the same time. The block of ice is the same mass as the solid cylinder and slides without friction. At any given point on the slope, the speed of the block of ice is greater than the speed of the solid cylinder. Outline why, using the answer to (b)(i). (1)

(d) The solid cylinder is replaced by a hollow cylinder of the same mass and radius. Suggest how this change will affect, if at all, the acceleration in (b)(i). (2)

(Total 10 marks)

7. (a) When a car accelerates forward, it tends to rotate about its center of mass. When will the car nose upward? (1)

A   When the driving force is imposed by the rear wheels (for front-wheel drive, the car would nose downward).

B   Whether the driving force is imposed by the rear or the front wheels.

(b) Explain your answer. (1)

(Total 2 marks)

8. (a) Pull a swing or pendulum to one side, let it go and it will swing back and forth by itself. As it goes back and forth, what is the swing conserving? (1)

A   Angular and linear momentum

B   Only angular momentum

C   Only linear momentum

D   Neither angular momentum nor linear momentum

(b) Explain your answer. (1)

(Total 2 marks)

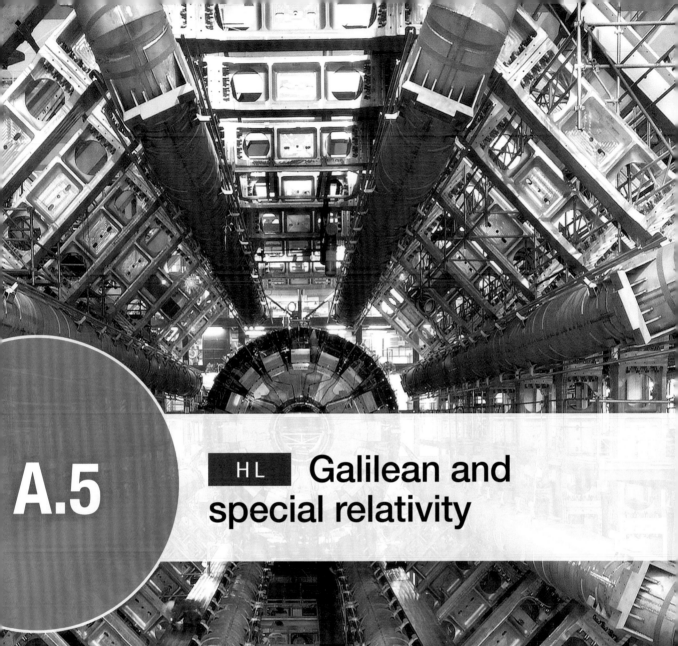

A.5 HL Galilean and special relativity

Rockets only travel close to the speed of light in science fiction and physics exams. However, at CERN, relativity becomes reality.

HL

Guiding Questions

How do observers in different reference frames describe events in terms of space and time?

How does special relativity change our understanding of motion compared to Galilean relativity?

How are space–time diagrams used to represent relativistic motion?

So far, everything has been quite intuitive. A ball starts to move when thrown and stops when caught, a rocket moves forward when burning gas is ejected, and a plank of wood rotates when one end is lifted. You do not need physics to be aware of these things. However, this is about to change.

Clocks that are moving tick more slowly. Lengths contract. $E = mc^2$.

You may be uneasy about the validity of these effects because you have not personally observed them. But instead of doubting the consequences, focus on the postulates. If you accept the bases for relativity, the outcomes follow.

Nature of Science

Accepting that the velocity of light in a vacuum is $3 \times 10^8\,\mathrm{m\,s^{-1}}$, traveling at $2 \times 10^8\,\mathrm{m\,s^{-1}}$ toward the source requires a totally different approach than that accepted by Galilean relativity. Such a radical change is called a **paradigm shift**. Maxwell's equations turned up a result that did not fit in with the classical view of relativity. If two theories do not predict the same outcomes, then at least one of them must be wrong.

Students should understand:

reference frames
Newton's laws of motion are the same in all inertial reference frames and this is known as Galilean relativity
in Galilean relativity the position x' and time t' of an event are given by $x' = x - vt$ and $t' = t$
Galilean transformation equations lead to the velocity addition equation as given by $u' = u - v$
the two postulates of special relativity
the postulates of special relativity lead to the Lorentz transformation equations for the coordinates of an event in two inertial reference frames as given by $x' = \gamma(x - vt)$ $t' = \gamma\left(t - \frac{vx}{c^2}\right)$ where $\gamma = \dfrac{1}{\sqrt{1 - \frac{v^2}{c^2}}}$
Lorentz transformation equations lead to the relativistic velocity addition equation as given by $u' = \dfrac{u - v}{1 - \frac{uv}{c^2}}$

the space–time interval Δs between two events is an invariant quantity as given by $(\Delta s)^2 = (c\Delta t)^2 - \Delta x^2$
proper time interval and proper length
time dilation as given by $\Delta t = \gamma \Delta t_0$
length contraction as given by $L = \dfrac{L_0}{\gamma}$
the relativity of simultaneity
space–time diagrams
the angle between the world line of a moving particle and the time axis on a space–time diagram is related to the particle's speed as given by $\tan\theta = \dfrac{v}{c}$
muon decay experiments provide experimental evidence for time dilation and length contraction.

Reference frames

During this course, we have sometimes used the term **relative**; for example, when quoting a velocity, it is very important to say what the velocity is measured relative to.

Relative velocity

Consider the example shown in Figure 1. A, B, and C measure each other's velocity but they do not agree.

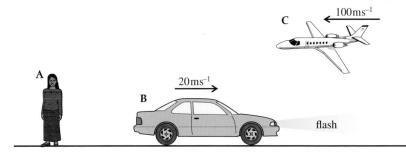

A.5 Figure 1 A, B, and C measure each other's velocity.

Measured by A:

velocity of car = 20 m s^{-1}
velocity of plane = -100 m s^{-1}

Measured by B:

velocity of woman = -20 m s^{-1}
velocity of plane = -120 m s^{-1}

They do not agree because velocity is relative.

When considering an example like this, there are some useful terms worth defining.

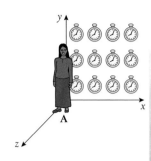

A.5 Figure 2 A in her frame of reference with some of her clocks. We will not always draw the clocks but remember they are there.

A.5 Figure 3 It takes 5 minutes for the light to reach **A** from this event. If she had used her own clock, she would have got the wrong time.

A.5 Figure 4 Three observers; three frames of reference.

An inertial frame of reference is a frame of reference within which Newton's laws of motion apply.

Event

An event is some change that takes place at a point in space at a particular time. If **B** were to flash his headlights, this would be an event.

Observer

An observer is someone who measures some physical quantity related to an event. In this case, **A** measures the time and position when **B** flashed his lights, so **A** is an observer.

Frames of reference

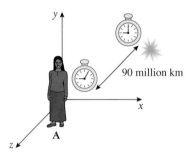

90 million km

Figure 2 shows a frame of reference. This is a coordinate system covered in clocks that an observer uses to measure the time and position of an event. It is covered in clocks so we can measure the time that an event took place *where* it took place. If we used our own clock, we would always measure a time that was a little late, since it takes time for light to get from the event to us (Figure 3). An observer can only make measurements in their own frame of reference.

In the example of Figure 1, we have three observers with three different frames of reference. As we can see in Figure 4, the frames of reference are moving relative to each other with constant velocity.

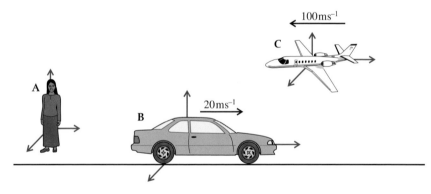

When we look at Figure 4, we can see that the car and plane are moving, but the woman is standing still. This is because we often measure velocity relative to the Earth and she is stationary relative to the Earth. But, according to **B**, **A** is moving with a velocity of 20 m s⁻¹ to the left, so who is moving: **A** or **B**? Is there an experiment that we can do to prove which one of them is moving and which one is stationary?

The presence of the Earth can be confusing, since we always think of the person standing as stationary. Also the Earth's gravitational field and the fact that it is spinning complicate matters. For that reason, we will now move our observers into space, as in Figure 5, and ask the question again: Is there any experiment that **A** or **B** could do to prove who is moving and who is stationary?

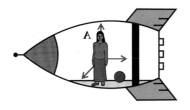
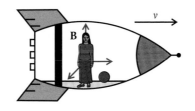

A.5 Figure 5 Two observers; **B** is traveling at velocity *v* relative to **A**.

Let us try a simple experiment. In Figure 5, **A** and **B** take a ball (red) and place it on the floor. If they apply a force to the ball, it will accelerate, and if they do not, it will remain at rest. Newton's laws of motion apply in each frame of reference. There is in fact no experiment that **A** and **B** can do to show who is moving: they are moving relative to each other but there is no absolute movement. We call these **inertial frames of reference**. Let us compare this to the situation in Figure 6 where **B**'s rocket is accelerating.

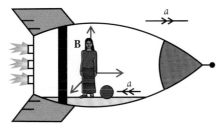

A.5 Figure 6 Observer **B** fires the rocket engines and accelerates.

If observer **B** now places a ball on the ground, it will start to roll toward her even though there is no force acting on the ball, Newton's laws do not apply! Watching from the outside, we can see what is happening. There is a force acting on the rocket pushing it past the ball and the ball is stationary. Newton's laws have not been broken, but inside the rocket it appears that they have. Accelerating frames of reference like this are called **non-inertial frames of reference**.

Galilean relativity

Galileo did many experiments with both live and dead objects and came to the conclusion that the basic laws of physics are the same in all inertial frames of reference.

Coordinate transformations

If observers **A** and **B** in Figure 7 both measure the position of the blue balloon floating weightlessly in the spaceship of **B**, then they will get different values. So if **B** were to tell **A** where the ball was, it would not make sense unless **A** knew how to transform **B**'s measurement into her own frame of reference. The classical way of doing this is called a **Galilean transformation**.

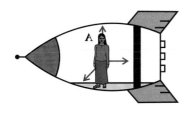
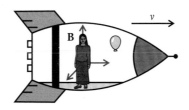

A.5 Figure 7 A blue balloon is at rest in **B**'s rocket.

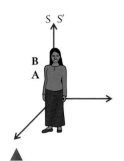

A.5 Figure 8 At time $t = 0$, **A** and **B** are coincident, as can be seen from the magenta dress.

A.5 Figure 9 The balloon pops when **A** and **B** have moved apart.

> The transformations work in both directions. Sometimes, you might see examples where the event is at rest in **A**'s frame of reference rather than **B**'s.

Galilean transformations

At some time, as **A** and **B** are flying apart, the balloon bursts. **A** and **B** measure the time and position of this event. To do this, they will use clocks and tape measures in their own frames of reference, but to transform the measurements, they need to have some reference point. So let us assume that at the time when they started their clocks, the two frames were at the same place. This is not really possible with the rockets but we can imagine it.

If we call **A**'s frame of reference S and **B**'s S′ (S and S dash), we can then distinguish between **A** and **B**'s measurement by using a dash.

Figure 9 shows the moment that the balloon bursts.

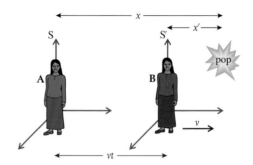

A and B record the coordinates and time for this event and get the results in Table 1.

We can see that only the coordinate in the direction of motion (x) is changed.

Notice that the time is the same in both frames of reference.

A (S)	B (S′)	Transformation
x	x'	$x = x' + vt$
y	y'	$y = y'$
z	z'	$z = z'$
t	t'	$t = t'$

A.5 Table 1

Transformation for length

A length is simply the difference between two positions so consider an object stationary in the frame of reference S′ with one end at x_1' and the other at x_2'. It will have length, $L' = x_2' - x_1'$ measured in S′. An observer in S will measure the ends to be at positions x_1 and x_2 where $x_1 = x_1' + vt$ and $x_2 = x_2' + vt$, so its length measured in S will be:

$$L = x_2 - x_1 = (x_2' + vt) - (x_1' + vt) = x_2' - x_1'$$

As you would expect, according to the Galilean transformations, the length of objects are the same as measured in all inertial frames of reference.

Galilean transformations for velocity

Galilean transformations can also be used to transform velocities. Consider a small bird flying in the x-direction in **B**'s spacecraft as shown in Figure 10.

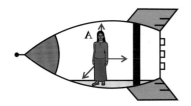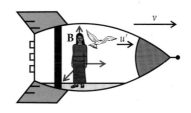

A (S)	B (S')	Transformation
u	u'	$u = u' + v$

The velocity of the bird as measured by A (u) is equal to the velocity of the bird as measured by B (u') minus the velocity of A's frame of reference relative to B ($-v$):

$$u = u' - (-v) = u' + v$$

Galilean transform for acceleration

If the bird in the previous example accelerates from velocity u_1' to u_2' in a time Δt, then the acceleration measured by **B** is:

$$a' = \frac{u_2' - u_1'}{\Delta t}$$

The acceleration measured by **A** will therefore be:

$$a' = \frac{(u_2' + v) - (u_1' + v)}{\Delta t} = \frac{u_2' - u_1'}{\Delta t}$$

So the acceleration is the same in each frame of reference. This means that Newton's second law applies in the same way in all inertial frames of reference.

How are equations of linear motion adapted in relativistic contexts? (A.1)

Exercise

Q1. A train travels through a station at a constant velocity of 8 m s^{-1}. One observer sits on the train and another sits on the platform. As they pass each other, they start their stopwatches and take measurements of a passenger on the train who is walking in the same direction as the train.

Before starting to answer the following questions, make sure you understand what is happening; drawing a diagram will help.

(a) The train observer measures the velocity of the passenger to be 0.5 m s^{-1}. What is the velocity to the platform observer?

(b) After 20 s, how far has the walking passenger moved according to the observer on the train?

(c) After 20 s, how far has the walking passenger moved according to the observer on the platform?

James Clerk Maxwell.

A.5 Figure 11 The rockets have a relative velocity close to the speed of light.

The nature of light

Light travels so fast that it appears to take no time at all. However, if the distances are long enough, time taken is noticeable. The first measurement of the speed of light was undertaken by Danish astronomer Ole Rømer, who in 1776 observed that the timing of the eclipses of Jupiter by its moon Io were not as expected. When the Earth is at its furthest from Jupiter, the eclipse is late, and when the Earth is closest, the eclipse occurs early. The reason for this, he concluded, was due to the time taken for light to travel from Jupiter to the Earth. Others used his measurements to calculate the velocity of light and by 1809, with improved instrumentation, the value stood at $3 \times 10^8 \, \text{m s}^{-1}$, quite close to today's value. At this time, scientists knew that light had wave-like properties but not what sort of wave it was. It was not until 1864 that James Clerk Maxwell deduced that light was an electromagnetic wave.

The speed of light

Maxwell showed how light could propagate through a vacuum and also gave a value for the speed at which the changing fields (regions in which forces are experienced) would travel. This turned out to be:

$$v = \frac{1}{\sqrt{\varepsilon_0 \mu_0}} = \frac{1}{\sqrt{8.85 \times 10^{-12} \times 4\pi \times 10^{-7}}} = 3.00 \times 10^8 \, \text{m s}^{-1}$$

which was the same as the accepted speed of light. Maxwell therefore concluded that light is an electromagnetic wave. One strange thing about this result was that there was nothing in the solution that made it possible to calculate what the velocity would be if the observer, source, or medium was moving. The constants are the permittivity and permeability of a vacuum. It seemed, therefore, that the speed of light is always the same.

This causes some problems when we try to transform the velocity of light using the Galilean transformations. Imagine that observer B in Figure 11 measured the speed of a photon of light traveling along the x-axis to be $3 \times 10^8 \, \text{m s}^{-1}$.

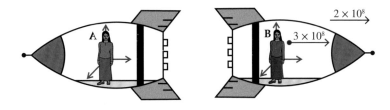

According to the Galilean transformations (and intuition), the velocity of the photon measured by B would be $2 \times 10^8 \, \text{m s}^{-1} + 3 \times 10^8 \, \text{m s}^{-1} = 5 \times 10^8 \, \text{m s}^{-1}$. This is not the case according to Maxwell's equations, which say the speed should be $3 \times 10^8 \, \text{m s}^{-1}$. To test whether the velocity of light is indeed the same as measured by all inertial observers, many experiments have been carried out. Some involve moving light sources such as gamma rays emitted by pions traveling at $0.999c$ and others use the motion of the Earth through space. All results verify that the speed of light in a vacuum is independent of the relative motion of observers.

Momentum of light

Maxwell not only predicted the velocity of light but also that a pulse of light should have momentum equal to $\frac{E}{c}$ where E is the energy of the pulse and c the velocity of light. This does not make sense according to classical mechanics, since light has zero mass and momentum = mass × velocity so should also be zero. In short, electricity and magnetism seem to cause some problems for classical relativity. Einstein showed how it could all fit together.

Special relativity places a limit on the speed of light. What other limits exist in physics? (NOS)

Special relativity

Nature of Science

Einstein developed the theory of relativity by applying mathematics to two clear and experimentally verifiable statements of fact. Anyone wanting to disprove the theory would need to show that the postulates were incorrect or find a fault in the mathematics.

By multiplying by a constant, Einstein got the Galilean transformations to agree with Maxwell's equations, but it is no good to simply fix the equations: there must be some theoretical explanation.

The two postulates of special relativity

Einstein's theory of relativity extends the Galilean idea that Newton's laws apply in all inertial frames of reference to take into account the nature of light. His theory is based on two statements of fact, or postulates:

First postulate:
The laws of physics are the same in all inertial frames of reference.

Second postulate:
The speed of light in a vacuum is the same as measured by all inertial observers.

Light clock experiments

At first, the far-reaching consequences of the constancy of the velocity of light may not be apparent, but a couple of interesting thought experiments will show how this simple statement will make us see the Universe in a totally different way. For this, we need to consider a special type of clock called a light clock. To understand how it works, first we will consider a rubber ball clock.

A rubber ball clock is made using a perfectly elastic ball and two perfectly elastic metal plates as in Figure 12(a). This cannot be made in reality but could be simulated with a program such as Algodoo®.

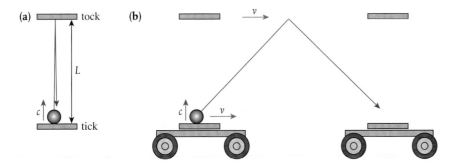

◄ **A.5 Figure 12** The rubber ball clock.

125

The rubber ball bounces back and forth between the metal plates, counting off the seconds as it ticks and tocks. If the velocity of the ball is c, then the time between ticks is $\frac{2L}{c}$. Now, if the clock is rolled past us on a trolley as shown in Figure 12(b), the ball moves to the right as it bounces so will be seen to follow the much longer path shown. However, the clock still ticks at the same rate since although the ball is traveling further, it is also traveling faster. The velocity of the ball is now the vector sum of c and v.

Time dilation

Now we are ready for the light clock. This is the same except that it has a photon of light bouncing between two mirrors as shown in Figure 13(a).

A.5 Figure 13 The light clock.

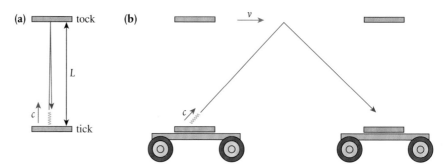

This time, the clock will tick much more quickly but it still ticks at a constant rate. If this clock is now moved past on a trolley, the photon also follows a longer path. But light always has the same velocity, so even though the clock moves on the trolley, the light will travel along this long path with the same velocity as the stationary clock. This means the moving clock must tick more slowly than it does when stationary. The problem is that if we compared this light clock with another type of clock, a wrist watch, for example, we would be able to tell the difference between the moving slow-ticking clock and the stationary fast-ticking one. This is not possible. According to the principle of relativity, we are not supposed to be able to distinguish between inertial frames of reference. So to satisfy this condition, **all** moving clocks tick slowly, including the rubber ball clock and the watch on your wrist.

Proper time

Remember that an inertial frame of reference is a coordinate system covered with clocks so you have to use the clock at the position of the event. Let us say you want to measure the time between two flashes of a light. This could be done with one clock placed by the light. However, if the light was moving past on a trolley as in Figure 14, you would have to use two clocks, one for the first flash and one for the second. A time interval measured by a clock at the same point in space is called the **proper time**.

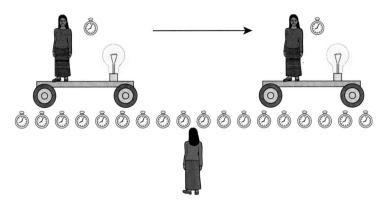

◀ **A.5 Figure 14** Proper time measured by the woman in the red dress.

You may think that the red woman's clock is moving so how can her clock be in the same point in space? Well, in her frame of reference, the clock is in the same point in space; it is the blue woman's clocks that are moving.

Length contraction

The slowing down of time has further consequences. For example, consider a spaceship traveling from Earth to a distant planet at a velocity v as shown in Figure 15. Observers on Earth will see the rocket moving away at a velocity v and the observers on the rocket will see the Earth moving away at velocity $-v$.

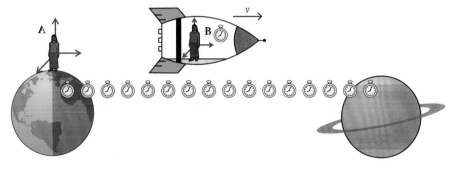

◀ **A.5 Figure 15** Traveling to a distant planet.

Now, if we try to measure the velocity of the rocket, we could divide the distance by the time of flight. The **proper time** between take-off and landing is measured by the observer in the rocket. Since she can use the same clock to measure both leaving the Earth and arriving at the planet, she will actually see the Earth and the planet move relative to the clock, which stays in the same position. This time will be shorter than the time measured by an observer on Earth, who would have to use two separate clocks to measure the same event, one at the start and one at the end. The problem is that they will not agree on the velocity since they will measure the same distance but different times. That is, unless the distance measured by the astronaut is shorter. This is called **length contraction** and only happens along the direction of motion. The distance measured by an observer at rest relative to the length is called the **proper length**.

Clock synchronization

Stranger still is that events that happen at the same time for one observer do not happen at the same time for another. Imagine trying to synchronize two clocks so that they show the same time. The clocks are separated by some distance so you stand exactly in the middle and ask two friends to start the clocks when you raise your hand. When you do this, light will travel from you to your helpers. Because they are the same distance from you, they will receive the light at the same time and start their watches together. As you are doing this, a third friend moves past on a trolley as shown in Figure 16.

i If simultaneous events occur at the same point in space, they will be simultaneous in all frames of reference. So if the woman in the middle raised her hand and smiled at the same time, they will all agree that these two events were simultaneous.

127

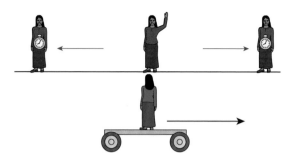

A.5 Figure 16 Synchronize your watches.

The third friend will see the three of you moving past to the left as in Figure 17.

The observer on the trolley will see **A** moving away from the light beam and **B** moving toward it. This means that the light traveling to **A** has to travel further than light traveling to **B**. The light therefore arrives at **B** first, so **B** will start her clock before **A**. So according to the observer on the trolley, the clocks are not synchronized.

A.5 Figure 17 The red women move relative to the blue woman.

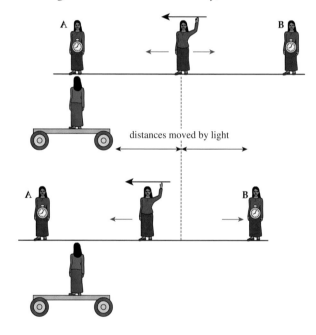

distances moved by light

Events at two different points in space that are simultaneous in one frame of reference are not simultaneous in all frames of reference.

Lorentz transformations

To relate the measurements in one frame of reference to another in Galilean relativity, we use the Galilean transformations. These, however, do not give the correct answer when we try to transform the velocity of light. If we make the assumption that the equations can be corrected by multiplying by some constant γ, then, given that the velocity of light must be the same in all inertial frames of reference, we get the following transformations between measurements taken in S and S′:

A.5 Figure 18 Inertial frames of reference S and S′.

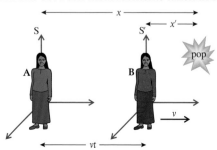

$$x' = \gamma(x - \nu t)$$

$$y' = y$$

$$z' = z$$

$$t' = \gamma\left(t - \frac{\nu x}{c^2}\right)$$

where

$$\gamma = \frac{1}{\sqrt{1 - \frac{\nu^2}{c^2}}} \quad \text{(the Lorentz factor)}$$

These can also be written in terms of the time interval, Δt, and the distance between two events, Δx.

$$\Delta x' = \gamma(\Delta x - \nu \Delta t)$$

$$\Delta t' = \gamma\left(\Delta t - \frac{\nu \Delta x}{c^2}\right)$$

Worked example

Two inertial frames, S and S', coincident at time 0 s, move apart with relative velocity $0.9c$ as shown in Figure 18. An observer in S sees a balloon pop at $x = 5$ m at time 10^{-8} s. When and where did the balloon pop as measured by an observer in S'?

Solution

The relative speed of the two reference frames $= 0.9c$ so:

$$\gamma = \frac{1}{\sqrt{1 - \frac{\nu^2}{c^2}}} = \frac{1}{\sqrt{1 - \frac{0.9^2 c^2}{c^2}}} = 2.3$$

Using the Lorentz transform for x:

$$x' = \gamma(x - \nu t) = 2.3(5 - 0.9 \times 3 \times 10^8 \times 10^{-8}) = 5.29 \, \text{m}$$

and for t:

$$t' = \gamma\left(t - \frac{\nu x}{c^2}\right) = 2.3\left(10^{-8} - \frac{0.9c \times 5}{c^2}\right) = -1.15 \times 10^{-8} \, \text{s}$$

This is before the clocks were started.

Exercise

Q2. An event takes place at a position of $x = 100$ m at a time 4×10^{-8} s as measured by an observer in frame of reference S. A second observer traveling at a speed of 2×10^8 m s^{-1} relative to the first along the line of the x-axis also measures the position and time for the event.

(a) Calculate the Lorentz factor between the two reference frames.

(b) Calculate the time and position measured in the second frame of reference.

Simultaneity

Using these transformations, we can also show that events that are simultaneous in one frame are not simultaneous in another.

Worked example

Consider two trees observed in an inertial frame S at rest relative to the trees. One tree (an oak) is at the origin of the frame of reference. The other (a fir) is 5 km along the *x*-axis as shown in Figure 19. At a time of 4 μs, both trees get hit simultaneously by lightning.

For a second observer flying past in a rocket at a speed of 0.9c, the lightning strikes will not be simultaneous. Calculate the time between the two lightning strikes.

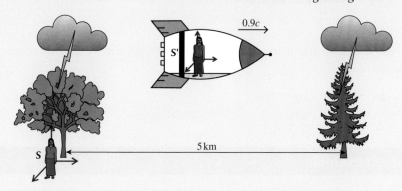

A.5 Figure 19 The frames of reference at the time of the lightning strikes. At time $t = 0$, the two reference frames were coincident.

Solution

To calculate the time between the strikes, we need to transform the time of each strike in S to S′. Taking the oak tree first:

$t' = \gamma(t - \frac{vx}{c^2})$ where $\gamma = 2.3$ as in the previous problem

The oak tree got struck at $t_1 = 4$ μs at position $x = 0$ so $t_1' = 2.3(4 \times 10^{-6} - 0) = 9.2$ μs

The fir tree was struck at $t_2 = 4$ μs but at position $x = 5$ km

So:
$$t_2' = 2.3(4 \times 10^{-6} - \frac{0.9c \times 5 \times 10^3}{c^2})$$

$$t_2' = -25.3 \, \mu s$$

To the observer in the rocket, the fir tree was hit 34.5 μs before the oak tree. Remember that to the observer in the rocket, the trees are moving to the left, so the fir tree is moving toward the observer and the oak tree away. The light therefore travels a shorter distance from the fir tree to the red observer, which is why the fir tree is observed as being hit first.

Exercise

Q3. Repeat the Worked example above with the trees separated by 100 m and a speed of 0.8c.

Time dilation

To measure time between two events, we can use the Lorentz transformations to transform the times of the start and finish. Let us consider two events occurring at the same place measured by an observer in frame S. For example, a light at rest relative to an observer in S flashes at time t_1 then again at time t_2 so the time between flashes measured in S is $\Delta t = t_2 - t_1$. This is the **proper time** since the two events occur at the same point in space so can be measured by the same clock. A second observer moving past at velocity v also measures the time between the flashes as t_1' and t_2', giving a time between flashes of $\Delta t' = t_2' - t_1'$. To the second observer, the flashing light is moving so the two flashes do not occur at the same point in space so she will have to use different clocks to measure each flash.

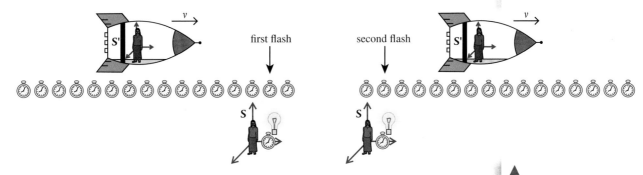

Transforming these times we get:

$$t_1' = \gamma\left(t_1 - \frac{vx}{c^2}\right)$$

$$t_2' = \gamma\left(t_2 - \frac{vx}{c^2}\right)$$

$$t_2' - t_1' = \gamma\left(t_2 - \frac{vx}{c^2}\right) - \gamma\left(t_1 - \frac{vx}{c^2}\right)$$

The position measured in S (x) is the same for each flash so:

$$t_2' - t_1' = \gamma(t_1 - t_2)$$

$$\Delta t' = \gamma \Delta t$$

γ is always greater than 1, which means that the observer moving relative to the flashing light will measure a longer time between flashes. The time has **dilated**.

This can also be written as:

$$T = \gamma T_0$$

where T_0 = proper time (observer measures time with the same clock) and T = time measured by an observer who has to use two clocks to measure the time.

> **A.5 Figure 20** The light flashes twice as the rocket flies past. The line of clocks belong to S′ and are moving with the rocket. The clock used to measure each flash is as labeled.

> ⓘ This can also be solved using the interval transformation:
> $$\Delta t' = \gamma\left(\Delta t - \frac{v\Delta x}{c^2}\right)$$
> where the events in S are happening at the same place so $\Delta x = 0$.

Worked example

A woman in a rocket claps her hands once every second as she flies past an observer on the Earth at a speed of $0.8c$. What is the time between hand claps for the Earth observer?

Solution

In this example, the proper time is the time measured by the woman in the rocket since she can use the same clock to measure each clap, so $T_0 = 1$ s.

The relative speed of the two frames of reference = $0.8c$ so $\gamma = \dfrac{1}{\sqrt{1 - \dfrac{v^2}{c^2}}} = \dfrac{1}{\sqrt{1 - \dfrac{0.8^2 c^2}{c^2}}} = 1.7$.

$$T = \gamma T_0' = 1.7 \times 1 = 1.7 \text{ s}$$

Q4. Two spaceships, A and B, pass in space at relative velocity $0.7c$. An observer on A measures the time between swings of a pendulum he is holding to be 2 s. What will the time period be to an observer in B?

Q5. The half-life of the decay of some radioactive isotope is 30 s. The nucleus is accelerated to a speed of $0.99c$ relative to some observer. What will the half-life be to that observer?

Q6. A rocket travels between the Earth and some distant point at a constant speed of $0.8c$. The time between these events is measured by an observer on the Earth and an observer on the rocket. The rocket observer measures the time to be 2 years.

(a) Which observer measures the proper time?

(b) What time will the Earth observer measure?

Length contraction

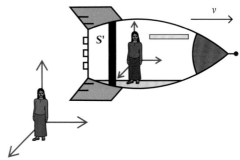

A.5 Figure 21 A rod at rest in inertial frame S'.

The length of a body is the difference in position of its ends, so a metal rod at rest in some inertial frame S' extending from x_1' to x_2' in the x-axis will have length $\Delta x' = x_2' - x_1'$. This is the **proper length** since the rod is not moving relative to the observer in S'. The rod and observer are moving at a velocity v relative to a second observer in frame of reference S, as shown in Figure 21. This observer measures the length of the rod as it passes. To do this, the observer must use a method that enables her to measure each end at the same time, t. If she does not, the rod will move between measurements, resulting in a false value. Observer A measures the length to be $x_2 - x_1$.

Transforming these positions we get:

$$x_1' = \gamma (x_1 - vt)$$

$$x_2' = \gamma (x_2 - vt)$$

$$x_2' - x_1' = \gamma (x_2 - vt) - \gamma (x_2 - vt)$$

But the observer in S measured both ends at the same time so:

$$x_2' - x_1' = \gamma (x_2 - x_1)$$

$$(x_2 - x_1) = \frac{1}{\gamma}(x_2' - x_1')$$

This means that the length measured by the observer moving relative to the rod is shorter than its proper length.

This can also be solved using the interval transformation:
$$\Delta x' = \gamma (\Delta x - v\Delta t)$$
where the time measurements in S are the same so $\Delta t = 0$. In this case, the proper length is $\Delta x'$.

This can also be written: $L = \dfrac{L_0}{\gamma}$

where:

L_0 = proper length

L = length as measured by an observer moving relative to the rod.

Proper length L_0 is said to be **invariant**. This is because all observers will agree that the length measured by an observer at rest with respect to the object would be L_0. If they were moving relative to the object, they would measure it to be shorter, but if they applied the Lorentz transformation, they would agree that the length measured by an observer at rest would be L_0. Proper time is also invariant.

Worked example

A 1 m ruler is lying next to an observer on the Earth. How long will the ruler be if measured by a second observer traveling at a constant velocity of $0.9c$ along the line of the ruler?

Solution

In this case, the proper length, $L_0 = 1$ m.

The relative speed of the two reference frames = $0.9c$ so:

$$\gamma = \dfrac{1}{\sqrt{1 - \dfrac{v^2}{c^2}}} = \dfrac{1}{\sqrt{1 - \dfrac{0.9^2 c^2}{c^2}}} = 2.3$$

According to the length contraction formula: $L = \dfrac{L_0}{\gamma} = \dfrac{1}{2.3} = 0.43$ m

Exercise

Q7. Two spaceships, A and B, pass in space at relative velocity $0.7c$. If an observer in B measures the length of a metal rod he is holding to be 2 m, what is the length of the rod as measured by an observer in A?

Q8. A nucleus decays 2×10^{-8} s (measured in the nucleus' frame of reference) after passing an observer standing on the Earth traveling at a speed of $0.99c$.

(a) Calculate how far the nucleus traveled in the nucleus' frame of reference.

(b) Calculate the time of flight as measured by the Earth observer.

(c) Calculate the distance traveled as measured by the Earth observer.

(d) Which observer measured the proper time?

(e) Which observer measured the proper distance?

Q9. A rocket travels to a distant point, fixed relative to the Earth, at a speed of $0.8c$. The distance to the point measured by an observer on the Earth is 5 light hours (one light hour is the distance traveled by light in 1 hour).

(a) Calculate the time period of the flight measured by an observer on the Earth.

(b) Calculate the distance traveled as measured by an observer on the rocket.

(c) Calculate the time taken measured by the rocket observer.

How would the length and time period of a pendulum change if the maximum velocity was close to the speed of light? (C.1)

Addition of velocity

So far, we have dealt with the Lorentz transformations for position and time but there is also a transformation for velocity. So if a velocity u is measured in frame of reference S, then the velocity measured in frame S' traveling at velocity v relative to S will be given by the equation:

$$u' = \dfrac{u - v}{1 - \dfrac{uv}{c^2}}$$

A.5 Figure 22 Observer in S measures velocity of bird to be *u*. ▶

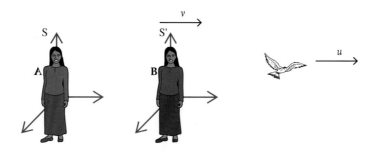

Worked example

An observer in some frame of reference S measures the velocity of a particle moving along the *x*-axis to be 0.9*c*. What would the velocity of the particle be if measured by an observer in S′ moving at 0.5*c* relative to S (along the *x*-axis)?

Solution

Here we can simply substitute the values into the Lorentz velocity transformation:

$$u' = \frac{u - v}{1 - \dfrac{uv}{c^2}}$$

where:
$u = 0.9c$
$v = 0.5c$

$$u' = \frac{0.9c - 0.5c}{1 - \dfrac{0.9c \times 0.5c}{c^2}} = 0.7c$$

Worked example

Two rockets approach an astronaut at speeds of 0.8*c* from the left and 0.9*c* from the right. At what speed will the rockets approach one another from the frame of reference of one of the rockets?

Solution

This typical problem is slightly confusing because there are three possible frames of reference: two rockets and one floating astronaut. The velocities 0.8*c* and 0.9*c* are measured in the frame of reference of the astronaut. What you have to do is transform the velocity of one of the rockets into the frame of reference of the other. A diagram always helps (see the figure below).

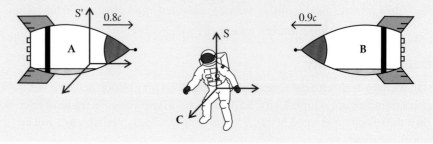

$$u' = \frac{u - v}{1 - \dfrac{uv}{c^2}}$$

If we substitute the values from the rocket example, we get:

$$u = -0.9c$$

$$v = 0.8c$$

$$u' = \frac{-0.9c - 0.8c}{1 - \dfrac{-0.9c \times 0.8c}{c^2}}$$

$$u' = \frac{-1.7c}{1.72}$$

$$u' = -0.988c$$

So, the rockets do not approach each other faster than the speed of light. If the velocities are small, then $\frac{uv}{c^2}$ is approximately zero, so the equation will be the same as the Galilean transform, $u' = u - v$.

Measuring the speed of light

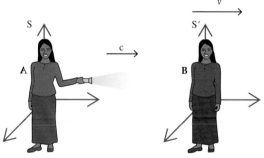

A.5 Figure 23 Two observers measure the speed of light.

An observer A in frame of reference S shines a light and measures the speed at which it propagates in the x-direction to be c. We can now use the Lorentz transform to find the velocity of light as measured by a second observer moving at speed v relative to A.

Using $\qquad u' = \dfrac{u - v}{1 - \dfrac{uv}{c^2}}$ where $u = c$:

$$u' = \frac{c - v}{1 - \dfrac{cv}{c^2}}$$

$$u' = \frac{c - v}{1 - \dfrac{v}{c}}$$

$$u' = \frac{c\left(1 - \dfrac{v}{c}\right)}{\left(1 - \dfrac{v}{c}\right)} = c$$

So, as expected, the Lorentz transformations always return a value of c for the velocity of light, independent of the relative velocity of the observers.

Why is the equation for the Doppler effect for light so different from that for sound? (C.5)

Q10. Two subatomic particles are collided head on in a particle accelerator. Each particle has a velocity of $0.9c$ relative to the Earth. Calculate the velocity of one of the particles, as measured in the frame of reference of the other particle.

Q11. An observer on Earth sees a meteorite traveling at $0.5c$ on collision course with a spaceship traveling at $0.6c$. What is the velocity of the meteorite as measured by the spaceship?

Q12. A relativistic fly flies at $0.7c$ in the same direction as a car traveling at $0.8c$. According to the driver of the car, how quickly will the fly approach the car?

The muon experiment

Rockets traveling close to the speed of light seem a little far-fetched, so it is worth looking at some results from an actual experiment with particles that really do travel that fast. Muons are produced in the upper atmosphere as a result of the decay of pions produced by cosmic rays. They travel at $0.98c$ and have a half-life of $1.6\,\mu s$. They can be detected using two GM tubes, one on top of the other. They travel so fast that they appear to pass through both tubes at the same time.

Let us consider 100 muons at a height of $480\,m$. As they travel to the Earth, they will decay, so less than 100 muons will be detected at ground level. Traveling at almost the speed of light, the muons will take $1.6\,\mu s$ to travel to Earth. According to non-relativistic physics, in this time, half should have decayed.

But according to special relativity, the time in the muons' frame of reference is dilated so the half-life will be longer:

$$T = \gamma T_0$$

where:

$$\gamma = 5$$

T_0 = the proper time for the half-life measured in the inertial frame of the muon
T = the half-life measured from the Earth.

So the half-life measured from the Earth = $8\,\mu s$.

In traveling the $1.6\,\mu s$ down to Earth, less than half of them will decay.

The actual number can be found using $N = N_0 e^{-\lambda t}$, which gives 87 remaining.

We can also look at this in the reference frame of the muons. They will be decaying with a half-life of $1.6\,\mu s$ but the distance to the Earth will be contracted:

$$L = \frac{L_0}{\gamma}$$

So the length measured by the muons = $\frac{480}{5} = 96\,m$.

The time taken traveling at $0.98c = 0.32\,\mu s$, which is much less than one half-life.

It is obviously not possible to measure the same muons at $480\,m$ and at ground level. What is actually measured is the *average* incidence of muons over a long period of time.

The actual number can again be found from $N = N_0 e^{-\lambda t}$, which gives 87 remaining.

The results from muon experiments agree with those predicted by the Lorentz transformations.

Worked example

A very common type of problem is the 'sending a signal home problem'.

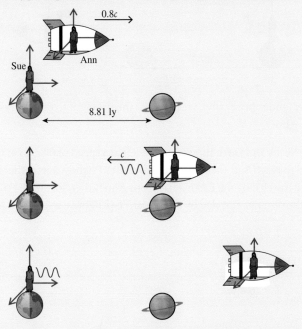

Ann and Sue are twins. Sue remains on Earth. Ann travels to the star Sirius in a spaceship moving at a speed of 0.8c relative to Sue. The distance between Earth and Sirius is 8.8 ly, as measured by Sue. As Ann approaches Sirius, she sends a radio message back to Sue. Determine the time, as measured by Ann, that it takes for the signal to reach Sue.

Solution

The problem here is that neither Ann or Sue can measure the time for the signal to travel from the start to Earth with one clock, so what is the proper time? Sue can, however, measure the time between Ann leaving and the signal arriving with one earthbound clock so this will be the proper time.

T_0 = time for Ann to get to Sirius + time for signal to reach Earth

$T_0 = \frac{8.8}{0.8} + 8.8$ (note that $\frac{\text{distance in light years}}{\text{speed in } c}$ = time in years)

$T_0 = 19.8$ yrs

This is the proper time measured by Sue, so the time measured by Ann will be:
$\gamma T_0 = 1.67 \times 19.8 = 33$ yrs

But this includes the time taken to get to Sirius. For Ann, this was a contracted distance of $\frac{8.8}{1.67} = 5.27$ ly, so the time taken at a speed of 0.8c was $\frac{5.27}{0.8} = 6.6$ yrs.

time for signal to get to Earth = 33 − 6.6 = 26.4 yrs

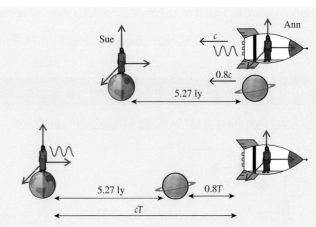

Alternatively (and more easily), we can take the problem from the frame of reference of Ann.

Ann will see Sirius and Earth moving to the left at a speed of 0.8c, so in the time taken for the signal to reach Sue, the Earth has moved 0.8T light years further away. The total distance traveled by the Earth away from Ann is therefore 0.8T + 5.27, which must equal the distance traveled by the signal, cT (T light years).

So: $T = 0.8T + 5.27$

           $(1 - 0.8)T = 5.27$

                    $T = 26.4 \text{ yrs}$

Space–time interval

We have seen that the position and time of an event varies from one frame of reference to another, but the space–time interval does not; we say it is **invariant**. To understand what the space–time interval is and why it is invariant, we will bring back the light clock used earlier in this chapter.

A.5 Figure 24 The light clock.

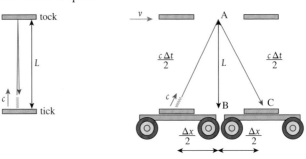

The light clock in Figure 24 is moving relative to the stationary light clock so the light is following a longer path and thus will take a longer time. The interval between ticks is therefore not the same. However, the distance between the mirrors *is* the same. Applying Pythagoras to the triangle ABC:

$$L^2 = \left(\frac{c\Delta t}{2}\right)^2 - \left(\frac{\Delta x}{2}\right)^2$$

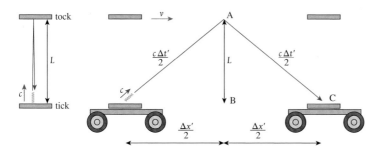

▲ **A.5 Figure 25** A faster light clock.

The light clock in Figure 25 is traveling faster than the previous one so the distance traveled by the trolley and light is longer. Applying Pythagoras again gives:

$$L^2 = \left(\frac{c\Delta t'}{2}\right)^2 - \left(\frac{\Delta x'}{2}\right)^2$$

So:

$$\left(\frac{c\Delta t}{2}\right)^2 - \left(\frac{\Delta x}{2}\right)^2 = \left(\frac{c\Delta t'}{2}\right)^2 - \left(\frac{\Delta x'}{2}\right)^2$$

$$(c\Delta t)^2 - (\Delta x)^2 = (c\Delta t')^2 - (\Delta x')^2$$

The quantity $(ct)^2 - x^2$ is invariant. This is called the **space–time interval**. Note that this is also the same for the stationary light clock where $\Delta x = 0$.

We can try this out for the example considered earlier in the chapter.

Worked example

Two inertial frames, S and S′, coincident at time 0 s, move apart with relative velocity $0.9c$ as shown in Figure 29. An observer in S sees a balloon pop at $x = 5$ m at time 10^{-8} s. What is the space–time interval to the event for each observer?

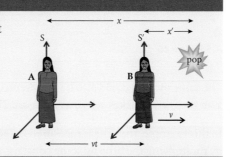

A negative space–time interval is said to be **space-like** since distance is the dominant quantity; a positive space–time interval is **time-like**.

Solution

The relative speed of the two reference frames $= 0.9c$ so $\gamma = \dfrac{1}{\sqrt{1 - \dfrac{v^2}{c^2}}} = \dfrac{1}{\sqrt{1 - \dfrac{0.9^2 c^2}{c^2}}} = 2.3.$

Using the Lorentz transformation for x:

$$x' = \gamma(x - vt) = 2.3(5 - 0.9 \times 3 \times 10^8 \times 10^{-8}) = 5.29\,\text{m}$$

And for t:

$$t' = \gamma\left(t - \frac{vx}{c^2}\right) = 2.3\left(10^{-8} - \frac{0.9c \times 5}{c^2}\right) = -1.15 \times 10^{-8}\,\text{s}$$

So: space–time interval in S $= (ct)^2 - (x)^2 = (3 \times 10^8 \times 10^{-8})^2 - 5^2 = -16$

And: space–time interval in S′ $= (ct')^2 - (x')^2 = (3 \times 10^8 \times -1.15 \times 10^{-8})^2 - 5.29^2 = -16$

Exercise

Q13. Find the space–time interval for the two observers in Q2 on page 129.

Space–time diagrams

Nature of Science

A theory is not very useful if no one can understand it. Space–time diagrams provide a visual representation, making difficult concepts easier to understand (hopefully).

The way that space and time are connected is rather difficult to comprehend. However, things can be made easier by using space–time diagrams, which give a visual representation of one dimension of space and time. To help understand what a space–time diagram represents, we can start with a simple displacement–time graph as in Figure 26. This graph is drawn the usual way with t on the x-axis and displacement on the y-axis (slightly confusing since the y-axis represents x displacement).The gradient of the line will equal the velocity.

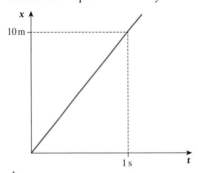

A.5 Figure 26 Displacement–time graph for a body moving at $10\,\text{m}\,\text{s}^{-1}$.

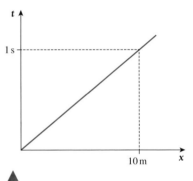

A.5 Figure 27 Swap the axes.

The same motion could also be represented with time t on the y-axis and displacement x on the x-axis (makes sense) as in Figure 27. The velocity is now $\frac{1}{\text{gradient}}$.

In this example, the time is measured in seconds but we could change the scale by multiplying the time by some constant value (e.g., $10\,\text{m}\,\text{s}^{-1}$) to give the graph in Figure 28. Multiplying the time by some constant velocity turns the time into distance; $10\,\text{m}$ is equivalent to $1\,\text{s}$ of the body's movement.

The gradient of this graph is 1 (no units) because the body travels $10\,\text{m}$ in $10\,\text{m}$ worth of time. If we use the same axes to draw the graph for a body traveling at $5\,\text{m}\,\text{s}^{-1}$, we get Figure 29. Here the body is traveling only $5\,\text{m}$ in $10\,\text{m}$ worth of time ($1\,\text{s}$, because the constant value in the previous paragraph is unchanged). The gradient of this line = 2, which is $\frac{10\,\text{m}}{5\,\text{m}}$, so we can calculate the velocity from $(\frac{1}{\text{gradient}}) \times 10\,\text{m}\,\text{s}^{-1}$.

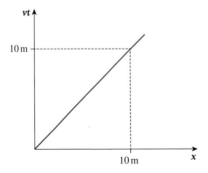

A.5 Figure 28 Change the scale.

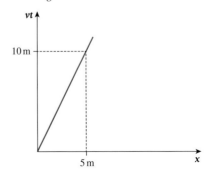

A.5 Figure 29 Same axes displaying a slower body.

Space–time diagrams used in relativity are drawn in the same way except that the constant speed that is used is the speed of light, c. If we draw the position of a photon on these axes, then the line will have gradient = 1, as shown in Figure 30.

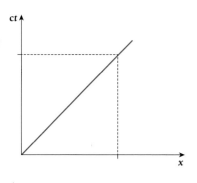

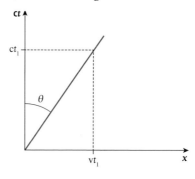

▲ **A.5 Figure 30** Gradient = 1.

▲ **A.5 Figure 31** Gradient = 2.

A body traveling with velocity 0.5c will therefore have a line that is steeper as in Figure 31. A line representing the position of a body at different times is called a **worldline**. The gradient of this line is 2, so we can easily calculate the velocity in multiples of c from $\frac{1}{\text{gradient}} = \frac{v}{c} = \beta$.

In a time t_1 the body will have traveled distance = vt_1 and the value of $ct = ct_1$ so the angle of the line is given by the equation: $\tan\theta = \frac{vt_1}{ct_1} = \frac{v}{c}$

Frames of reference in space–time diagrams

Every point on a space–time diagram represents an event, for example the position of a moving body at different times. The axes of the graph represent the coordinate system used for the time and place of an event. A moving body is represented by a line as in Figure 31. If that body was an observer, then the observer's frame of reference would be tilted along the same line as in Figure 32.

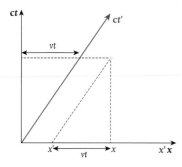

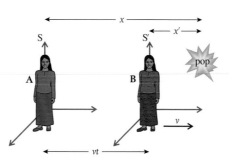

◀ **A.5 Figure 32** ct' represents the position axis for an observer in S′.

The yellow dot represents an event observed by both observers. To record the time of the event, a line is drawn from the event parallel to the x-axis until it coincides with the time axes ct and ct' (the black dotted line). This is the same for both observers. The position measured in S is found by taking a line parallel to the ct-axis to where it crosses the x-axis (the blue dashed line). This is the normal way you would read the position from a graph. The position measured in S′ is found in a similar way but taking the line parallel to the ct'-axis (the red dashed line). This results in two different positions: x and x′. We can see from the geometry of the lines that $x' = x - vt$. This is the Galilean transformation.

We can also use the same method to transform the velocity of a moving object. The green line in Figure 33 represents an object moving with velocity v starting from the origin in the same direction as S'. The velocity u measured in S is $\frac{x}{t}$ and in S' is $u' = \frac{x'}{t}$. Since $x' = x - vt$, we can deduce that $u' = u - v$.

A.5 Figure 33 An object moving with velocity v.

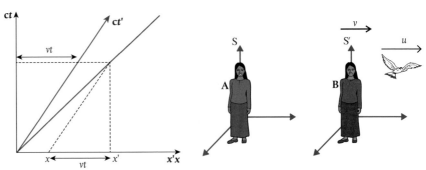

If we use the graph in Figure 33 to measure the speed of light, we run into a problem as shown in Figure 34(a). The gradient of the line is different for each set of axes, meaning that light will have a different velocity for each observer. This cannot be the case but it can be corrected by tilting the x-axis to match the y-axis as in Figure 34(b). We can now use the space–time diagrams to illustrate the consequences of relativity.

A.5 Figure 34 By tilting the x'-axis, we can make the gradient of the green line 1 for both sets of coordinates.

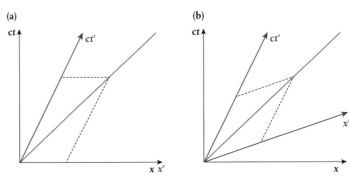

Simultaneity

Using a space–time diagram, we can show that simultaneous events in one frame are not simultaneous in all frames. Consider two events occurring at time 0 in a frame of reference S. These are represented by the two lightning strikes in Figure 35(a). They are on either side of the origin so one occurs on the left, the other on the right. A second frame of reference traveling in a positive direction with respect to S is shown in Figure 35(b). If we plot the time and position of these events in S', we can clearly see that the right-hand event occurred before the in-between one. This is in agreement with the previous explanation in terms of the distance traveled by light being shorter for the one where the observer is moving toward the event.

A.5 Figure 35 Simultaneous events in S.

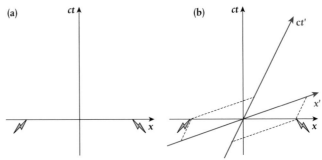

Note that if each flash occurred on the x'-axis, the observer in S′ would see the two flashes simultaneously but the observer in S would not.

Time dilation

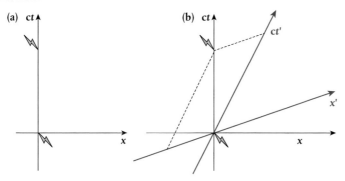

A.5 Figure 36 Time dilation.

Consider two events occurring at $x = 0$ in the S frame. These can be represented by the two lightning flashes on the space–time diagram Figure 36(a). If these flashes are observed in S′, then plotting on the space–time diagram Figure 36(b) we can clearly see that the time between flashes is longer; time has dilated. Notice in this diagram that the position of the second flash is negative. This is because it took place to the left of the observer.

Length contraction

Consider a rod measured in S that is moving at the same velocity as the S′ reference frame. The ends of the rod will have the worldlines shown by the parallel black lines in Figure 37(a). To measure the length of the rod, an observer will need to devise a method to simultaneously measure each end at the same time. Simultaneous events in S occur along the x-axis so the length will be L as shown.

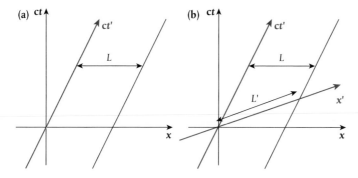

A.5 Figure 37 Length contraction.

If we now observe the rod in the reference frame S′, then the rod would be stationary. Simultaneous events in S′ occur along the x'-axis so the measure of length in S′ would give L' as in Figure 37(b). This is longer than L so we can deduce that the lengths of objects contract when measured by observers moving relative to the rod (in this case, the observer in S). Notice that the measurements simultaneous in one frame of reference are not simultaneous in the other.

Faster than the speed of light

We can use a space–time diagram to show that it is not possible to travel faster than the speed of light. Consider a spaceship leaving planet A and traveling to planet B faster than the speed of light, as illustrated on the space–time diagram of Figure 38. By tracing back the axis, we can see that according to an observer in S, the rocket left at time t_A and arrived some time later at t_B. However, in S', the rocket left at time t_A' and arrived some time earlier at t_B' and this is not possible.

A GeoGebra worksheet linked to this topic is available in the eBook.

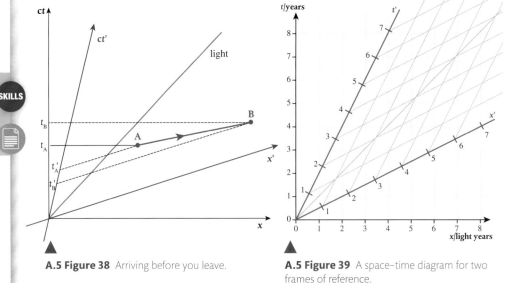

A.5 Figure 38 Arriving before you leave.

A.5 Figure 39 A space–time diagram for two frames of reference.

Exercise

Figure 39 is a space–time diagram for two frames of reference traveling at $0.5c$ relative to each other. Use it to estimate the answers to the following questions.

Q14. Estimate the position and time in S' of an event that takes place at a position 4 light years from the origin at a time of 6.5 years from the time when the clocks were started in S. Check your solution using the Lorentz transformations and find the space–time interval for the event in each frame.

Q15. Estimate the position and time in S of an event measured in S' that takes place at a position 5 light years from the origin at a time of 1 year from the time when the clocks were started. Check your solution using the Lorentz transformations and find the space–time interval for the event in each frame.

Q16. Two events, stationary with respect to S, occur at a distance 4 light years from the origin at times 3 years and 6.5 years. Estimate the time between the events as measured by an observer in S'. Check your solution with the time dilation formula.

Q17. A very long rod at rest in the S' frame is measured at time 0 to extend from 2 ly to 5 ly. An observer in S manages to simultaneously measure the position of each end of the rod. Estimate the length it will measure. Check your solution with the length contraction formula.

Q18. A rocket travels from Earth at a speed of 0.5c. After traveling for 4 years (measured by the rocket so take the rocket as S), a radio signal is sent back to the Earth. How long after the rocket left the Earth will the Earth receive the signal (as measured on Earth)? By following the method in the earlier example, check your solution by calculation.

Q19. Assuming the origin represents the year 2000, a rocket leaves Earth in 2003 and travels at 4c for 2 years. When will the rocket depart and arrive as measured by an observer traveling at 0.5c relative to the Earth? What if the rocket traveled at twice the speed of light? Why can you not check this with a calculation?

Guiding Questions revisited

How do observers in different reference frames describe events in terms of space and time?

How does special relativity change our understanding of motion compared to Galilean relativity?

How are space–time diagrams used to represent relativistic motion?

In this chapter, we have expanded our considerations of motion to include light itself and how:

- Identifying events, observers and frames of reference allows physicists to evaluate the relative nature of quantities.
- Newton's laws apply in inertial (stationary or constant velocity) frames of reference but do not apply in non-inertial (accelerating) frames of reference.
- Galilean transformations enable conversions of the velocity and acceleration measured by two observers in two inertial reference frames.
- The first postulate of special relativity is that the laws of physics are the same in all inertial frames of reference.
- The second postulate of special relativity is that the speed of light in a vacuum is the same as measured by all inertial observers.
- The Lorentz transformations enable calculations of relative quantities using a constant that depends on the ratio of the square of the ratio of the frame of reference velocity to the speed of light.
- Space–time diagrams are graphs of time multiplied by the speed of light against displacement, in which lines represent how the position of a body (e.g. an observer) varies with time and points represent events.
- The non-absolute nature of simultaneity at a distance, time dilation, length contraction and the speed of light as an upper limit for the speed of any body can be represented on space–time diagrams.

Practice questions

1. **(a)** Define the terms *proper time* and *proper length*.   (2)

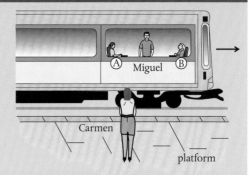

In the diagram, Miguel is in a railway carriage that is traveling in a straight line with uniform speed relative to Carmen who is standing on the platform.

Miguel is midway between two people sitting at opposite ends A and B of the carriage.

At the moment that Miguel and Carmen are directly opposite each other, the person at end A of the carriage strikes a match as does the person at end B of the carriage.

According to Miguel these two events take place simultaneously.

(b) **(i)** Discuss whether the two events will appear to be simultaneous to Carmen.   (4)

(ii) Miguel measures the distance between A and B to be 20.0 m. However, Carmen measures this distance to be 10.0 m. Determine the speed of the carriage relative to Carmen.   (2)

(iii) Explain which of the *two* observers, if either, measures the correct distance between A and B?   (2)

(Total 10 marks)

2. **(a)** State what is meant by an *inertial* frame of reference.   (1)

An observer S in a spacecraft sees a flash of light. The light is reflected from a mirror, distance D from the flash, and returns to the source of the flash as shown in the diagram. The speed of light is c.

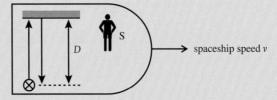

observer E

(b) Write down an expression, in terms of D and c, for the time T_0 for the flash of light to return to its original position, as measured by the observer S who is at rest relative to the spaceship.   (1)

The spaceship is moving at speed v relative to the observer labeled E in the diagram. The speed of light is c.

(c) (i) Copy the diagram and draw the path of the light as seen by observer E. Label the position F from where the light starts and the position R where the light returns to the source of the flash. (1)

(ii) The time taken for the light to travel from F to R, as measured by observer E, is T. Write down an expression, in terms of the speed v of the spacecraft and T, for the distance FR. (1)

(iii) Using your answer in (ii), determine, in terms of v, T, and D, the length L of the path of light as seen by observer E. (2)

(iv) Hence derive an expression for T in terms of T_0, v, and c. (4)

(Total 10 marks)

3. (a) State the **two** postulates of the special theory of relativity. (2)

(b) Two identical spacecraft are moving in opposite directions, each with a speed of $0.80\,c$ as measured by an observer at rest relative to the ground. The observer on the ground measures the separation of the spacecraft as increasing at a rate of $1.60\,c$.

0.80c     0.80c

ground

(i) Explain how this observation is consistent with the theory of special relativity. (1)

(ii) Calculate the speed of one spacecraft relative to an observer in the other. (3)

(Total 6 marks)

4. (a) The space–time diagram shows two inertial frames of reference and the position and time of an event A.

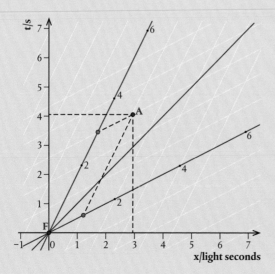

(i)   Show that the relative velocity of the two frames of reference is 0.5c.   (2)

(ii)   Calculate the Lorentz factor for this situation.   (2)

(iii)   Using the diagram, deduce the time and position of event A as measured in the frame of reference at rest with the event.   (1)

(iv)   Using the diagram, deduce the time and position of event A as measured in the frame of reference moving at 0.5c relative to the event.   (1)

(v)   Show that your answers to (iii) and (iv) are in agreement with the values obtained using the Lorentz transformations.   (2)

(b)   A spaceship travels from the origin to point A.

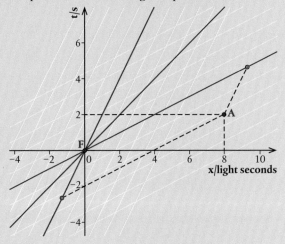

(i)   Calculate the velocity of the spaceship.   (1)

(ii)   Using the diagram, explain why it is not possible to travel faster than the speed of light.   (2)

(c)   A spaceship travels from 0 to A at 0.5c. Use the space–time diagram to determine:

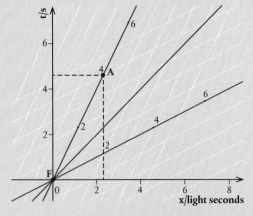

(i)   the time taken according to a observer on the spaceship   (1)

(ii)   the time taken according to an observer at rest at 0.   (1)

(iii)   State which of these values is the proper time.   (1)

(iv)   Use these results to explain what is meant by time dilation.   (2)

(Total 16 marks)

5. Muons are unstable particles with a proper lifetime of 2.2 μs. Muons are produced 2.0 km above the ground and move downward at a speed of 0.98c relative to the ground. For this speed, $\gamma = 5.0$. Discuss, with suitable calculations, how this experiment provides evidence for time dilation. (3)

(Total 3 marks)

6. If you were traveling with respect to the stars at a speed close to the speed of light, you could detect your speed because of which reason?

A Your mass would increase.

B Your heart would slow down.

C You would shrink.

D All of the above.

E You could never tell your speed by changes in you. *(Total 1 mark)*

7. There is a person from the mythology of physics who can run faster than light. Of course, that can never be. But why can they never run faster than light? The following reason is sometimes given:

As the person runs faster and faster, their mass increases, so they find they have become a very massive person as they approach the speed of light. They find, also, that their muscles can no longer cope with the increased mass of their body. Try as they will, they cannot go any faster.

Discuss why this explanation is not sound.

(Total 2 marks)

8. **(a)** A spear, 10 meters long, is thrown at a relativistic speed through a pipe, which is also 10 meters long. These dimensions are measured when the spear and the pipe are at rest. When the spear passes through the pipe, which of the following statements best describes what is observed? (1)

A The spear shrinks so that the pipe completely covers it at some point.

B The pipe shrinks so that the spear extends from both ends at some point.

C The spear and the pipe shrink equally so the pipe just covers the spear at some point.

D Any of these, depending on the motion of the observer.

(b) Explain your answer. (2)

(Total 3 marks)

THEME **B** **The particulate nature of matter**

B The particulate nature of matter

Klaus Hasselmann, based at the Max Planck Institute for Meteorology in Hamburg, Germany, was one of the winners of the Nobel Prize for Physics in 2021 'for the physical modeling of Earth's climate, quantifying variability and reliably predicting global warming' (Nobel Prize, 2021). Our awareness today that global warming is caused by increasing CO_2 in the atmosphere is in part a result of Hasselmann's advocacy and modeling work.

A model is a representation of reality. During your study of matter, you will consider the energies and momentum changes of individual particles (in which assumptions are made about elasticity, size and forces experienced) and the collective properties of substances in macroscopic containers. Part of what made Hasselmann so successful was his acknowledgement that short-term weather fluctuations and broader climate trends cannot be fully understood in terms of individual spheres behaving as perfect ballistic projectiles.

These chapters commence by defining density, temperature, internal energy, heat transfer, phase, luminosity, brightness and emissivity, and getting to know how particles are involved in conduction, convection and radiation. These abstract concepts are then brought into sharp real-world focus through the greenhouse effect, which describes the processes by which radiation from the Sun interacts with the Earth and its atmosphere and the enhanced impact that human activities are making. Armed with an awareness of the very small and the very big, you will be well-placed to grapple with theory, problems, experiments and modeling of your own on the gas laws, which in turn have implications for engines via the laws of thermodynamics.

The chapters on Thermal energy transfers (B.1), the Greenhouse effect (B.2) and Gas laws (B.3) relate to particles that, in general, do not interact with each other except when touching or when radiation passes between them. Another property of particles is charge. Differences in charges across spaces lead to electric fields, and electric fields are regions in which electric forces are exerted. In a similar way to how a ball falls when released from a height, charged particles move in a net direction when constrained within electrical conductors that form a complete circuit. They release the energy provided by a source of emf in resistors.

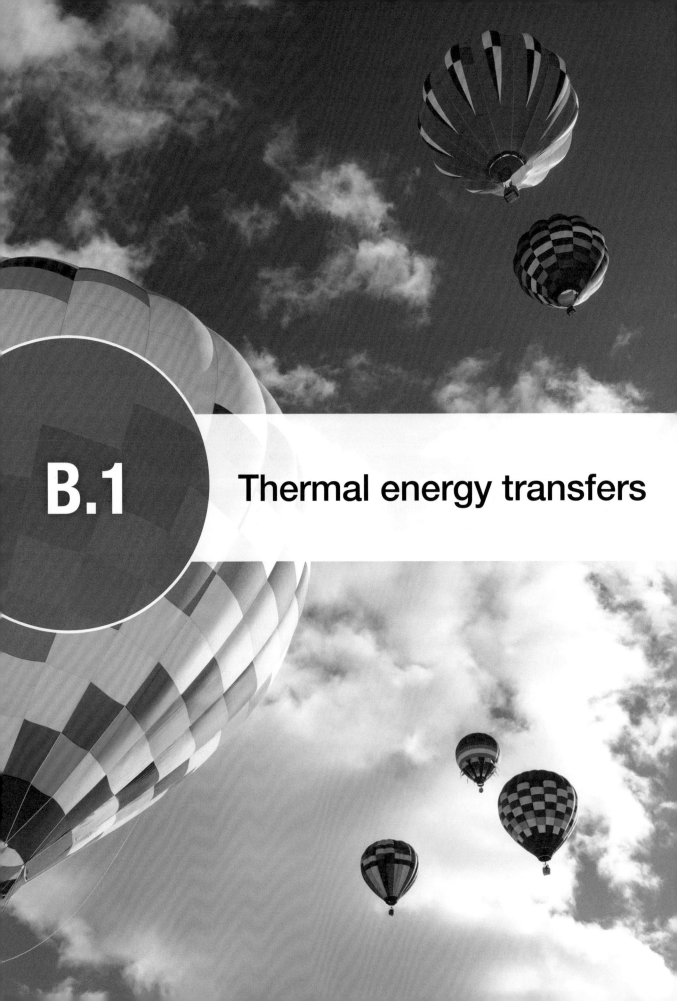

B.1 Thermal energy transfers

◀ When a gas is heated, it expands making it less dense than the surrounding air. If the balloon is large enough, the buoyancy it experiences will be sufficient to lift several people.

Guiding Questions

How do macroscopic observations provide a model of the microscopic properties of a substance?

How is energy transferred within and between systems?

How can observations of one physical quantity be used to determine the other properties of a system?

When a box is pulled along a rough horizontal surface at constant velocity, it gains no kinetic energy or potential energy. However, the point of application of the pulling force that overcomes friction moves in the direction of the force. Work is done.

If work is done, energy must be transferred. So, if energy is not transferred to the box, where does this energy go?

We can answer this question if we think of the box as being made of particles. It is the particles that gain kinetic energy and potential energy, not the box.

Our role as physicists is to observe our physical surroundings, take measurements and think of ways to explain what we see. Up to this point in the course, we have been dealing with the motion of bodies. We can describe bodies in terms of their mass and volume, and if we know their speed and the forces that act on them, we can calculate where they will be at any given time. We even know what happens if two bodies hit each other. However, knowledge of relative motion, forces and energy is not enough to describe all the differences between objects. For example, by simply holding different objects we can feel that some are hot and some are cold.

In this chapter, we will develop a model to explain these differences, but first of all we need to know what is inside matter.

Nature of Science

So far we have dealt with the motion of particles and, given their initial conditions, can predict their speed and position at any time. Once we realize that all matter is made up of particles, we can use this knowledge to build a model of the way those particles interact with each other. So even though we cannot see these particles, we can make predictions related to them.

Students should understand:

molecular theory in solids, liquids and gases
density ρ as given by $\rho = \frac{m}{V}$
Kelvin and Celsius scales are used to express temperature
the change in temperature of a system is the same when expressed with the Kelvin or Celsius scales

Kelvin temperature is a measure of the average kinetic energy of particles as given by $\overline{E_k} = \frac{3}{2} k_B T$
the internal energy of a system is the total intermolecular potential energy arising from the forces between the molecules plus the total random kinetic energy of the molecules arising from their random motion
temperature difference determines the direction of the resultant thermal energy transfer between bodies
a phase change represents a change in particle behavior arising from a change in energy at constant temperature
quantitative analysis of thermal energy transfers Q with the use of specific heat capacity c and specific latent heat of fusion and vaporization of substances L as given by $Q = mc\Delta T$ and $Q = mL$
conduction, convection and thermal radiation are the primary mechanisms for thermal energy transfer
conduction in terms of the difference in the kinetic energy of particles
quantitative analysis of rate of thermal energy transfer by conduction in terms of the type of material and cross-sectional area of the material and the temperature gradient as given by $\frac{\Delta Q}{\Delta t} = kA\frac{\Delta T}{\Delta x}$
qualitative description of thermal energy transferred by convection due to fluid density differences
quantitative analysis of energy transferred by radiation as a result of the emission of electromagnetic waves from the surface of a body, which, in the case of a black body, can be modeled by the Stefan–Boltzmann law as given by $L = \sigma A T^4$ where L is the luminosity, A is the surface area and T is the absolute temperature of the body
the concept of apparent brightness b
luminosity L of a body as given by $b = \frac{L}{4\pi d^2}$
the emission spectrum of a black body and the determination of the temperature of the body using Wien's displacement law as given by $\lambda_{max} T = 2.9 \times 10^{-3}$ m K where λ_{max} is the peak wavelength emitted.

The nature of radiation is covered in C.2.

The particle model of matter

Ancient Greek philosophers spent a lot of time thinking about what would happen if they took a piece of cheese and kept cutting it in half.

B.1 Figure 1 Can we keep cutting the cheese forever?

They did not think it was possible to keep halving it forever, so they suggested that there must exist a smallest part, which they called the **atom**.

Atoms are too small to see (about 10^{-10} m in diameter) but we can think of them as very small perfectly elastic balls. This means that when they collide, both momentum and kinetic energy are conserved.

Elements and compounds

We might ask: 'If everything is made of atoms, why is everything not the same?' The answer is that there are many different types of atom.

There are 118 different types of atom, and a material made of just one type of atom is called an **element**. There are, however, many more than 118 different types of material. The other types of matter are made of atoms that have joined together to form molecules. Materials made from molecules that contain more than one type of atom are called **compounds**.

The three states of matter

Solid

A solid has a fixed shape and volume so the molecules must have a fixed position.

This means that if we try to pull the molecules apart, there will be a force pulling them back together. This force is called the **intermolecular force**. This force is due to a property of the molecules called **charge** which will be dealt with properly in B.5. This force does not only hold the particles together but also stops the particles getting too close. It is this force that is responsible for the tension in a string (as the molecules are pulled apart they pull back) and the normal reaction (when the surfaces are pushed together the molecules push back).

attractive forces between molecules pulled apart

repulsive forces between molecules pushed together

Molecules of a solid are not free to move about but they can vibrate.

Liquid

A liquid does not have a fixed shape but does have a fixed volume so the molecules are able to move about but still have an intermolecular force between then. This force is quite large when you try to push the molecules together (a liquid is very difficult to compress) but not so strong when pulling molecules apart (if you throw a bucket of water in the air it does not stay together).

When a liquid is put into a container, it presses against the sides of the container. This is because of the intermolecular forces between the liquid and the container. At the bottom of the container, the molecules are forced together by the weight of liquid above. This results in a bigger force per

B.1 Figure 5 Molecules in a liquid.

unit area on the sides of the container. This can be demonstrated by drilling holes in the side of the container and watching the water squirt out as shown in Figure 6. The pressure under a solid block also depends on the height of the block but this pressure only acts downward on the ground not outward or upward as it does in a fluid.

B.1 Figure 6 Higher pressure at the bottom of the container forces the water out at higher speed. ▶

○ hydrogen atom

● gold atom

▲
B.1 Figure 2 Gold is made of gold atoms and hydrogen is made of hydrogen atoms.

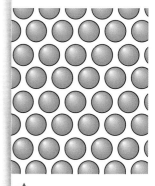

▲
B.1 Figure 3 Molecules in a solid.

◀ **B.1 Figure 4** Intermolecular forces.

𝑖

The molecules of a liquid are often drawn further apart than molecules of a solid but this is not always the case; water is a common but atypical example of the opposite. When ice turns to water, it contracts, which is why water is more dense than ice and ice floats on water.

The pressure in a container of water increases with depth but, because of the effect of gravity, and hence lower potential energy, this does not mean that the lowest water jet has the longest range, as demonstrated in Figure 6.

So the force per unit area, or **pressure**, increases with depth. If you have ever been diving, you may have felt this as the water pushed against your ears. The increase in pressure with depth is also the reason why submerged objects experience a buoyant force. If you consider the submerged cube shown in Figure 7, the bottom surface is deeper than the top so experiences a greater force, resulting in a resultant upward force. Note that this is only the case if the water is in a gravitational field, e.g. on the Earth, as it relies on the weight of the water pushing down on the water below.

Gas

A gas does not have fixed shape or volume; it simply fills whatever container it is put into. The molecules of a gas are completely free to move about without any forces between the molecules except when they are colliding.

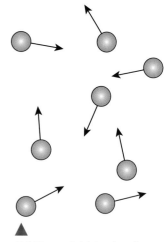

Since the molecules of a gas are moving, they collide with the wall of the container. The change in momentum experienced by the gas molecules means that they must be subjected to an unbalanced force, resulting in an equal and opposite force on the container. This results in gas pressure. If the gas is on the Earth, then the effect of gravity will cause the gas nearest the ground to

B.1 Figure 8 Molecules of a gas.

be compressed by the gas above. This increases the density of the gas so there are more collisions between the gas molecules and the container, resulting in a higher pressure. The difference between the pressure at the top of an object and the pressure at the bottom results in a buoyant force is illustrated in Figure 9.

A car moving through air will collide with the air molecules. As the car hits the air molecules, it increases their momentum so they must experience a force. The car experiences an equal and opposite force which we call air resistance or drag.

Brownian motion

The explanation of the states of matter supports the theory that matter is made of particles but it is not completely convincing. More evidence was found by Robert Brown when, in 1827, he was observing a drop of water containing pollen grains under a microscope. He noticed that the pollen grains had an unusual movement. The particles moved around in an erratic zigzag pattern similar to Figure 10. The explanation for this is that they are being hit by the invisible molecules of water that surround the particles. The reason we do not see this random motion in larger objects is because they are being bombarded from all sides so the effect cancels out.

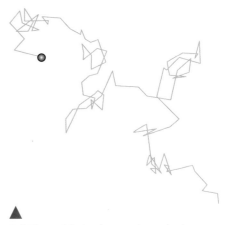

B.1 Figure 10 Smoke particles jiggle about.

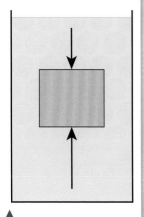

B.1 Figure 7 The force at the bottom is greater than at the top, resulting in a buoyant force.

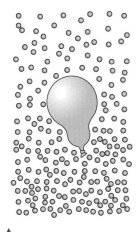

B.1 Figure 9 The density of air molecules is greater at the bottom of the atmosphere.

Density

Density is defined as the ratio of the mass of a body to its volume:

$$\rho = \frac{m}{V}$$

where ρ is density in $kg\,m^{-3}$, m is mass in kg and V is volume in m^3.

The density of a gas is lower than the density of a solid. But not all liquids are less dense than solid bodies. On a macroscopic level, densities cause bodies to float or sink. On a microscopic level, density is related to the spacing of particles.

Because density is unrelated to the dimensions of a body, it is a material property. A given material will always have the same density, irrespective of whether it is a gigantic sphere or stretched out into a sheet.

Informally, some refer to density as 'the amount of stuff in a thing'.

Internal energy

Consider a car moving along the road at constant velocity.

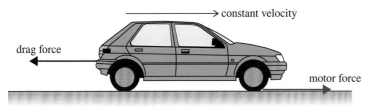

The force of the motor is applied via the friction between the tires and the road. We know that energy is transferred from the petrol to the car so the motor is doing work, but since the car is not getting any faster or going up a hill, it is not gaining any kinetic or potential energy. If we consider the air to be made up of particles, we can answer the question of what is happening to the energy and thus gain a better understanding of drag force.

As the car moves through the air, it collides with the molecules of air as in Figure 12. When a collision is made, the car exerts a force on the molecules so, according to Newton's third law, the car must experience an equal and opposite force. This is the drag force.

As the car moves forward hitting the air molecules, it gives them kinetic energy; this is where all the energy is going. We call this **internal energy**, and since gas molecules have no forces between them (and have no net increase in height), this energy is all kinetic energy.

Another example we could consider is a block sliding down a slope at a constant speed as in Figure 13. As it slides down the slope, it is losing potential energy, but where is the energy going? This time it is the friction between the block and the slope that provides the answer. As the surfaces rub against each other, energy is transferred to the molecules of the block and slope; the rougher the surfaces, the more the molecules get bumped about. The effect of all this bumping is to increase the kinetic energy of

SKILLS

A GeoGebra worksheet linked to this topic is available in the eBook.

How does density relate to gases, standing waves, gravitation, fuels and thermal energy transfer? (B.3, C.4, D.1, A.3, B.1)

How has international collaboration helped to develop the understanding of the nature of matter? (NOS)

B.1 Figure 11 Forces on a car moving with constant velocity.

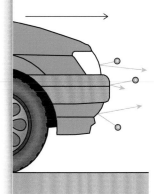

B.1 Figure 12 The front of the car collides with air molecules.

the molecules. Solid molecules cannot fly about, they can only vibrate, and as they do this, they move apart. This moving apart requires energy because the molecules have a force holding them together. The result is an increase in both kinetic energy and potential energy.

B.1 Figure 13 A block slides down a slope at constant speed.

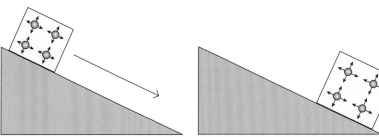

What role does the molecular model play in understanding other areas of physics? (NOS)

Worked example

A 4 kg block slides down the slope at a constant speed of $1\,\text{m s}^{-1}$ as in the figure given below. What is the work done against friction?

Solution

The loss of $E_p = mgh = 4 \times 10 \times 3 = 120\,\text{J}$.

This energy has not been transferred to kinetic energy since the speed of the block has not increased. The energy has been given to the internal energy of the slope and block. The work done against friction (friction force × distance traveled in direction of the force) is therefore 120 J. The block is losing energy so this should be negative.

$$\text{friction} \times 5 = -120\,\text{J}$$

$$\text{friction} = -24\,\text{N}$$

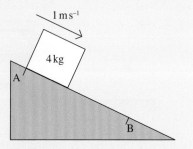

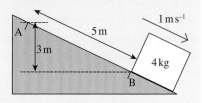

Worked example

A car of mass 1000 kg is traveling at $30\,\text{m s}^{-1}$. If the brakes are applied, how much thermal energy is transferred to the brakes?

Solution

When the car is moving, it has kinetic energy. This must be transferred to the brakes when the car stops.

$$E_k = \frac{1}{2}mv^2$$
$$= \frac{1}{2} \times 1000 \times 30^2$$
$$= 450\,\text{kJ}$$

thermal energy transferred to the brakes = 450 kJ

When a car slows down using its brakes, kinetic energy will be transferred to internal energy in the brake pads and disks.

This thermogram of a car shows how the wheels have become hot due to friction between the road and the tires, and the brakes pads and disks.

Exercise

Q1. A block of metal, mass 10 kg, is dropped from a height of 40 m.
   (a) How much energy does the block have before it is dropped?
   (b) How much thermal energy do the block and floor gain when it hits the floor?

Q2. If the car in the second Worked example on page 158 was traveling at $60 \, \text{m s}^{-1}$, how much thermal energy would the brakes receive?

Q3. A 75 kg free fall parachutist falls at a constant speed of $50 \, \text{m s}^{-1}$. Calculate the amount of energy given to the surroundings per second.

Q4. A block, starting at rest, slides down the slope as shown. Calculate the amount of work done against friction and the size of the friction force.

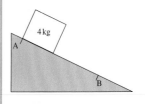

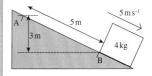

Internal energy and the three states of matter

Internal energy is the sum of the energies of the molecules of a body. In solids and liquids, there are forces between the molecules so to move them around requires work to be done (like stretching a spring). The internal energy of solids and liquids is therefore made up of the total kinetic + potential energy. There is no force between molecules of a gas so changing their position does not require work to be done. The internal energy of a gas is therefore only the total kinetic energy.

Temperature (*T*)

If we rub our hands together, we are doing work since there is movement in the direction of the applied force. If work is done, then energy must be transferred but we are not increasing the kinetic energy or potential energy of our hands; we are increasing their **internal energy**. As we do this, we notice that our hands get hot.

TOK

We perceive how hot or cold something is with our senses but to quantify this we need a measurement

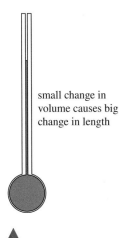

small change in
volume causes big
change in length

B.1 Figure 14 A simple
thermometer.

This is a sensation that we perceive through our senses and it seems from this simple experiment to be related to energy. The harder we rub, the hotter our hands become. Before we can go further, we need to define a quantity we can use to measure how hot or cold a body is.

Temperature is a measure of how hot or cold a body is.

When we defined a scale for length, we simply took a known length and compared other lengths to it. With temperature it is not so easy. First we must find some directly measurable physical quantity that varies with temperature. One possibility is the length of a metal rod. As the internal energy of a solid temperature increases, the molecules move faster, causing them to move apart. The problem is that the length does not change very much so it is not easy to measure. A better alternative is to use the change in volume of a liquid. This also is not very much, but if the liquid is placed into a container with a thin tube attached as in Figure 14, then the change can be quite noticeable.

To define a scale, we need two fixed points. In measuring length, we use two of the positions along a ruler. In this case, we will measure the length of the liquid at two known temperatures: the boiling and freezing points of pure water. But how do we know that these events always take place at the same temperatures before we have made our thermometer? What we can do is place a tube of liquid in many containers of freezing and boiling water to see if the liquid always has the same lengths. If it does, then we can deduce that the freezing and boiling temperatures of water are always the same. Having defined our fixed points, we can make the scale by marking the tube at the highest and lowest points and dividing the range into 100 equal units.

Nature of Science

It is important to use pure water at normal atmospheric pressure. Otherwise, the temperatures will not be quite right.

B.1 Figure 15 Calibrating a
thermometer.

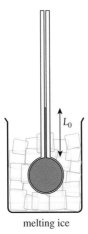

L_0

melting ice

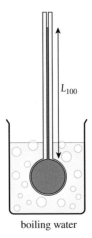

L_{100}

boiling water

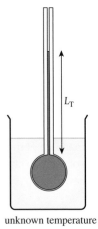

L_T

unknown temperature

So if we place the thermometer into water at an unknown temperature, resulting in length L_T, then the temperature can be found from:

$$T = \frac{L_T - L_0}{L_{100} - L_0} \times 100$$

This is how the Celsius scale is defined.

Temperature and kinetic energy

The reason that a liquid expands when it gets hot is because its molecules vibrate more and move apart. Higher temperature implies faster molecules so the temperature is directly related to the average kinetic energy of the molecules. However, since the kinetic energy of the particles is not zero when ice freezes, the average kinetic energy cannot be directly proportional to the temperature in °C. The lowest possible temperature is the point at which the kinetic energy of molecules becomes zero. This happens at −273 °C.

Kelvin scale

An alternative way to define a temperature scale would be to use the pressure of a constant volume of gas. As temperature increases, the kinetic energy of the molecules increases so they move faster. The faster moving molecules hit the walls of the container harder and more often, resulting in an increased pressure. As the temperature gets lower and lower, the molecules slow down until at some point they stop moving completely. This is the lowest temperature possible or **absolute zero**. If we use this as the zero in our temperature scale, then the average kinetic energy is directly proportional to temperature. In defining this scale, we then only need one fixed point in addition to absolute zero. This point could be the freezing point of water but the triple point is more precisely defined. This is the temperature at which water can be solid, liquid, and gas in equilibrium, which in degrees Celsius is 0.01 °C. If we make this 273.16 in our new scale, then a change of 1 unit will be the same as 1 °C. This is called the Kelvin scale.

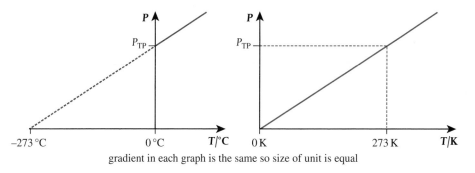

gradient in each graph is the same so size of unit is equal

Because the size of the unit is the same, to convert from °C to K, we simply add 273. So:

$$10 °C = 283 K$$

$$50 °C = 323 K$$

A change from 10 °C to 50 °C is 50 − 10 = 40 °C.

A change from 283 K to 323 K is 323 − 282 = 40 K, so $\Delta °C = \Delta K$.

Temperature and molecular speed

Now we have a temperature scale that begins at absolute zero, we can say that, for an ideal gas, the average kinetic energy of the molecules is directly proportional to its temperature in kelvin.

$$\overline{E_k} = \frac{3}{2} k_B T$$

where k_B is the Boltzmann constant, $1.38 \times 10^{-23} \, m^2 \, kg \, s^{-2} \, K^{-1}$.

The lowest theoretical temperature is −273.15 °C, but for conversion purposes, we usually find it more convenient to use −273 °C.

Not all countries use the same units of temperature when describing the weather but the agreed SI unit is the kelvin.

B.1 Figure 16 Pressure vs temperature in Celsius and kelvin.

B.1 Figure 17
Molecular velocity distribution for a gas.

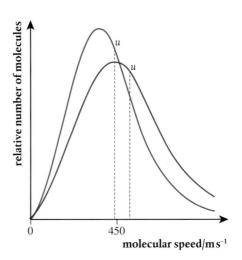

> This is the *average* kinetic energy of the molecules. The different molecules of a gas travel at *different random* velocities, some faster and some slower. The range of velocities can be represented by the velocity distribution curve in Figure 17. Because the curve is not symmetrical, the mean value is to the *right* of center.

Figure 17 shows the distribution of molecular velocities for different temperatures – the blue line describes molecular speed distribution of molecules in the air at about 0 °C, and the red line is at about 100 °C.

Exercise

Q5. The length of a column of liquid is 30 cm at 100 °C and 10 cm at 0 °C. At what temperature will its length be 12 cm?

Q6. The average molar mass of air is 29 g mol^{-1}. Calculate:

  (a) the average kinetic energy of air molecules at 20 °C

  (b) the average mass of one molecule of air

  (c) the average speed of air molecules at 20 °C.

Thermal energy

We know that the temperature of a body is related to the average kinetic energy of its molecules and that the kinetic energy of the molecules can be increased by doing work, for example, against friction. The internal energy of a body can also be increased by putting it in contact with a hotter body. Energy transferred in this way is called **thermal energy** (or sometimes **heat**).

B.1 Figure 18 Heat flows from hot to cold until thermal equilibrium is established.

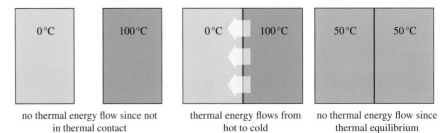

no thermal energy flow since not in thermal contact

thermal energy flows from hot to cold

no thermal energy flow since thermal equilibrium

When bodies are in thermal contact, heat will always flow from a high temperature to a low temperature until the bodies are at the same temperature. Then we say they are in **thermal equilibrium**.

Thermal energy transfer

Three ways that thermal energy can be transferred from one body to another are conduction, convection, and radiation.

Conduction

Conduction takes place when bodies are in contact with each other. The vibrating molecules of one body collide with the molecules of the other. The fast-moving hot molecules lose energy and the slow-moving cold ones gain it.

Metals are particularly good conductors of thermal energy because not only are their atoms well connected but metals contain some free particles (electrons) that are able to move freely about, helping to pass on the energy.

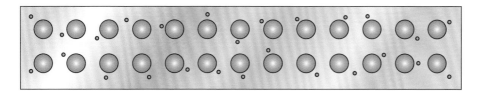

Gases are not very good conductors of thermal energy because their molecules are far apart. However, heat is often transferred to a gas by conduction. This is how heat would pass from a room heater into the air of a room, for example. For a conducting material, the rate of flow of thermal energy is proportional to the temperature gradient:

$$\frac{\Delta Q}{\Delta t} = kA \frac{\Delta T}{\Delta x}$$

where k is material conductivity ($\text{W K}^{-1}\text{m}^{-1}$) and A is cross-sectional area (m^2).

Convection

Convection is the way that heat is transferred through fluids by collections of fast-moving molecules moving from one place to another. When heat is given to air, the molecules move around faster. This causes an increase in pressure in the hot air, which enables it to expand, pushing aside the colder surrounding air. The hot air has now displaced more than its own weight of surrounding air so experiences an unbalanced upward force, resulting in motion in that direction.

As the hot air rises, it will cool and then come back down (this is also the way that a hot air balloon works). The circular motion of air is called a **convection current** and is the way that heat is transferred around a room.

Radiation

Radiation is the mechanism by which thermal energy can pass directly between two bodies without increasing the temperature of the material in-between. In fact, there does not even have to be a material between since radiation can pass through a vacuum. The name of this radiation is **infrared** and it is a part of the electromagnetic spectrum. The amount of radiation emitted and absorbed by a body depends on its color. Dark, dull bodies both **emit** and **absorb** radiation better than light shiny ones. When you stand in front of a fire and feel the heat, you are feeling radiated heat.

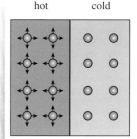

hot   cold

▲ **B.1 Figure 19** Energy passed from fast molecules to slow molecules.

◀ **B.1 Figure 20** Electrons pass energy freely.

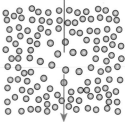

buoyant force = weight of air displaced

weight of hot air

▲ **B.1 Figure 21** Hot air expands.

Although the first ever engine was probably a steam turbine, cylinders of expanding gas are the basis of most engines.

Emissivity is a measure of how effectively a body radiates heat.

It is measured from 0 to 1, with 1 being a perfect radiator and 0 being the opposite; a perfect reflector.

When a rod of metal is heated to around 1000 K, it starts to glow red. Although the most intense part of the spectrum is not in the visible region, there is enough visible red light to make the rod glow.

This experiment involves electrical quantities and apparatus, which we will discuss fully in B.5.

B.1 Figure 22 Finding the relationship between radiated power and temperature.

The *y*-axis is the potential difference measured across the thermopile. This is proportional to the power absorbed per unit area of the sensor.

B.1 Figure 23 Graph of potential difference across thermopile vs temperature.

Black-body radiation

A black body is a perfect radiator of thermal energy. The power radiated per unit area of the body is related to the temperature of the body. To determine how it is related we can do an experiment using a tungsten lightbulb as a source of radiation.

SKILLS

It is not easy to measure the temperature of the filament, but we do know that the resistance of tungsten changes with temperature. In Figure 22, we measure the potential difference across the bulb and the current through it so we can calculate the resistance and hence determine the temperature. The power radiating from the filament is measured using a thermopile. This absorbs the radiation, causing its temperature to rise and leading to a potential difference that is measured with the voltmeter.

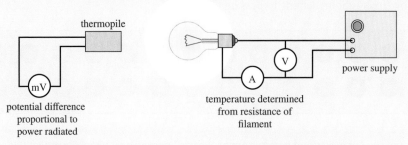

thermopile

mV

potential difference proportional to power radiated

power supply

temperature determined from resistance of filament

Figure 23 shows the results from this experiment. The potential difference measured across the thermopile is on the *y*-axis and the filament temperature on the *x*-axis.

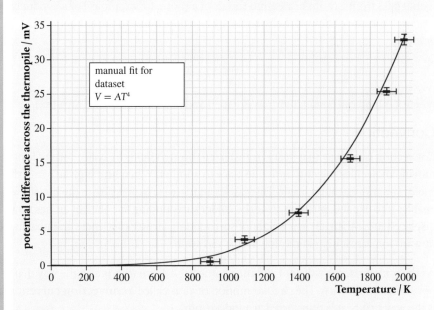

manual fit for dataset
$V = AT^4$

This is obviously not a straight line but the curve $y = 2.09 \times 10^{-12}x^4$ is a good fit. Since the thermopile potential difference is proportional to the power absorbed per unit area, this shows that:

$$\frac{P}{A} \propto T^4$$

By calibrating the thermopile to W m^{-2} instead of mV, we find the constant of proportionality $\sigma = 5.67 \times 10^{-8}$ W m^{-2} K^{-4}. This is known as the **Stefan–Boltzmann constant**. So for a perfect black-body radiator:

$$P = A\sigma T^4$$

The black-body spectrum

If we view the visible spectrum of a filament lamp, we see that it is made up of a continuous spread of visible wavelengths as shown in Figure 24.

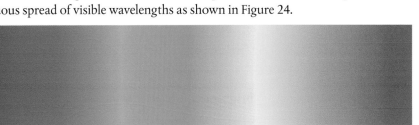

What applications does the Stefan–Boltzmann equation have in astrophysics and in the use of solar energy? (E.5, B.2)

B.1 Figure 24 The visible spectrum.

By using sensors, we can measure the intensity of the different colors and also electromagnetic radiation outside the visible range to produce the complete **black-body spectrum** as shown in Figure 25. Notice how the peak in the spectrum is actually in the infrared region, which is why a light bulb gives out more heat than light.

Wien's displacement law

A body at room temperature emits infrared radiation but not visible light. However, the spectrum of infrared radiation will be a similar shape to that of the tungsten bulb in Figure 25, except that the peak will be moved to the right and, as the total power emitted is less, the area under the graph will be smaller. Figure 26 shows the black-body spectra for a range of different temperatures.

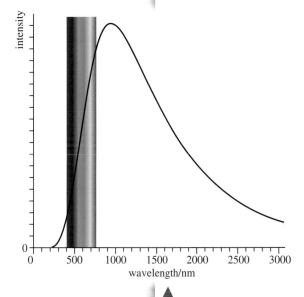

B.1 Figure 25 The black-body spectrum for a tungsten filament lamp.

Bodies with the same temperature but lower emissivity will emit radiation with the same peak wavelength but less power.

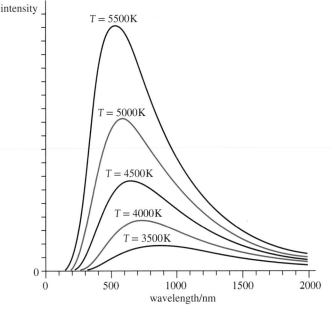

B.1 Figure 26 The intensity distribution for a black body at different temperatures.

The relationship between the peak wavelength in meters and the temperature in kelvin is given by **Wien's displacement law**, which states that peak wavelength and absolute temperature are inversely proportional:

$$\lambda_{peak} = \frac{0.0029}{T}$$, where 0.0029 is a constant with the unit m K (not millikelvin)

How can observations of one physical quantity allow for the determination of another? (NOS)

Inverse square law

The radiation from a spherical body spreads out radially in all directions so the power per unit area decreases as distance from the source increases. At distance r, the radiation from a source emitting a total power P has spread out to cover a sphere of area $4\pi r^2$ so the power per unit area, I, is given by the formula:

$$I = \frac{P}{4\pi r^2}$$

In the case of stars, we refer to **brightness**, b, in place of intensity.

Where do inverse square law relationships appear in other areas of physics? (NOS)

Exercise

Q7. A black ball of radius 2 cm can be considered to be a perfect black-body radiator. If the temperature of the ball is 500 K, calculate:

   (a) the peak wavelength of the spectrum of electromagnetic radiation emitted

   (b) the power per unit area emitted from the ball

   (c) the total power emitted from the ball

   (d) the intensity of radiation received at a distance of 1 m.

Preventing heat loss

In cold countries, houses are insulated to prevent thermal energy from escaping. Are houses in hot countries insulated to stop thermal energy entering?

In everyday life, as well as in the physics lab, we often concern ourselves with minimizing heat loss. Insulating materials are often made out of fibrous matter that traps pockets of air. The air is a poor conductor and when it is trapped it cannot convect. Covering something with silver-colored paper will reduce radiation.

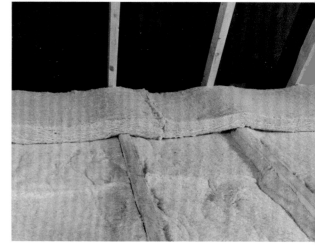

Roof insulation in a house.

Thermal capacity (C)

This applies not only when things are given heat, but also when they lose heat.

If thermal energy is added to a body, its temperature rises, but the actual increase in temperature depends on the body.

The thermal capacity, C, of a body is the amount of heat needed to raise its temperature by one unit. Typical units: $\text{J}\,^\circ\text{C}^{-1}$ or $\text{J}\,\text{K}^{-1}$.

If the temperature of a body increases by an amount ΔT when quantity of heat Q is added, then the thermal capacity is given by the equation:

$$C = \frac{Q}{\Delta T}$$

Worked example

If the thermal capacity of a quantity of water is 5000 J K^{-1}, how much heat is required to raise its temperature from 20°C to 100°C?

Solution

Thermal capacity:	$C = \dfrac{Q}{\Delta T}$
So:	$Q = C\Delta T$
Therefore:	$Q = 5000 \times (100 - 20)$ J
So the heat required:	$Q = 400$ kJ

Worked example

How much heat is lost from a block of metal of thermal capacity 800 J K^{-1} when it cools down from 60°C to 20°C?

Solution

Thermal capacity:	$C = \dfrac{Q}{\Delta T}$
So:	$Q = C\Delta T$
Therefore:	$Q = 800 \times (60 - 20)$ J
So the heat lost:	$Q = 32$ kJ

Exercise

Q8. The thermal capacity of a 60 kg human is 210 kJ K^{-1}. How much heat is lost from a body if its temperature drops by 2°C?

Q9. The temperature of a room is 10°C. In 1 hour the room is heated to 20°C by a 1 kW electric heater.

    (a) How much heat is delivered to the room?

    (b) What is the thermal capacity of the room?

    (c) Does all this heat go to heat the room?

Specific heat capacity (c)

The thermal capacity depends on the size of the object and what it is made of. The **specific heat capacity** depends only on the material. Raising the temperature of 1 kg of water requires more heat than raising the temperature of 1 kg of steel by the same amount, so the specific heat capacity of water is higher than that of steel.

Remember, power is energy per unit time.

The specific heat capacity of water is quite high, so it takes a lot of energy to heat up the water for a shower.

It is possible to buy a special shower head that uses less water. In some countries, this is used to save energy; in others, to save water.

It takes 4200 J of energy to raise the temperature of water by 1°C. This is equivalent to lifting 420 kg a height of 1 m. This makes water a good medium for transferring energy but also makes it expensive to take a shower. Oil would be cheaper to heat but not so good to wash in!

The specific heat capacity of a material is the amount of heat required to raise the temperature of a unit mass of the material by one unit. Typical units: $J\,kg^{-1}\,°C^{-1}$ or $J\,kg^{-1}\,K^{-1}$.

If a quantity of heat Q is required to raise the temperature of a mass m of material by ΔT, then the specific heat capacity c of that material is given by the following equation:

$$c = \frac{Q}{m\Delta T}$$

Worked example

The specific heat capacity of water is $4200\,J\,kg^{-1}\,K^{-1}$. How much heat will be required to heat 300 g of water from 20°C to 60°C?

Solution

Specific heat capacity:

$$c = \frac{Q}{m\Delta T}$$

So:

$$Q = cm\Delta T$$

Therefore:

$Q = 4200 \times 0.3 \times 40$ (Note: Convert g to kg)

$Q = 50.4\,kJ$

Worked example

A metal block of mass 1.5 kg loses 20 kJ of thermal energy. As this happens, its temperature drops from 60°C to 45°C. What is the specific heat capacity of the metal?

Solution

Specific heat capacity:

$$c = \frac{Q}{m\Delta T}$$

So:

$$c = \frac{20\,000}{1.5(60 - 45)}$$

$$c = 888.9\,J\,kg^{-1}\,K^{-1}$$

Substance	Specific heat capacity ($J\,kg^{-1}\,K^{-1}$)
Copper	380
Steel	440
Aluminum	900
Water	4200

B.1 Table 1

Exercise

Use the data in Table 1 to solve the problems:

Q10. How much heat is required to raise the temperature of 250 g of copper from 20°C to 160°C?

Q11. The density of water is $1000\,kg\,m^{-3}$.

(a) What is the mass of 1 liter of water?

(b) How much energy will it take to raise the temperature of 1 liter of water from 20°C to 100°C?

(c) A water heater has a power rating of 1 kW. How many seconds will this heater take to boil 1 liter of water?

Q12. A 500 g piece of aluminum is heated with a 500 W heater for 10 minutes.

    (a) How much energy will be given to the aluminum in this time?

    (b) If the temperature of the aluminum was 20 °C at the beginning, what will its temperature be after 10 minutes?

Q13. A car of mass 1500 kg traveling at 20 m s^{-1} brakes suddenly and comes to a stop.

    (a) How much kinetic energy does the car lose?

    (b) If 75% of the energy is given to the front brakes, how much energy will they receive?

    (c) The brakes are made out of steel and have a total mass of 10 kg. By how much will their temperature rise?

Q14. The water comes out of a showerhead at a temperature of 50 °C at a rate of 8 liters per minute.

    (a) If you take a shower lasting 10 minutes, how many kg of water have you used?

    (b) If the water must be heated from 10 °C, how much energy is needed to heat the water?

Phase change

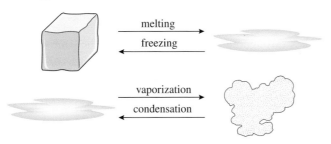

B.1 Figure 27 When matter changes from liquid to gas, or solid to liquid, it is changing state.

When water boils, this is called a **change of state** (or **change of phase**). As this happens, the temperature of the water does not change – it stays at 100 °C. In fact, we find that while the state of a material is changing, the temperature stays the same provided that no particles are added or lost. We can explain this in terms of the particle model.

An iceberg melts as it floats into warmer water.

B.1 Figure 28 Molecules gain potential energy when the state changes.

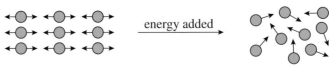

Solid molecules have kinetic energy since they are vibrating.

Liquid molecules are now free to move about but have the same kinetic energy as before.

When matter changes state, the energy is needed to enable the molecules to move more freely. To understand this, consider the example below.

B.1 Figure 29 A ball-in-a-box model of change of state.

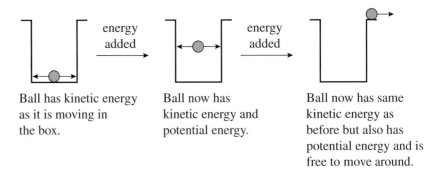

Ball has kinetic energy as it is moving in the box.

Ball now has kinetic energy and potential energy.

Ball now has same kinetic energy as before but also has potential energy and is free to move around.

Boiling and evaporation

These are two different processes by which liquids can change to gases.

Boiling takes place throughout the liquid and always at the same temperature (for a given pressure). **Evaporation** takes place only at the free surface of the liquid and can happen at all temperatures.

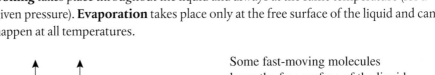

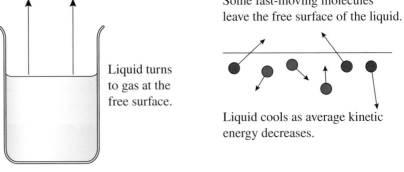

Liquid turns to gas at the free surface.

Some fast-moving molecules leave the free surface of the liquid.

Liquid cools as average kinetic energy decreases.

B.1 Figure 30 A microscopic model of evaporation.

How can the phase change of water be used in the process of electricity generation? (B.5, E.4)

When a liquid evaporates, the fastest-moving particles leave the free surface. This means that the average kinetic energy of the remaining particles is lower, resulting in a drop in temperature for the liquid that remains.

The rate of evaporation can be increased by:

- increasing the free surface area. This increases the number of molecules near the free surface, giving more of them a chance to escape
- blowing across the free surface. After molecules have left the free surface, they form a small 'vapor cloud' above the liquid. If this is blown away, it allows further molecules to leave the free surface more easily
- raising the temperature. This increases the kinetic energy of the liquid molecules, enabling more to escape.

Specific latent heat (*L*)

The **specific latent heat** of a material is the amount of heat required to change the state of a unit mass of the material without change of temperature.

Typical unit: $J\,kg^{-1}$

Latent means *hidden*. This name is used because when matter changes state, the heat added does not cause the temperature to rise, but seems to disappear.

If it takes an amount of energy Q to change the state of a mass m of a substance, then the specific latent heat of that substance is given by the equation:

$$L = \frac{Q}{m}$$

Worked example

The specific latent heat of fusion of water is $3.35 \times 10^5\,J\,kg^{-1}$. How much energy is required to change $500\,g$ of $0\,°C$ ice into $0\,°C$ water?

Solution

Latent heat of fusion: $\qquad L = \frac{Q}{m}$

So: $\qquad Q = mL$

Therefore: $\qquad Q = 0.5 \times 3.35 \times 10^5\,J$

So the heat required: $\qquad Q = 1.675 \times 10^5\,J$

Worked example

The amount of heat released when $100\,g$ of steam turns to water is $2.27 \times 10^5\,J$. What is the specific latent heat of vaporization of water?

Solution

Specific latent heat of vaporization: $\qquad L = \frac{Q}{m}$

Therefore: $\qquad L = \frac{2.27 \times 10^5}{0.1}\,J\,kg^{-1}$

So the specific latent heat of vaporization: $\qquad L = 2.27 \times 10^6\,J\,kg^{-1}$

People sweat to increase the rate at which they lose thermal energy. When you get hot, sweat comes out of your skin onto the surface of your body. When the sweat evaporates, it cools you down. In a sauna, there is so much water vapor in the air that the sweat does not evaporate.

Solid → liquid (or vice versa) Specific latent heat of fusion
Liquid → gas (or vice versa) Specific latent heat of vaporization

This equation ($L = \frac{Q}{m}$) can also be used to calculate the heat lost when a substance changes from gas to liquid, or liquid to solid.

Specific latent heat of vaporization	$2.27 \times 10^6 \, \text{J kg}^{-1}$
Specific latent heat of fusion	$3.35 \times 10^5 \, \text{J kg}^{-1}$

B.1 Table 2 Specific latent heats of water.

Before the invention of the refrigerator, people would collect ice in the winter and store it in well-insulated rooms so that it could be used to make ice cream in the summer. The reason it takes so long to melt is because to melt 1 kg of ice requires 3.3×10^5 J of energy; in a well-insulated room, this could take many months.

In this example, we are ignoring the heat given to the kettle and the heat lost.

B.1 Figure 31 Temperature–time graph for 1 kg of water being heated in an electric kettle.

Exercise

Use the data about water in Table 2 to solve the following problems.

Q15. If the mass of water in a cloud is 1 million kg, how much energy will be released if the cloud turns from water to ice?

Q16. A water boiler has a power rating of 800 W. How long will it take to turn 400 g of boiling water into steam?

Q17. The ice covering a 1000 m² lake is 2 cm thick.

    (a) If the density of ice is 920 kg m⁻³, what is the mass of the ice on the lake?

    (b) How much energy is required to melt the ice?

    (c) If the Sun melts the ice in 5 hours, what is the power delivered to the lake?

    (d) How much power does the Sun deliver per m²?

Graphical representation of heating

The increase of the temperature of a body can be represented by a temperature–time graph. Observing this graph can give us a lot of information about the heating process.

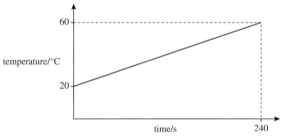

From this graph, we can calculate the amount of heat given to the water per unit time (power).

$$\text{gradient of the graph} = \frac{\text{temperature rise}}{\text{time}} = \frac{\Delta T}{\Delta t}$$

We know from the definition of specific heat capacity that:

$$\text{heat added} = mc\Delta T$$

$$\text{rate of adding heat} = P = \frac{mc\Delta T}{\Delta t}$$

So:
$$P = mc \times \text{gradient}$$

$$\text{gradient of this line} = \frac{(60 - 20)}{240} \, °\text{C s}^{-1} = 0.167 \, °\text{C s}^{-1}$$

So:
$$\text{power delivered} = 4200 \times 0.167 \, \text{W} = 700 \, \text{W}$$

B.1 Figure 32 A graph of temperature vs time for boiling water. When the water is boiling, the temperature does not increase any more.

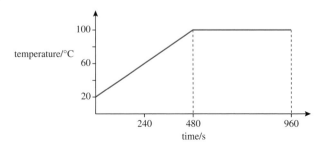

If we continue to heat this water, it will begin to boil.

If we assume that the heater is giving heat to the water at the same rate, then we can calculate how much heat was given to the water while it was boiling.

$$\text{power of the heater} = 700\,\text{W}$$

$$\text{time of boiling} = 480\,\text{s}$$

$$\text{energy supplied} = \text{power} \times \text{time} = 700 \times 480\,\text{J} = 3.36 \times 10^5\,\text{J}$$

From this, we can calculate how much water must have turned to steam.

$$\text{heat added to change state} = \text{mass} \times \text{latent heat of vaporization}$$

$$\text{specific latent heat of vaporization of water} = 2.27 \times 10^6\,\text{J kg}^{-1}$$

$$\text{mass changed to steam} = \frac{3.36 \times 10^5}{2.27 \times 10^6} = 0.15\,\text{kg}$$

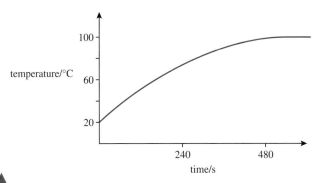

▲ **B.1 Figure 33** Heat loss.

The amount of thermal energy loss is proportional to the difference between the temperature of the kettle and its surroundings. For this reason, a graph of temperature against time is actually a curve, as shown in Figure 33. The fact that the gradient decreases tells us that the amount of heat given to the water gets less with time. This is because as it gets hotter, more and more of the heat is lost to the room.

Measuring thermal quantities by the method of mixtures

The method of mixtures can be used to measure the specific heat capacity and specific latent heat of substances.

A metal sample is first heated to a known temperature. The most convenient way of doing this is to place it in boiling water for a few minutes; after this time it will be at 100 °C. The hot metal is then quickly moved to an insulated cup containing a known mass of cold water. The hot metal will cause the temperature of the cold water to rise; the rise in temperature is measured with a thermometer. Some example temperatures and masses are given in Figure 34.

As the specific heat capacity of water is 4180 J kg^{-1} K^{-1}, we can calculate the specific heat capacity of the metal.

$$\Delta T \text{ for the metal} = 100 - 15 = 85\,°\text{C}$$

$$\Delta T \text{ for the water} = 15 - 10 = 5\,°\text{C}$$

Applying the formula $Q = mc\Delta T$ we get:

$$(mc\Delta T)_{\text{metal}} = 0.1 \times c \times 85 = 8.5c$$

$$(mc\Delta T)_{\text{water}} = 0.4 \times 4180 \times 5 = 8360$$

If no heat is lost, then the heat transferred from the metal = heat transferred to the water:

$$8.5c = 8360$$

$$c_{\text{metal}} = 983\,\text{J kg}^{-1}\,\text{K}^{-1}$$

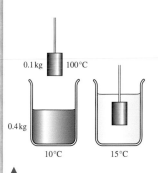

▲ **B.1 Figure 34** Measuring the specific heat capacity of a metal.

A worksheet with full details of how to carry out this experiment is available in the eBook.

To measure the latent heat of vaporization, steam is passed into cold water. Some of the steam condenses in the water, causing the water temperature to rise.

thermal energy from the steam = thermal energy to the water

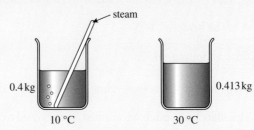

In Figure 35, 13 g of steam have condensed in the water, raising its temperature by 20 °C. The steam condenses then cools down from 100 °C to 30 °C.

$$\text{heat from steam} = ml_{\text{steam}} + mc\Delta T_{\text{water}}$$
$$0.013 \times L + 0.013 \times 4.18 \times 10^3 \times 70 = 0.013L + 3803.8$$
$$\text{heat transferred to cold water} = mc\Delta T_{\text{water}} = 0.4 \times 4.18 \times 10^3 \times 20$$
$$= 33\,440\,J$$

Since: heat from steam = heat to water
$$0.013L + 3803.8 = 33\,440$$
So: $$L = \frac{33\,440 - 3803.8}{0.013}$$
$$L = 2.28 \times 10^6\,J\,kg^{-1}$$

When melting sugar to make confectionary, be very careful: liquid sugar takes much longer to cool down than you might think. This is because as it changes from liquid to solid it is giving out heat but does not change temperature. You should wait a long time before trying to pick up any of your treats with your fingers.

Thermal energy loss

In both of these experiments, some of the heat coming from the hot source can be lost to the surroundings. To reduce heat loss, the temperatures can be adjusted, so you could start the experiment below room temperature and end the same amount above (e.g. if room temperature is 20 °C, then you can start at 10 °C and end at 30 °C).

Guiding Questions revisited

How do macroscopic observations provide a model of the microscopic properties of a substance?

How is energy transferred within and between systems?

How can observations of one physical quantity be used to determine the other properties of a system?

In this chapter, we have considered the evidence for and historical development of the particle model and conservation of energy on different scales to appreciate how:

- The macroscopic properties of solids, liquids and gases (e.g. hardness, rigidity) can be explained by considering the arrangement of particles.
- Density is a material property that describes the ratio of mass to volume, irrespective of a body's dimensions.
- Internal energy is the sum of the kinetic and potential energies of particles in liquids and solids and the sum of the kinetic energies of particles in gases.
- Temperature, a macroscopic property associated with whole bodies or containers, is proportional to the mean kinetic energy of particles when measured in kelvin.
- The Kelvin scale is a translation of the Celsius scale for temperature.

- Specific heat capacity is a measure of the thermal energy required to change the temperature of a unit mass of a material.
- Specific latent heat is a measure of the thermal energy required to change the state of a unit mass of a body.
- Thermal energy flows in solids via conduction, at a rate affected by conductivity, cross-sectional area and temperature gradient (the variation in temperature with distance), until the mean kinetic energy of particles becomes evenly distributed.
- Thermal energy is transferred in fluids by convection currents, with high temperature, low-density collections of particles rising, and low temperature, high-density collections of particles sinking and filling any low-pressure regions.
- Emissivity is a measure of how effectively a body radiates heat. It is measured from 0 to 1, with 1 being a perfect radiator and 0 being the opposite; a perfect reflector.
- All bodies that are hotter than absolute zero radiate thermal energy, with power (or luminosity) dependent on emissivity, cross-sectional area and absolute temperature, and with peak wavelength inversely proportional to absolute temperature.
- Intensity (or brightness) of radiation is inversely proportional to the square of the distance from the source.
- Knowledge of a body's brightness and distance away can be used to determine its luminosity, which in turn can be used to determine its surface temperature and, hence, peak wavelength (or vice versa).

Practice questions

1. A steel hacksaw blade is used to cut a short cylindrical bar of steel in half.

   (a) To do this, the blade is moved forward and backward 100 times through a distance of 30 cm. The temperature of the blade and metal bar rises from 20 °C to 26 °C.

   Length of steel bar = 10 cm

   Radius of bar = 0.5 cm

   Density of steel = 8000 kg m^{-3}

   Mass of blade = 30 g

   Specific heat capacity of steel = 470 J kg^{-1} K^{-1}

   Specific latent heat of fusion of steel = 270 kJ kg^{-1}

   Melting temperature of steel = 1500 °C

   (i) Explain why the metal bar gets hot. (3)

   (ii) Calculate the mass of the bar. (2)

   (iii) Assuming no heat is lost, calculate the average force exerted on the saw blade. Note that the blade only cuts on the way forward. (3)

(b) One piece of the cut bar is melted by passing an electric current through it. The electrical power is calculated to be 1 kW.

  (i) Assuming no heat is lost, calculate the time taken to melt the bar. (2)

  (ii) In reality the bar only reaches 1000 °C and does not melt. State the heat loss per second at this temperature. (1)

(c) An electric water boiler has a heating element at the bottom.

  (i) Give **two** reasons why the element is at the bottom. (2)

  (ii) Explain why the element gets dangerously hot if there is no water in the boiler. (2)

(Total 15 marks)

2. An electric current is passed through a 1 cm long tungsten wire situated in a vacuum. The temperature of the wire rises then remains constant at 2500 °C. The electric power dissipated is 30 W.

  (i) State the temperature in kelvin to an appropriate number of significant figures. (2)

  (ii) Explain why we can assume that the wire radiates 30 J of energy per second. (2)

  (iii) Calculate the diameter of the wire. (3)

(Total 7 marks)

3. A quantity of crushed ice is removed from a freezer and placed in a calorimeter. Thermal energy is supplied to the ice at a constant rate. To ensure that all the ice is at the same temperature, it is continually stirred. The temperature of the contents of the calorimeter is recorded every 15 seconds. The diagram shows the variation with time t of the temperature θ of the contents of the calorimeter. (Uncertainties in the measured quantities are not shown.)

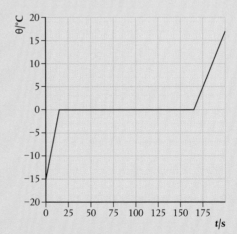

(a) Copy the graph and mark with an X the data point at which all the ice has just melted. (1)

(b) Explain, with reference to the energy of the molecules, the constant temperature region of the graph. (3)

The mass of the ice is 0.25 kg and the specific heat capacity of water is $4200\,J\,kg^{-1}\,K^{-1}$.

(c) Use these data and data from the graph to:

    (i) deduce that energy is supplied to the ice at the rate of about 530 W (3)

    (ii) determine the specific heat capacity of ice (3)

    (iii) determine the specific latent heat of fusion of ice. (2)

(Total 12 marks)

4. When running, a person generates thermal energy but maintains approximately constant temperature.

(a) Explain what thermal energy and temperature mean. Distinguish between the two concepts. (4)

The following simple model may be used to estimate the rise in temperature of a runner, assuming no thermal energy is lost.

A closed container holds 70 kg of water, representing the mass of the runner. The water is heated at a rate of 1200 W for 30 minutes. This represents the energy generation in the runner.

(b) (i) Show that the thermal energy generated by the heater is 2.2×10^6 J. (2)

    (ii) Calculate the temperature rise of the water, assuming no energy losses from the water.

The specific heat capacity of water is $4200\,J\,kg^{-1}\,K^{-1}$. (3)

(c) The temperature rise calculated in (b)(ii) would be dangerous for the runner. Outline **three** mechanisms, other than evaporation, by which the container in the model would transfer energy to its surroundings. (6)

A further process by which energy is lost from the runner is the evaporation of sweat.

(d) (i) Describe, in terms of molecular behavior, why evaporation causes cooling. (3)

Percentage of generated energy lost by sweating: 50%

Specific latent heat of vaporization of sweat: 2.26×10^6 J kg^{-1}

    (ii) Using the information above, and your answer to (b)(i), estimate the mass of sweat evaporated from the runner. (3)

    (iii) State and explain **two** factors that affect the rate of evaporation of sweat from the skin of the runner. (4)

(Total 25 marks)

5. A black body at temperature T emits radiation with peak wavelength λ_p and power P. What is the temperature of the black body and the power emitted for a peak wavelength of $\frac{\lambda_p}{2}$?

	Temperature of the black body	Power emitted by the black body
A	$\frac{T}{2}$	$\frac{P}{16}$
B	$\frac{T}{2}$	$\frac{P}{4}$
C	$2T$	$4P$
D	$2T$	$16P$

(Total 1 mark)

6. A black-body radiator emits a peak wavelength of λ_{max} and a maximum power of P_0. The peak wavelength emitted by a second black-body radiator with the same surface area is $2\lambda_{max}$. What is the total power of the second black-body radiator?

   A $\frac{1}{16}P_0$ **B** $\frac{1}{2}P_0$ **C** $2P_0$ **D** $16P_0$

(Total 1 mark)

7. A bicycle of mass M comes to rest from speed v using the back brake. The brake has a specific heat capacity of c and a mass m. Half of the kinetic energy is absorbed by the brake. What is the change in temperature of the brake?

   A $\frac{Mv^2}{4mc}$ **B** $\frac{Mv^2}{2mc}$ **C** $\frac{mv^2}{4Mc}$ **D** $\frac{mv^2}{2Mc}$

(Total 1 mark)

8. An object can lose energy through:

   I. conduction   **II.** convection   **III.** radiation

(Total 1 mark)

   Which are the principal means for losing energy for a hot rock resting on the surface of the Moon?

   A I and II only   **C** II and III only

   B I and III only   **D** I, II and III

(Total 1 mark)

9. (a) Show that the brightness, $b \propto \frac{AT^4}{d^2}$, where d is the distance of the object from Earth, T is the surface temperature of the object and A is the surface area of the object. (1)

   (b) Two of the brightest objects in the night sky seen from Earth are the planet Venus and the star Sirius. Explain why the relationship $b \propto \frac{AT^4}{d^2}$ is applicable to Sirius but not to Venus. (2)

(Total 3 marks)

10. The masses of different material samples are recorded and mass is plotted against density for the sample. The samples are labeled 1 to 5. Which two samples have the same volume?

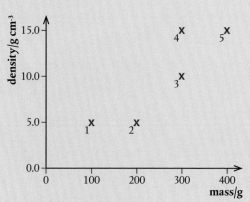

A   1 and 2

B   4 and 5

C   3 and 4

D   1 and 4

E   None of them

(Total 1 mark)

11. The thermal power flowing by conduction through a surface is proportional to the temperature difference across the surface, $\Delta\theta$, and the area of the surface, A, and inversely proportional to the thickness, Δx. The constant of proportionality is known as the thermal conductivity. A 60 cm composite rod, of constant cross-section, is made of 20 cm lengths of steel, copper and aluminum joined together. The rod is well insulated. The tip of the steel end of the rod is maintained at 100 °C and the tip of the aluminum end at 0 °C. What are the temperatures at each of the two junctions of dissimilar metals? Thermal conductivities: steel 60 W m^{-1} K^{-1}; copper 400 W m^{-1} K^{-1}; aluminum 240 W m^{-1} K^{-1}.

(5)

(Total 5 marks)

B.2

Greenhouse effect

◀ This photograph of the Earth was taken in 1968 during the Apollo 8 astronauts' return from orbiting the Moon. It reveals a lot on the greenhouse effect that has relevance in this chapter: not all of the Earth is illuminated by the Sun at a time; there are variations in how reflective different surfaces are; and the Earth is surrounded by swirling gases known as the atmosphere. What this snapshot cannot reveal is the impact that humans are having on the balance of thermal energy and temperature overall.

Guiding Questions

How does the greenhouse effect help to maintain life on Earth and how does human activity enhance this effect?

How is the atmosphere as a system modeled to quantify the Earth–atmosphere energy balance?

The Sun can be considered to be a black-body radiator so we can use Wein's law to find the temperature of its surface and the Stefan–Boltzmann law to find the power emitted per unit area and hence the total power. This radiation spreads out according to the inverse square relationship and is incident on the Earth, but before it can reach us on the ground, it must pass through the atmosphere. Here, we will investigate how the different wavelengths interact with the gases in the atmosphere.

Nature of Science

We can model the behavior of a small amount of ideal gas trapped in a cylinder quite accurately but modeling the whole atmosphere is quite different. To model the atmosphere, computer programs are used in an attempt to make predictions. However, we cannot run experiments to see if these predictions are correct. In 1988, the Intergovernmental Panel on Climate Change (IPCC) was established by the United Nations Environmental Programme and the World Meteorological Organization to provide the decision-makers of the world with a clear scientific view on the current state of knowledge in climate change and its potential environmental and socio-economic impacts.

There are problems associated with the use of energy. It will be up to the imagination and creativity of future scientists to solve them.

Students should understand:

conservation of energy
emissivity as the ratio of the power radiated per unit area by a surface compared to that of an ideal black surface at the same temperature as given by emissivity $= \dfrac{\text{power radiated per unit area}}{\sigma T^4}$
albedo as a measure of the average energy reflected off a macroscopic system as given by albedo $= \dfrac{\text{total scattered power}}{\text{total incident power}}$
Earth's albedo varies daily and is dependent on cloud formations and latitude
the solar constant S
the incoming radiative power is dependent on the projected surface of a planet along the direction of the path of the rays, resulting in a mean value of the incoming intensity being $\dfrac{S}{4}$

methane CH_4, water vapor H_2O, carbon dioxide CO_2, and nitrous oxide N_2O are the main greenhouse gases, and each has origins that are both natural and created by human activity

absorption of infrared radiation by the main greenhouse gases in terms of the molecular energy levels and the subsequent emission of radiation in all directions

the greenhouse effect can be explained in terms of a resonance model and molecular energy levels

the augmentation of the greenhouse effect due to human activities is known as the enhanced greenhouse effect.

Radiation from the Sun

B.2 Figure 1
Electromagnetic spectrum of sunlight.

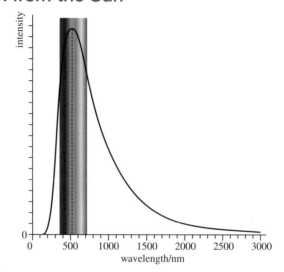

The spectrum of light arriving at the edge of the Earth's atmosphere, as shown in Figure 1, has a peak value at about 500 nm. This is in the green region of the visible spectrum. Putting this value into Wien's displacement law equation gives a temperature of 5800 K.

Now we know the temperature of the Sun, we can use the Stefan–Boltzmann law to calculate the power radiated per unit area:

$$P = A\sigma T^4$$

$$\frac{P}{A} = 5.67 \times 10^{-8} \times 5800^4 = 6.42 \times 10^7 \, W$$

The radius of the Sun = 6.9×10^8 m so the surface area = $4\pi r^2$ = 6.0×10^{18} m², which means that the total power radiated = 3.9×10^{26} W.

The Earth is 1.5×10^{11} m from the Sun so the power per unit area (also known as intensity or brightness) at the Earth will be:

$$I = \frac{P}{4\pi r^2} = \frac{3.9 \times 10^{26}}{4\pi (1.5 \times 10^{11})^2} = 1400 \, W \, m^{-2}$$

This quantity is called the **solar constant**, S.

Although the peak is in the visible region, sunlight also contains ultraviolet (10%) and infrared (50%).

Interaction between solar radiation and the atmosphere

Electromagnetic radiation spreads out as a wave, rather like a water wave, but in three dimensions. A water wave has peaks and troughs that spread across the water; electromagnetic waves have peaks and troughs in electromagnetic fields that spread through a vacuum at speed v. The distance between the peaks is called the wavelength λ, and the number of oscillations per second is called the frequency f. These quantities are related by the equation:

$$v = f\lambda$$

The electromagnetic radiation from the Sun can be split into three regions according to their wavelength: ultraviolet, visible and infrared. Ultraviolet has the shortest wavelength and highest frequency, infrared has the longest wavelength and lowest frequency, and visible light is in-between. These types of radiation interact with the atmosphere in different ways.

Ultraviolet radiation

Ultraviolet radiation has the highest energy, enough to split molecules. Ultraviolet radiation is absorbed by oxygen molecules (O_2) and splits them into two oxygen atoms (2O). The atoms can then join with O_2 molecules to form ozone (O_3), which forms a thin layer in the upper atmosphere where it absorbs ultraviolet radiation as it splits it into O_2 and O. Most of the higher energy ultraviolet radiation is absorbed in this way. This can be represented by the absorption spectrum in Figure 3, where 100% means that all of the radiation of that wavelength is absorbed.

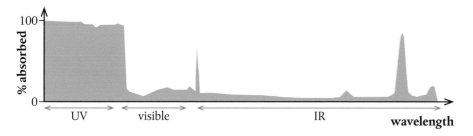

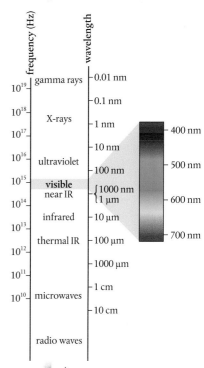

B.2 **Figure 2** The electromagnetic spectrum. Waves can be classified in terms of their wavelength. Each range of wavelengths has a different name, mode of production and uses.

B.2 **Figure 3** Absorption of electromagnetic radiation by ozone.

Visible light

Visible light excites atoms. In other words, the atoms absorb the energy of the light. An excited atom is unstable so will re-emit the light a short time later. You might expect that the atmosphere would absorb all the visible light, but it does not. If it did, the atmosphere would not be transparent. The reason for this is that the atoms can only be excited by certain specific energies, so only these are absorbed, leaving the rest to pass through. This gives rise to dark lines in the spectrum of light received on the Earth because the re-emitted radiation does not come out in the same direction as the original. You will find out more about this when you study atomic models and learn about electron energy levels.

Infrared radiation

Infrared radiation is produced by the vibration of molecules. The molecules of a solid vibrate in bodies that have thermal energy and infrared radiation is emitted. So, infrared radiation is associated with hot bodies. This is the radiation that we can feel coming from a fire and is detected in night vision cameras. When infrared radiation is absorbed, it makes molecules vibrate, which is why you get hot standing in front of a fire.

Certain molecules present in the atmosphere (e.g. carbon dioxide CO_2, water vapor H_2O, nitrous oxide N_2O, and methane CH_4) can be made to vibrate by infrared radiation, but they only absorb the frequency of radiation that is the same as their natural frequency. This is called resonance and is similar to someone pushing a child on a swing. If the frequency of pushing matches the natural frequency of the swing, maximum energy is transferred from the pusher to the swing. Once the molecule starts to vibrate, it will re-emit the radiation in a random direction (including toward space), which results in less radiation reaching the Earth. The absorption spectrum for CO_2 is shown in Figure 4.

Notice the three distinct peaks; each is due to a different mode of vibration of the molecule.

B.2 Figure 4 Absorption of electromagnetic radiation by carbon dioxide.

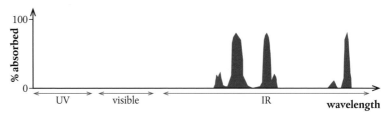

Although CO_2 is the most talked about absorber of infrared radiation, water has the greatest effect. This is not only due to its absorption characteristics, shown in Figure 5, but also because it is more abundant in the atmosphere.

B.2 Figure 5 Absorption of electromagnetic radiation by water.

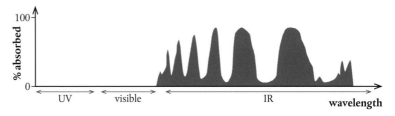

What relevance do simple harmonic motion and resonance have to climate change? (C.1, C.4)

Radiation reaching the Earth

By the time the black-body radiation from the Sun reaches the surface of the Earth, a lot of wavelengths have been absorbed by the different gases in the atmosphere. The spectrum now looks something like Figure 6. The actual amount of energy absorbed will vary with latitude since the sunlight has to pass through more atmosphere to reach the surface at the poles than at the equator.

B.2 Figure 6 Spectrum of light reaching the Earth's surface.

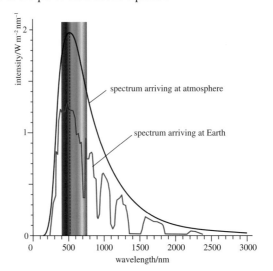

Interaction between light and solids

When the radiation hits the surface of the Earth, some is reflected and some is absorbed. The molecules in solids are much closer than the molecules in a gas. This results in electron energy **bands** rather than the discrete levels of a single atom. Solids can therefore emit and absorb many more wavelengths of light than low-pressure gases. When electromagnetic radiation is incident on a solid surface, there are three main possibilities:

- It may be absorbed then re-emitted in the opposite direction. This is **reflection** or **scattering**.
- It may be absorbed, transferring energy to electrons, which in turn pass the energy to molecular kinetic energy, resulting in an increase in temperature.
- It may not be absorbed but pass through the material.

When dealing with the surface of the Earth, we can discount the last option since no radiation will pass through the Earth. Whether a particular wavelength photon (particle of light) is absorbed or reflected depends upon the material of the surface; it is this that makes different materials have different colors. A red object appears red because it reflects red photons but absorbs the other colors, a white object reflects all colors and a black object reflects none.

Albedo

The ratio of scattered to total incident power is called the **albedo**:

$$\text{albedo} = \frac{\text{total scattered power}}{\text{total incident power}}$$

Different surfaces have different albedos; for example, the albedo of snow is 0.9 since almost all of the light is reflected, whereas the albedo of asphalt is 0.04. If we consider the whole Earth, then we should also take into account the light scattered by the clouds. This gives an average albedo of 0.3.

Emissivity (e)

Not all bodies are perfect black-body radiators so they emit less radiation than predicted by the equation for a perfect black-body radiator. The ratio of the energy radiated by a body to the energy radiated by a perfect black body at the same temperature is called the **emissivity**. So, for a body with emissivity e, the power radiated is given by the adapted equation:

$$P = eA\sigma T^4$$

The emissivity of a black body is 1; other materials will have emissivity less than 1 (see Table 1).

Note that the total power radiated from a star is known as the **luminosity**, L.

Emissivity is the ratio of actual thermal energy emitted to that which would be emitted by a black body. We can express this in terms of energy or power by the equation:

$$\text{emissivity} = \frac{\text{power radiated per unit area}}{\sigma T^4}$$

The greenhouse effect

Since the atmosphere does not absorb much visible light, the visible light reflected from the Earth's surface will pass back out through the atmosphere.

The absorbed light will cause the temperature of the surface to rise so it will emit radiation at a peak wavelength given by $\frac{0.0029}{T}$. The Earth is not very hot so this peak is in the infrared region, which means that the radiation is absorbed by carbon dioxide (CO_2), water vapor (H_2O), nitrous oxide (N_2O), and methane (CH_4) in the atmosphere. These molecules then

Using the color picker tool in photo-editing software, you can measure the amount of red, green, and blue light reflected on different surfaces in a digital photograph.

If more light is reflected from the Earth, then less energy is absorbed, so the temperature will be less. One way of increasing the amount of radiation reflected would be to paint surfaces white. If asphalt was white, how much difference would that make?

Material	Emissivity (at 0°C)
Polished aluminum	0.02
Brick	0.85
Dull black paper	0.94

B.2 Table 1 Emissivity of different materials.

re-emit the radiation in random directions so some of it returns to the Earth. The net result is that the energy leaving the Earth is reduced, resulting in a lower emissivity. This effect is called the **greenhouse effect** due to its similarity with the way glass traps thermal energy in a greenhouse. The gases that cause the effect are called **greenhouse gases**.

Energy balance

Imagine we could build the solar system by taking a Sun at 5800 K and putting a 0 K Earth in orbit at a distance of 1.5×10^{11} m. The Earth would absorb energy from the Sun causing its temperature to rise. As the Earth's temperature rose above 0 K, it would begin to radiate energy itself. The amount of energy radiated would increase until the amount of energy radiated equaled the amount of energy absorbed and equilibrium was reached.

Earth without atmosphere

To understand the principle of energy balance, we will first consider the simplified version of a perfectly black Earth without atmosphere. We have already calculated the power received per unit area at the Earth as 1400 W m^{-2}. This radiation only lands on one side of the Earth, so treating the Earth as a disk of radius 6400 km, we can calculate the total power incident at the surface:

$$\text{incident power} = \frac{\text{power}}{\text{area}} \times \text{area of Earth} = 1400 \times \pi r^2 = 1400 \times \pi \times (6.4 \times 10^6)^2$$

$$= 1.8 \times 10^{17}\,\text{W}$$

To find the average power per unit area over the whole Earth, we must divide by the surface area of the Earth ($4\pi r^2$). Note that, this time, we take the Earth to be a sphere, which results in an average intensity of $\frac{S}{4}$.

As the temperature of the Earth increases, it will emit radiation which, according to the Stefan–Boltzmann law, will be proportional to the fourth power of the temperature in kelvin. So when the temperature is T, the power radiated per unit area is given by:

$$\text{power per unit area} = \sigma T^4$$

The total power radiated can be found by multiplying this value by the area of the Earth. This time, we must use the total area of the sphere since energy is radiated by all parts, not just the side facing the Sun:

$$\text{power radiated from Earth} = \sigma T^4 \times 4\pi r^2$$

$$\text{power} = 2.9 \times 10^7 \times T^4\,\text{W}$$

The temperature of the Earth will rise until the incident power = radiated power:

$$1.8 \times 10^{17}\,\text{W} = 2.9 \times 10^7 \times T^4$$

$$T = \sqrt[4]{\frac{1.8 \times 10^{17}}{2.9 \times 10^7}} = 280\,\text{K}$$

Earth with atmosphere

The previous example is not realistic because not all the radiation is absorbed by the ground: some is scattered by the atmosphere (particularly clouds) and some is reflected off the surface. The average albedo of the Earth is 0.3, which means that only $\frac{7}{10}$ of the power will be absorbed by the ground. So the power absorbed = $0.7 \times 1.8 \times 10^{17} = 1.26 \times 10^{17}$ W.

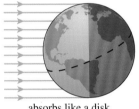

absorbs like a disk

▲ **B.2 Figure 7** To an imaginary observer on the Sun, the Earth would look like a circle.

The radiation from the Sun hits the Earth from one direction so the Earth absorbs the same amount of radiation as if it were a disk. In this case, power absorbed = $1400\pi r^2$. However, the Earth is actually a sphere so:

power absorbed per unit area

$= \frac{1400\pi r^2}{4\pi r^2} = \frac{1400}{4}$

$= 350\,\text{W m}^{-2}$

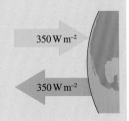

350 W m^{-2}

350 W m^{-2}

Also, the power radiated is not as high because the Earth is not a black body and the greenhouse gases re-radiate some of the infrared radiation emitted by the Earth back to the ground. This results in an emissivity of around 0.6.

So: power radiated from Earth $= 0.6 \times 2.9 \times 10^7 \times T^4$ W
$$= 1.74 \times 10^7 \times T^4 \text{ W}$$

Equilibrium will therefore be reached when $1.26 \times 10^{17} = 1.74 \times 10^7 \times T^4$ which gives a value of $T = 292$ K.

We can see that the atmosphere has two competing effects: the clouds raise the albedo, resulting in less power reaching the Earth, but the greenhouse gases make the emissivity lower, resulting in a higher overall temperature.

Figure 8 is a very simplified picture of the energy flow. Figure 9 is a more complete (but still simplified) representation showing some of the detail of the exchange of energy between the ground and the atmosphere. Notice that the atmosphere absorbs energy in two ways: $358\,\text{W m}^{-2}$ is absorbed by greenhouse gases and $105\,\text{W m}^{-2}$ due to convection and the energy used to turn water into water vapor.

$105\,\text{W m}^{-2}$

$350\,\text{W m}^{-2}$

$245\,\text{W m}^{-2}$

B.2 Figure 8 Energy flow with some reflected energy.

110   350
      68

79

31

172

199

41   358

399   105   332

B.2 Figure 9 The numbers in this diagram represent power per unit area in W m^{-2}.

Challenge yourself

1. Estimate the average surface temperature of the Moon.

What limitations are there in using a resonance model to explain the greenhouse effect? (C.4)

Exercise

Q1. Use Figure 10 to answer the following questions.

(a) What is the total power per square meter absorbed by the atmosphere?

(b) How much of the incident radiation is reflected from the surface?

(c) What is the albedo of the Earth?

(d) How much power per square meter is re-radiated by the atmosphere?

(e) What percentage of the energy radiated from the Earth passes straight through the atmosphere?

SKILLS

We cannot do experiments to see the effect of changing the albedo of the Earth, but we can set up computer models to see what might happen. Using a spreadsheet, we can make a simple model to predict the final temperature of the Earth given certain starting conditions. We are going to start the Earth off with an initial temperature of, say, 20 K and calculate how much the temperature will rise in a certain time interval.

The first thing we need to do is enter the variables and constants. These can be entered into a table on the side of the spreadsheet as in Table 2.

B.2 Table 2 ▶

	I	J
Solar constant S/W m^{-2}	350
Albedo α	0.3
Emissivity e	0.6
Time interval/yrs	2
Initial temp T/K	20

Next we will add some formulae to calculate the initial conditions of the Earth using the headers in Table 3 (Note: all energy and power values are for 1 m^2 of Earth).

	A	B	C	D	E	F	G
1	Time/yr	Energy in/J m^{-2}	Power radiated/W m^{-2}	Energy out/J m^{-2}	Energy added/J m^{-2}	Change in temperature/K	New temperature/K
2	0	15 452 640 000	0.005 443 2	343 313.5	1 545 229 668	38.630 741 7	20

▲

B.2 Table 3

The columns are as follows:

Time: a starting value, 0; subsequent times will be this plus the time interval.

Energy in: solar constant × (1 − albedo) × time interval in seconds (= J$1*(1−J$2)*J$4*3600*24*365).

Power radiated: emissivity × σ × T^4 where T is the initial temperature, in this case, 20 K (= J$3*0.000 000 056 7*J$5^4).

Energy out: power out × time interval in seconds (= C2*J$4*3600*24*365).

Energy added: energy in − energy out (= B2 − D2).

Change in temp: calculated using the surface heat capacity, which is 4 × 10^8 J m^{-2} K^{-1}; this is rather like the specific heat capacity with area instead of mass, so $\Delta T = \frac{E}{C_s}$ where E is the heat received per unit area (= $\frac{E2}{400\,000\,000}$).

New temperature: this is simply the initial temperature; subsequent values will be calculated from the previous temperature + the change in temperature. Enter = J5 to take the temperature from the table.

Once this row has been fixed, the formulae are put in the next row to calculate the heat lost at the new temperature (Table 4).

	A	B	C	D	E	F	G
1	Time/ yr	Energy in/J m^{-2}	Power radiated/W m^{-2}	Energy out/J m^{-2}	Energy added/J m^{-2}	Change in temp/K	New temp/K
2	0	15 452 640 000	0.005 443 2	343 313.5104	15 452 296 686	38.630 741 72	20
3	2	15 452 640 000	0.402 009 08	25 355 516.7	15 427 284 483	38.568 211 21	58.630 741 72

B.2 Table 4

Time: previous time + time interval (= A2+J$4).

Energy in: solar constant × (1– albedo) × time interval in seconds (= J$1*(1–J$2)*J$4*3600*24*365).

Power radiated: emissivity × σ × T^4 where T is the new temperature; this is the temperature of the last time interval + the change in temperature (= J$3*0.0000000567*G3^4).

Energy out: power out × time interval in seconds (= C3*J$4*3600*24*365).

Energy added: energy in – energy out (= B3 – D3).

Change in temperature: (= $\frac{E3}{400\,000\,000}$).

New temperature: change in temperature + previous temperature (= F2 + G2).

These formulae are now copied down for fifty rows by highlighting row 3 and dragging down. (Note: it is row 3 you copy down, not row 2, which just contains the starting values.)

To plot a graph, highlight column A, then while pressing control, highlight column G. This will leave both A and G highlighted. Now insert scatter graph, giving Figure 11.

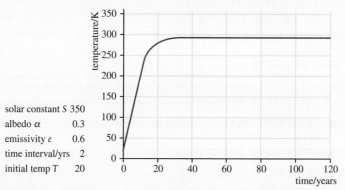

solar constant S 350
albedo α 0.3
emissivity e 0.6
time interval/yrs 2
initial temp T 20

Although this simulation aids our understanding of the underlying physics, it is a much simplified version of reality and should not be taken to represent reality. There are, however, some much more complex computer simulations that are thought to come close.

◀ **B.2 Figure 10** Graph of temperature against time.

You can now find out what happens if you change the values of albedo, emissivity, etc.

Global warming

The spreadsheet simulation shows that if the energy in is not balanced with energy out, then the average temperature of the Earth will change. A rise in average temperature is called **global warming** and can be caused by four factors: solar constant, albedo, emissivity and quantity of greenhouse gases.

Increase in solar constant

The amount of energy reaching the Earth depends on how much energy the Sun is giving out and the distance between the Earth and the Sun. Neither of these is constant. The Earth's orbit is elliptical so the orbital radius changes. There are also some changes due to the change of angle of the Earth's axis in relation to the Sun; these variations are called Milankovitch cycles.

The increase in temperature as a result of many sunspots leads to solar flares, which are jets of gas flying out from the Sun's surface like huge flames.

The Sun's surface is an ever-changing swirling mass of gas that sometimes sends out flares many tens of thousands of kilometers high. More easily viewed from the Earth are the number of sunspots, which can be seen if an image of the Sun is projected onto a screen (never look directly at the Sun). The number of sunspots present gives an indication of the amount of energy emitted; more sunspots implies more energy. This seems the wrong way round but it is the gas around the sunspot that emits the energy, not the sunspot itself.

Reduced albedo

Albedo is the ratio of reflected to total incident radiation. If the albedo is low, then more energy is absorbed by the Earth, resulting in a higher equilibrium temperature. Snow has a high albedo so a reduction in the amount of snow present at the poles and on glaciers would reduce the albedo.

Reduced emissivity

The emissivity is related to the greenhouse effect, which reduces the amount of radiation leaving the Earth at a given temperature. Increasing the amount of greenhouse gases in the atmosphere will result in a higher equilibrium temperature.

Enhanced greenhouse effect

There is a lot of evidence that the burning of fossils fuels, which produce carbon dioxide at a faster rate than the living trees remove it, has led to an increase in the percentage of CO_2 in the atmosphere. This has enhanced the greenhouse effect, causing more energy to be radiated back to Earth. If more radiation returns to the Earth, then more must be emitted, resulting in a higher average temperature. This effect is made worse by deforestation because fewer trees are available to store the carbon dioxide.

How is the understanding of systems applied to other areas of physics? (e.g. B.4, D.4, E.4, E.5)

What can be done

To reduce the enhanced greenhouse effect, the levels of greenhouse gases must be reduced, or at the very least, the rate at which they are increasing must be slowed down. There are several ways that this can be achieved:

Global warming is an international problem that requires an international solution. What measures are the government and people making in the country where you live?

1 Greater efficiency of power production

In recent years, the efficiency of power plants has been increasing significantly. According to the second law of thermodynamics, they can never be 100% efficient but some of the older, less efficient ones could be replaced. This would mean that producing the same amount of power would require less fuel, resulting in reduced CO_2 emission.

2 Replacing the use of coal and oil with natural gas

Gas-fired power stations are more efficient than oil and gas and produce less CO_2.

3 Use of combined heating and power systems (CHP)

Using the excess thermal energy from power stations to heat homes would result in a more efficient use of fuel.

4 Increased use of renewable energy sources and nuclear power

Replacing fossil fuel burning power stations with alternatives such as wave power, solar power and wind power would reduce CO_2 emissions.

5 Use of electric vehicles

A large amount of the oil used today is used for transport and, even without global warming, there will be a problem when the oil runs out. Cars that run on electricity are already in production. Airplanes will also have to use a different fuel.

6 Carbon dioxide capture and storage

A different way of reducing greenhouse gases is to remove CO_2 from the waste gases of power stations and store it underground.

An international problem

Global warming is an international problem, and if any solution is going to work, then it must be a joint international solution. Before working on the solution, the international community had to agree on pinpointing the problem and it was to this end that the Intergovernmental Panel on Climate Change (IPCC) was formed.

Guiding Questions revisited

How does the greenhouse effect help to maintain life on Earth and how does human activity enhance this effect?

How is the atmosphere as a system modeled to quantify the Earth–atmosphere energy balance?

In this chapter, we have studied different variables that enable us to understand that:

- The Sun's radiation forms a spectrum including ultraviolet (which is absorbed by oxygen molecules), visible light (which passes through the atmosphere except for a few discrete wavelengths) and infrared (which is absorbed by methane, water, carbon dioxide and nitrous oxide molecules before being re-emitted in all directions).
- Energy is incident on the Earth from the Sun with an intensity referred to as the solar constant.
- On average, 30% of the incident energy is reflected because of the Earth's albedo.
- The energy that is absorbed by the Earth is later emitted in the infrared region, which means that greenhouse gas molecules in the atmosphere will absorb and re-emit some of this radiation back toward the Earth's surface (and, in turn, reduce the Earth's overall emissivity).
- In the absence of any greenhouse effect, the Earth's surface temperature would be cooler than is needed to maintain life.
- The burning of fossil fuels (e.g. to heat houses, to travel and to generate electricity) and deforestation (e.g. wood for manufacturing) are contributing to greenhouse gases in the atmosphere, which in turn is enhancing the greenhouse effect.
- The enhanced greenhouse effect and its associated reduction in emissivity is contributing to global warming, which is an increase in the Earth's average temperature, which in turn reduces albedo (due to the melting of snow at the poles and on glaciers).

How do different methods of electricity production affect the energy balance of the atmosphere? (B.5, E.4)

How are developments in science and technology affected by climate change and other important international concerns? (NOS)

Practice questions

There is a general expectation that you will become familiar with mechanisms for electricity generation and heating over the course of IB physics. These questions may require you to draw upon your existing knowledge or to carry out research (e.g. in a library or online).

1. (a) Fossil fuels are being used continuously for electricity production. Outline why fossil fuels are classed as non-renewable. (2)

    (b) Some energy consultants suggest that the solution to the problem of carbon dioxide pollution is to use nuclear energy for the generation of electrical energy. Identify **two** disadvantages of the use of nuclear fission when compared to the burning of fossil fuels for the generation of electrical energy. (2)

    (Total 4 marks)

2. (a) By reference to energy transfers, distinguish between a solar panel and a solar cell. (2)

    Some students carry out an investigation on a solar panel. They measure the output temperature of the water for different solar input powers and for different rates of extraction of thermal energy. The results are shown on the graph.

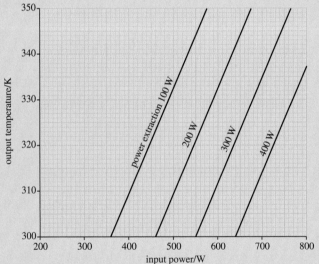

    (b) Use the data from the graph to answer the following.

        (i) The solar panel is to provide water at 340 K while extracting energy at a rate of 300 W when the intensity of the sunlight incident normally on the panel is $800 \, W \, m^{-2}$. Calculate the effective surface area of the panel that is required. (2)

        (ii) Deduce the overall efficiency of the panel for an input power of 500 W at an output temperature of 320 K. (3)

    (Total 5 marks)

3. (a) The intensity of the Sun's radiation at the position of the Earth is approximately $1400 \, W \, m^{-2}$.
        Suggest why the average power received per unit area of the Earth is $350 \, W \, m^{-2}$. (2)

(b) The diagram shows a simplified model of the energy balance of the Earth's surface. It shows radiation entering or leaving the Earth's surface only.

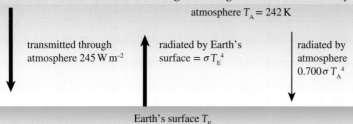

atmosphere $T_A = 242$ K

transmitted through atmosphere 245 W m^{-2}

radiated by Earth's surface $= \sigma T_E^{\,4}$

radiated by atmosphere $0.700\sigma T_A^{\,4}$

Earth's surface T_E

The average equilibrium temperature of the Earth's surface is T_E and that of the atmosphere is $T_A = 242$ K.

(i) Using the data from the diagram, state the emissivity of the atmosphere. (1)

(ii) Show that the intensity of the radiation radiated by the atmosphere toward the Earth's surface is 136 W m^{-2}. (1)

(iii) By reference to the energy balance of the Earth's surface, calculate T_E. (2)

(c) (i) Outline a mechanism by which part of the radiation radiated by the Earth's surface is absorbed by greenhouse gases in the atmosphere. (3)

(ii) Suggest why the incoming solar radiation is not affected by the mechanism you outlined in (c)(i). (2)

(iii) Carbon dioxide (CO_2) is a greenhouse gas. State **one** source and **one** sink (object that removes CO_2) of this gas. (2)

(Total 13 marks)

4. (a) State the Stefan–Boltzmann law for a black body. (2)

(b) The following data relate to the Earth and the Sun.

Earth–Sun distance $= 1.5 \times 10^{11}$ m
Radius of Earth $= 6.4 \times 10^6$ m
Radius of Sun $= 7.0 \times 10^8$ m
Surface temperature of Sun $= 5800$ K

(i) Use the data to show that the power radiated by the Sun is about 4×1^{26} W. (1)

(ii) Calculate the solar power incident per unit area at a distance from the Sun equal to the Earth's distance from the Sun. (2)

(iii) The average power absorbed per unit area at the Earth's surface is 240 W m^{-2}.
State **two** reasons why the value calculated in (b)(ii) differs from this value. (2)

(iv) Show that the value for power absorbed per unit area of 240 W m^{-2} is consistent with an average equilibrium temperature for Earth of about 255 K. (2)

(c) Explain, by reference to the greenhouse effect, why the average temperature of the surface of the Earth is greater than 255 K. (3)

(d) Suggest why the burning of fossil fuels may lead to an increase in the temperature of the surface of the Earth. (3)

(Total 15 marks)

5. **(a)** A white plate has a black pattern on it.

   (i) Explain why, when red hot, the black pattern is brighter than the white parts of the plate. (2)

   (ii) State which part of the plate has the highest emissivity. (1)

   (iii) State which part of the plate has the highest albedo. (1)

(b) There is a scheme to paint the roofs of houses white. Explain how this would reduce the enhanced greenhouse effect. (3)

(c) The diagram shows the absorption spectrum for the atmosphere of a planet.

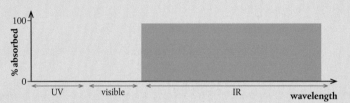

   (i) Explain why burning fossil fuels would not cause any change in the temperature of the planet. (2)

   (ii) Explain why life would not be possible on this planet. (2)

   (iii) Water vapor is an effective greenhouse gas. Explain why the presence of water vapor in the atmosphere does not necessarily lead to global warming. (2)

   (Total 13 marks)

6. A 2 cm diameter metal ball with emissivity 0.8 is heated to 1000 K.

   (a) **(i)** Calculate the total power of the radiation emitted by the ball. (2)

   (ii) Calculate the intensity of the radiation at a distance of 2 m. (2)

   (b) A thermal radiation sensor with area of 0.5 cm² is placed 2 m from the ball.

   (i) Calculate the energy absorbed by the sensor in 1 minute. (2)

   (ii) State **two** assumptions that you have made about the sensor. (2)

   (Total 8 marks)

7. In a simple climate model for a planet, the incoming intensity is 400 W m^{-2} and the radiated intensity is 300 W m^{-2}. The temperature of the planet is constant. What is the reflected intensity from the planet and the albedo of the planet?

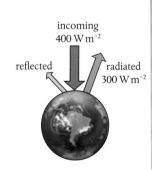

incoming
400 W m^{-2}

reflected   radiated
300 W m^{-2}

	Reflected intensity from the planet	Albedo of the planet
A	100 W m^{-2}	0.25
B	100 W m^{-2}	0.75
C	300 W m^{-2}	0.25
D	300 W m^{-2}	0.75

(Total 1 mark)

8. What is the main role of carbon dioxide in the greenhouse effect?

   A It absorbs incoming radiation from the Sun.

   B It absorbs outgoing radiation from the Earth.

   C It reflects incoming radiation from the Sun.

   D It reflects outgoing radiation from the Earth.   *(Total 1 mark)*

9. The average temperature of the surface of a planet is five times greater than the average temperature of the surface of its moon. The emissivities of the planet and the moon are the same. The average intensity radiated by the planet is I. What is the average intensity radiated by its moon?

   A $\dfrac{I}{25}$ B $\dfrac{I}{125}$ C $\dfrac{I}{625}$ D $\dfrac{I}{3125}$

   (Total 1 mark)

10. The orbital radius of the Earth around the Sun is 1.5 times that of Venus. What is the intensity of solar radiation at the orbital radius of Venus?

    A $0.6\,\mathrm{kW\,m^{-2}}$ C $2\,\mathrm{kW\,m^{-2}}$ B $0.9\,\mathrm{kW\,m^{-2}}$ D $3\,\mathrm{kW\,m^{-2}}$

    (Total 1 mark)

11. A photovoltaic panel of area S has an efficiency of 20%. A second photovoltaic panel has an efficiency of 15%. What is the area of the second panel so that both panels produce the same power under the same conditions?

    A $\dfrac{S}{3}$ B $\dfrac{3S}{4}$ C $\dfrac{5S}{4}$ D $\dfrac{4S}{3}$

    (Total 1 mark)

12. The three statements give possible reasons why an average value should be used for the solar constant.

    I. The Sun's output varies during its 11-year cycle.

    II. The Earth is in an elliptical orbit around the Sun.

    III. The plane of the Earth's spin on its axis is tilted to the plane of its orbit about the Sun.

    Which are the correct reasons for using an average value for the solar constant?

    A I and II only   C II and III only   B I and III only   D I, II and III

    (Total 1 mark)

13. It is suggested that the solar power incident at a point on the Earth's surface depends on:

    I. daily variations in the Sun's power output

    II. the location of the point

    III. the cloud cover at the point

    Which suggestion(s) is/are correct?

    A III only   C II and III only   B I and II only   D I, II and III

    (Total 1 mark)

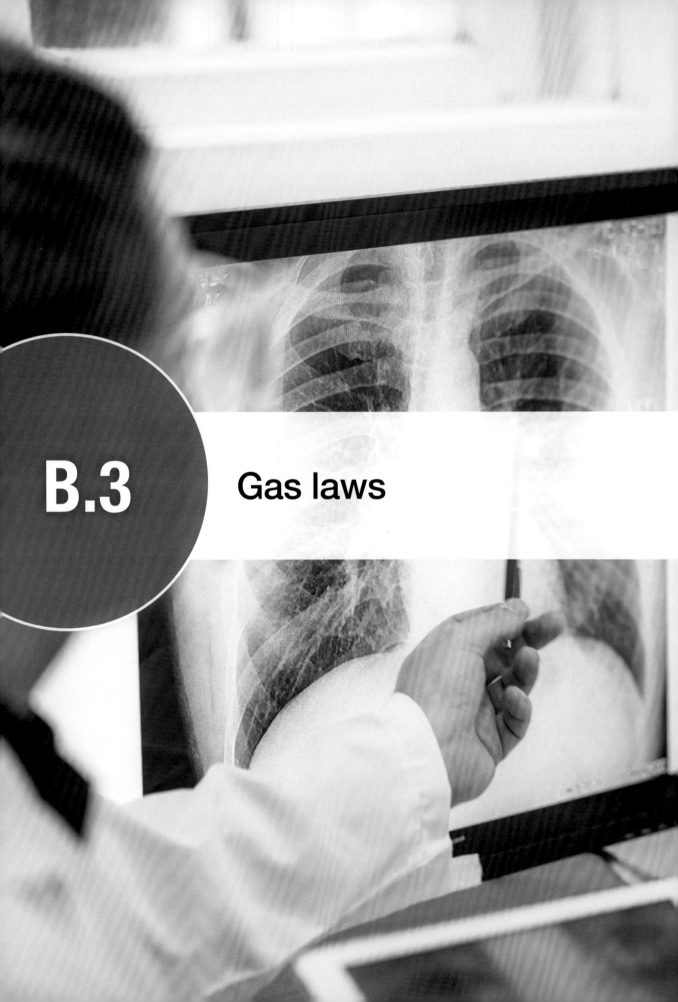

B.3 Gas laws

◀ During the study of gas quantities, such as volume, pressure and temperature, it can be helpful to remember that we all carry our own gas container at all times: our lungs. Breathing in involves a contraction of the diaphragm, which increases the volume of the cavity. Air moves in to reduce the pressure difference between the atmosphere and that inside. Breathing out involves a relaxation of the diaphragm, which reduces the volume and increases the internal pressure so air moves out. The rib cage supports the process in both directions using the intercostal muscles.

Guiding Questions

How are the macroscopic characteristics of a gas related to the behavior of individual molecules?

What assumptions and observations lead to universal gas laws?

How can models be used to help explain observed phenomena?

In the chapters on mechanics, we first defined the quantities associated with particles (e.g. mass, momentum and energy), then we looked at how they interact when they collide. We are going to follow a similar procedure with gases by first defining the relevant quantities (pressure, volume, temperature and number of particles), then using those quantities to model interactions between two gases.

When considering a gas in a container, it is easy to forget that we are actually looking at two gases: the gas held within the container and the air that surrounds the container. This is made clearer if we consider two gases separated by a shared piston.

What we want to know is how the gas on one side affects the gas on the other. Which way will thermal energy flow and which way will the piston move?

To answer this, we could look at the mechanical properties of the individual particles or the relationships between pressure, volume and temperature. The only problem is that there are too many variables, all of which affect the others. We will instead relate just two variables at a time, keeping the others constant (for example, pressure is inversely proportional to volume if temperature and the number of particles remain the same).

The macroscopic relationships involving pressure, volume, temperature and amount of gas discussed in this chapter form the ideal gas laws and can be used in place of the microscopic kinetic theory explanations of gas transformations.

What happens to the pressure of a gas when its volume is reduced? The kinetic theory explanation would be that the reduction in volume leads to an increase in density, which results in more collisions with the walls per unit time, which increases the total rate of change of momentum of particles hitting the walls, leading to a greater force over a smaller area and, therefore, a higher pressure.

Or, using the ideal gas laws, the pressure of a fixed mass of gas is inversely proportional to its volume at constant temperature. Reducing the volume will result in a higher pressure.

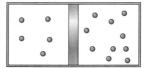

▲
B.3 Figure 1 Two chambers within a container, separated by a piston.

Students should understand:

pressure as given by $P = \frac{F}{A}$ where F is the force exerted perpendicular to the surface
the amount of substance n as given by $n = \frac{N}{N_A}$ where N is the number of molecules and N_A is the Avogadro constant
ideal gases are described in terms of the kinetic theory and constitute a modeled system used to approximate the behavior of real gases
the ideal gas law equation can be derived from the empirical gas laws for constant pressure, constant volume and constant temperature as given by $\frac{PV}{T}$ = constant
the equations governing the behavior of ideal gases as given by $PV = Nk_B T$ and $PV = nRT$
the change in momentum of particles due to collisions with a given surface gives rise to pressure in gases and, from that analysis, pressure is related to the average translational speed of molecules as given by $P = \frac{1}{3}\rho v^2$
the relationship between the internal energy U of an ideal monatomic gas and the number of molecules or amount of substance as given by $U = \frac{3}{2}Nk_B T$ or $U = \frac{3}{2}RnT$
the temperature, pressure and density conditions under which an ideal gas is a good approximation of a real gas.

TOK

This is a good example of how models are used in physics. Here we are modeling something that we cannot see, the atom, using a familiar object, a perfectly elastic ball.

How do we know atoms have different masses?

The answer to that question is thanks originally to the chemists, John Dalton in particular. Chemists make compounds from elements by mixing them in very precise proportions. This is quite complicated but we can consider a simplified version as shown in Figure 2. An atom of A joins with an atom of B to form molecule AB.

B.3 Figure 2 Atoms join to make a molecule.

A + B = A B

We first try by mixing the same masses of A and B but find that when the reaction has finished there is some B left over; we must have had too many atoms of B. If we reduce the amount of B until all the A reacts with all the B to form AB, we know that there must have been the same number of atoms of A as there were of B as shown in Figure 3. We can therefore conclude that the mass of an A atom is larger than an atom of B. In fact, the ratio of:

B.3 Figure 3 To combine completely there must be equal numbers of atoms.

$$\frac{\text{mass of atom A}}{\text{mass of atom B}} = \frac{\text{total mass A}}{\text{total mass B}}$$

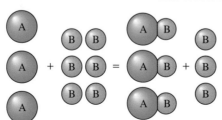

equal masses of A and B

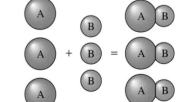

equal number of atoms of A and B

By finding out the ratios of masses in many different reactions, the atomic masses of the elements relative to each other were measured. Originally, everything was compared to oxygen since it reacts with so many other atoms. Later, when physicists started to measure the mass of individual atoms, the standard atom was changed to carbon-12. This is taken to have an atomic mass of exactly 12 unified mass units (u). The size of 1 u is therefore equal to $\frac{1}{12}$ of the mass of a carbon-12 atom, which is approximately the mass of the smallest atom, hydrogen.

Avogadro's hypothesis

The simplified version of chemistry given here does not give the full picture: for one thing, atoms do not always join in pairs. Maybe one A joins with two Bs. Without knowing the ratio of how many atoms of B join with one atom of A, we cannot calculate the relative masses of the individual atoms. Amedeo Avogadro solved this problem by suggesting that equal volumes of all gases at the same temperature and pressure will contain the same number of molecules. So if one atom of A and one atom of B join to give one molecule AB, then the number of molecules of AB is equal to the number of atoms of A or B and the volume of AB = $\frac{1}{2}$(A + B). But if one atom of A joins with two atoms of B, then the volume of B atoms is twice the volume of the A atoms, and the volume of AB_2 = the volume of A. This is illustrated in Figure 4.

The volume of 1 mole of any gas at normal atmospheric pressure (101.3 kPa) and a temperature of 0 °C is 22.4 liters (L), or 0.0224 m³.

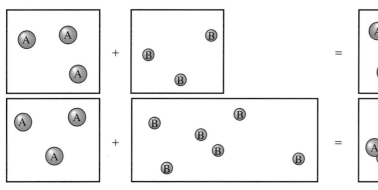

B.3 Figure 4 Equal volumes of gases contain the same number of molecules.

The mole and Avogadro's constant

It can be shown that 12 g of carbon-12 contains 6.02×10^{23} atoms. This amount of material is called a mole. This number of atoms is called Avogadro's constant (N_A) (named after him but not calculated by him). If we take 6.02×10^{23} molecules of a substance that has molecules that are four times the mass of carbon-12 atoms, it would have relative molecular mass of 48.

A rough calculation of how big a grain of sand is compared to volume of sand reveals that there are approximately 6×10^{23} (Avogadro's constant) grains of sand in the Sahara desert.

Moles of different compounds have different masses.

6.02×10^{23} molecules of this substance would therefore have a mass four times the mass of the same number of carbon-12 atoms, 48 g. So to calculate the mass of a mole of any substance, we simply express its relative molecular mass in grams. This gives us a convenient way of measuring the amount of substance in terms of the number of molecules rather than its mass.

Worked example

If a mole of carbon has a mass of 12 g, how many atoms of carbon are there in 2 g?

Solution

One mole contains 6.02×10^{23} atoms.

2 g is $\frac{1}{6}$ of a mole so contains $\frac{1}{6} \times 6.02 \times 10^{23}$ atoms = 1.00×10^{23} atoms

Be careful with the units. For example, do all volume calculations using m^3.

Worked example

The density of iron is 7874 kg m^{-3} and the mass of a mole of iron is 55.85 g. What is the volume of 1 mole of iron?

Solution

$$\text{density} = \frac{\text{mass}}{\text{volume}}$$

$$\text{volume} = \frac{\text{mass}}{\text{density}}$$

$$\text{volume of 1 mole} = \frac{0.05585}{7874} \, m^3$$

$$= 7.093 \times 10^{-6} \, m^3$$

$$= 7.093 \, cm^3$$

Exercise

Q1. The mass of 1 mole of copper is 63.54 g and its density 8920 kg m^{-3}.

   (a) What is the volume of one mole of copper?

   (b) How many atoms does one mole of copper contain?

   (c) How much volume does one atom of copper occupy?

Q2. If the density of aluminum is 2700 kg m^{-3} and the volume of 1 mole is 10 cm^3, what is the mass of one mole of aluminum?

Nature of Science

Experiments in the physics lab are sometimes designed to reinforce theory. However, real science is not always like that. Theories are often developed as a result of observation and experiment. Given that gases are made of randomly-moving tiny particles, it is not difficult to explain their properties. Deducing that gases are made of particles from the properties of the gas is a much more difficult proposition that will fortunately never have to be done again.

The gas laws are examples of how we investigate the relationship between two variables while controlling all other factors.

The ideal gas

Of the three states of matter, the gaseous state has the simplest model. This is because the forces between the molecules of a gas are very small, so they are able to move freely. We can therefore use information about the motion of particles in A.1, A.2 and A.3 sections to study gases in more detail.

According to our simple model, a gas is made up of a large number of perfectly elastic, tiny spheres moving in random motion.

This model makes some assumptions:
- The collisions between molecules are perfectly elastic.
- The molecules are spheres.
- The molecules are identical.
- There are no forces between the molecules (except when they collide). This means that the molecules move with constant velocity between collisions.
- The molecules are very small; that is, their total volume is much smaller than the volume of the container.

Some of these assumptions are not true for all gases, especially when the gas is compressed (when the molecules are so close together that they experience a force between them). The gas then behaves as a liquid. However, to keep things simple, we will only consider gases that behave like our model. We call these gases **ideal** gases.

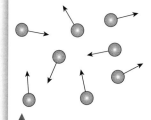

▲ **B.3 Figure 5** Molecules of gas in random motion.

Defining the state of a gas

To define the state of an amount of matter, we need to give enough information so that another person could obtain the same material with the same properties. If we were to describe a 100 g cube of copper at 300 K, we have stated how much and how hot it is, and even its shape. Someone else would be able to take an identical piece of copper and it would behave in the same way as ours. If, on the other hand, we were to take 100 g of helium gas, then we would also need to define its volume since density can vary depending on the container, and different volumes cause the pressure exerted by the gas on its sides to vary.

Volume

The volume of a gas is simply the volume of the container. If we want to vary the volume, we can place the gas in a cylinder with a movable end (a piston) as in Figure 6.

▲ **B.3 Figure 6** Gas molecules trapped in an adjustable container.

Temperature

Since gas molecules have no forces between them (except when colliding), no work is done when they move around, which means that there is no energy associated with their position. In other words, the molecules have no potential energy. The temperature of a gas in kelvin is therefore directly proportional to the average kinetic energy of a molecule.

$$E_{k,\,mean} = \frac{3}{2}kT$$

where k is the Boltzmann constant, $1.38 \times 10^{-23}\,\text{J K}^{-1}$.

If there are N molecules then the total kinetic energy of the gas $= N \times E_{k,\,mean} = \frac{3}{2}NkT$

A more convenient expression (because we are more often presented with the total number of moles than the number of molecules) is $E_{k,\,total} = \frac{3}{2}nRT$ where R is the molar gas constant and n is the number of moles. Higher kinetic energy implies higher velocity so the molecules of a gas at high temperature will have a higher average velocity than molecules of the same gas at a low temperature, as shown in Figure 7.

B.3 Figure 7 Kinetic energy is related to temperature.

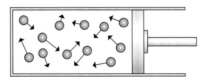 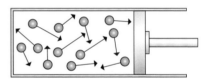

low temperature, small average kinetic energy   high temperature, large average kinetic energy

How does a consideration of the kinetic energy of molecules relate to the development of the gas laws? (A.3)

Let us compare two gases, A and B, with molecules of different mass at the same temperature. Applying $E_{k,\,total} = \frac{3}{2}nRT$, we can deduce that if temperature is the same, then the average kinetic energy of the molecules will be the same. But $E_k = \frac{1}{2}mv^2$ so:

$$\frac{1}{2}m_A v_A^2 = \frac{1}{2}m_B v_B^2$$

$$\frac{m_A}{m_B} = \frac{v_B^2}{v_A^2}$$

This means that if A has lighter molecules, the molecules in gas A must have higher velocity, as represented by the red balls in Figure 8.

B.3 Figure 8 Same temperature, different gases.

temperature = T   temperature = T

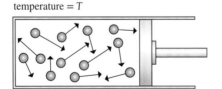 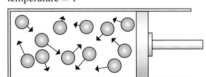

low mass molecules, high velocity   high mass molecules, low velocity

Pressure

Pressure at a surface is defined as the force per unit area, where the force acts perpendicular to the surface:

$$P = \frac{F}{A}$$

Pressure is measured in pascals (Pa), where 1 Pa is equivalent to $1\,\mathrm{N\,m^{-2}}$.

When gas molecules collide with the sides of the container, their momentum changes. This is because they have experienced an unbalanced force from the wall. According to Newton's third law, the wall must experience an equal and opposite force so will be pushed out by the gas. This is why the piston must be held in place by the man in Figure 9. The collective force from all collisions between particles and the walls is responsible for the pressure a gas exerts on its container.

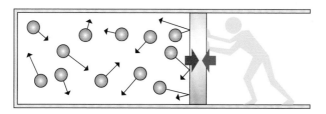

◄ **B.3 Figure 9** Gas pushes the piston to the right so something must push it to the left.

To understand how the pressure is related to the motion of the molecules, we can consider the simplified version shown in Figure 10 where one molecule is bouncing rapidly between the piston and the far wall of the cylinder.

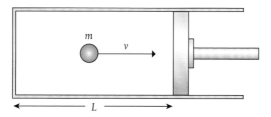

◄ **B.3 Figure 10** One molecule of gas.

When the molecule hits the piston, it bounces off elastically. The magnitude of change in momentum is therefore $2mv$. The force exerted on the piston is equal to the rate of change of momentum, which in this case is the change in momentum × rate of hitting the wall. The rate at which the molecule hits the wall depends on how long it takes for the molecule to travel to the other end of the cylinder and back:

$$\text{time for molecule to travel to other end and back} = \frac{2L}{v}$$

$$\text{number of hits per unit time} = \frac{1}{\left(\frac{2L}{v}\right)} = \frac{v}{2L}$$

$$\text{force on wall} = \text{rate of change of momentum} = 2mv \times \frac{v}{2L} = \frac{mv^2}{L}$$

Since pressure is force per unit area, $P_{\text{piston}} = \frac{mv^2}{AL}$, where A is the surface area of the piston.

Because density is mass per unit volume, $P_{\text{piston}} = \rho v^2$, this allows us to scale up this derivation to containers with more than one molecule. So far, we have only considered the forces acting on the piston. To relate the overall pressure across the entire container to the average speed of the molecules, we must account for three dimensions and not just one:

$$P = \frac{1}{3}\rho v^2$$

How can gas particles with high kinetic energy be used to perform work? (A.3)

So the pressure is directly related to the square of the average speed of the particles and therefore the temperature of the gas.

The force exerted by the gas on the piston would cause the piston to move outward unless there was a force opposing it. In the lab, this force is normally provided by the air on the outside, which is also made of molecules in random motion as shown in Figure 11.

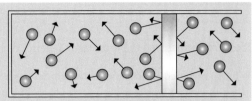

B.3 Figure 11 Piston pushed by trapped gas on one side and air on the other.

When you make changes to the state of a gas, all three quantities will change unless one is kept constant. This is a rather artificial condition but makes modeling the gas easier.

Relationships between *P*, *V*, *T* and *n*

When dealing with relationships, we generally are concerned with two quantities, e.g. distance and time, force and area, mass and volume. Here we have four variables, each depending on each other. To make life easier, we can keep two constant and look at the relationships between pressure and volume, pressure and temperature, and volume and temperature. This will give three different relationships that are known as the gas laws.

The Boyle–Mariotte law (constant temperature and number of moles)

The Boyle–Mariotte law states that the pressure of a fixed mass of gas at constant temperature is inversely proportional to its volume.

$$P \propto \frac{1}{V}$$

As the volume of a gas is reduced, the gas will become more dense, because the molecules are pushed together. The molecules will therefore hit the walls more often, increasing the rate of change of momentum (while the area of the walls decreases) and hence the pressure as shown in Figure 12.

Keeping the temperature constant is quite difficult because when you push in the piston, you do work on the gas, increasing the kinetic energy of the molecules and hence increasing the temperature. If the compression is slow, then the temperature will have time to return to the temperature of the surroundings.

B.3 Figure 12 Reducing the volume increases the pressure.

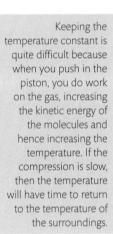

The easiest way to test the relationship between pressure and volume is to compress a gas in a syringe that is connected via some rubber tubing to a pressure sensor as in Figure 13. The range of pressure will be limited to how much force you can apply but should be enough to show the relationship.

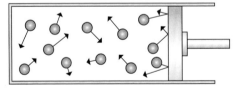

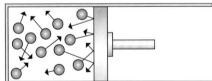

pressure sensor   gas syringe

B.3 Figure 13 Apparatus to measure *P* and *V*

A worksheet with full details of how to carry out this experiment is available in the eBook.

Boyle–Mariotte law apparatus.

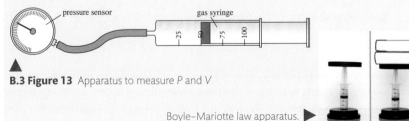

Graphical representation of the Boyle–Mariotte law

Since the pressure of a fixed mass of gas at constant temperature is inversely proportional to its volume, a graph of pressure against volume will be a curve as shown in Figure 14.

If the experiment was now repeated with the same amount of gas at different temperatures, the set of blue lines shown in Figure 15 would be achieved. Each line is called an **isotherm**. The effect of increasing the temperature of a fixed volume of gas is to increase the pressure so we can see that the curves further away from the origin are for higher temperatures. The orange lines on the graph represent the following gas transformations:

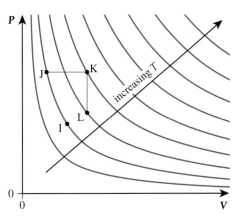

IJ   constant temperature (isothermal)

JK   constant pressure (isobaric)

KL   constant volume (isovolumetric).

We can describe each isotherm (at a constant number of moles) with the equation: $P_1V_1 = P_2V_2$. This is a simple rewriting of the inverse proportion relationship above, but can be a more direct way to solve problems with just these two variables.

Gay-Lussac's law (constant volume and number of moles)

Gay-Lussac's law states that the pressure of a fixed mass of gas with constant volume is directly proportional to its temperature in kelvin.

$$P \propto T$$

As the temperature of a gas is increased, the average kinetic energy of the molecules increases. The change in momentum as the molecules hit the walls is therefore greater and they hit the walls more often as shown in Figure 16. According to Newton's second law, the force exerted = rate of change of momentum, so the force on the walls increases and hence the pressure increases.

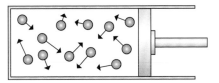 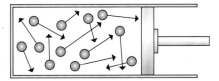

increased temperature   →   increased kinetic energy   →   increased pressure

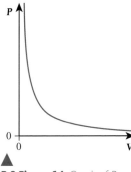

▲ **B.3 Figure 14** Graph of P vs V. A straight line through the origin would be seen if, instead, P was plotted against 1/V.

◄ **B.3 Figure 15** P vs V for different T.

When a gas is compressed, work is done on it. Work done = force × distance, which is the area under the P–V graph. This makes this graph particularly useful when investigating the energy changes that a gas undergoes when transformed.

◄ **B.3 Figure 16** Pressure increases with temperature.

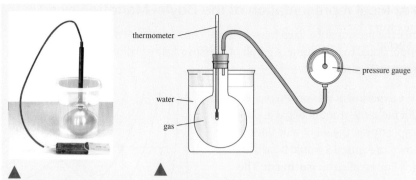

Gay-Lussac's law apparatus.   **B.3 Figure 17** Apparatus to measure P and T.

A pressure sensor can be used in an experiment to show the relationship between pressure and temperature. A flask of fixed volume is placed in a water bath as shown in Figure 17. A temperature sensor measures the temperature of the gas while a pressure sensor measures its pressure. The temperature of the gas is changed by heating the water, and the pressure and temperature are recorded simultaneously.

Graphical representation of Gay-Lussac's law

Since pressure is proportional to temperature, a graph of pressure vs temperature for a fixed mass of gas at constant volume will be a straight line as shown in Figure 18.

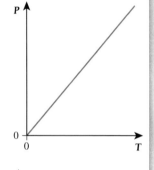

B.3 Figure 18 Graph of P vs T.

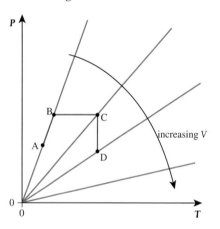

B.3 Figure 19 P vs T for different V.

If the experiment was repeated with different volumes of the same amount of gas, then the set of lines shown in Figure 19 would be achieved, each line representing a different volume. Increasing the volume at constant temperature (line CD) will make the pressure lower so the less steep lines are for larger volumes. The orange lines on the graph represent the following gas transformations:

AB   constant volume (isovolumetric)

BC   constant pressure (isobaric)

CD   constant temperature (isothermal).

We can describe each isovolume (at a constant number of moles) with the equation: $P_1T_2 = P_2T_1$. This is equivalent to the proportional relationship above.

Charles' law (constant pressure)

Charles' law states that the volume of a fixed mass of gas at a constant pressure is directly proportional to its temperature in kelvin.

$$V \propto T$$

As the temperature of a gas is increased, the molecules move faster, causing an increase in pressure. However, if the volume is increased in proportion to the increase in temperature, the pressure will remain the same. This is shown in Figure 20.

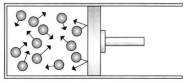 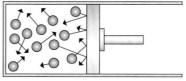 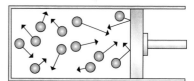

increased temperature → increased kinetic energy → increased pressure → increased volume reduces pressure to original

SKILLS

To test the relationship between volume and temperature, you need a narrow tube with the top end open and a small amount of liquid that traps a sample of dry air, as shown in Figure 21. Traditionally, concentrated sulfuric acid was used to trap the air because it absorbs water. However, this might be against the safety regulations in your country. If so, oil will do the job but might not give such good results. The temperature of the sample of air is changed by placing it in a water bath which is heated. The temperature of the gas is then assumed to be the same as the temperature of the water, which is measured using a thermometer. If we assume the tube has a uniform cross-section then, as the temperature is changed, the volume is measured by measuring the length of the cylinder of gas. At the start of the experiment, the pressure of the gas = the pressure of the surrounding air. As the temperature increases, the pressure also increases, pushing the bead up the tube increasing the volume, causing the pressure to reduce until it is again equal to the pressure of the surrounding air. We therefore assume the gas pressure is constant.

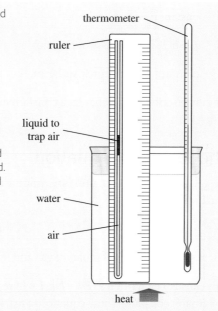

thermometer
ruler
liquid to trap air
water
air
heat

B.3 Figure 20 Constant pressure expansion.

The liquid will add a little bit to the pressure.

B.3 Figure 21 Apparatus to measure *T* and (indirectly) *V*.

Graphical representation of Charles' law

Since volume is proportional to temperature, a graph of volume vs temperature for a fixed mass of gas at constant pressure will be a straight line as shown in Figure 22.

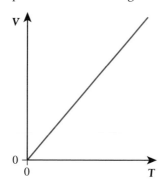

How does the concept of force and momentum link mechanics and thermodynamics? (A.2)

B.3 Figure 22 Graph of *V* vs *T*.

If the experiment was repeated with the gas at different constant pressures, then the set of lines shown in Figure 23 would be achieved, each line representing a different pressure. Reducing the volume at a constant temperature (line GH) will result in a higher pressure so the less steep lines have higher pressure. The orange lines on the graph represent the following gas transformations:

EF   constant pressure (isobaric)

FG   constant volume (isovolumetric)

GH   constant temperature (isothermal).

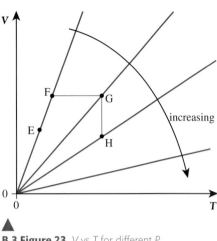

B.3 Figure 23 V vs T for different P.

We can describe each isobar (at a constant number of moles) with the equation: $V_1T_2 = V_2T_1$.

The ideal gas equation

The relationship between pressure, volume and temperature can be expressed in one equation:

$$PV = nRT$$

where n = the number of moles of gas and R = the molar gas constant ($8.31\,\mathrm{J\,mol^{-1}\,K^{-1}}$).

This can also be written as $PV = Nk_BT$ where N is the number of molecules and k_B is the Boltzmann constant. These equations are used to determine one of the four variables from the other three at a given instant.

When working with a changing container of gas ('before and after'), we can say that $P_1V_1T_2n_2 = P_2V_2T_1n_1$ (and simply cancel any quantities that remain constant).

Graphical representation of the ideal gas equation

This relationship can be represented on a graph with three axes as in Figure 24. The shaded area represents all the possible states of a fixed mass of gas. No matter what you do to the gas, its P, V, and T will always be on this surface for a given number of moles. This is quite difficult to draw so the 2-dimensional views shown before are used instead.

B.3 Figure 24 P, V and T in three dimensions.

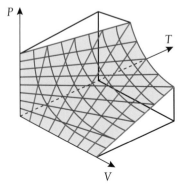

Worked example

The pressure of a gas inside a cylinder is 300 kPa. If the gas is compressed to half its original volume and the temperature rises from 27 °C to 327 °C, what will its new pressure be?

Solution

Using the ideal gas equation:

$$PV = nRT$$

Rearranging:

$$\frac{PV}{T} = \text{constant}$$

So:

$$\frac{PV}{T} \text{ at the beginning} = \frac{PV}{T} \text{ at the end}$$

$$\frac{PV}{T} \text{ at the beginning} = \frac{300\,000 \times V}{300}$$

$$\frac{PV}{T} \text{ at the end} = \frac{P \times \frac{V}{2}}{600}$$

Equating:

$$300\,000 \times \frac{V}{300} = \frac{P \times \frac{V}{2}}{600}$$

$$P = 300\,000 \times 600 \times \frac{2}{300}$$

$$P = 1200\,\text{kPa}$$

Temperatures must be changed to kelvin because we are working with absolute values rather than changes.

Exercise

Q3. The pressure of 10 m³ of gas in a sealed container at 300 K is 250 kPa. If the temperature of the gas is changed to 350 K, what will the pressure be?

Q4. A container of volume 2 m³ contains 5 moles of gas. If the temperature of the gas is 293 K:
   (a) what is the pressure exerted by the gas?
   (b) what is the new pressure if half the gas leaks out?

Q5. A piston contains 250 cm³ of gas at 300 K and a pressure of 150 kPa. The gas expands, causing the pressure to go down to 100 kPa and the temperature drops to 250 K. What is the new volume?

Q6. A sample of gas trapped in a piston is heated and compressed at the same time. This results in a doubling of temperature and a halving of the volume. If the initial pressure was 100 kPa, what will the final pressure be?

Challenge yourself

1. Two identical flasks, each full of air, are connected by a thin tube on a day when the temperature is 300 K and the atmospheric pressure 100 kPa. One of the flasks is then heated to 400 K while the other one is kept at 300 K. What is the new pressure in the flasks?

Internal energy of a gas (U)

We think of an ideal gas as being made of a large number of perfectly elastic spheres moving in random motion. When the molecules collide, momentum and energy are conserved, but between collisions, there is no force acting between them. This means that no work needs to be done to change the position of a molecule. The molecules therefore have no potential energy. The total kinetic energy of all the molecules is called the **internal energy** of the gas.

Challenge yourself

2. Try to derive these relationships between kinetic energy and temperature by combining the pressure–velocity relationship and ideal gas law described previously in this chapter.

We know that the average kinetic energy of the molecules of a gas is proportional to the temperature in kelvin.

$$E_{k,\, mean} \text{ of a molecule} = \frac{3}{2}kT$$

where k = Boltzmann's constant = $1.38 \times 10^{-23}\, \text{J K}^{-1}$.

A mole of gas contains N_A (Avogadro's constant) molecules so the $E_{k,\, total} = \frac{3}{2}N_A kT$. $N_A k$ is also a constant, the universal gas constant = $8.31\, \text{J mol}^{-1}\, \text{K}^{-1}$.

So, the total kinetic energy of one mole of gas = $\frac{3}{2}RT$.

Since the internal energy of a gas is the total kinetic energy, then for n moles we can say:

$$\text{internal energy, } U = \frac{3}{2}nRT$$

When heat is transferred to a *fixed volume* of gas, it will increase the internal energy and hence the temperature of the gas. If no heat is lost, we can say that:

$$Q = \Delta U$$

i $U = \frac{3}{2}nRT$ is only true for monatomic gases such as helium, neon and argon.

Worked example

500 J of thermal energy are transferred to 2 g of helium gas kept at constant volume in a cylinder. Calculate the temperature rise of the gas.

Solution

The molar mass of helium is 4 g so $n = 0.5$.

$$\text{increase in internal energy} = \frac{3}{2}nR\Delta T = \text{thermal energy added} = 500\, \text{J}$$
$$\Delta T = \frac{2 \times 500}{3 \times 0.5 \times 8.31} = 80\, \text{K}$$

Exercise

Q7. Calculate the internal energy of 100 g of argon (nucleon number 40) at 300 K.

Q8. Calculate the average kinetic energy of atoms of helium at 400 K.

Real gases

The assumptions we made when developing the model for an ideal gas do not fully apply to real gases except when the pressure is low and the temperature high. At high pressures and densities, the molecules can be close together so the assumption that the volume of the molecules is negligible does not apply, nor does the assumption about there being no forces between the molecules. What can also happen at low temperatures is the gas can change into a liquid which, for obvious reasons, does not behave like a gas. However, although no gas behaves exactly as an *ideal* gas, air at normal room temperature and pressure comes pretty close, as experiments show.

What other simplified models are relied upon to communicate the understanding of complex phenomena? (NOS)

Nitrogen becomes a liquid at low temperatures.

Guiding Questions revisited

How are the macroscopic characteristics of a gas related to the behavior of individual molecules?

What assumptions and observations lead to universal gas laws?

How can models be used to help explain observed phenomena?

In this chapter, we have combined the macroscopic properties of a gas in a container with a microscopic understanding of kinetic theory to explain that:

- Pressure (the force per unit area at a surface) is increased when the average speed of particles increases and/or the total container surface area decreases.
- Real gases at low pressures and densities and high temperatures can be modeled as ideal gases, for which certain assumptions in kinetic theory apply.
- The Boyle–Mariotte law states that pressure and volume are inversely proportional for a constant temperature and number of moles of gas.

- Gay-Lussac's law states that pressure and absolute temperature are proportional for a constant volume and number of moles of gas.
- Charles' law states that volume and absolute temperature are proportional for a constant pressure and number of moles of gas.
- The ideal gas law is comprised of three empirical relationships and relates pressure, volume, absolute temperature and number of moles.
- The internal energy of an ideal gas is related to the kinetic energy, but not to the potential energy, of particles in a gas and therefore to the temperature of the gas.

Practice questions

1. A hot air balloon works because the density of hot air is less than the surrounding cold air.

   (a) The volume of a hot air balloon is 2500 m³ and its mass is 250 kg (without passengers). The temperature of the air in the balloon is 100 °C.

   Density of air at 20 °C = 1.21 kg m⁻³

   Average relative molecular mass of air = 29 g mol⁻¹

   (i) Calculate the volume of 1 kg of at 20 °C. (2)

   (ii) Calculate the volume of 1 kg of air at 100 °C and hence show that the density of air at 100 °C is 0.95 kg m⁻³. (3)

   (iii) Calculate the mass of hot air in the balloon. (1)

   (iv) Calculate the mass of cold air displaced by the balloon. (1)

   (v) Determine whether the balloon will leave the ground or not. (2)

   (b) (i) Calculate the average velocity of air molecules in the balloon (assume the molecules behave as spheres). (2)

   (ii) Calculate the number of moles in the balloon. (2)

   (iii) Calculate the internal energy contained by the hot gas. (2)

   (Total 15 marks)

2. (a) The atoms or molecules of an ideal gas are assumed to be identical hard elastic spheres that have negligible volume compared with the volume of the containing vessel.

   (i) State **two** further assumptions of the kinetic theory of an ideal gas. (2)

   (ii) Suggest why only the average kinetic energy of the molecules of an ideal gas is related to the internal energy of the gas. (3)

cylinder
ideal gas
piston

   (b) An ideal gas is contained in a cylinder by means of a frictionless piston.

   At temperature 290 K and pressure 4.8 × 10⁵ Pa, the gas has volume 9.2 × 10⁻⁴ m³.

   (i) Calculate the number of moles of the gas. (2)

(ii) The gas is compressed isothermally to a volume of $2.3 \times 10^{-4}\,\text{m}^3$. Determine the pressure P of the gas. (2)

(iii) The gas is now heated at constant volume to a temperature of 420 K. Show that the pressure of the gas is now $2.8 \times 10^6\,\text{Pa}$. (1)

(c) Sketch a pressure–volume (P–V) diagram for the changes in (b)(ii) and (b)(iii). (3)

(Total 13 marks)

3. The equipment shown in the diagram was used by a student to investigate the variation with volume, of the pressure p of air, at constant temperature. The air was trapped in a tube of constant cross-sectional area above a column of oil. The pump forces oil to move up the tube, decreasing the volume of the trapped air.

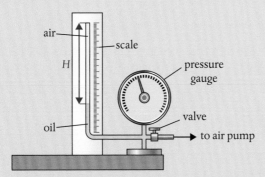

(a) The student measured the height H of the air column and the corresponding air pressure p. After each reduction in the volume, the student waited for some time before measuring the pressure. Outline why this was necessary. (1)

(b) The following graph of p versus $\frac{1}{H}$ was obtained. Error bars were negligibly small. Error bars were negligibly small. The equation of the line of best fit is $p = a + \frac{b}{H}$. Determine the value of b including an appropriate unit. (3)

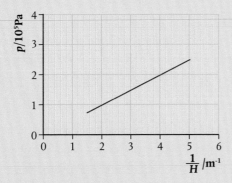

(c) Outline how the results of this experiment are consistent with the ideal gas law at constant temperature. (2)

(d) The cross-sectional area of the tube is $1.3 \times 10^{-3} \, m^2$ and the temperature of air is 300 K. Estimate the number of moles of air in the tube. (2)

(e) The equation in (b) may be used to predict the pressure of the air at extremely large values of $\frac{1}{H}$. Suggest why this will be an unreliable estimate of the pressure. (2)

(Total 10 marks)

4. Which aspect of thermal physics is best explained by the molecular kinetic model?

   A   The equation of state of ideal gases

   B   The difference between Celsius and kelvin temperature

   C   The value of the Avogadro constant

   D   The existence of gaseous isotopes   *(Total 1 mark)*

5. A quantity of 2.00 mol of an ideal gas is maintained at a temperature of 127 °C in a container of volume $0.083 \, m^3$. What is the pressure of the gas?

   A   8 kPa   B   25 kPa   C   40 kPa   D   80 kPa

   (Total 1 mark)

6. Two ideal gases, X and Y, are at the same temperature. The mass of a particle of gas X is larger than the mass of a particle of gas Y. Which is correct about the average kinetic energy and the average speed of the particles in gases X and Y?

	Average kinetic energy	Average speed
A	larger for Y	larger for Y
B	same	larger for Y
C	same	same
D	larger for Y	same

(Total 1 mark)

7. A substance in the gas state has a density about 1000 times less than when it is in the liquid state. The diameter of a molecule is d. Which is the best estimate of the average distance between molecules in the gas state?

   A d B $10d$ C $100d$ D $1000d$

   (Total 1 mark)

8. An ideal gas and a solid of the same substance are at the same temperature. The average kinetic energy of the gas molecules is E_g and the average kinetic energy of the solid molecules is E_s. What is the comparison between E_g and E_s?

   A E_g is less than E_s

   B E_g equals E_s

   C E_g is greater than E_s

   D   The relationship between E_g and E_s cannot be determined   *(Total 1 mark)*

9. Two flasks, P and Q, contain an ideal gas and are connected with a tube of negligible volume compared to that of the flasks. The volume of P is twice the volume of Q. P is held at a temperature of 200 K and Q is held at a temperature of 400 K. What is ratio of $\frac{\text{mass of gas in P}}{\text{mass of gas in Q}}$?

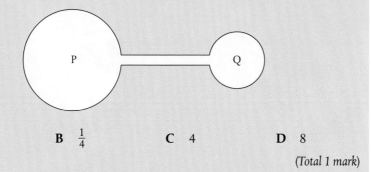

A $\frac{1}{8}$ **B** $\frac{1}{4}$ **C** 4   **D** 8

(Total 1 mark)

10. Q and R are two rigid containers of volumes 3V and V respectively, containing molecules of the same ideal gas initially at the same temperature. The gas pressures in Q and R are p and $3p$ respectively. The containers are connected through a valve of negligible volume that is initially closed. The valve is opened in such a way that the temperature of the gases does not change. What is the change of pressure in Q?

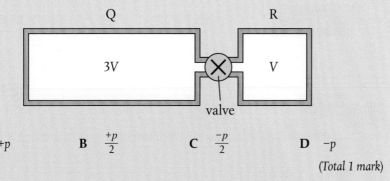

A $+p$ **B** $\frac{+p}{2}$ **C** $\frac{-p}{2}$ **D** $-p$

(Total 1 mark)

11. **(a)** A closed box of fixed volume $0.15\,\text{m}^3$ contains $3.0\,\text{mol}$ of an ideal monatomic gas. The temperature of the gas is 290 K. Calculate the pressure of the gas. (1)

(b) When the gas is supplied with $0.86\,\text{kJ}$ of energy, its temperature increases by 23 K. The specific heat capacity of the gas is $3.1\,\text{kJ}\,\text{kg}^{-1}\,\text{K}^{-1}$. Calculate, in kg, the mass of the gas. (1)

(c) Calculate the average kinetic energy of the particles of the gas. (1)

(d) Explain, with reference to the kinetic model of an ideal gas, how an increase in temperature of the gas leads to an increase in pressure. (3)

(Total 6 marks)

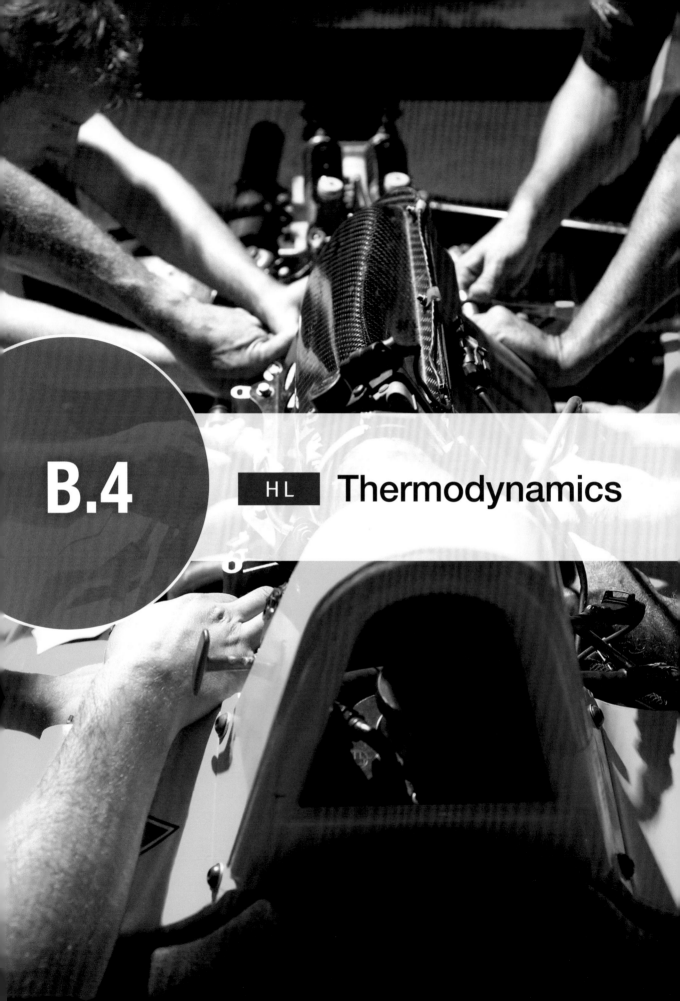

B.4

HL Thermodynamics

Formula 1 is the highest class of international car racing. The cars entered are subject to rules, including those that govern the type and size of engine. Engines (in any context) are designed to transfer energy and work so that a body moves.

Guiding Questions

How can energy transfers and energy storage within a system be analyzed?

How can the future evolution of a system be determined?

In what way is entropy fundamental to the evolution of the Universe?

A helium balloon can be used to power a toy car. As the balloon rises (due to its low density), its string could turn an axle, moving the car forward.

The problem is that when the string becomes unwound, the balloon would have to be pulled down again, and you would use more energy doing this than the work that was done on the car.

An alternative is to use a hot air balloon. A burner fills the balloon with hot air and it rises. When the string runs out, the burner is turned off and the air cools, allowing the balloon to come down. The chemical energy in the fuel has been transferred to work.

This is the principle of all heat engines. They do work when hot and are reset when cold. There is no way around this; thermal energy must be lost for the engine to work.

B.4 Figure 1 Underneath the balloon is a flame that heats the air in the balloon and this drives the engine.

Students should understand:

the first law of thermodynamics as given by $Q = \Delta U + W$ results from the application of conservation of energy to a closed system and relates the internal energy of a system to the transfer of energy as heat and as work
the work done by or on a closed system as given by $W = P\Delta V$ when its boundaries are changed can be described in terms of pressure and changes of volume of the system
the change in internal energy as given by $\Delta U = \frac{3}{2}Nk_B\Delta T = \frac{3}{2}nR\Delta T$ of a system is related to the change of its temperature
entropy S is a thermodynamic quantity that relates to the degree of disorder of the particles in a system
entropy can be determined in terms of macroscopic quantities, such as thermal energy and temperature as given by $\Delta S = \frac{\Delta Q}{T}$, and also in terms of the properties of individual particles of the system as given by $S = k_B \ln \Omega$ where k_B is the Boltzmann constant and Ω is the number of possible microstates of the system
the second law of thermodynamics refers to the change in entropy of an isolated system and sets constraints on possible physical processes and on the overall evolution of the system
processes in real isolated systems are almost always irreversible and consequently the entropy of a real isolated system always increases
the entropy of a non-isolated system can decrease locally, but this is compensated for by an equal or greater increase of the entropy of the surroundings
isovolumetric, isobaric, isothermal and adiabatic processes are obtained by keeping one variable fixed
adiabatic processes in monatomic ideal gases can be modeled by the equation as given by $PV^{\frac{5}{3}} = $ constant

cyclic gas processes are used to run heat engines
a heat engine can respond to different cycles and is characterized by its efficiency as given by $\eta = \dfrac{\text{useful work}}{\text{input energy}}$
the Carnot cycle sets a limit for the efficiency of a heat engine at the temperatures of its heat reservoirs as given by $\eta_{\text{Carnot}} = 1 - \dfrac{T_c}{T_h}$.

Nature of Science

To study the microscopic motion of all the atoms involved even in a simple system such as gas in a cylinder is very complex. However, we do not need to understand the motion of every particle to be able predict the behavior of the system.

Nature of Science

A lot of the original work on thermodynamics was done to improve the efficiency of engines, which would in turn increase the profits of factory and mine owners.

When Carnot developed his idea of the most efficient heat engine, he imagined heat to be a flowing fluid, which, like water, could be used to turn a wheel. This is no longer thought to be the case. However, his predictions still turn out to be true.

What paradigm shifts enabling change to human society, such as harnessing the power of steam, can be attributed to advancements in physics understanding? (NOS)

When a gas does work, it is pushing the piston out; work is positive. If work is done on the gas, then something must be pushing the piston in. Work is taken to be negative.

Thermodynamic systems

When work is done, energy is transferred, so to do work requires a source of energy. When fuel is burned, chemical energy is transferred to thermal energy but to do work we need an engine. In this section, we will consider a simple engine: an ideal gas trapped in a cylinder by a piston. Before we can understand the principle of its operation, we need to investigate the relationship between the system and energy.

In the previous chapter, we saw how internal energy, $U = \dfrac{3}{2}nRT$.

So if the temperature changes by ΔT, the internal energy change, $\Delta U = \dfrac{3}{2}nR\Delta T$

Work done by a gas

Work is done when the point of application of a force moves in the direction of the force. If the pressure of a gas pushes a piston out, then the force exerted on the piston is moving in the direction of the force, so work is done. The example in Figure 2 is of a gas expanding at constant pressure. In this case, the force exerted on the piston = $P \times A$. The work done when the piston moved distance Δd is therefore given by:

$$\text{work done} = P \times A \times \Delta d$$

B.4 Figure 2 A gas expands at constant pressure.

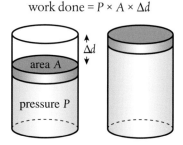

But $A\Delta d$ is the change in volume ΔV, so:

$$\text{work done} = P\Delta V$$

Figure 3 is the P–V graph for this constant pressure expansion. From this, we can see that the work done is given by the area under the graph. This is true for all processes.

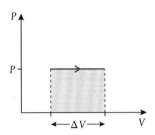

◀ **B.4 Figure 3** A gas expands at constant pressure.

The first law of thermodynamics

According to the law of conservation of energy, energy can neither be created nor destroyed, so the amount of thermal energy (or 'heat'), Q, added to a gas must equal the work done by the gas, W, plus the increase in internal energy, ΔU. This is so fundamental to the way physical systems behave that it is called the **first law of thermodynamics**. This can be written in the following way:

$$Q = \Delta U + W$$

This would be nice and easy if the only things a gas could do were gain thermal energy, get hot, and do work. However, thermal energy can be added and lost, internal energy can increase and decrease and work can be done by the gas and on the gas. To help us understand all the different possibilities, we will use the P–V diagram to represent the states of a gas.

The first law (simple version) states that: If a gas expands and gets hot, heat must have been added.

Using *P–V* diagrams in thermodynamics

We have seen how a P–V diagram enables us to see the changes in P, V, and T that take place when a gas changes from one state to another. It also tells us what energy changes are taking place. If we consider the transformation represented in Figure 4, we can deduce that when the gas changes from A to B:

1 Since the volume is increasing, the gas is doing work (W is positive).

2 Since the temperature is increasing, the internal energy is increasing (ΔU is positive).

We can imagine that the *P–V* diagram is covered in a set of curves (isotherms) representing the gas at different temperatures. A transformation from A to B will imply a rise in temperature.

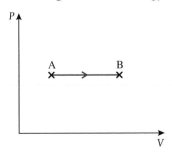

◀ **B.4 Figure 4** An isobaric transformation.

If we then apply the first law $Q = \Delta U + W$, we can conclude that if both ΔU and W are positive, then Q must also be positive, so heat must have been added.

This is a typical example of how we use the *P–V* diagram with the first law. We use the diagram to find out how the temperature changes and whether work is done *by* the gas or *on* the gas, and then use the first law to deduce whether heat is added or lost.

Constant pressure compression (isobaric)

The previous example was an expansion at constant pressure. Now we will consider the constant pressure (isobaric) compression shown in Figure 5.

B.4 Figure 5 An isobaric transformation.

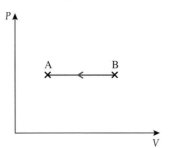

1 Temperature decrease implies that the internal energy decreases (ΔU = negative).

2 Volume decrease implies that work is done on the gas (W = negative).

Applying the first law, $Q = \Delta U + W$, tells us that Q is also negative, so heat is lost.

Constant volume increase in temperature (isovolumetric)

Figure 6 is the *P–V* graph for a gas undergoing a constant volume transformation. From the graph, we can deduce that:

B.4 Figure 6 An isovolumetric transformation.

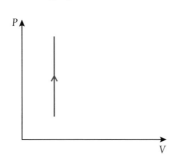

1 The volume is not changing, so no work is done (W = 0).

2 The gas changes to a higher isotherm so the temperature is increasing. This means that the internal energy is increasing (ΔU = positive).

Applying the first law $Q = \Delta U + W$, we can conclude that $Q = \Delta U$, so if ΔU is positive, then Q is also positive – heat has been added.

Isothermal expansion

B.4 Figure 7 An isothermal expansion.

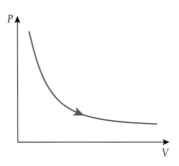

For an ideal gas, $PV = nRT$, so if the temperature is constant, PV = constant, which implies that $P = \frac{\text{constant}}{V}$ so the *P–V* graph follows the curve shown in Figure 7 ($y = \frac{k}{x}$).

From this *P–V* diagram, we can deduce that:

1 The temperature does not change so there is no change in internal energy (ΔU = 0).

2 The volume increases so work is done by the gas (W = positive).

Applying the first law, $Q = \Delta U + W$, we conclude that $Q = W$ so heat must have been added. The heat added enables the gas to do work.

Adiabatic expansion

An adiabatic process is when there is no exchange of thermal energy between the system and the surroundings. To understand how this will be on a P–V graph, let us compare an adiabatic expansion with an isothermal expansion between the same two volumes.

During an isothermal expansion, work is done by the gas and the internal energy stays constant so heat must have been added. To do the same amount of work without adding heat, the internal energy must decrease, resulting in a reduction in temperature leading to the curve in Figure 8.

It can be shown that for an adiabatic transformation where the gas is monatomic and ideal, $PV^{\frac{5}{3}}$ = constant so the shape of this curve is of the form $y = \frac{1}{x^{\frac{5}{3}}}$. Notice that the gradient is always larger than the adjacent isotherms.

From the P–V diagram, we can deduce that:

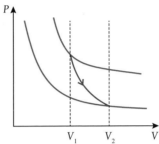

1   The volume is increased so work is done by the gas (W = positive).

2   The temperature decreases so the internal energy is reduced (ΔU = negative).

◀ **B.4 Figure 8** An adiabatic transformation (shown in red with two isotherms in blue).

We also know that $Q = 0$, so if we apply the first law, $Q = \Delta U + W$, we get:

$$0 = -\Delta U + W$$

$$W = \Delta U$$

So the energy required for the gas to do work comes from its internal energy.

Worked examples

In the following worked examples, we will consider a cylinder containing 1.203×10^{-3} moles of a monatomic gas. This makes $nR = 0.01$ J K^{-1} so $\frac{PV}{T} = 10$ kPa cm^3 K^{-1}.

1.   Isobaric expansion

The gas is kept at a constant pressure of 100 kPa as it expands from 100 cm^3 to 150 cm^3. As this happens, the temperature rises from 1000 K to 1500 K. The process can be represented by the P–V graph shown in the figure on the right.

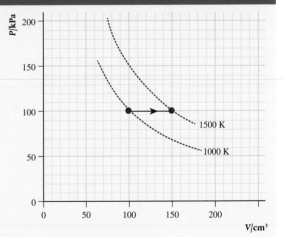

When a gas expands at constant pressure, work is done by the gas and it gets hot. First let us work out the work done:
$W = P\Delta V = 100 \times 10^3 \times 50 \times 10^{-6} = 5$ J
increase in internal energy $= \frac{3}{2}nR\Delta T = 1.5 \times 0.01 \times 500 = 7.5$ J
heat added = increase in internal energy + work done ($Q = \Delta U + W$) = 12.5 J

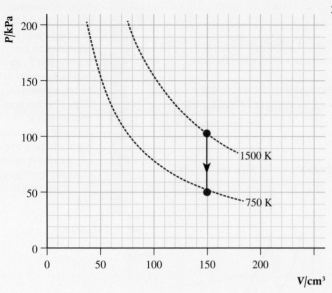

2. Isovolumetric fall in temperature

The gas is now cooled at constant volume until its temperature reaches 750 K. The pressure of a gas at constant volume is proportional to the temperature so the pressure will fall to 50 kPa as shown in the P–V graph on the left.

The volume is constant so no work is done on or by the gas. The reduction in internal energy is therefore equal to the loss of heat ($Q = \Delta U$).

$$\Delta U = \tfrac{3}{2}nR\Delta T = 1.5 \times 0.01 \times 750 = 11.25 \text{ J}$$

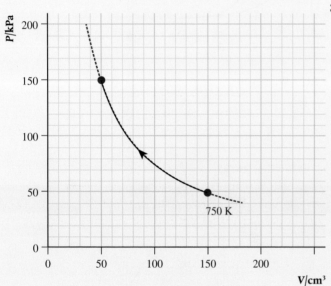

3. Isothermal compression

The gas is compressed from 150 cm^3 to 50 cm^3 at a constant temperature of 750 K. The volume is reduced by $\frac{1}{3}$ so the pressure must be three times the original pressure shown in the P–V graph on the left.

The volume of the gas is reduced, which means work is done on the gas. However, the temperature of the gas does not increase so the work done must equal the loss of heat ($Q = W$).

The work done on the gas = the area under the curve. We can find this by counting the squares. There are approximately 82 squares: each square represents 0.1 J
work done on the gas = 8.2 J

4. Adiabatic expansion

After the previous expansion, the gas at pressure 150 kPa and volume 50 cm³ expands adiabatically until the volume is 100 cm³. For an adiabatic process, $PV^{\frac{5}{3}} = $ constant so we can calculate the final pressure.

$$P_1V_1^{\frac{5}{3}} = P_2V_2^{\frac{5}{3}}$$

$$P_2 = P_1 \left(\frac{V_1}{V_2}\right)^{\frac{5}{3}}$$

$$P_2 = 150 \times \left(\frac{50}{100}\right)^{\frac{5}{3}} = 47\,\text{kPa}$$

The final temperature can be found from $PV = nRT$:

$$T = \frac{PV}{nR} = \frac{47 \times 10^3 \times 100 \times 10^{-6}}{0.01} = 470\,\text{K}$$

This process is represented by the P–V graph on the right.

The work done on the gas can be found by counting the squares under the line.

There are approximately 42 squares each representing 0.1 J so the work done = 4.2 J.

This should be the same as the loss of internal energy of the gas, which can be found from $\frac{3}{2}nR\Delta T$:

$$\Delta U = 1.5 \times 0.01 \times (750 - 470) = 4.2\,\text{J}$$

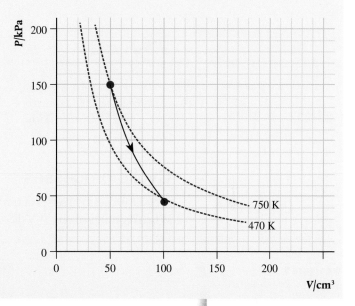

Exercise

In the following questions, we will consider a cylinder containing 1.203×10^{-3} moles of a monatomic gas, which will make $nR = 0.01\,\text{JK}^{-1}$. This means you can use a simulation to confirm your answers but do not forget nR is *not* always 0.01.

Q1. The gas is compressed from 100 cm³ to 50 cm³ at a constant pressure of 70 kPa. Calculate:

(a) the initial temperature
(b) the final temperature
(c) the change in internal energy
(d) the work done on the gas
(e) the heat lost to the surroundings.

Q2. The gas is compressed isothermally as illustrated by the P–V graph in Figure 13. Calculate:

(a) the temperature of the gas
(b) the work done on the gas
(c) the heat lost to the surroundings.

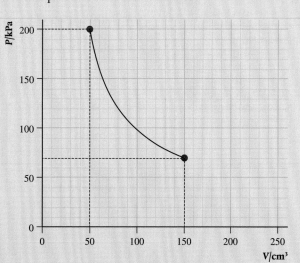

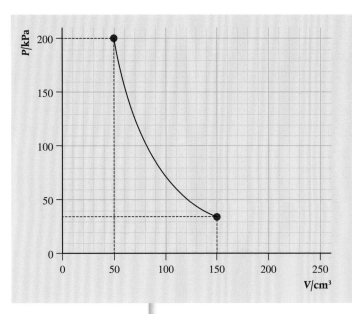

Q3. The gas expands adiabatically from $52\,cm^3$ at a pressure of $200\,kPa$ to a volume of $150\,cm^3$.

(a) Show by calculation that the new pressure is in agreement with the P–V graph of Figure 14.

Calculate:

(b) the work done by the gas

(c) the initial temperature

(d) the final temperature

(e) the change of internal energy

(f) heat lost/heat gained.

Cyclic processes

A cyclic process is a series of transformations that take a gas back to its original state. When represented on a P–V diagram, they form a closed loop such as the one shown in Figure 9.

B.4 Figure 9 A thermodynamic cycle.

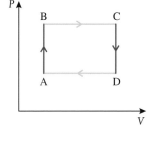

In this example, the cycle is clockwise so the sequence of transformations is:

A–B isovolumetric temperature rise

B–C isobaric expansion

C–D isovolumetric temperature drop

D–A isobaric compression.

The net work done during a cycle is the difference between the work done by the gas and the work done on the gas. This is equal to the area enclosed by the cycle on the P–V diagram.

In the process of completing this cycle, work is done on the gas from D to A and the gas does work from B to C. It is clear from the diagram that the work done by the gas is greater than the work done on the gas (since the area under the graph is greater from B to C than from D to A) so net work is done. What we have here is an engine; heat is added and work is done. Let us look at this cycle more closely.

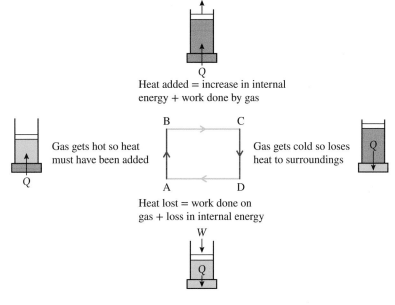

Heat added = increase in internal
energy + work done by gas

Gas gets hot so heat
must have been added

Gas gets cold so loses
heat to surroundings

Heat lost = work done on
gas + loss in internal energy

B.4 Figure 10 An example
of a thermodynamic cycle.
The red and blue rectangles
placed under the piston
represent hot and cold
bodies used to add and take
away heat.

The secret to the operation of all heat engines is that the gas is cooled down before it is compressed back to its original volume. The cold gas is easier to compress than a hot one, so when the gas is hot, it does work, but it is reset when it is cold.

The balloon engine from the chapter introduction operates on the same principle; when the gas is hot the balloon goes up, doing work. The balloon is then allowed to cool so that pulling it down does not use as much energy as was gained when it went up.

Energy flow diagram

The principle of a heat engine can be represented by an energy flow diagram as in Figure 11. Heat flows from a hot source to a cold one through the engine, which converts some of it into work.

The **thermal efficiency** η of an engine is defined as the ratio of the work it does to the amount of heat energy put in:

$$\eta = \frac{W}{Q_h}$$

But $W = Q_h - Q_c$ so:

$$\eta = \frac{Q_h - Q_c}{Q_h} = 1 - \frac{Q_c}{Q_h}$$

The Carnot cycle

When heat is added in the previous example, the source of heat is much hotter than the gas. A more efficient process would be to transfer heat from a source that is similar in temperature to the gas, although the proof of this is outside the scope of the course. The most efficient cycle possible is the Carnot cycle as represented in Figure 12. As Figure 12 shows, the Carnot cycle consists of two isothermal transformations when heat is transferred at the same temperature as the surroundings (thereby making these transformations reversible and efficient), and two adiabatic processes when the volume is

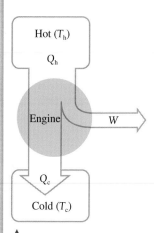

B.4 Figure 11 Energy flow for a heat engine.

How are efficiency considerations important in motors and generators? (A.3, D.3, D.4)

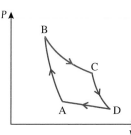

B.4 Figure 12 The Carnot cycle.

If a gas can be returned to its original state without any change to the environment then the process is said to be *reversible* (e.g. an adiabatic process where no heat is transferred).

changed, resulting in a change in temperature without exchanging heat to the surroundings (which, in turn, also maximizes efficiency). This is an idealized process that would have to take place very slowly but sets the limit on what is possible.

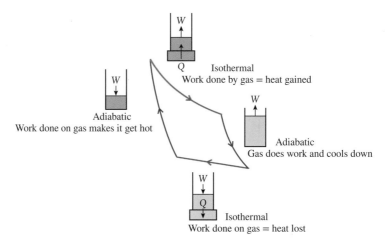

B.4 Figure 13 The Carnot cycle in detail. Notice that during the adiabatic transformations, the cylinder is isolated from its surroundings.

The amount of heat transferred in and out of the gas during the isothermal processes is directly proportional to the temperature in kelvin, so:

$$Q_h \propto T_h \text{ and } Q_c \propto T_c$$

The efficiency of a Carnot cycle is therefore:

$$\eta = 1 - \frac{Q_c}{Q_h} = 1 - \frac{T_c}{T_h}$$

No engine can have a higher efficiency than this.

Why is there an upper limit on the efficiency of any energy source or engine? (A.3)

We can see that the efficiency depends on the difference between the temperatures of the hot and cold parts of the cycle. If the cold part was absolute zero (0 K), then no work would have to be done to push back the piston and the efficiency would be 1.

The reverse cycle

Let us consider what would happen if the Carnot cycle was operated in reverse. The details of this are shown in Figure 14.

B.4 Figure 14 The reverse Carnot cycle.

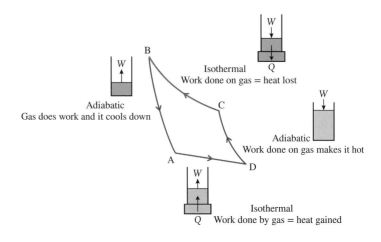

The interesting thing about this cycle is that heat is lost to the hot body during the isothermal compression (C to B) and gained from the cold body during the isothermal expansion (A to D). So heat has been taken from something cold and given to something hot. This is what a refrigerator does – it takes thermal energy from the cold food inside and gives it to the warm room. To make this possible, work must be done on the gas (D to C) so that it gets hot enough to give heat to the hot body.

In cold countries, a heat pump is used to extract thermal energy from the cold air outside and give it to the inside of a house. It works in the same way as a refrigerator.

In warm countries, the same principle is used to take heat from the cool inside to the warmer outside.

Exercise

Q4. 250 cm³ of gas at 300 K exerts a pressure of 100 kPa on its container (state A). It undergoes the following cycle of transformations:

1. an isobaric expansion to 500 cm³ (state B)
2. an isovolumetric transformation to a pressure of 200 kPa (state C)
3. an isobaric contraction back to 250 cm³ (state D)
4. an isovolumetric transformation back to state A.

(a) Sketch a P–V diagram representing this cycle, labeling the states A, B, C, and D.

(b) Use the ideal gas equation to calculate the temperature at B, C, and D.

(c) Calculate the amount of work done by the gas.

(d) Calculate the amount of work done on the gas.

(e) What is the net work done during one cycle?

Q5. This figure on the right represents a Carnot cycle. The areas of the colored regions are as follows:

A – 50 J

B – 45 J

C – 40 J

D – 35 J

E – 150 J

If the cycle is performed clockwise, how much work is done:

(a) during the isothermal expansion?

(b) during the adiabatic compression?

(c) by the gas?

(d) on the gas?

(e) in total?

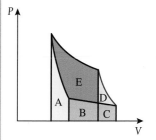

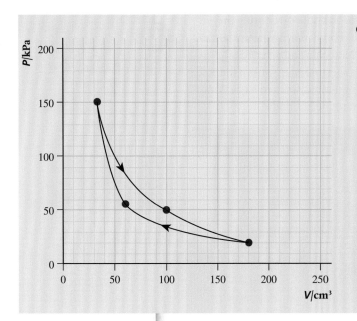

Q6. The graph in Figure 22 shows a Carnot cycle performed with 1.203×10^{-3} moles of a monatomic gas ($nR = 0.01\ \mathrm{J\,K^{-1}}$).

Use information from the graph to calculate:

(a) the temperature of the gas when expanding isothermally (T_h)

(b) the temperature of the gas when being compressed isothermally (T_c)

(c) the thermal efficiency of the engine

(d) the net work done (W)

(e) the amount of heat added during the isothermal expansion (Q_h).

(f) Use the last two answers to get the value for the efficiency.

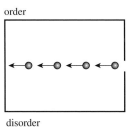

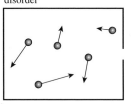

▲ **B.4 Figure 15** From an orderly entrance to disordered random motion.

▲ A Stirling engine will work if you put it on top of a mug of hot coffee or on a bowl of ice. Provided there is a temperature difference, work can be done.

The second law of thermodynamics

We have seen that we can use our simple thermodynamic system of a cylinder of gas to convert heat into work, but to do this, we must transfer heat from a hot body to a cold one. This means that we will always lose some heat. It would be even better if we could take a source of heat and transfer all the energy to work without losing any to a cold body. According to the first law of thermodynamics, this should be possible since energy would be conserved. However, it cannot be done. The reason for this is fundamental to the way matter behaves.

To understand why, let us first consider a seemingly unconnected example in which a gas is pumped into a container as in Figure 15. The molecules of gas flow into the container in an orderly fashion through a small opening, all traveling in the same direction with the same speed (not really possible but this is a thought experiment). The molecules travel across the container and hit the other side, at which point things start to get messy. The molecules hit each other. They no longer have the same energy and direction but move about in random motion, some moving fast and others moving slowly, just like the way we know the molecules of gas behave.

The different kinetic energies of each molecule are called microstates. Ω is the number of possible microstates. The entropy, S, of a system is defined as $k_B \ln \Omega$. More microstates implies higher entropy, so we can see that, in this process, entropy increases. No matter how long we wait, the molecules will never line up with the same speed again (at least, it is extremely unlikely) even though, according to the law of conservation of energy, this would be perfectly conceivable.

Just as physical systems tend to a position of lowest potential energy, they also tend to a state where the energy is most disordered. When a metal block is held above the ground, each molecule has approximately the same amount of potential energy, but when it is dropped onto the floor, that energy goes to increase the kinetic energy and potential energy of the molecules of the block and ground. Those molecules interact with neighboring molecules as the energy is spread out. This energy will never collect together again to allow the block to return to its original position.

So an engine that, for example, took heat energy from the random motion of molecules in the air and transferred it all to work to lift a mass from the ground would be creating an ordered form of energy (the potential energy of the mass) out of the random spread of energy in the air, and that is not possible. However, if some of the energy was put into a cold body, the net effect could be a more disordered form of energy overall, so this could be possible.

The second law of thermodynamics states this in a concise way:

It is not possible for a heat engine working in a cycle to absorb thermal energy and transfer it all to work.

This is the Kelvin–Planck statement of the law.

Entropy

The second law of thermodynamics is about the spreading out of energy. This can be quantified by using the quantity **entropy**.

The change of entropy is ΔS, when a quantity of heat flows into a body at temperature T is equal to $\frac{Q}{T}$:

$$\Delta S = \frac{Q}{T}$$

The unit of entropy is $J\,K^{-1}$.

For example, consider the situation of a 1 kg block of ice melting in a room that is at a constant temperature 300 K. To melt the block of ice, it must gain 3.35×10^5 J of energy. Ice melts at a constant 273 K so:

gain in entropy of the ice $= \dfrac{3.35 \times 10^5}{273} = 1.23 \times 10^3\,J\,K^{-1}$

loss of entropy by the room $= \dfrac{3.35 \times 10^5}{300} = 1.12 \times 10^3\,J\,K^{-1}$

We can see from this that the entropy has increased.

Entropy always increases in any transfer of thermal energy since heat always flows from hot bodies to cold bodies. We can therefore rewrite the second law in terms of entropy:

In any cyclic process, the entropy will either stay the same or increase.

Entropy is a measure of how spread out or disordered the energy has become. Saying entropy has increased implies that the energy has become more spread out, which is always the case during irreversible changes for real, isolated systems.

This statement also implies that heat cannot spontaneously flow from a cold object to a hot object. We have seen that this is possible by reversing the cycle of a heat engine but then work must be done. A third way of stating the second law is therefore:

It is not possible for heat to be transferred from a cold body to a warmer one without work being done.

This is the Clausius statement of the law.

What are the consequences of the second law of thermodynamics to the Universe as a whole? (E.5)

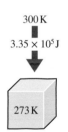

300 K

3.35×10^5 J

273 K

▲ **B.4 Figure 16** Heat moves from the room to the ice.

There can be no change in entropy in a reversible process.

Although a reduction in entropy is conceivable at a local level for a non-isolated system, the entropy increase in the surroundings will more than compensate, and entropy will increase overall.

Exercise

Q7. 500 J of heat flows from a hot body at 400 K to a colder one at 250 K.

(a) Calculate the entropy change in:

(i) the hot body

(ii) the cold body.

(b) What is the total change in entropy?

Q8. Use the second law to explain why heat is released when an electric motor is used to lift a heavy load.

Guiding Questions revisited

How can energy transfers and energy storage within a system be analyzed?

How can the future evolution of a system be determined?

In what way is entropy fundamental to the evolution of the Universe?

In this chapter, we have reconsidered the implications of the conservation of energy and introduced the concept of entropy to discuss how:

- The thermal energy added to a container of gas is equal to the sum of the increase in internal energy and the work done by the gas on the surroundings.
- Adiabatic transformations involve no change in thermal energy, isothermal transformations involve no change in temperature and isovolumetric transformations involve no work done by or on the system, and these simplifications enable calculations to be performed to predict the future variable quantities of a system.
- Pressure–volume diagrams can be used to show thermodynamic processes, with a closed loop representing a cyclic process.
- A cyclic process in which net work is done on the surroundings is known as an engine, with the reverse known as a heat pump.
- The Carnot cycle, involving two isothermal and two adiabatic processes, is the most efficient possible for an engine in which the maximum and minimum volumes are predetermined.
- Systems, from the smallest engine to the whole Universe, always tend toward disorder, energy always tends to spread out and entropy always increases (or stays the same).
- Not all thermal energy can be transferred to work.
- Heat cannot be transferred from a cold to a hot body without work being done.

Practice questions

1. A student undertakes an investigation into the relationship between the volume of air V in a measuring cylinder and its depth h in a swimming pool. The diagram shows the cylinder just below the surface. They measure the depth from the surface to the bottom of the cylinder.

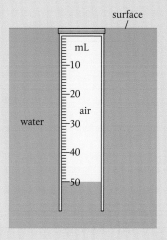

(i) Explain why the student expects there to be a relationship between depth and volume. (2)

(ii) If the atmospheric pressure is A and the density ρ, show that
$$h = \frac{nRT}{\rho g V} - \frac{A}{\rho g}.$$ (3)

(iii) The student decides to plot h vs $\frac{1}{V}$. Explain why they choose to plot this. (2)

(iv) The uncertainty in volume is estimated to be ± 2 cm³. Calculate the length of the error bar plotted on the graph. (2)

(v) The number of moles is calculated from the graph. State what additional measurement must be made to enable the student to calculate this. (1)

(vi) To check the validity of the results, the student compares the atmospheric pressure calculated from the graph with the atmospheric pressure measured with a barometer. They find that their calculated value is different from the measured value. Explain why this is the case. (2)

(Total 12 marks)

2. In an idealized heat engine, a fixed mass of a gas undergoes various changes of temperature, pressure and volume. The P–V cycle (A → B → C → D → A) for these changes is shown in the diagram.

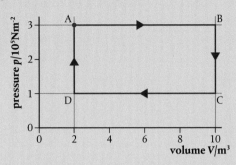

(a) Use the information from the diagram to calculate the work done during one cycle. (2)

(b) During one cycle, a total of 1.8×10^6 J of thermal energy is ejected into a cold reservoir. Calculate the efficiency of this engine. (2)

(c) Copy the diagram below and on the axes sketch the P–V changes that take place in the fixed mass of an ideal gas during one cycle of a Carnot engine. (Note this is a sketch graph – you do not need to add any values.) (2)

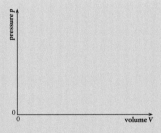

(d) (i) State the names of the **two** types of change that take place during one cycle of a Carnot engine. (2)

(ii) Add labels to your graph drawn in (c) to indicate which parts of the cycle refer to which type of change. (2)

(Total 10 marks)

3. When a balloon filled with helium at 300 K pops, the gas expands from 200 cm³ to 205 cm³ in 0.01 s. The pressure of the gas is 104 kPa before the balloon pops.

(a) Explain why this can be considered to be an adiabatic expansion. (2)

(b) Calculate the number of moles of helium in the balloon. (2)

(c) The pressure of the gas is 100 kPa after it pops. Calculate the change in temperature of the gas. (2)

(d) Calculate the change in internal energy of the helium. (2)

(e) (i) Copy the graph below and plot the expansion. (1)

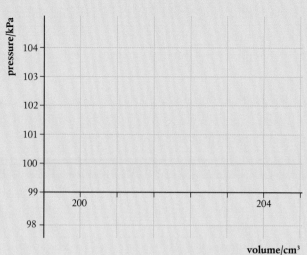

(ii) State how the graph can be used to find the work done by the helium and show that this is consistent with your answer to (i). (3)

(f) A student wishes to perform an investigation measuring the temperature change for different size balloons. Explain why it might be difficult to do this experiment. (2)

(Total 14 marks)

4. A heat engine uses a fixed mass of an ideal gas as a working substance. The diagram shows the changes in pressure and volume of the gas during one cycle, ABCA, of operation of the engine.

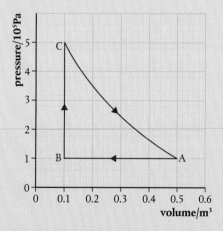

(a) For the part A → B of the cycle, explain whether:

(i) work is done **by** the gas or work is done **on** the gas (1)

(ii) thermal energy (heat) is absorbed **by** the gas or is ejected **from** the gas to the surroundings. (1)

(**b**) Calculate the work done during the change A → B.   (2)

(**c**) Use the diagram to estimate the total work done during one cycle.   (2)

(**d**) The total thermal energy supplied to the gas during one cycle is 120 kJ. Estimate the efficiency of this heat engine.   (2)

(Total 8 marks)

5. A system consists of a refrigerator with its door open operating in a thermally insulated room. What are the change in the entropy of the system and the change in temperature of the room?

	Entropy change of system	Temperature change of room
A	decreases	decreases
B	decreases	increases
C	increases	decreases
D	increases	increases

(Total 1 mark)

6. The graph shows how the volume of a system varies with pressure during a cycle ABCA. What is the work done in joules during the change AB?

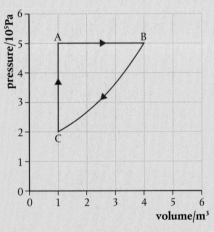

A 15×10^5 **B** 9.0×10^5 **C** 4.5×10^5 **D** 0

(Total 1 mark)

7. (**a**) An ideal nuclear power plant can be modeled as a heat engine that operates between a hot temperature of 612 °C and a cold temperature of 349 °C. Calculate the Carnot efficiency of the nuclear power plant.   (2)

(**b**) Explain, with a reason, why a real nuclear power plant operating between the stated temperatures cannot reach the efficiency calculated in (**a**).   (2)

(**c**) The nuclear power plant works at 71.0% of the Carnot efficiency. The power produced is 1.33 GW. Calculate how much waste thermal energy is released per hour.   (3)

(**d**) Discuss the production of waste heat by the power plant with reference to the first law and the second law of thermodynamics.   (3)

(Total 10 marks)

8. A cylinder is fitted with a piston. A fixed mass of an ideal gas fills the space above the piston. The gas expands isobarically. The following data are available.

Amount of gas   = 243 mol
Initial volume of gas   = 47.1 m³
Initial temperature of gas = −12.0°C
Final temperature of gas = +19.0°C
Initial pressure of gas   = 11.2 kPa

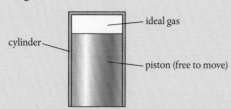

(a) Show that the final volume of the gas is about 53 m³. (2)

(b) Calculate, in J, the work done by the gas during this expansion. (2)

(c) Determine the thermal energy that enters the gas during this expansion. (3)

(d) The gas returns to its original state by an adiabatic compression followed by cooling at constant volume. Copy the *pV* diagram below and sketch the complete cycle of changes for the gas, labeling the changes clearly. The expansion shown in (a) and (b) is drawn for you. (2)

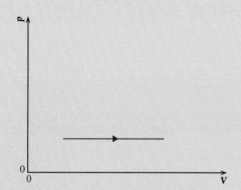

(e) Outline the change in entropy of the gas during the cooling at constant volume. (1)

(f) There are various equivalent versions of the second law of thermodynamics. Outline the benefit gained by having alternative forms of a law. (1)

(Total 11 marks)

HL end

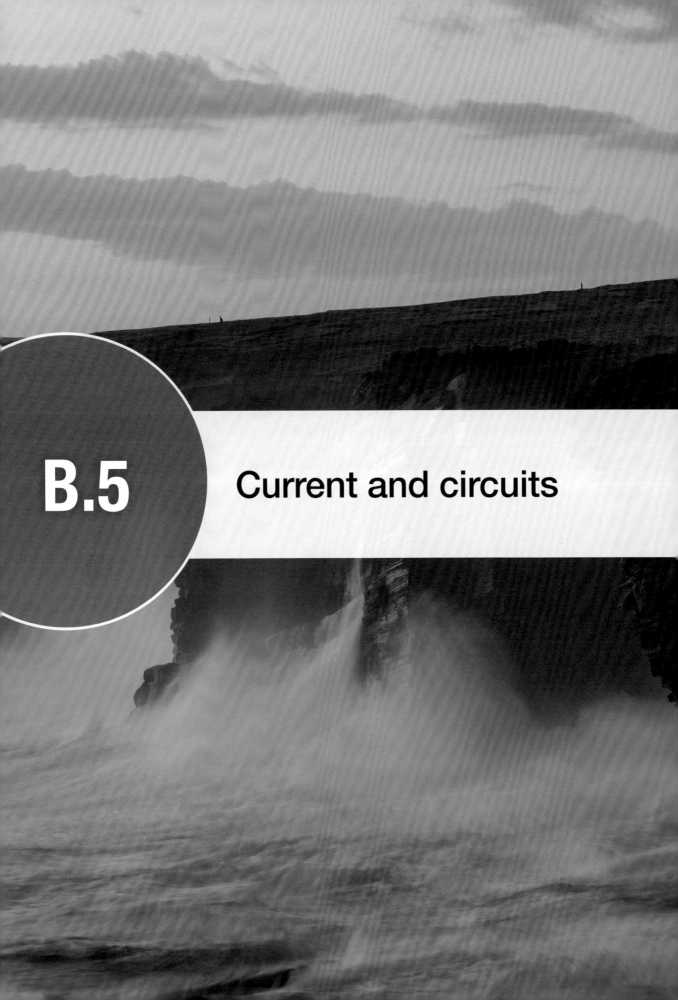

B.5

Current and circuits

◀ The Orkney islands experience high winds, waves and ocean tides, and there is ample land (and sunlight in summer) available for solar cells. Residents require less electricity than can be generated by these renewable means. The only problem is reliability.
The solution? Large-scale batteries have been installed to store excess electrical energy and even to generate income for residents who sell the electricity generated on their land to the UK's national grid.

Guiding Questions

How do charge particles flow through materials?

How are the electrical properties of materials quantified?

What are the consequences of resistance in conductors?

It may seem strange to include electrical circuits in the section on particles. Whole circuits do not fly through the air in parabolic paths, reside in the atmosphere or interact with pressure and volume (or at least they should not). However, electrical current in metals is due to the flow of electrons. Electrons are particles that follow parabolic paths when traveling in uniform fields.

There are two main themes in physics: waves and particles. When a stone is thrown into a pond, a water wave spreads out in a circle. Unless the energy is focused (e.g. laser light), the amount of energy transferred from the stone to a point at the edge of the pond is much less than the original energy transferred to the water. In contrast, if a particle travels between A and B without making a collision, all of the energy it had at A will arrive at B.

When studying gases, we looked at the microscopic motion of the molecules, then at a more general picture of energy changes. We will do the same thing here – first looking at the microscopic interaction between electrons and lattice atoms and then at some rules to solve problems that relate to the macroscopic quantities of current, resistance and potential.

Students should understand:

cells provide a source of emf
chemical cells and solar cells as the energy source in circuits
circuit diagrams represent the arrangement of components in a circuit
direct current (DC) I as a flow of charge carriers as given by $I = \dfrac{\Delta q}{\Delta t}$
the electric potential difference V is the work done per unit charge on moving a positive charge between two points along the path of the current as given by $V = \dfrac{W}{q}$
the properties of electrical conductors and insulators in terms of mobility of charge carriers
electric resistance and its origin
electrical resistance R as given by $R = \dfrac{V}{I}$
resistivity as given by $\rho = \dfrac{RA}{L}$
Ohm's law

the ohmic and non-ohmic behavior of electrical conductors, including the heating effect of resistors
electrical power P dissipated by a resistor as given by $P = IV = I^2R = \dfrac{V^2}{R}$

the combinations of resistors in series and parallel circuits	
Series circuits	**Parallel circuits**
$I = I_1 = I_2 = ...$	$I = I_1 + I_2 + ...$
$V = V_1 + V_2 + ...$	$V = V_1 = V_2 = ...$
$R_s = R_1 + R_2 + ...$	$\dfrac{1}{R_p} = \dfrac{1}{R_1} + \dfrac{1}{R_2} + ...$

electric cells are characterized by their emf ε and internal resistance r as given by $\varepsilon = I(R + r)$
resistors can have variable resistance.

Simple circuits

The charged particles responsible for current in dissolved salts are ions. Can you think of any other alternatives?

What are the parallels in the models for thermal and electrical conductivity? (NOS)

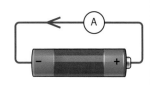

Electrical current in metals is due to the flow of small particles called electrons. Electrons have charge, which causes them to repel each other (because the charges are the same type). Pushing electrons together requires work, and this work is converted to electrical potential energy (which you will learn more about in D.2). However, you do not need to know anything about charges and fields to understand circuits if we use an analogy.

A simple circuit consists of a wire connecting the terminals of a cell (this is actually a short circuit). When the wire is connected to the cell, a circuit is formed and current flows in the wire. **Current** is defined as the rate of flow of charge and can be measured in amperes (A):

$$I = \frac{\Delta q}{\Delta t}$$

Let us consider an analogous situation. Here, we have a container full of balls and an empty container. To get the balls to move from A to B, we need to connect the containers with a pipe, and lift container A (or lower container B) so that the balls roll down the hill.

The height difference between A and B is an analogy for the potential difference in the circuit. This difference is key to understanding the relationship between potential difference and current; a great height is meaningless if the balls are no higher than the next container. **Potential difference** is defined as the work done per unit charge passing between two points and is measured in volts (V):

$$V = \frac{W}{q}$$

How are the fields in other areas of physics similar to and different from each other? (A.1, B.5, D.1, D.2)

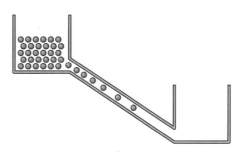

B.5 Figure 2 Balls flow from box A to box B when there is a height difference between them.

In what ways can an electrical circuit be described as a system like the Earth's atmosphere or a heat engine? (B.2, B.4)

To produce a continuous flow of particles, we need to lift the balls back up to the start. Let us imagine that we do this with an escalator.

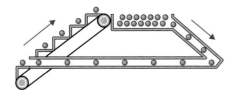

B.5 Figure 3 An escalator ensures that the balls can continue to flow.

So, the pipe represents the wires and the escalator the cell. The current is the rate of flow of the particles and is dependent on the height difference between the two containers and the diameter of the pipe. But there is more to it than that. If we make an electric circuit, the wire gets hot, which means that the atoms in the wire are being made to vibrate more. If we add atoms to our wire analogy, we can see how this takes place.

As the particles move through the wire, they collide with the atoms (green balls), increasing their kinetic energy. More green balls will mean more collisions and less current. We say that the wire has **resistance** to the flow of particles. From an energy perspective, the particles are given potential energy by the battery, which is transferred to the kinetic energy of the atoms.

In the analogy, we can represent this by replacing the pipe with stairs. As the balls bounce down the stairs, they lose energy. If you have ever fallen down the stairs, you will know that you do not accelerate; you roll down with constant velocity. This is the same for the electrons. Although they accelerate between collisions, their average speed is constant all along the wire. The current leaving the cell equals the current into it.

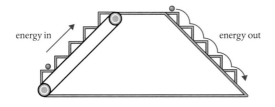

energy in   energy out

B.5 Figure 5 Average speed is constant across both energy changes.

The current varies with both the cross-sectional area and length of the pipe and is also affected by the nature and density of the atoms. We represent resistors with a rectangular symbol.

A complete circuit consists of at least one source of electrical energy (e.g. a cell) and one resistor.

B.5 Figure 4 Atoms impede the flow of charged particles.

How does a particle model allow electrical resistance to be explained? (NOS)

B.5 Figure 6 Resistor symbol.

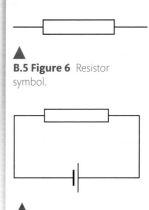

B.5 Figure 7 Circuit diagram consisting of a cell and resistor.

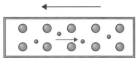

B.5 Figure 8 Conventional current and electron flow are in opposite directions.

When the current is in one direction, as in this case, it is termed **direct current** (DC). When it changes direction, it is termed **alternating current** (AC).

A resistor is a component with a known resistance. You can work out the resistance from the colors.

$$\text{Area} = \pi r^2$$
where radius
$$= \frac{1}{2} \times 2\,mm$$
$$= 0.001\,m$$

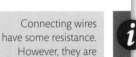

Connecting wires have some resistance. However, they are made of a good conductor, so this resistance is normally much less than the resistance of other components and can normally be ignored.

Because electrons have a negative charge, they flow from low potential to high potential. This is like particles rolling uphill. However, it is easier to consider particles rolling downhill, which in electricity is equivalent to the flow of charges (current) being from high potential to low potential. We refer to this as 'conventional current'. We will always consider conventional current in our calculations, but remember that the electrons are actually flowing in the opposite direction. There is no mathematical difference to worry about, as current has a constant value in a single loop of circuit.

Resistance

The amount of current flowing through a particular conductor is related to the potential difference across it. A good conductor will allow a large current for a given potential difference whereas a poor conductor will only allow a small current. The ratio $\frac{V}{I}$ is defined as the **resistance**.

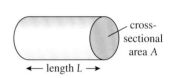

B.5 Figure 9 Dimensions of a conductor.

The unit of resistance is the ohm (Ω).

Resistivity

Resistance R is related to cross-sectional area A, length L, and the material:

$$R \propto \frac{L}{A}$$

The constant of proportionality is called the **resistivity** (ρ).

So:

$$R = \rho \frac{L}{A}$$

By rearranging $R = \rho \frac{L}{A}$, we get $\rho = \frac{RA}{L}$ (units $\Omega\,m$).

From this, we can deduce that if the length of a sample of material is 1 m and the cross-sectional area is 1 m^2, then $\rho = R$. You can probably imagine that the resistance of such a piece of metal will be very small – that is why the values of resistivity are so low (e.g. for copper $\rho = 1.72 \times 10^{-8}\,\Omega\,m$).

Worked example

The resistivity of copper is $1.72 \times 10^{-8}\,\Omega\,m$. What is the resistance of a 1 m length of 2 mm diameter copper wire?

Solution

$$\text{cross-sectional area} = \pi\,(0.001)^2$$
$$= 3.14 \times 10^{-6}\,m^2$$

$$R = \rho \frac{L}{A}$$

$$= 1.72 \times 10^{-8} \times \frac{1}{3.14 \times 10^{-6}}$$

$$= 0.0055\,\Omega$$

Exercise

Q1. Nichrome has a resistivity of $1.1 \times 10^{-6}\,\Omega\,m$. Calculate the diameter of a 2 m long nichrome wire with resistance $5\,\Omega$.

Q2. Calculate the resistance of a copper cable (resistivity $1.7 \times 10^{-8}\,\Omega\,m$) of length 2 km and diameter 0.2 cm.

Ohm's law

Provided that temperature remains constant, the resistance of many conductors is constant. Such conductors are said to be ohmic and obey Ohm's law:

The current through an ohmic conductor is directly proportional to the potential difference across it provided that the temperature remains constant.

This means that if we know the resistance of a component, we can calculate the potential difference V required to cause a current I to flow through it using the equation:

$$V = IR$$

Worked example

If the potential difference across a $3\,\Omega$ resistance is 9 V, what current will flow?

Solution

From Ohm's law: $\quad V = IR$

Rearranging: $\quad I = \dfrac{V}{R}$

$$I = \dfrac{9}{3}$$

$$= 3\,A$$

Graphical treatment of Ohm's law

Ohmic conductor

Since $V \propto I$ for an ohmic conductor, a graph of I against V will be a straight line.

In Figure 10, resistance $= \dfrac{V}{I} = 2\,\Omega$

Non-ohmic conductors

Not all conductors obey Ohm's law. I–V graphs for these conductors will not be straight. A light bulb filament is an example of a non-ohmic conductor.

In Figure 11, the resistance at the start is $\dfrac{1}{3}\,\Omega$ ($0.33\,\Omega$), and at the end it is $\dfrac{4}{7}\,\Omega$ ($0.57\,\Omega$).

The reason for this is that as current flows through the light bulb, electrical energy is transferred to thermal energy, resulting in an increased temperature and leading to an increase in resistance. Electrical energy is of course transferred to thermal energy in all resistors, but if the rate of thermal energy production is not too high, it will be transferred to the surroundings rather than causing significant temperature change.

The reason that temperature affects resistance is because an increase in temperature means increased lattice vibrations, resulting in more collisions between electrons and the lattice. However, increasing the temperature of a semiconductor leads to the liberation of more free electrons, which results in a *lower* resistance.

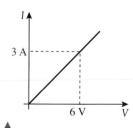

▲ **B.5 Figure 10** Ohm's law is obeyed.

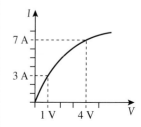

▲ **B.5 Figure 11** Ohm's law is not obeyed.

Resistance is defined as the ratio of $\frac{V}{I}$. Note that this is not the same as a rate of change.

In A.1 Kinematics, we determined acceleration from a velocity–time graph using the gradient. To find the resistance of a component from a current–potential difference graph, however, we divide the potential difference value by the current value at a given point.

Exercise

Q3. If a potential difference of 9 V causes a current of 3 mA to flow through a wire, what is the resistance of the wire?

Q4. A current of 1 µA flows through a 300 kΩ resistor. What is the potential difference across the resistor?

Q5. If the potential difference across a 600 Ω resistor is 12 Ω, how much current flows?

Q6. Table 1 shows the potential difference and current through a device called a thermistor. Calculate the resistance at these different potential differences.

 Table 1

Potential difference/V	Current/A
1.0	0.01
10.0	0.10
25.0	1.00

Nature of Science

When an ammeter or voltmeter is connected to a circuit, it changes the current flowing. It is important to take this into account when analyzing data from experiments.

When drawing electrical circuits, we use universal symbols, with one or two alternatives, that can be understood throughout the world.

Electric cells and batteries

An electric cell is a device that uses the energy stored in chemicals to arrange charges in such a way that a potential difference is created that can be used to cause a current to flow in a conductor. Chemical energy is the energy associated with molecules. When chemical reactions take place, this energy can be transferred to other stores; for example, when coal burns, the molecules of coal and oxygen recombine in a way that has less potential energy. This potential energy is transferred to the kinetic energy of the new molecules, resulting in a rise in temperature. Chemical energy is actually electrical potential energy: molecules are simply charged bodies arranged in such a way that they do not fly apart. If they are reorganized so that their potential energy decreases, then energy is released. What is happening in a cell is that, by reorganizing molecules, energy is used to separate the positive and negative charges. One way this can be done is in the so-called simple cell (Figure 12), which can consist of zinc and copper plates dipped into acid, for example.

What are the advantages of cells as a source of electrical energy? (A.3)

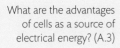

zinc   copper

acid

B.5 Figure 12 A simple cell.

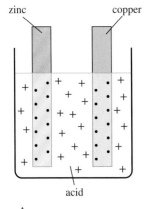

When metals are put into acid (e.g. sulfuric acid), they react to give bubbles of hydrogen. While this is happening, the metal reacts with the acid. When the metal atoms react, they do not take all their electrons with them: this leaves them positively charged (they are called **positive ions**). The electrons remaining give the metal plates a negative charge. Since zinc reacts faster than copper, it gets more charge, resulting in a potential difference. The chemical energy of the metal and acid has been transferred to electrical potential energy. A zinc plate is more negative than a copper plate and so will have a lower potential. When connecting a battery, we often refer to the ends as positive and negative. However, they could be both negative, as in this case. The important thing is to realize that they have different potentials, not different charge.

long side –
high potential

short side –
low potential

◀ **B.5 Figure 13** The symbol for a cell makes a lot of sense; the side that is at the highest potential has the longest line.

The simple Cu–Zn cell produces a potential difference of about 1 V. Even though each side is at a negative potential, it is convenient to take the low side to be 0 V, called ground or **earth**, and the high side to be 1 V. A battery is a row of cells. However, the word **battery** is often used to mean anything that transfers chemical energy to electrical energy. There are many different types and sizes of cell but all work on the same principle.

A cell that cannot be recharged is termed a primary cell; a rechargeable cell is a secondary cell.

Internal resistance

The internal components of a cell have resistance. This internal resistance is due to the resistive components of the cell but there is also resistance due to the changing chemical composition. This means that the internal resistance can vary depending on how long the cell is used for.

When current flows from the cell, some energy is lost. This resistance is represented in circuit diagrams by placing a small resistor next to the symbol for a cell. In some examples, we will consider cells with zero internal resistance, but this is not possible in reality.

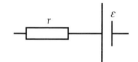

▲ **B.5 Figure 14** The symbol for a cell includes a resistor r next to the cell to show internal resistance.

Emf (ε) and terminal potential difference

The emf of a cell is the work done per unit charge taking the charges from the low potential to the high potential. The energy to do this work has been transferred from the chemical energy so the emf of the cell is the amount of chemical energy transferred to electrical potential energy per unit charge.

The unit of emf is the volt (V). Note that 'emf' stands for electromotive force, but as this is misleading in terms of units, we tend to use the abbreviation.

If no current flows, then there will be no potential drop across the internal resistance. The potential difference across the terminals of the cell (**terminal potential difference**) will equal the emf. However, if current flows, the terminal potential difference will be *less* than the emf.

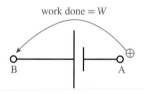

work done = W

B   A

▲ **B.5 Figure 15** $\varepsilon = \dfrac{W}{q}$

Discharge of a cell

If a simple cell is not connected to a circuit, charges will be moved to the position of high potential until the force pushing them is balanced by the repulsive force of the charges already there. This occurs at about 1 V. If the cell is now connected to a resistance, charge will flow from high potential to low potential. As this happens, more charges will be added to the high potential, maintaining the terminal potential difference at 1 V. The cell is said to be discharging. This continues until all the chemical energy is used up, at which point the potential difference starts to get less.

As a cell discharges (runs out) its internal resistance increases, resulting in a lower terminal potential difference.

You can measure the terminal potential difference of a cell as it discharges through a resistor by using a voltmeter connected across the terminals of the cell. However, the cell should not be discharged too rapidly, which means the experiment might take a long time. If you use a voltage sensor and computer interface, you could leave the cell to discharge on its own.

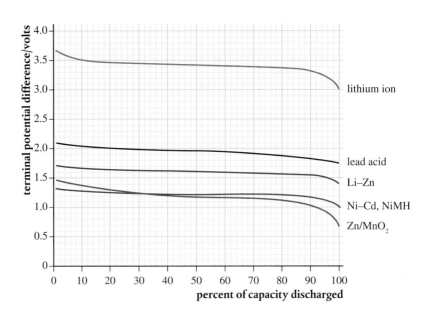

B.5 Figure 16 Potential difference vs time for different cells discharging showing how the terminal potential difference is almost constant for the life of the cell.

Generators

A generator consists of a coil rotating in a magnetic field. As the coil rotates, the magnetic flux enclosed by the coil changes, which induces an emf in the coil (see D.4). The current flowing in the coil also creates a magnetic field that interacts with the field causing it. This causes a force that opposes the motion of the coil, which means that to move the coil at constant speed, some force has to be applied, for example, by someone turning the handle in the diagram. As the handle rotates, the point of application of the force moves in a circle; work is therefore done. In this way, mechanical energy is transferred to electrical energy. The amount of energy transferred per unit charge is the emf of the generator.

B.5 Figure 17 A simple AC generator.

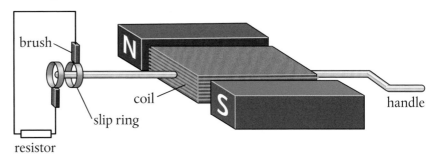

The force applied usually comes from a turbine, which in turn could be a result of combustion of fossil fuels to boil water (B.2), hydroelectric power (A.3) or nuclear energy (E.4).

Photovoltaic cell

The photovoltaic cell is a semiconductor device based on a PN junction that transfers solar energy directly to electrical energy. A PN junction is a slice of semiconductor where one side has been doped with impurities with holes and the other side impurities with electrons. At the junction between the N and P, the electrons and holes migrate to give a region of electric field as shown in Figure 18.

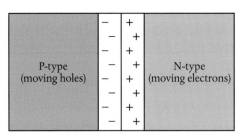

 B.5 Figure 18 A PN junction.

depletion layer
(electrons fill holes)

Solar panels are positioned so that they absorb maximum sunlight in the middle of the day. On the equator, the Sun is directly overhead at midday, so the panels are placed horizontally, but in other countries, the position of the Sun changes with the seasons, so a compromise has to be made. In countries with less sunshine (lots of clouds), the position is not so important because the sunlight does not come from one direction (it is said to be diffuse).

Light incident on the N-type semiconductor can cause more free electrons to be emitted by a process known as the **photovoltaic effect**. The electric field at the junction traps the electrons on this side, creating a potential difference across the slice. If connected to a circuit, current will flow. Figure 19 shows the different parts of the photovoltaic cell. Notice the top contact does not cover the whole top, leaving space for the light to get to the PN junction. Silicon is very shiny so an anti-reflective layer is added to reduce energy loss by reflection. Light reflecting off the top layer of this very thin film undergoes a phase change of π whereas light reflected off the bottom layer does not. The reflected light therefore interferes destructively, reducing energy loss.

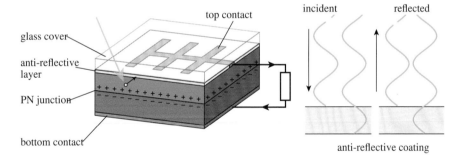

B.5 Figure 19 A photovoltaic cell with detail of anti-reflective coating.

Simplest circuit

When a resistor is connected across the terminals of a cell, current will flow from high potential to low potential. Charges are given electrical energy by the cell, which is then transferred to thermal energy in the resistor. We can think of this as being like a shopping center with stairs and escalators. Shoppers are taken up from ground level to the first floor (given potential energy) by the escalators and then come down (lose potential energy) by the stairs.

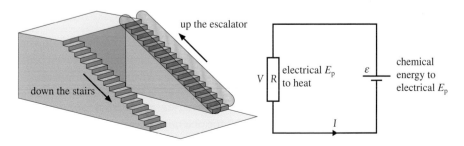

B.5 Figure 20 The simplest circuit and equivalent shopping center.

Since energy is conserved, we can say that when a unit charge flows around the circuit, the amount of chemical energy transferred to electrical energy = the amount of electrical energy transferred to thermal energy. In other words, emf = potential difference. In the shopping center, this would translate to 'the height you go up by the escalator = the height you come down by the stairs'.

We can write this statement as an equation: $\varepsilon = V = IR$

Worked example

If the emf of a battery is 9 V, how much energy is transferred from chemical to electrical when 2 C of charge flow?

Solution

emf = energy transferred from chemical to electrical per unit charge

$$\text{energy converted} = 2 \times 9\,\text{J}$$
$$= 18\,\text{J}$$

Worked example

What is the potential difference across a resistor if 24 J of heat are produced when a current of 2 A flows through it for 10 s?

Solution

2 A for 10 s = 2 × 10 C = 20 C of charge

If 20 C of charge flows, then the energy per unit charge = $\dfrac{24}{20}$ V = 1.2 V

B.5 Figure 21 A circuit with internal resistance.

 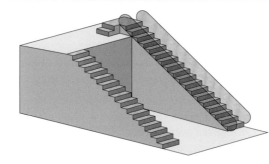

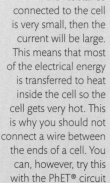

If the resistance connected to the cell is very small, then the current will be large. This means that most of the electrical energy is transferred to heat inside the cell so the cell gets very hot. This is why you should not connect a wire between the ends of a cell. You can, however, try this with the PhET® circuit construction kit.

The not-so-simple circuit

The previous example did not include the internal resistance of the cell. In the shopping center, the escalator is still the same height but now there is a short stairway down before you reach the first floor. In terms of the circuit, some energy is lost as heat before the charge leaves the cell.

Applying Ohm's law to the internal resistance, the potential difference across it will be Ir.

From the law of conservation of energy, when a certain charge flows, the amount of energy transferred from chemical to electrical equals the amount transferred from electrical to thermal.

$$\varepsilon = IR + Ir$$

Rearranging this formula, we can get an equation for the current from the cell.

$$I = \frac{\varepsilon}{R + r}$$

Worked example

A cell of emf 9 V with an internal resistance 1 Ω is connected to a 2 Ω resistor, as shown in Figure 22.

How much current will flow?

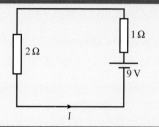

Solution

$$I = \frac{\varepsilon}{R + r}$$

$$I = \frac{9}{2 + 1} = 3\,\text{A}$$

What is the potential difference across the 2 Ω resistor?

$$V = IR$$

$$V = 3 \times 2 = 6\,\text{V}$$

Exercise

Q7. A current of 0.5 A flows when a cell of emf 6 V is connected to an 11 Ω resistor. What is the internal resistance of the cell?

Q8. A 12 V cell with internal resistance 1 Ω is connected to a 23 Ω resistor. What is the potential difference across the 23 Ω resistor?

Electrical power

Just as in mechanics, electrical power is the rate at which energy is transferred from one store to another.

Power delivered

In a perfect cell, the power is the amount of chemical energy transferred to electrical energy per unit time.

If the emf of a cell is ε, then if a charge q flows, the amount of energy transferred from chemical to electrical is εq.

If this charge flows in a time t, then the power delivered $= \frac{\varepsilon q}{t}$

But: $\frac{q}{t}$ = current, I

So: power delivered $= \varepsilon I$

In a real cell, the actual power delivered will be a bit lower, since there will be some power dissipated in the internal resistance.

Always start by drawing a circuit showing the quantities you know and labeling the ones you want to find.

The electrical energy used at home is measured in kilowatt-hours. This is the amount of energy used by a 1 kilowatt heater switched on for 1 hour. When the electricity bill comes, you have to pay for each kilowatt-hour of energy that you have used.

Since 1 W = 1 J s⁻¹

1 kilowatt-hour = 1000 × 60 × 60 J

1 kWh = 3.6 × 10⁶ J

The kilowatt-hour is often wrongly called 'kilowatts per hour'. This is because people are used to dealing with quantities that are rates of change like kilometers per hour.

How can the heating of an electrical resistor be explained using other areas of physics?
(A.3, B.1)

Power dissipated

The power dissipated in the resistor is the amount of electrical energy transferred to thermal energy per unit time.

Consider a resistance R with a potential difference V across it. If a charge q flows in time t, then the current, $I = \frac{q}{t}$.

The potential difference, V, is defined as the energy transferred to heat per unit charge q, so the energy transferred to heat in this case $= Vq$.

Power is the energy used per unit time, so $P = \dfrac{Vq}{t}$

But $\frac{q}{t} = I$, so $P = VI$.

Worked example

If a current of 2 A flows through a resistor that has a potential difference of 4 V across it, how much power is dissipated?

Solution

$$P = VI \text{ where } V = 4\,V \text{ and } I = 2\,A$$

$$P = 4 \times 2\,W$$
$$= 8\,W$$

Worked example

What power will be dissipated when a current of 4 A flows through a resistance of 55 Ω?

Solution

Using Ohm's law: $V = IR$
$$= 4 \times 55$$
$$= 220\,V$$

$$P = VI$$
$$= 220 \times 4$$
$$= 880\,W$$

Alternative ways of writing $P = VI$

In the second worked example, we had to calculate the potential difference before finding the power. It would be convenient if we could solve this problem in one step. We can write alternative forms of the equation by substituting for I and V from Ohm's law.

We have shown that: power, $P = VI$

But from Ohm's law: $V = IR$

If we substitute for V, we get: $P = IR \times I = I^2R$

We can also substitute for $I = \dfrac{V}{R}$

So: $P = V \times I = \dfrac{V \times V}{R} = \dfrac{V^2}{R}$

$P = VI$
$P = I^2R$
$P = \dfrac{V^2}{R}$

Q9. 5 A flows through a 20 Ω resistor.

    (a) How much electrical energy is transferred to heat per second?

    (b) If the current flows for one minute, how much energy is released?

Q10. If a cell has an internal resistance of 0.5 V, how much power will be dissipated in the cell when 0.25 A flows?

Q11. A current of 0.5 A flows from a cell of emf 9 V. If the power delivered is 4 W, how much power is dissipated in the internal resistance?

Electric kettle (water boiler)

An electric kettle transfers the heat produced when current flows through a wire element to the water inside the kettle.

Worked example

A current of 3 A flows through an electric kettle connected to the 220 V mains. What is the power of the kettle and how long will it take to boil 1 liter of water?

This is quite a low-powered electric kettle.

Solution

power of the kettle = VI = 220 × 3 = 660 W

To calculate energy needed to boil the water, we use the formula:

    heat required = mass × specific heat capacity × temperature change

specific heat capacity of water = 4180 J kg^{-1} °C^{-1}

The mass of 1 liter of water is 1 kg, so if we assume that the water was at room temperature, 20 °C, then to raise it to 100 °C the energy required is:

    1 × 4180 × 80 = 334 400 J

$$\text{power} = \frac{\text{energy}}{\text{time}}, \text{ so the time taken} = \frac{\text{energy}}{\text{power}}$$

$$= \frac{334\,400}{660}$$

    time = 506.67 s

        = 8 minutes 27 seconds

The electric circuits in a house are organized in rings. Each ring consists of three cables going around the house and back to the supply. Sockets are connected to the live cable in parallel with each other. Each ring has its own circuit breaker, which will cut the power if the current gets too big.

The light bulb

If the power dissipated in a wire is large, then a lot of heat is produced per second. When heat is added quickly, the wire does not have time to lose this heat to the surroundings. The result is that the temperature of the wire increases, and if the temperature is high enough, the wire will begin to glow, giving out light. Only about 10% of the energy dissipated in an incandescent light bulb is transferred to light – the rest is heat.

Fluorescent tubes and light-emitting diodes (LEDs) are much more efficient than incandescent light bulbs, transferring most of the electrical energy to light.

The electric motor

A motor transfers electrical energy to mechanical energy. This could be potential energy if something is lifted by the motor, or kinetic energy if the motor is accelerating something like a car.

Worked example

An electric motor is used to lift 10 kg through 3 m in 5 seconds. If the potential difference across the motor is 12 V, how much current flows (assuming no energy is lost)?

Solution

$$\text{work done by the motor} = mgh$$
$$= 10 \times 10 \times 3 \quad = 300\,\text{J}$$

$$\text{power} = \frac{\text{work done}}{\text{time}}$$
$$= \frac{300}{5}$$
$$= 60\,\text{W}$$

$$\text{electrical power, } P = IV \text{ so } I = \frac{P}{V}$$
$$= \frac{60}{12} = 5\,\text{A}$$

Exercise

Q12. An electric car of mass 1000 kg uses twenty-five 12 V cells connected together to create a potential difference of 300 V. The car accelerates from rest to a speed of 30 m s^{-1} in 12 seconds.

(**a**) What is the final kinetic energy of the car?

(**b**) What is the power of the car?

(**c**) How much electrical current flows from the battery?

Q13. What assumptions have you made in calculating Q12. (a)–(c)?

Q14. A light bulb for use with the 220 V mains is rated at 100 W.

(**a**) What current will flow through the bulb?

(**b**) If the bulb transfers 20% of the energy to light, how much light energy is produced per second?

Q15. A 1 kW electric heater is connected to the 220 V mains and left on for 5 hours.

(**a**) How much current will flow through the heater?

(**b**) How much energy will the heater release?

▲
A battery-powered car.

Combinations of components

In practical situations, resistors and cells are often joined together in combinations, for example, fairy lights, flashlight batteries.

There are many ways of connecting a number of components. Here we will consider two simple arrangements: series and parallel.

Resistors in series

In a series circuit, the same current flows through each resistor.

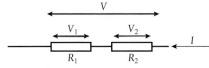

The combination could be replaced by one resistor.

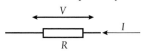

Applying the law of conservation of energy, the potential difference across R_1 plus the potential difference across R_2 must be equal to the potential difference across the combination.

$$V_1 + V_2 = V$$

Applying Ohm's law to each resistor: $IR_1 + IR_2 = IR$

Dividing by I: $R_1 + R_2 = R$

Worked example

What is the total resistance of a $4\,\Omega$ and an $8\,\Omega$ resistor in series?

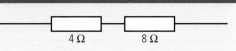

Solution

$$\begin{aligned}\text{total resistance} &= R_1 + R_2\\ &= 4 + 8\\ &= 12\,\Omega\end{aligned}$$

Resistors in parallel

In a parallel circuit, the current splits in two.

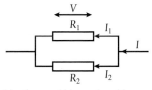

The combination could be replaced by one resistor.

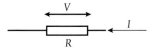

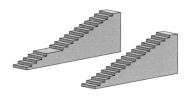

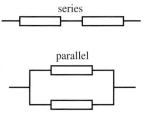

B.5 Figure 22 Two simple combinations of resistors.

◄ **B.5 Figure 23** Two resistors in series are similar to two flights of stairs.

▲ These colored lights are connected in series. If you take one out, they all go out.

◄ **B.5 Figure 24** Resistors in parallel are similar to stairs side by side.

Applying the law of conservation of charge, we know that the current going into a junction must equal the current coming out.

$$I = I_1 + I_2$$

Applying Ohm's law to each resistor gives: $\dfrac{V}{R} = \dfrac{V}{R_1} + \dfrac{V}{R_2}$

Dividing by V: $\dfrac{1}{R} = \dfrac{1}{R_1} + \dfrac{1}{R_2}$

Worked example

What is the total resistance of a $4\,\Omega$ and an $8\,\Omega$ resistor in parallel?

Solution

$$\frac{1}{R} = \frac{1}{R_1} + \frac{1}{R_2}$$

$$\frac{1}{R} = \frac{1}{4} + \frac{1}{8}$$

$$= \frac{2+1}{8} = \frac{3}{8}$$

$$R = \frac{8}{3}\,\Omega$$

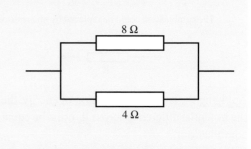

The total resistance of two equal resistors in parallel is half the resistance of one of them.

Worked example

What is the total resistance of two $8\,\Omega$ resistors in parallel?

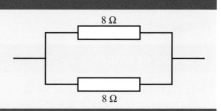

Solution

$$\frac{1}{R} = \frac{1}{R_1} + \frac{1}{R_2}$$

$$\frac{1}{R} = \frac{1}{8} + \frac{1}{8} = \frac{2}{8}$$

$$R = \frac{8}{2} = 4\,\Omega$$

Exercise

Q16. Calculate the total resistance for the circuits in the figure below.

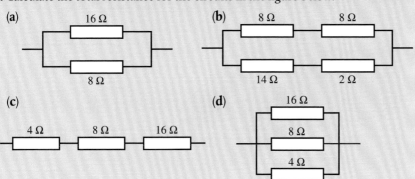

(a) 16 Ω, 8 Ω

(b) 8 Ω, 8 Ω, 14 Ω, 2 Ω

(c) 4 Ω, 8 Ω, 16 Ω

(d) 16 Ω, 8 Ω, 4 Ω

Investigating combinations of resistors in parallel and series circuits

The simplest way to measure the electrical resistance of a component is to use a multimeter. This actually measures the current through the component when a known potential difference is applied to it; the resistance is then calculated and displayed. Try connecting various combinations of resistors in different combinations. Calculate the combined resistance using the equations for series and parallel resistors then check your calculation by measurement. You need to make sure that the resistors are connected properly; this is best done by soldering. If you do not have the necessary equipment, then you can simulate the circuit with a custom built sim such as the PhET circuit construction kit.

Note that not all combinations of resistors can be solved using the series and parallel equations.

Cells in series

Cells are often added in series to obtain a larger emf.

$$\varepsilon = \varepsilon_1 + \varepsilon_2$$

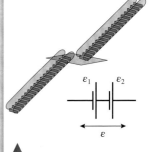

B.5 Figure 25 Cells in series are similar to two flights of escalators.

Worked example

Two 12 V cells are connected in series to a 10 Ω resistor. If each cell has an internal resistance of 1 Ω, how much current will flow?

Solution

total emf for two cells in series

= 12 + 12 = 24 V

total resistance = 1 + 1 + 10 = 12 Ω

Applying Ohm's law:

$$I = \frac{V}{R}$$

$$= \frac{24}{12}$$

$$= 2\,A$$

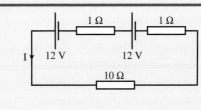

Cells in parallel

When two identical cells are connected in parallel, the emf across the combination is the *same* as the emf of one cell (Figure 26).

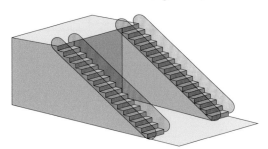
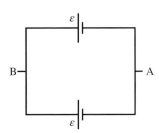

If the cells are different, then it is not so simple. In fact, if we were to connect two cells with different emfs and zero internal resistance, we would not be able to solve the problem of how you can have two different emfs depending on which cell you consider. This would be like having two different height escalators going between the same two floors, and that is not possible unless you connect a short staircase between them. This is where internal resistance comes in.

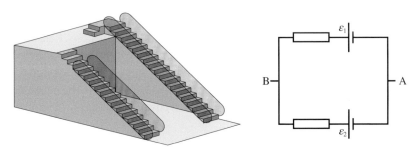

As can be seen in the shopping center model in Figure 27, the potential drops over the internal resistance of the big escalator, and goes up over the internal resistance of the small one. This means that a current is being forced to flow into the smaller cell. This is what happens when a rechargeable cell is being recharged.

Electrical measurement

Measurement of potential difference

The potential difference can be measured using a voltmeter. Voltmeters can have either a numerical display (digital) or a moving pointer (analog).

The potential difference is the difference in potential between two points. To measure the potential difference between A and B, one lead of the voltmeter must be connected to A, the other to B.

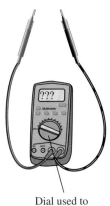

Dial used to change from voltmeter to ammeter.

▲

B.5 Figure 28 A multimeter is a common instrument that can measure both potential difference and current. It can also measure resistance.

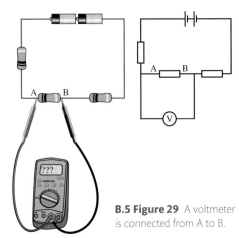

B.5 Figure 29 A voltmeter is connected from A to B.

Measurement of current

To measure the current flowing through a resistor, the ammeter must be connected so that the same current will flow through the ammeter as flows through the resistor. This means disconnecting one of the wires and connecting the ammeter (Figure 30).

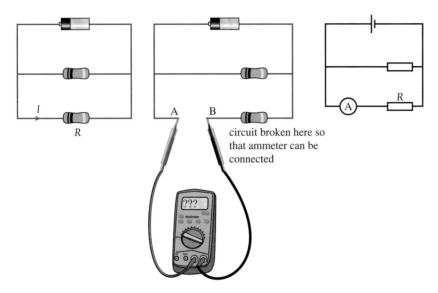

circuit broken here so
that ammeter can be
connected

B.5 Figure 30 The ammeter is connected to measure the current through R.

An ideal ammeter has zero resistance so that it does not change the current in the circuit.

SKILLS

By measuring the current through and potential difference across a conductor, it is possible to calculate its resistance. Measuring the resistance of different lengths of nichrome wire, the relationship between length and resistance can be found.

Nichrome is a metal that has a resistivity ρ of $1.5 \times 10^{-6}\,\Omega$ m. It is suitable to use in this experiment as it is a good conductor, but not so good that it will short circuit the battery. When performing this experiment one should use:

- quite a thin wire so that the resistance is not too small

- a low voltage power supply so that the current is not too big; a high current will make the wire hot

- the least sensitive range on the ammeter (10 A range) so that the ammeter fuse does not blow

- a technique to prevent loops of wire touching as this would shorten the effective length of the sample.

A length of about 1 m of wire is measured with a ruler then connected in a circuit as shown in Figure 31.

The current, potential difference, and length are recorded in an appropriate table and the measurements repeated with at least five more different lengths.

The data can be processed in a spreadsheet using Ohm's law to calculate the resistance of each length. The uncertainty in the resistance should also be found. The easiest way of doing this is to use the percentage uncertainties in V and I, which simply add when you calculate $\frac{V}{I}$.

> So if $V = 1.5 \pm 0.1$ V, the percentage uncertainty = 7%.
> And if $I = 0.54 \pm 0.01$ A, the percentage uncertainty = 2%.
> So the percentage uncertainty in $R = \frac{V}{I}$ is 9%.
> But $\frac{V}{I} = 2.8\,\Omega$ so the absolute uncertainty = 0.3 Ω.
> The final value is $2.8 \pm 0.3\,\Omega$.

This is quite a lot of steps but you only have to write the calculation once if a spreadsheet is used.

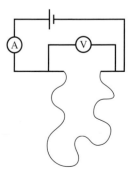

▲
B.5 Figure 31 Measuring the resistance of nichrome.

The theoretical relationship between resistance R and length L is given by the equation:

$$R = \frac{\rho L}{A}$$

where ρ is the resistivity and A is the cross-sectional area. This means that a graph of R vs L will be a straight line with gradient $\frac{\rho}{A}$.

In this case, the uncertainties are quite small, but as with all experiments, the results never exactly match the theoretical model. Some factors that might be worth taking into consideration are:

- the changing temperature of the wire
- the resistance of the meters
- the uniformity of cross-section.

If multimeters are used to measure **V** and **I**, it is quicker to simply switch to measure resistance in ohms, then measure the resistance directly just using one multimeter. In this setting, the multimeter uses its internal power supply (usually a 9 V cell) to pass a small current through the wire. By measuring this current and the potential difference across the wire, the multimeter can give a value for resistance.

Variable resistors

The resistance of a wire is proportional to its length so we can make a variable resistor by varying the length of the wire. This can be done by adding a sliding contact.

B.5 Figure 32 Variable resistor.

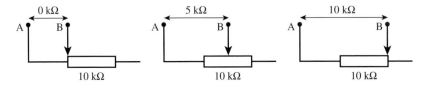

If a current is passed through the resistor, then moving the slider would give a varying potential difference. This is a potentiometer.

B.5 Figure 33 Variable potential.

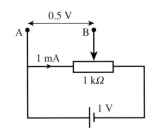

This is how lights are dimmed with a dimmer switch or whenever a varying potential difference is required; for example, to measure the *I–V* characteristics of a component such as a light bulb.

Light-dependent resistor

A light-dependent resistor (LDR) has a resistance that varies as the light intensity falling on it changes. This is a semiconductor device. Light causes the release of more charge carriers in the semiconductor, resulting in a lowering of resistance.

If you have used a light gate, it might have included an LDR. The change of resistance when the light is prevented from reaching the LDR is detected by the computer interface, enabling the computer to record the time that it happened.

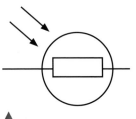

B.5 Figure 34 The symbol for an LDR.

Thermistor

A thermistor is also a semiconductor device. Unlike conductors, whose resistance increases with rising temperature due to the increased lattice vibrations, the resistance of a semiconductor decreases as its temperature rises. This is due to the increase in charge carriers.

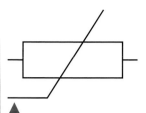

B.5 Figure 35 The symbol for a thermistor.

Exercise

Q17. The slider in the circuit is set at the middle of the variable resistor. Calculate the potential difference across the 2 kΩ resistor.

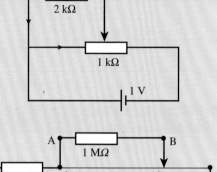

Q18. A 1 m length of nichrome resistance wire ($\rho = 1.5 \times 10^{-6}\,\Omega\,m$) of diameter 0.1 mm is connected to the circuit. Calculate the potential difference from A to B.

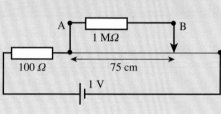

The circuit used to measure the current flowing through a component for different values of *V* is shown in Figure 36. Here the potential divider (a circuit in which the terminal potential difference is split up) is providing the changing potential difference that is measured by the voltmeter. Note that, in this circuit, it is assumed that the voltmeter has a much higher resistance than the bulb so it will not draw much current.

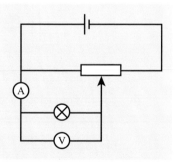

A worksheet with full details of how to carry out this experiment is available in the eBook.

B.5 Figure 36 Circuit used to measure the characteristics of a light bulb.

Guiding Questions revisited

How do charge particles flow through materials?

How are the electrical properties of materials quantified?

What are the consequences of resistance in conductors?

In this chapter, we have used an analogy and topic-specific terminology to understand that:

- Like all bodies, charged particles (such as those found in metal wires) require a force to accelerate, which in electrical circuits can be supplied by a cell.
- The rate of flow of charge (current) is the same for all points in a series circuit, but shared in a parallel circuit.
- The energy transferred per unit charge between two points (potential difference) is shared among components in a series circuit, but is the same across sections in parallel.

- The ratio of potential difference to current is known as resistance.
- Resistance can be conceptualized as being due to collisions between electrons and lattice atoms.
- The power output of a resistor can be desirable (e.g. for household lighting, sound, heating or movement) or undesirable (e.g. wasted thermal energy).
- Resistivity is a property that enables quantitative comparisons between materials of equal dimensions.
- Ohmic conductors have constant resistance because temperature is constant, whereas non-ohmic conductors, such as lightbulb filaments, light-dependent resistors and thermistors, do not.
- Terminal potential difference across a power supply (irrespective of whether a chemical cell, solar cell or generator) is equal to the emf minus the potential lost across the internal resistance.
- Cells and resistors can be combined when solving complex circuit problems.

Practice questions

1. The diagram shows a screenshot from a two-dimensional simulation of electric current. The black circles represent the atomic lattice and the white circles represent the charge carrier, which each have charge $-q$ and travel with average velocity v.

    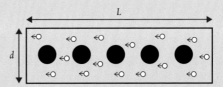

    (a) Copy the diagram and draw the direction of the electric field.   (1)

    (b) Use this model to explain:

       (ii) resistance   (2)

       (iii) electric heating   (2)

       (iv) the change of resistance of a light bulb when it is switched on.   (2)

    (c) Show that the current is $15\frac{vq}{L}$.   (3)

    (Total 10 marks)

2. A 20 cm long nichrome wire of diameter 0.1 mm is connected to a 9 V cell of internal resistance 1 Ω.

    Resistivity of nichrome = 1.1×10^{-6} Ω m

    Heat capacity of a 9 V cell = 10 J K^{-1}

    (a) Calculate:

       (i) the resistance of the wire   (2)

       (ii) the current flowing through the wire   (2)

       (iii) the power dissipated in the wire   (2)

       (iv) the temperature of the wire.   (2)

(b) Explain why your answer to **(a)(iii)** is different from the actual value. (1)

(c) By making an appropriate calculation, deduce whether the power dissipation in the cell is likely to cause damage. (3)

(Total 12 marks)

3. Two conductors, *S* and *T*, have the *V*/*I* characteristic graphs shown below.

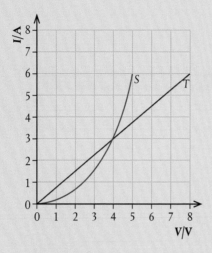

When the conductors are placed in the circuit below, the reading of the ammeter is 6.0 A. What is the emf of the cell?

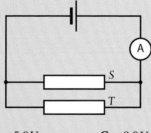

A 4.0 V	**B** 5.0 V	**C** 8.0 V	**D** 13 V

(Total 1 mark)

4. The diagram shows two cylindrical wires, X and Y. Wire X has length *l*, diameter *d*, and resistivity ρ. Wire Y has length $2l$, diameter $\frac{d}{2}$ and resistivity $\frac{\rho}{2}$. What is $\dfrac{\text{resistance of X}}{\text{resistance of Y}}$?

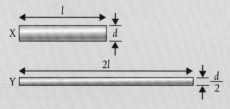

A 4	**B** 2	**C** 0.5	**D** 0.25

(Total 1 mark)

5   In the circuits shown below, the cells have the same emf and zero internal resistance. All the resistors are identical. What is the order of increasing power dissipated in each circuit?

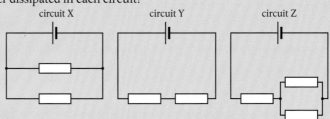

circuit X   circuit Y   circuit Z

	Lowest → Highest power dissipated
A	Y, Z, X
B	X, Y, Z
C	X, Z, Y
D	Y, X, Z

(Total 1 mark)

6. Four resistors of $4\,\Omega$ each are connected as shown. What is the effective resistance between P and Q?

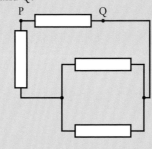

A $1.0\,\Omega$	**B** $2.4\,\Omega$	**C** $3.4\,\Omega$	**D** $4.0\,\Omega$

(Total 1 mark)

7. A cell with negligible internal resistance is connected as shown. The ammeter and the voltmeter are both ideal. What changes occur in the ammeter reading and in the voltmeter reading when the resistance of the variable resistor is increased?

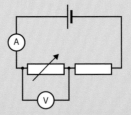

	Change in ammeter reading	Change in voltmeter reading
A	increases	increases
B	increases	decreases
C	decreases	increases
D	decreases	decreases

(Total 1 mark)

8. A photovoltaic cell is supplying energy to an external circuit. The photovoltaic cell can be modeled as a practical electrical cell with internal resistance. The intensity of solar radiation incident on the photovoltaic cell at a particular time is at a maximum for the place where the cell is positioned. The following data are available for this particular time:

Operating current = 0.90 A

Output potential difference to external circuit = 14.5 V

Output emf of photovoltaic cell = 21.0 V

Area of panel = 350 mm × 450 mm

(a) Explain why the output potential difference to the external circuit and the output emf of the photovoltaic cell are different. (2)

(b) Calculate the internal resistance of the photovoltaic cell for the maximum intensity condition using the model for the cell. (3)

The maximum intensity of sunlight incident on the photovoltaic cell at the place on the Earth's surface is 680 W m^{-2}.

A measure of the efficiency of a photovoltaic cell is the ratio:

$$\frac{\text{energy available every second to the external circuit}}{\text{energy arriving every second at the photovoltaic cell surface}}$$

(c) Determine the efficiency of this photovoltaic cell when the intensity incident on it is at a maximum. (3)

(d) State **two** reasons why future energy demands will be increasingly reliant on sources such as photovoltaic cells. (2)

(Total 10 marks)

9. A girl rides a bicycle that is powered by an electric motor. A battery transfers energy to the electric motor. The emf of the battery is 16 V, and it can deliver a charge of 43 kC when discharging completely from a full charge. The maximum speed of the girl on a horizontal road is 7.0 m s^{-1} with energy from the battery alone. The maximum distance that the girl can travel under these conditions is 20 km.

(a) Show that the time taken for the battery to discharge is about 3 × 10^3 s. (1)

(b) Deduce that the average power output of the battery is about 240 W. (2)

(c) Friction and air resistance act on the bicycle and the girl when they move. Assume that all the energy is transferred from the battery to the electric motor. Determine the total average resistive force that acts on the bicycle and the girl. (2)

(d) The bicycle and the girl have a total mass of 66 kg. The girl rides up a slope that is at an angle of 3.0° to the horizontal. Calculate the component of weight for the bicycle and girl acting down the slope. (1)

The battery continues to give an output power of 240 W. Assume that the resistive forces are the same as in (**c**).

(**e**) Calculate the maximum speed of the bicycle and the girl up the slope. (2)

(**f**) On another journey up the slope, the girl carries an additional mass. Explain whether carrying this mass will change the maximum distance that the bicycle can travel along the slope. (2)

The bicycle has a meter that displays the current and terminal potential difference (pd) for the battery when the motor is running. The diagram shows the meter readings at one instant. The emf of the cell is 16 V.

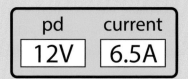

(**g**) Determine the internal resistance of the battery. (2)

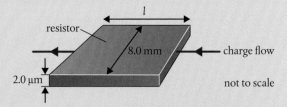

The battery is made from an arrangement of ten identical cells as shown.

(**h**) Calculate the emf of **one** cell. (1)

(**i**) Calculate the internal resistance of **one** cell. (2)

(Total 15 marks)

10. Electrical resistors can be made by forming a thin film of carbon on a layer of an insulating material. A carbon film resistor is made from a film of width 8.0 mm and thickness 2.0 μm. The diagram shows the direction of charge flow through the resistor.

(**a**) The resistance of the carbon film is 82 Ω. The resistivity of carbon is $4.1 \times 10^{-5}\ \Omega\,\text{m}$. Calculate the length l of the film. (1)

(**b**) The film must dissipate a power less than 1500 W from each square meter of its surface to avoid damage. Calculate the maximum allowable current for the resistor. (2)

(c) State why knowledge of quantities such as resistivity is useful to
     scientists.   (1)

The current direction is now changed so that charge flows vertically
through the film.

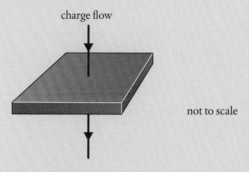

charge flow

not to scale

(d) Deduce, without calculation, the change in the resistance.   (2)

(e) Draw a circuit diagram to show how you could measure the resistance
     of the carbon-film resistor using a variable resistor to limit the
     potential difference across the resistor.   (2)

                                                           (Total 8 marks)

THEME C Wave behavior

C Wave behavior

◀ Gravitational waves are transmitted at the speed of light by astronomical bodies moving at high speeds or orbiting one another. Their existence was first postulated by Albert Einstein in 1916 as part of his general theory of relativity. It was not until 2015 that they were detected – at the Laser Interferometer Gravitational-Wave Observatory – having been released by two colliding black holes more than a billion years ago.

Traveling waves transfer energy. They are made up of oscillations of particles; the particles vibrate about fixed points while the energy propagates through space. These chapters commence, therefore, with simple harmonic motion, a special type of oscillation in which the acceleration of a particle is proportional to its displacement from an equilibrium position and always in the opposite direction. As with gases, the assumptions required for a model do not always apply, but many systems can be approximated as being like masses on springs or simple pendula and described by equations.

Armed with an understanding of oscillations, you will be well-placed to tackle a variety of categories of waves, which often overlap: transverse and longitudinal, mechanical and electromagnetic, high and low frequency, and high and low amplitude. Yet, all waves share common behaviors; they reflect, refract, diffract and interfere. You can expect to use a protractor to investigate angles and vernier calipers for measuring the separation of bright light 'fringes' on a screen.

It is precisely the phenomena of reflection and interference that turn these chapters on their head. Not all waves travel. Standing waves form when two identical waves at particular frequencies travel toward one another. Nodes and antinodes form in precise positions that will be familiar to wind and string musicians, and engineers can exploit both resonance and damping to good effect in their designs.

And finally, no course on wave behavior would be complete without pausing to consider the Doppler effect, which is the change in an observed frequency caused by relative motion between the source of a wave and its observer. The Doppler effect in the context of sound is an everyday occurrence for anyone living near a road; you can hear when a car is approaching and receding ('neeeaooooowwww'). With light, the effect is even more profound; it is most noticeable when the relative speed is comparable to the speed of light, and therefore in the motion of stars and galaxies. These detections provide us with evidence for the expansion of the Universe.

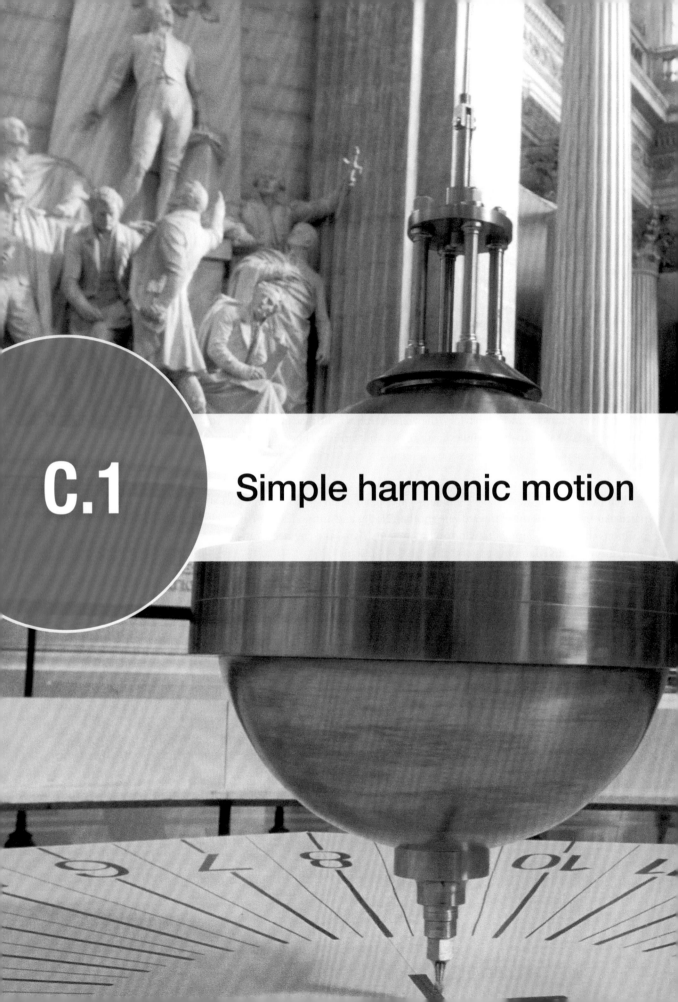

C.1

Simple harmonic motion

◀ A pendulum is a hanging mass that oscillates back and forth. The time period of these oscillations depends on the length of the string and the gravitational field strength where the mass is located, but not on the amplitude of the swings. This 'Foucault pendulum', which gives evidence of the Earth's rotation, is located in the Paris Panthéon.

Guiding Questions

What makes the harmonic oscillator model applicable to a wide range of physical phenomena?

Why must the defining equation of simple harmonic motion take the form it does?

How can the energy and motion of an oscillation be analyzed graphically and algebraically?

An oscillating body does not move in a circle but the mathematical model representing one component of circular motion is the same as that which describes a simple oscillation. In physics, it is quite common that the same model can be used in different applications. As the mathematics becomes more complicated, it becomes more difficult to relate the equations to the motion. This is where computer simulations and visual representations help our understanding.

Students should understand:

conditions that lead to simple harmonic motion
the defining equation of simple harmonic motion as given by $a = -\omega^2 x$
a particle undergoing simple harmonic motion can be described using time period T, frequency f, angular frequency ω, amplitude, equilibrium position and displacement
the time period in terms of frequency of oscillation and angular frequency as given by $T = \dfrac{1}{f} = \dfrac{2\pi}{\omega}$
the time period of a mass–spring system as given by $T = 2\pi\sqrt{\dfrac{m}{k}}$
the time period of a simple pendulum as given by $T = 2\pi\sqrt{\dfrac{l}{g}}$
a qualitative approach to energy changes during one cycle of an oscillation
HL   a particle undergoing simple harmonic motion can be described using phase angle
HL   problems can be solved using the equations for simple harmonic motion as given by $x = x_0 \sin(\omega t + \phi)$ $v = \omega x_0 \cos(\omega t + \phi)$ $v = \pm\omega\sqrt{x_0^2 - x^2}$ $E_T = \dfrac{1}{2}m\omega^2 x_0^2$ $E_p = \dfrac{1}{2}m\omega^2 x^2$

A swing is an example of oscillatory motion.

C.1 Figure 1
A swinging pendulum.

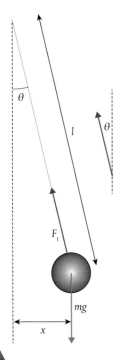

C.1 Figure 2 Pendulum with small displacement.

The simple pendulum

A simple pendulum consists of a small mass, called a bob, hanging on an inextensible (non-stretchy) string. If left alone, the mass will hang at rest so that the string is vertical, but if pushed to one side, it will oscillate about its equilibrium position. The reason that the pendulum oscillates is because there is always a force acting toward the center, as shown in Figure 1.

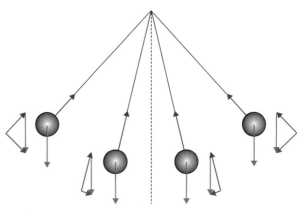

The forces drawn are the tension and weight. These forces add to give a resultant that is always directed toward the equilibrium position. Furthermore, the size of the resultant (blue arrow) gets bigger as the distance from the equilibrium position increases. Notice how the tension decreases as the pendulum swings up toward the horizontal position, and is greatest when it swings through the bottom. At the bottom of the swing, the tension will actually be greater than the weight, allowing the bob to accelerate as it changes direction. Due to the changing angles, this motion is rather difficult to analyze. To make this simpler, we will consider only very small displacements.

Simple harmonic motion

If the swings are kept small, then force becomes directly proportional to the displacement from equilibrium. Consider the small angle shown in Figure 2.

Since the angle is very small, we can make some approximations.

1. The displacement is horizontal (in reality, the bob moves slightly up).
2. The force acting toward the equilibrium position is the horizontal component of the tension, $F_t \sin \theta$ (in reality, it is the **resultant** of the weight and the tension, which is not shown on this diagram).
3. The weight is approximately the same as the tension (in reality, the tension is greatest at the bottom of the swing) $F_t = mg$ so the restoring force = $F_t \sin \theta = mg \sin \theta$.

Since the motion has been reduced to a one-dimensional problem, we can write $F = -mg \sin \theta$ to take into account the direction of this force.

Looking at the triangle made by the string and the equilibrium position, we can see that:

$$\sin \theta = \frac{x}{l}$$

So: $$F = -\frac{mgx}{l}$$

Substituting for $F = ma$ from Newton's second law gives: $ma = -\frac{mgx}{l}$

$$a = -\frac{gx}{l}$$

So we can deduce that, for this motion, the acceleration is directly proportional to the displacement from a fixed point and always directed toward that point. This kind of motion is called **simple harmonic motion** (SHM), with the displacement determining the magnitude of the acceleration and the minus sign indicating the opposite direction of these vector quantities.

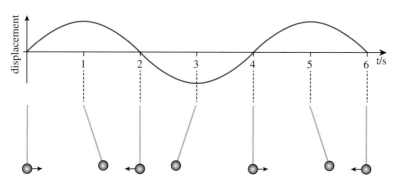

C.1 Figure 3 Sinusoidal motion.

SHM and the sine function

Later in this chapter, we will do a more thorough graphical analysis of SHM, but by observing the motion of the pendulum, we can already see that as the bob swings back and forth its displacement will be described by a sine curve. In other words, the displacement of the bob is sinusoidal. Let us analyze the motion represented by the sine curve in Figure 3.

0 s   At the start, the displacement is zero and the velocity is maximum and positive (this can be deduced from the gradient). The pendulum bob is swinging through the center, moving right.

1 s   The displacement is maximum and positive and the velocity has reduced to zero. The pendulum bob is at the top of its swing to the right.

2 s   The displacement is zero again and the velocity maximum and negative. The pendulum bob is swinging through the center, traveling left.

3 s   The displacement is maximum and the velocity is zero.

4 s   Back to the start.

Terms and quantities

Before going any further, we shall define the terms and quantities used to describe oscillatory motion with reference to the pendulum in Figure 4 swinging backward and forward between A and B.

Cycle
One complete 'there and back' swing, from A to B and back to A.

Amplitude (x_0)
The maximum displacement from the equilibrium position; the swing of the pendulum is symmetric so this could be the distance OA or OB.

The unit of amplitude is the meter.

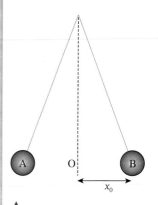

C.1 Figure 4 A swinging pendulum.

Time period (T)
The time taken for one complete cycle from A to B and back to A.

The unit of time period is the second.

Frequency (f)

The number of complete cycles per second.

$$f = \frac{1}{T}$$

The unit of frequency is the hertz (Hz).

Equations for SHM

Standard Level students need not learn the equations for how displacement, velocity or acceleration vary with time. However, understanding these will provide greater clarity about how the motion varies and where the defining equation for SHM comes from.

C.1 Figure 5 Sine curve compared to SHM.

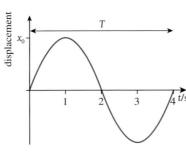

We have seen that simple harmonic motion is sinusoidal so the graph of displacement against time has the shape of a sine function, but what is the equation of the line? Let us compare the displacement–time graph against a sine curve as in Figure 5.

The equation of the sine curve is $y = A \sin \theta$.

Comparing the two curves, we can see that an angle of 2π is equivalent to one cycle of the oscillation so when $t = T$, $\theta = 2\pi$ so $\theta = \frac{2\pi t}{T}$. For the example with time period 4 s, the equivalent angle after 1 s will be:

$$\theta = \frac{(2\pi \times 1)}{4} = \frac{\pi}{2}$$

An alternative way of writing this is: $\theta = 2\pi f t$

So: $x = x_0 \sin (2\pi f t)$

Angular frequency (ω)

This is the angular equivalent to frequency.

$$\omega = 2\pi f$$

The unit of angular frequency is radian per second (rad s^{-1}) so the equation for the displacement of a body experiencing SHM can be written $x = x_0 \sin \omega t$.

If the time starts when the pendulum is at its highest point to the right (point B in Figure 4), then the equation would be $x = x_0 \cos \omega t$; this is still said to be sinusoidal.

Because the time taken for a pendulum to complete one cycle is always the same, even if the amplitude of the swing changes, pendulums used to be extensively used in clock mechanisms. Today, pendulums are not used so often but many clocks still use some sort of mechanical oscillation.

SKILLS

The equation for the displacement–time curve is $x = x_0 \sin (\omega t)$.

You can try plotting this in a graph plotting or spreadsheet program to see the effect of changing x_0 and f.

Set up the spreadsheet as below so that you can vary the values of f and x_0.

Notice the $ sign in the equation. This prevents the values of x_0 and f changing when you copy the equation down into all the cells.

Mass on a spring

If a mass hanging on the end of a spring is lifted up and released, it will bounce up and down as in Figure 6. The forces acting on the mass are weight and the tension in the spring. The weight is always the same but the tension depends on how far the spring is stretched. (Recall from Hooke's law that when you stretch a spring, the tension is proportional to the extension.)

U shows the unstretched length of the spring (Figure 6(i)).

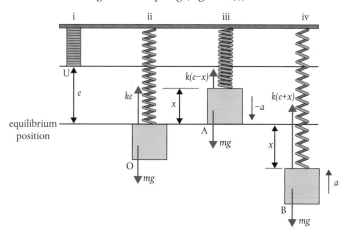

◀ **C.1 Figure 6** Spring and mass in four positions: (i) unstretched spring, (ii) spring in equilibrium with mass attached, (iii) mass above equilibrium position, (iv) mass below equilibrium position.

How can greenhouse gases be modeled as simple harmonic oscillators? (B.2)

When a mass m is hung on the string (Figure 6(ii)), it stretches to the length O with extension e. The mass is at rest so the elastic restoring force and weight are balanced:

$$ke = mg$$

This is the equilibrium position.

The mass is lifted up to position A (Figure 6(iii)), reducing the extension by x, and released. The weight is now larger than the elastic restoring force, so the mass accelerates downward. Applying Newton's second law to find the magnitude of the resultant force:

$$ma = mg - k(e - x) = mg - ke + kx$$

But recall that $ke = mg$.

So: $\qquad ma = mg - mg + kx = kx$

The mass accelerates downward, passing the equilibrium position to position B (Figure 6(iv)). Applying Newton's second law again to find the magnitude of the resultant force:

$$ma = k(e + x) - mg = ke + kx - mg$$

But recall, once more, that $ke = mg$.

So: $\qquad ma = mg + kx - mg = kx$

We can see that the acceleration is always proportional to the distance from the equilibrium position. However, if we consider the displacement instead of the distance, we can see that:

• When the displacement is down (−), the acceleration is up (+).

• When the displacement is up (+), the acceleration is down (−).

They are always in opposite directions, so we insert a minus sign: $ma = -kx$

So: $$a = -\frac{k}{m}x$$

Since acceleration is proportional to displacement but in the opposite direction, this is simple harmonic motion.

Exercise

Q1. State whether the following are examples of simple harmonic motion.
   (a) A ball rolling up and down on a track.
   (b) A cylindrical tube floating in water when pushed down and released.
   (c) A tennis ball bouncing back and forth across a net.
   (d) A bouncing ball.

Q2. A pendulum completes 20 swings in 12 s.
   What is:
   (a) the frequency?
   (b) the angular frequency?

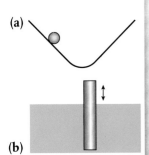

Graphical representation of SHM

When representing the motion of bodies in A.1 Kinematics, we drew displacement–time, velocity–time and acceleration–time graphs. Let us now do the same for the motion of a mass on a spring. The mass on the end of the spring is lifted to point A and released. In this example, we will start with the mass at its highest point (maximum positive displacement) so the equation for the displacement will be $x = x_0 \cos \omega t$.

Displacement–time

As before, O is the equilibrium position and we will take this to be our position of zero displacement. Above this is positive displacement and below is negative.

At A, the mass has maximum positive displacement from O.

At O, the mass has zero displacement from O.

At B, the mass has maximum negative displacement from O.

We can see that the shape of this displacement–time graph is a cosine curve.

C.1 Figure 7 Displacement–time graph.

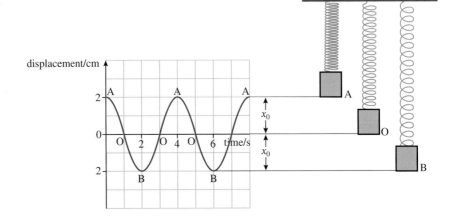

The equation of this line is $x = x_0 \cos \omega t$ where x_0 is the maximum displacement and ω is the angular frequency.

Velocity–time

From the gradient of the displacement–time graph (Figure 7), we can calculate the velocity.

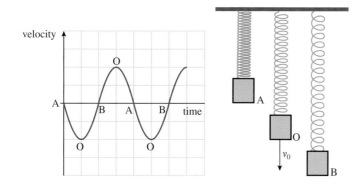

C.1 Figure 8 Velocity–time graph.

At A, gradient = 0 so velocity is zero.

At O, gradient is negative and maximum so velocity is down and maximum.

At B, gradient = 0 so velocity is zero.

The equation of this line is $v = -v_0 \sin \omega t$ where v_0 is the maximum velocity.

Acceleration–time

From the gradient of the velocity–time graph (Figure 8), we can calculate the acceleration.

At A, the gradient is maximum and negative so acceleration is maximum and downward.

At O, the gradient is zero so acceleration is zero.

At B, the gradient is maximum and positive so the acceleration is maximum and upward.

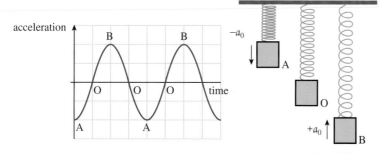

C.1 Figure 9 Acceleration–time graph.

The equation of this line is $a = -a_0 \cos \omega t$ where a_0 is this maximum acceleration.

So: $x = x_0 \cos \omega t$ and $a = -a_0 \cos t$

When displacement increases, the magnitude of acceleration increases proportionally. However, the direction of the acceleration is always opposite to the displacement, which introduces a minus sign to the relationship; in other words: $a \propto -x$

273

The constant of proportionality is ω^2, which means the general equation for SHM is $a = -\omega^2 x$.

We have confirmed that the acceleration of the body is directly proportional to the displacement of the body and always directed toward a fixed point.

Finding the frequencies of a pendulum and mass on a spring

We have seen that the acceleration of a **pendulum bob** is given by the equation:

$$a = -\frac{g}{l} x$$

We now know that:

$$a = -\omega^2 x$$

So:

$$\omega^2 = \frac{g}{l}$$

$$\omega = \sqrt{\frac{g}{l}}$$

$$2\pi f = \sqrt{\frac{g}{l}}$$

$$f = \frac{1}{2\pi}\sqrt{\frac{g}{l}}$$

Since time period is equal to the inverse of frequency:

$$T = 2\pi\sqrt{\frac{l}{g}}$$

For the **mass on a spring**:

$$a = -\frac{k}{m} x$$

So:

$$f = \frac{1}{2\pi}\sqrt{\frac{k}{m}}$$

$$T = 2\pi\sqrt{\frac{m}{k}}$$

When calculating $\cos \omega t$, you must have your calculator set on radians.

Worked example   HL

A mass on a spring is oscillating with a frequency 0.2 Hz and amplitude 3.0 cm. What is the displacement of the mass 10.66 s after it is released from the top?

Solution

Since this is SHM: $x = x_0 \cos \omega t$

Where: x = displacement

x_0 = amplitude = 3 cm

v = angular velocity = $2\pi f = 2\pi \times 0.2$
= 0.4π Hz

t = time = 10.66 s

$x = 0.03 \times \cos(0.4\pi \times 10.66)$

$x = 0.02$ m
= 2 cm

Q3. For the same mass on a spring in the Worked example on page 274, calculate the displacement after 1.55 s.

Q4. Draw a displacement–time sketch graph for this motion.

Q5. A long pendulum has time period 10 s. If the bob is displaced 2 m from the equilibrium position and released, how long will it take to move 1 m?

Q6. As a mass on a spring travels upward through the equilibrium position, its velocity is 0.5 m s^{-1}. If the frequency of the pendulum is 1 Hz, what will the velocity of the bob be after 0.5 s?

Representing SHM with circular motion

By applying the conservation of energy to (or simply observing) a swinging pendulum, it is clear that the maximum speed at which the bob swings past the equilibrium position is related to the maximum height of the swing and the frequency, but to find exactly how these quantities are related requires some more sophisticated mathematical analysis involving calculus. An alternative method is to use the horizontal component of circular motion. If you were to observe the ball in Figure 10 from below, you would only see one component of the motion, so the ball would appear to be moving up and down with an amplitude equal to the radius of the circle.

▲ **C.1 Figure 10** When a ball moving in a circle is viewed from below, it looks like it is moving with SHM.

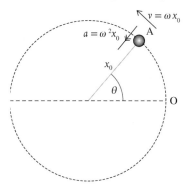

◀ **C.1 Figure 11** Circular analogy of SHM.

Let us consider a ball traveling in a circle of radius x_0 with constant speed v. The ball starts from point O and at some time, t, it is at position A as shown in Figure 11. In this time, the radius has swept out angle θ. From our previous study of circular motion, we know that:

$$\text{speed} = \frac{2\pi r}{T} = \omega r \text{ so in this case } v = \omega x_0$$

The centripetal acceleration = $\omega^2 x_0$ and is toward the center.

How can circular motion be used to visualize simple harmonic motion? (A.2)

We are only interested in one component of the motion. It does not matter which we take but this time we will consider the horizontal component.

Displacement

As can be seen in Figure 12, the horizontal component of displacement, $x = x_0 \cos \theta$.

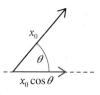

▲ **C.1 Figure 12** The horizontal component of displacement, $x = x_0 \cos \theta$.

Velocity

The velocity of the ball is directed perpendicular to the radius so the horizontal component of the velocity, $v_x = -\omega x_0 \sin \theta$. The negative sign is due to the fact that we are only considering one dimension, and this is in the opposite direction to the displacement, which was positive.

From Pythagoras' theorem, we know that $\sin^2 \theta + \cos^2 \theta = 1$

Rearranging gives: $\sin \theta = \sqrt{1 - \cos^2\theta}$

Substituting this in the equation for v_x gives: $v_x = -\omega x_0 \sqrt{1 - \cos^2\theta} = -\omega\sqrt{x_0^2 - x_0^2 \cos^2\theta}$

But: $x_0^2 \cos^2\theta = x$

So: $v_x = -\omega \sqrt{x_0^2 - x^2}$

which has its maximum value when: $x = 0$

Acceleration

The acceleration is *toward* the center so the horizontal component, $a_x = -\omega^2 x_0 \cos \theta$, again with a negative sign because of its direction.

So we have: $x = x_0 \cos \theta$

and $a_x = -\omega^2 x_0 \cos \theta$

So: $a_x = -\omega^2 x$

This shows that the horizontal motion is simple harmonic. We can also see that ω^2 is the constant of proportionality, relating the acceleration with displacement, which implies that if the acceleration increases a lot with a small displacement, then the frequency will be high. So a mass on a stiff spring will oscillate with higher frequency than a mass on a soft one, which seems to be as it is in practice.

Summary of equations

Standard Level students are not required to use these equations, but might find it helpful to have a quantitative understanding in addition to describing things qualitatively.

If a body is oscillating with SHM, starting from a position of maximum positive displacement, then its displacement, velocity and acceleration at any given time t can be found from the following equations:

$$x = x_0 \cos \omega t$$

$$v = -\omega x_0 \sin \omega t$$

$$a = -\omega^2 x_0 \cos \omega t$$

However, if the timing starts when the body is passing through the center, traveling in a positive direction, the equations of motion are:

$$x = x_0 \sin \omega t$$

$$v = \omega x_0 \cos \omega t$$

$$a = -\omega^2 x_0 \sin \omega t$$

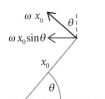

C.1 Figure 13 Finding a component of a vector.

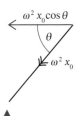

C.1 Figure 14 Finding a component of another vector.

If you have done differentiation in math, then you will understand that if:

displacement is

$x = x_0 \cos \omega t$

then velocity is

$\dfrac{dx}{dt} = -x_0 \omega \sin \omega t$

and acceleration is

$\dfrac{d^2x}{dt^2} = -x_0 \omega^2 \cos \omega t$

This implies that:

$a = -\omega^2 x$

This is a much shorter way of deriving the result!

At a given displacement x, the velocity and acceleration can be found from the following:

$$v = \omega \sqrt{x_0^2 - x^2}$$

$$a = -\omega^2 x$$

Worked example

A pendulum is swinging with a frequency of 0.5 Hz. What is the size and direction of the acceleration when the pendulum has a displacement of 2 cm to the right?

Solution

Assuming the pendulum is swinging with SHM, then we can use the following equation to calculate the acceleration:

$$a = -\omega^2 x$$

$$\omega = 2\pi f = 2\pi \times 0.5 = \pi$$

$$a = -\pi^2 \times 0.02 = -0.197 \, \text{m s}^{-2} \qquad \text{since } -\text{ve direction is to the left}$$

Worked example   HL

A pendulum bob is swinging with SHM at a frequency of 1 Hz and amplitude 3 cm. At what position will the bob be moving with maximum velocity and what is the size of the velocity?

Solution

Since the motion is SHM: $v = \omega \sqrt{x_0^2 - x^2}$

This is maximum when $x = 0$, which is when the pendulum swings through the central position.

maximum value = ωx_0 where $\omega = 2\pi f = 2 \times \pi \times 1 = 2\pi \, \text{rad s}^{-1}$

maximum $v = 2\pi \times 0.03 = 0.188 \, \text{m s}^{-1}$ $\qquad a = -\pi^2 \times 0.02 = -0.197 \, \text{m s}^{-2}$

Exercise   HL

Q7. A long pendulum swings with a time period of 5 s and an amplitude of 2 m.
   (a) What is the maximum velocity of the pendulum?
   (b) What is the maximum acceleration of the pendulum?

Q8. A mass on a spring oscillates with amplitude 5 cm and frequency 2 Hz. The mass is released from its highest point. Calculate the velocity of the mass after it has traveled 1 cm.

Q9. A body oscillates with SHM of time period 2 s. What is the amplitude of the oscillation if its velocity is $1 \, \text{m s}^{-1}$ as it passes through the equilibrium position?

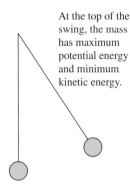

At the top of the swing, the mass has maximum potential energy and minimum kinetic energy.

At the bottom of the swing, the mass has maximum kinetic energy and minimum potential energy.

C.1 Figure 15 In the simple pendulum, energy is transferring from one store to another as the bob moves.

How does damping affect periodic motion? (C.4)

Energy transfers during SHM

Standard Level students must be able to describe the energy transfers during an oscillation. Higher Level students are also required to calculate the energy types at given displacements or velocities.

If we once again consider the simple pendulum, we can see that its energy is transferred as it swings.

Kinetic energy

We have already shown that the velocity of the mass is given by the equation:

$$v = \omega \sqrt{x_0^2 - x^2}$$

From definition: $E_k = \frac{1}{2}mv^2$

Substituting: $E_k = \frac{1}{2}m\omega^2(x_0^2 - x^2)$

Kinetic energy is a maximum at the bottom of the swing where $x = 0$.

So: $E_{k\,max} = \frac{1}{2}m\omega^2x_0^2$

At this point, the potential energy is zero.

Total energy

The total energy at any moment in time is given by:

$$\text{total energy} = E_k + E_p$$

So at the bottom of the swing:

$$\text{total energy} = \frac{1}{2}m\omega^2x_0^2 + 0 = \frac{1}{2}m\omega^2x_0^2$$

Since no work is done on the system, according to the law of conservation of energy, the total energy must be constant.

So: $\text{total energy} = \frac{1}{2}m\omega^2x_0^2$

Potential energy

Potential energy at any moment = total energy − kinetic energy

So: $E_p = \frac{1}{2}m\omega^2x_0^2 - \frac{1}{2}m\omega^2(x_0^2 - x^2)$

$E_p = \frac{1}{2}m\omega^2x^2$

HL So at the bottom of the swing:

$$\text{total energy} = \frac{1}{2}m\omega^2x_0^2 + 0 = \frac{1}{2}m\omega^2x_0^2$$

Since no work is done on the system, according to the law of conservation of energy, the total energy must be constant.

Graphical representation

Kinetic energy

From previous examples, we know that the velocity, $v = -v_0 \sin \omega t$.

So:
$$\frac{1}{2}mv^2 = \frac{1}{2}mv_0^2 \sin^2 \omega t$$

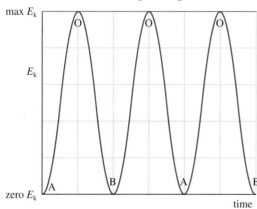

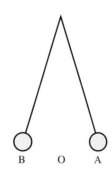

C.1 Figure 16 The graph of kinetic energy vs time is a $\sin^2$ curve.

Potential energy

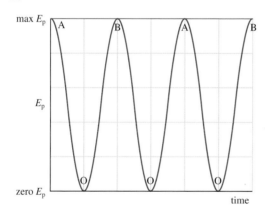

C.1 Figure 17 The graph of potential energy vs time is a $\cos^2$ curve.

The graph of potential energy can be found from $E_p = \frac{1}{2}m\omega^2 x^2$.

Since:
$$x = x_0 \cos \omega t$$
$$E_p = \frac{1}{2}m\omega^2 x_0^2 \cos^2 \omega t$$
$$= \frac{1}{2}mv_0^2 \cos^2 \omega t$$

Total energy

If these two graphs are added together, it gives a constant value, equal to the total energy.

This might remind you of Pythagoras: $1 = \cos^2\theta + \sin^2\theta$

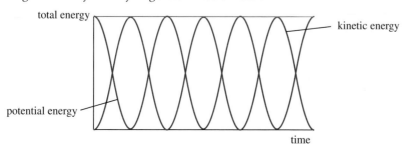

C.1 Figure 18 Total energy vs time.

What physical explanation leads to the enhanced greenhouse effect? (NOS)

Worked example   HL

A pendulum bob of mass 200 g is oscillating with amplitude 3 cm and frequency 0.5 Hz. How much kinetic energy will the bob have as it passes through the origin?

Solution

Since the bob has SHM: $E_{k\,max} = \frac{1}{2}m\omega^2 x_0^2$

Where: $x_0 = 0.03$ m and $\omega = 2\pi f = 2\pi \times 0.5 = \pi$

$$E_{k\,max} = \frac{1}{2} \times 0.2 \times \pi^2 \times (0.03)^2 = 8.9 \times 10^{-4}\,\text{J}$$

Exercise   HL

Q10. A pendulum bob of mass 100 g swings with amplitude 4 cm and frequency 1.5 Hz. Calculate:

  (a) the angular frequency of the pendulum

  (b) the maximum kinetic energy of the bob

  (c) the maximum potential energy of the bob

  (d) the kinetic energy of the bob when the displacement is 2 cm

  (e) the potential energy of the bob when the displacement is 2 cm.

SKILLS

At the start of this chapter, we analyzed the forces acting on the bob of a simple pendulum and found that, providing the displacement is small, the acceleration was given by $a = -\frac{gx}{l}$ (see Figure 19).

Then by comparing SHM with motion in a circle, we found that the acceleration for a body moving with SHM is $a = -\omega^2 x$.

So: $$\omega^2 x = \frac{gx}{l}$$
$$\omega^2 = \frac{g}{l}$$

Now: $$\omega^2 = \left(\frac{2\pi}{F_t}\right)^2 = \frac{g}{l}$$

Rearranging gives: $$F_t^2 = \frac{4\pi^2 l}{g} \text{ or } F_t = 2\pi\sqrt{\frac{l}{g}}$$

So if we measure the time period for different lengths of pendulum, we should find that F_t^2 is proportional to l. This means that a graph of F_t^2 vs l would be a straight line with gradient $= \frac{4\pi^2}{g}$.

The typical length of pendulum used in the lab has a time period of about 1 s so measuring one swing with a stop watch would be quite difficult. It is much better to time 10 swings then divide the time by 10 to give the time period. Alternatively, the pendulum could be made to swing through a photogate and a computer used to record the time period (see Figure 20).

The computer records every time the bob passes in and out of the photogate, but if you set the software to 'pendulum timing', then it will record the time between the first time the bob enters the gate and the third time. This will be the time period.

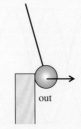
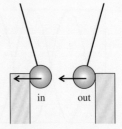
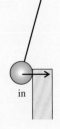

C.1 Figure 19 Pendulum with small displacement.

C.1 Figure 20 Pendulum and photogate.

Phase

If we take two identical pendulum bobs, displace each bob to the right and release them at the same time, then each will have the same displacement at the same time. We say the oscillations are **in phase**. If one is pulled to the left and the other to the right, then they are **out of phase** (see Figure 21).

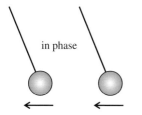
in phase

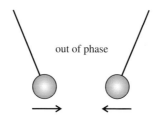
out of phase

This can be represented graphically.

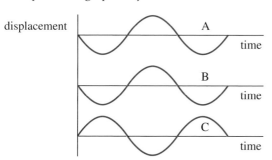

A and B represent motions that are in phase.

B and C represent motions that are out of phase.

C.1 Figure 21 The pendulum bobs are in phase when they swing together.

C.1 Figure 22
Displacement–time graphs for bodies in and out of phase.

When juggling balls, they go up and down at different times – they are out of phase.

Challenge yourself

1. A 12 cm long, narrow cylinder floats vertically in a bucket of water so that 2 cm is above the surface. It is then made to oscillate by pressing the end underwater. Calculate the time period of the motion.

HL

Phase difference

The phase difference is represented by an angle (usually in radians). We can see from graphs B and C in Figure 22 that if two oscillations are completely out of phase, then the graphs are displaced by an angle π. We say the **phase difference** is π.

If the displacement of an oscillating body A is $x_A = x_0 \sin(\omega t)$, then the equation for the displacement of body B with oscillations that are an angle φ out of phase will be:

$$x_B = x_0 \sin(\omega t + \varphi)$$

How can the understanding of simple harmonic motion apply to the wave model? (NOS)

C.1 Figure 23

Displacement–time graphs for bodies A and B.

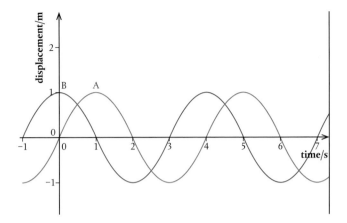

C.1 Figure 23

Displacement–time graphs for bodies A and B.

Figure 23 shows the displacement–time graphs of these two oscillations. We see that the sin curve is shifted to the right by 1 s, which is ¼ of a cycle. One complete cycle is 2π radians so the phase angle is $\frac{\pi}{2}$.

Note that A starts at the equilibrium position while moving toward its maximum positive displacement. B starts at the maximum positive displacement and is therefore ahead of A. If the phase angle was $-\frac{\pi}{2}$, B would lag behind A.

HL end

Guiding Questions revisited

What makes the harmonic oscillator model applicable to a wide range of physical phenomena?

Why must the defining equation of simple harmonic motion take the form it does?

How can the energy and motion of an oscillation be analyzed graphically and algebraically?

In this chapter, we have considered simple harmonic motion as a particular example of an oscillation. In doing so, we recognized:
- The inverse proportionality of the time period of a complete oscillation and the frequency of oscillations (the number per second).
- The amplitude in an oscillating system is the maximum displacement of the body from its equilibrium position.
- Simple harmonic oscillations as having acceleration proportional to displacement and in the opposite direction (the latter indicated by the minus sign).
- The magnitude of the constant of proportionality for simple harmonic motion is the square of angular frequency, where angular frequency is the product of 2π and the frequency of oscillations.
- Two common models for simple harmonic motion are mass–spring systems (with time period related to the mass and the spring constant) and simple pendula (with time period related to the length of the string and gravitational field strength).
- Many real-world scenarios as being relatable to these two models in design or problem-solving.
- Potential energy increases with displacement.
- Kinetic energy is maximum as the body passes through the equilibrium position.

- All energy and motion quantities can be analyzed through algebra (using the relationships between displacement and velocity, and velocity and acceleration) or graphically (using the gradients of graphs).
- The sinusoidal equation for displacement as a function of time can be differentiated to derive expressions for velocity and maximum velocity and, in turn, kinetic energy, total energy and potential energy.
- Phase angle as a means to compare to points relative to proportions of a 2π oscillation cycle.

HL end

Practice questions

1. A mass on a spring is displaced from its equilibrium position. Which graph represents the variation of acceleration with displacement for the mass after it is released?

A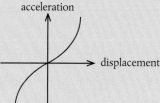

B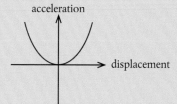

C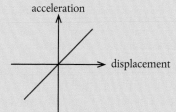

D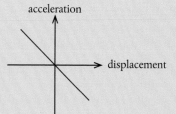

(Total 1 mark)

2. An object performs simple harmonic motion (SHM). The graph on the right shows how the velocity v of the object varies with time t. The displacement of the object is x and its acceleration is a. What is the variation of x with t and the variation of a with t?

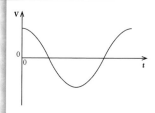

A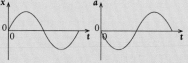

C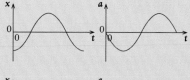

B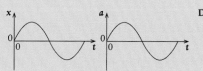

D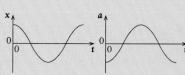

(Total 1 mark)

283

3. The bob of a pendulum has an initial displacement x_0 to the right. The bob is released and allowed to oscillate. The graph shows how the displacement varies with time. At which point is the velocity of the bob at its maximum magnitude directed toward the left?

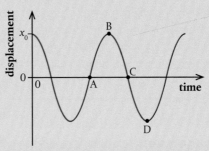

(Total 1 mark)

4. The four pendulums below have been cut from the same uniform sheet of board. They are attached to the ceiling with strings of equal length. Which pendulum has the shortest period?

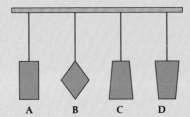

(Total 1 mark)

5. A mass at the end of a vertical spring and a simple pendulum both perform oscillations on Earth that are simple harmonic with time period T. The pendulum and the mass–spring system are taken to the Moon. The acceleration of free fall on the Moon is smaller than that on Earth. Which is correct about the time periods of the pendulum and the mass–spring system on the Moon?

	Simple pendulum	Mass–spring system
A	T	T
B	greater than T	T
C	greater than T	greater than T
D	T	greater than T

(Total 1 mark)

6. HL A particle of mass m oscillates with simple harmonic motion (SHM) of angular frequency ω. The amplitude of the SHM is A. What is the kinetic energy of the particle when it is halfway between the equilibrium position and one extreme of the motion?

A $\dfrac{mA^2\omega^2}{4}$ B $\dfrac{3mA^2\omega^2}{8}$

C $\dfrac{9mA^2\omega^2}{32}$ D $\dfrac{15mA^2\omega^2}{32}$

(Total 1 mark)

7. The graph shows the variation with time t of the acceleration a of an object X undergoing simple harmonic motion (SHM).

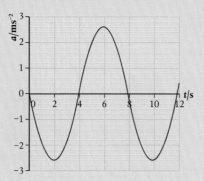

(a) Define simple harmonic motion (SHM). (2)

(b) X has a mass of 0.28 kg. Calculate the maximum force acting on X. (1)

(c) Determine the maximum displacement of X. Give your answer to an appropriate number of significant figures. (4)

(d) **HL** A second object Y oscillates with the same frequency as X but with a phase difference of $\frac{\pi}{4}$. Copy the graph and sketch how the acceleration of object Y varies with t. (2)

(Total 9 marks)

8. A student is investigating a method to measure the mass of a wooden block by timing the period of its oscillations on a spring.

(a) Describe the conditions required for an object to perform simple harmonic motion (SHM). (2)

(b) A 0.52 kg mass performs simple harmonic motion with a period of 0.86 s when attached to the spring. A wooden block attached to the same spring oscillates with a period of 0.74 s. Calculate the mass of the wooden block. (2)

frictionless surface

(c) **HL** In carrying out the experiment, the student displaced the block horizontally by 4.8 cm from the equilibrium position. Determine the total energy in the oscillation of the wooden block. (3)

(d) **HL** A second identical spring is placed in parallel, and the experiment in (b) is repeated. Suggest how this change affects the fractional uncertainty in the mass of the block. (3)

(Total 10 marks)

9. A vertical solid cylinder of uniform cross-sectional area A floats in water. The cylinder is partially submerged. When the cylinder floats at rest, a mark is aligned with the water surface. The cylinder is pushed vertically downward so that the mark is a distance x below the water surface.

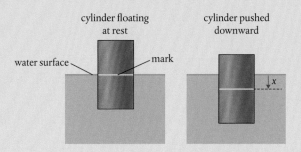

At time $t = 0$, the cylinder is released. The resultant vertical force F on the cylinder is related to the displacement x of the mark by:

$$F = -\rho A g x$$

where ρ is the density of water.

(a) Outline why the cylinder performs simple harmonic motion when released. (1)

(b) The mass of the cylinder is 118 kg and the cross-sectional area of the cylinder is 2.29×10^{-1} m². The density of water is 1.03×10^3 kg m⁻³. Show that the angular frequency of oscillation of the cylinder is about 4.4 rad s⁻¹. (2)

(c) **HL** The cylinder was initially pushed down a distance $x = 0.250$ m. Determine the maximum kinetic energy $E_{k\,max}$ of the cylinder. (2)

(d) **HL** Copy the axes below and sketch the graph to show how the kinetic energy of the cylinder varies with time during **one** period of oscillation T. (2)

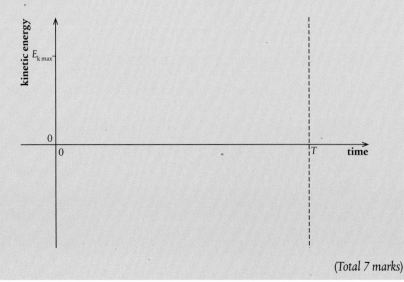

(Total 7 marks)

10. **HL** Two masses of 0.90 kg and 1.10 kg are hung vertically from identical springs on a common support, each with spring constant 39.48 N m^{-1}. Both are released simultaneously from a position of maximum extension to describe simple harmonic motion.

(a) Calculate the frequencies of the two masses. (2)

In acoustics, a 'beat' is an interference pattern between two sounds of similar frequencies. The 'beat frequency' is the difference in the two frequencies that interfere to produce the beats.

(b) Calculate the beat period and frequency. (2)

(Total 4 marks)

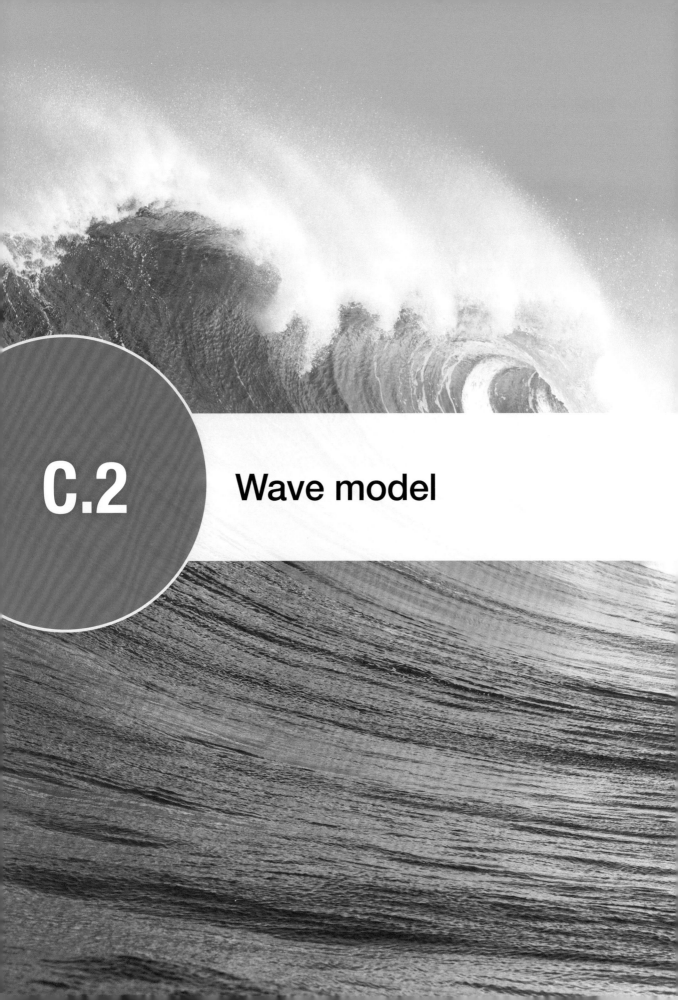

C.2

Wave model

◀ Waves change direction due to changing depth of the water. The change in wave speed can also cause the wave to break, resulting in the white foam visible in the photo.

Guiding Questions

What are the similarities and differences between different types of wave?

How can the wave model describe the transmission of energy as a result of local disturbances in a medium?

What effect does a change in the frequency of oscillation or medium through which the wave is traveling have on the wavelength of a traveling wave?

The wave created in a stadium is not a physical wave, but it can help us understand the nature of wave motion. To make a wave travel to the right, each person stands up just after the person to their left. Each person is out of phase with their neighbors. A stadium wave is just a pulse, but a continuous wave could be created if everyone stood up and sat down repeatedly.

What we see moving to the right is not people – they all stay in the same position – it is the disturbance.

The reason that this is not a real wave is that nothing is passed from one person to another. In real waves, energy is transmitted. Another reason this is not a real wave is that it exhibits none of the properties of a wave. If there is a wall in the stadium, the wave does not reflect off it, and if we have two waves traveling in opposite directions, they do not add together when they intersect. To make the wave go slower, we just have to tell the spectators to wait a short time before they stand up and sit down. Their frequency is the same but the wave speed is lower, which would mean that the distance between peaks becomes smaller.

▲
A 'stadium wave'.

Students should understand:

transverse and longitudinal traveling waves
wavelength λ, frequency f, time period T, and wave speed v applied to wave motion as given by $v = f\lambda = \dfrac{\lambda}{T}$
the nature of sound waves
the nature of electromagnetic waves
the differences between mechanical waves and electromagnetic waves.

Nature of Science

Complex models are often built of simple units. The complex motion of a wave becomes simple when we realize that each part is simply moving back and forth like a row of slightly out-of-step simple pendulums. In this chapter, you will be learning about waves in strings and springs, water waves, sound waves, and electromagnetic waves; all completely different things but with similar characteristics. When scientists started modeling the motion of ocean waves, they probably had no idea their work would one day be applied to light.

Waves

The word **wave** was originally used to describe the way that a water surface behaves when it is disturbed. We use the same model to explain sound, light, and many other physical phenomena. This is because they have some similar properties to water waves, so let us first examine the way water waves spread out.

If a stone is thrown into a pool of water, it disturbs the surface. The disturbance spreads out or **propagates** across the surface, and this propagating disturbance is called a **wave**. Observing water waves, we can see that they have certain basic properties (in other words, they do certain things).

Reflection

If a water wave hits a wall, the waves reflect.

Refraction

When sea waves approach a beach, they change direction because the difference in height of different parts of the sea floor leads to different wave speeds.

Interference

When two waves cross each other, they can add together, creating an extra big wave, or cancel out.

Diffraction

When water waves pass through a small opening, the waves spread out.

Anything that reflects, refracts, interferes and diffracts can also be called a wave.

Transfer of energy

Waves in the ocean are caused by winds that disturb the surface of the water. A big storm in the Atlantic Ocean can cause waves that break on the beaches of the west coast of Europe and the east coast of the Americas. The storm gives the water energy, which is then spread out in the form of water waves. So a wave is the transfer of energy through the disturbance of some medium.

Waves change direction as they approach a beach.

What happens when waves overlap or coincide? (C.3, C.4)

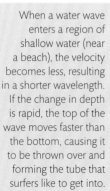

When a water wave enters a region of shallow water (near a beach), the velocity becomes less, resulting in a shorter wavelength. If the change in depth is rapid, the top of the wave moves faster than the bottom, causing it to be thrown over and forming the tube that surfers like to get into.

Although the water wave is the 'original' wave, it is not the simplest one to begin with. So to help understand how waves propagate, we will first consider two examples of one-dimensional waves: a wave in a string and a wave in a slinky spring.

Wave pulse in a string

If you take one end of a very long string and give it a flick (move it up and down once quickly), then you will see disturbance moving along the string. This is called a wave **pulse**. In lifting up the string and flicking it down, you have given the string energy. This energy is now being transferred along the string at a constant speed. This speed is called the **wave speed**.

To understand how the energy is transferred, consider the case where the rope is just lifted as shown in Figure 1. Here the string is represented by a line of balls, each joined to the next by an invisible string. When the end was lifted, the first ball lifted the next one, which lifted the next, etc. transferring energy from left to right.

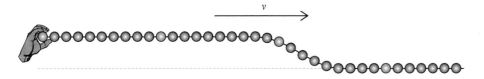

◀ **C.2 Figure 1** Energy transferred along the string.

If the end is moved up then down, then a pulse is sent along the string as in Figure 2.

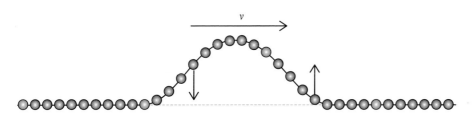

◀ **C.2 Figure 2** A pulse moves from left to right.

The particles at the front are moving up, and the ones at the back are moving down. As the pulse moves along the string, each part of the string has the same motion, up then down, but they do not all do it at the same time; they are **out of phase** (Figure 2). It is like a wave going around a stadium: the crowd all stand up then sit down at different times.

Reflection of a wave pulse

If the pulse meets a fixed end (e.g. a wall), it exerts an upward force on the wall. The wall being pushed up then pushes back down on the string, sending an inverted reflected pulse back along the string (Figure 3).

Can the wave model inform the understanding of quantum mechanics? (NOS)

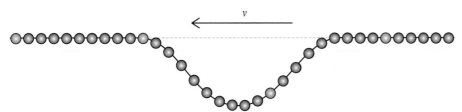

◀ **C.2 Figure 3** A wave pulse reflected off a fixed end.

If the end of the string is loose, then you also get a reflection, but this time it is reflected without phase change (Figure 4). It is just as if there is a hand at the end moving up then down like the one that made the original pulse.

C.2 Figure 4 A wave pulse reflected off a free end.

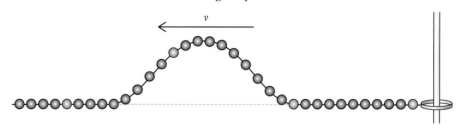

Even though the reflected wave from a free end looks the same as a wave resulting from the original hand movement (but moving in the opposite direction), note that the amplitude of the wave pulse at the free end is twice the amplitude of the movement of the hand.

A reflected wave is not only produced when the wave meets an end but whenever there is a change in the medium. If two different strings are joined together, there will be a reflection at the boundary between the strings. In this case, not all the wave is reflected; some is transmitted. If the second string is heavier, then the reflected wave is inverted as it is off a fixed end (Figure 3), but if the second string is lighter, then the wave is reflected as if off a free end (Figure 4).

Superposition of wave pulses

If two wave pulses are sent along a string from each end, they will coincide in the middle. When this happens, the displacements of each pulse add vectorially. This results in two peaks adding but a trough and a peak canceling out (Figure 5).

C.2 Figure 5 Wave pulses superpose.

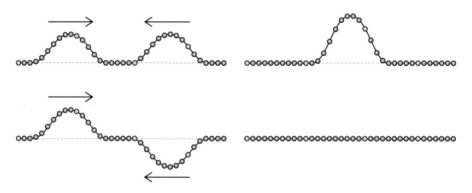

Note that when the waves cancel, it appears that the energy has disappeared, but if this was an animation, you would see the particles are actually moving up and down so the particles have kinetic energy.

Continuous wave in a string (transverse wave)

If the end ball on the string is moved up and down with simple harmonic motion, then a short time later, the next ball along the string will also move up and down with the same motion. This motion is passed along the string until all the parts of the string are moving with SHM, each with the same amplitude and frequency but different phase. In Figure 6, the wave is moving from left to right as a result of the end being disturbed. The green ball is just about to move downward; it is $\frac{3}{4}$ of a cycle $(\frac{3\pi}{2})$ out of phase with the end ball.

Amplitude (A)

The maximum displacement of the string from the equilibrium position.

Wave speed (v)

The distance traveled by the wave profile per unit time.

Wavelength (λ)

The distance between two consecutive crests or any two consecutive points that are in phase.

Frequency (f)

The number of complete cycles that pass a point per unit time.

Period (T)

Time taken for one complete wave to pass a fixed point $(T = \frac{1}{f})$.

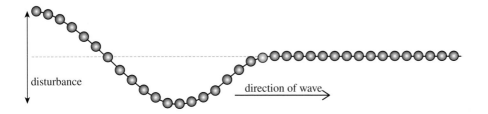

C.2 Figure 6 Forming a continuous wave.

We can see that after the end has completed one cycle, the front of the wave will be **in phase** with the original oscillation. The distance to this point depends on the speed of the wave and is called the **wavelength**, λ (Figure 7).

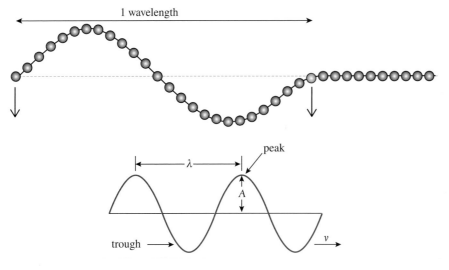

C.2 Figure 7 One complete cycle.

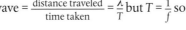

C.2 Figure 8 The quantities used to define a wave.

Relationship between v, f, and $λ$

If you observe a continuous wave moving along a string from a position at rest relative to the string, then you will notice that the time between one peak passing and the next is T, the period. In this time, the wave profile has progressed a distance equal to the wavelength, $λ$. The velocity of the wave $= \frac{\text{distance traveled}}{\text{time taken}} = \frac{λ}{T}$ but $T = \frac{1}{f}$ so:

$$v = fλ$$

How can the length of a wave be determined using concepts from kinematics? (A.1)

Worked example

The A string of a guitar vibrates at 110 Hz. If the wavelength is 153 cm, what is the velocity of the wave in the string?

Make sure to change cm to m.

Solution

$$v = fλ$$

$$f = 110 \, \text{Hz and } λ = 1.53 \, \text{m}$$

$$v = 110 \times 1.53 \, \text{m s}^{-1}$$

$$= 168.3 \, \text{m s}^{-1}$$

Worked example

A wave in the ocean has a period of 10 s and a wavelength of 200 m. What is the wave speed?

Solution

$$T = 10\,s$$
$$f = \frac{1}{T}\,Hz$$
$$= 0.1\,Hz$$
$$v = f\lambda$$
$$v = 0.1 \times 200\,m\,s^{-1}$$
$$= 20\,m\,s^{-1}$$

Exercise

Q1. Calculate the wave velocity of a tsunami with time period 30 min and wavelength 500 km.

Q2. Two strings are joined together as shown in the figure on the left.

   (a) If the wave velocity in the thin string is twice its velocity in the thick string, calculate the wavelength of the wave when it gets into the thick string.

   (b) When the wave meets the knot, part of it will be reflected. Explain whether the reflected wave will be inverted or not.

   (c) Why is the amplitude of the wave in the thick string smaller than in the thin string?

Q3. The end of a string is moved up and down with time period 0.5 s. If the wavelength of the wave is 0.6 m, what is the velocity of the wave?

0.4 m

> A wave in which the direction of disturbance is perpendicular to the direction of the transfer of energy is called a **transverse wave**.

Graphical representation of a transverse wave

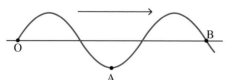

C.2 Figure 9 A snapshot of a transverse wave.

> Because the horizontal axis is time, the separation of the peaks represents the time period, not the wavelength.

> The event that will happen next is to the right on the graph but the part of the wave that will arrive next is to the left on the wave.

There are two ways we can represent a wave graphically: by drawing a displacement–time graph for one point on the wave, or a displacement–position graph for each point along the wave.

Displacement–time

Consider points A and B on the transverse wave in Figure 9.

Point A is moving up and down with SHM as the wave passes. At present, it is at its maximum negative displacement. As the wave progresses past A, this point will move up and then down (Figure 10).

We can also draw a graph for point B (Figure 11). This point starts with zero displacement then goes up.

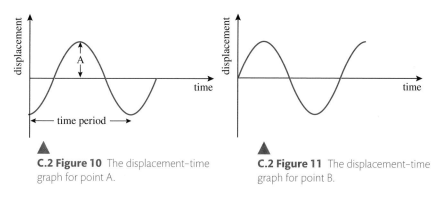

C.2 Figure 10 The displacement–time graph for point A.

C.2 Figure 11 The displacement–time graph for point B.

Displacement–position

To draw a displacement–position graph, we must measure the displacements of all the points on the wave at one moment in time.

Figure 12 shows the graph at the same time as the snapshot in Figure 10 was taken. The position is measured from point O.

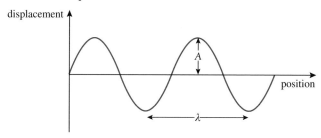

C.2 Figure 12 The displacement–position graph for all points at one time.

This is just like a snapshot of the wave – however, depending on the scale of the axis, it might not look quite like the wave.

The equation of a wave

We have seen that each point on the wave oscillates with simple harmonic motion. So if we take the end point of the string, its displacement is related to time by the equation: $y = A \sin \omega t$. If we now take a point a little bit further down the string, it will also be moving with SHM but a little behind the first one; let us say an angle θ behind. The equation for the displacement of this point is therefore: $y = A \sin(\omega t - \theta)$. This phase angle depends on how far along the string we go. In other words, θ is proportional to x or $\theta = kx$ where k is a constant.

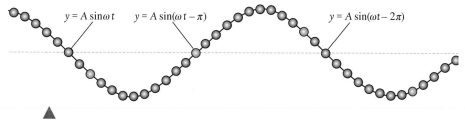

C.2 Figure 13 Different parts of the wave have different phase.

We can now write an equation for the displacement of any point, $y = A \sin(\omega t - kx)$.

If the point is one whole wavelength from the end, then the points will be in phase so $\theta = 2\pi$ or $k\lambda = 2\pi$, which means that $k = \frac{2\pi}{\lambda}$ (Figure 13).

The wave equation then becomes: $y = A \sin(\omega t - \frac{2\pi x}{\lambda})$

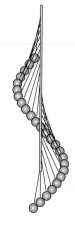

C.2 Figure 14 A wave profile is created when a row of pendulums are released at different times showing how a wave is made of a series of oscillations of different phase.

Waves in a slinky spring (longitudinal waves)

We can transfer energy along a long spring (a slinky) by moving the end up and down. We can also transfer energy if we move it in and out. Instead of peaks and troughs, compressions (squashed bits) and rarefactions (spread out bits) move along the spring, as seen in Figure 15. This doesn't look like a wave but we will find out that it has all the same properties as one.

This type of wave is called a longitudinal wave.

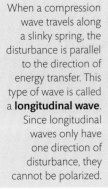

C.2 Figure 15 The difference between a compression wave in a spring and the transverse wave in a string is the direction of disturbance.

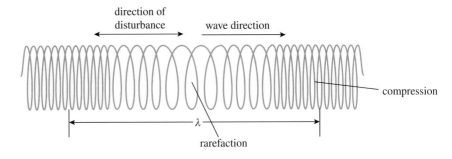

When a compression wave travels along a slinky spring, the disturbance is parallel to the direction of energy transfer. This type of wave is called a **longitudinal wave**. Since longitudinal waves only have one direction of disturbance, they cannot be polarized.

Reflection

When the wave in a spring meets a fixed end, the spring will push the wall so, according to Newton's third law, the wall will push back. This sends a reflected wave back along the spring (Figure 16).

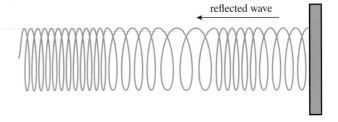

C.2 Figure 16 A wave in a spring is reflected off a wall.

Graphical representation of longitudinal waves

To get a better understanding of a longitudinal wave, let us consider a row of balls connected by springs as in Figure 17.

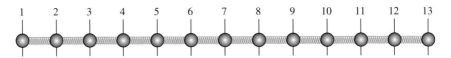

C.2 Figure 17 Balls connected by springs prior to any disturbance.

If the ball on the left is pulled to the left, then the spring connecting it to the next ball will be stretched, causing the next ball to move. In this way, the displacement is passed from one ball to the next. If the ball is moved with SHM, then a continuous wave is sent along the line. Each ball will move with the same frequency but a slightly different phase. The distance between two balls in phase is the wavelength. This is the same as the distance between two compressions. In Figure 18, it can be seen that the end balls are both at the same point in the cycle, at the equilibrium position moving left, so the distance between them is one wavelength.

The amplitude of the wave is the maximum displacement from the equilibrium position. This is marked by the letter A in Figure 18.

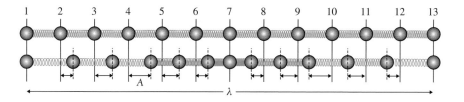

C.2 Figure 18 Showing the displaced positions of parts of a wave.

Displacement–time graph

A displacement–time graph shows how the displacement of one point varies with time.

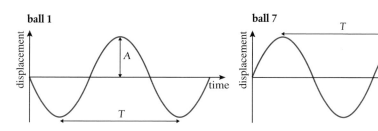

C.2 Figure 19 Balls moving in opposite directions to one another.

Figure 19 shows two graphs for two different points. Ball 1 is about to move to the left so its displacement will become negative. Ball 7 is about to move to the right so its displacement will become positive.

Displacement–position graph

A displacement–position graph represents the position of all the particles at one time. If the motion of the particles is sinusoidal, then the shape of the graph will be a sine curve as shown in Figure 20.

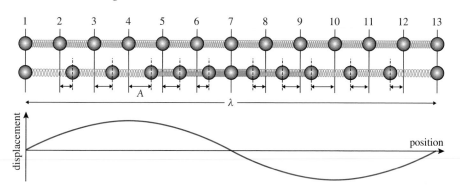

When looking at a wave traveling to the right, you must remember that the particles to the right, are lagging behind those to the left, so if we want to know what the displacement of particle 13 will be next, we look at the particle to the left. We can therefore deduce that particle 13 is moving to the left. When looking at a graph, time progresses from left to right so you can see what will happen next by looking to the right.

C.2 Figure 20 Check the axes in waves problems carefully.

A GeoGebra worksheet linked to this topic is available in the eBook.

In Figure 20, you can see how balls 1, 7, and 13 have zero displacement, ball 4 has maximum positive displacement, and ball 10 has maximum negative displacement.

Nature of Science

The speed of sound is approximately 340 m s^{-1}, which is larger than many everyday objects, so it proved quite difficult for early scientists to measure this accurately. The first methods assumed that light was instantaneous and measured the time difference between the light and sound arriving from an event such as a gunshot (you may have noticed this during a thunderstorm). An alternative approach is to use standing waves in pipes. With today's technology it is possible to get an accurate value by measuring the time taken for a sound to travel between two microphones on your computer.

Properties of sound waves

Sound is an example of a longitudinal wave. When a body moves through air, it compresses the air in front of it. This air then expands, compressing the next layer of air and passing the disturbance from one layer of air to the next (see Figure 21). If the body oscillates, then a continuous wave is propagated through the air. This is called a **sound wave**.

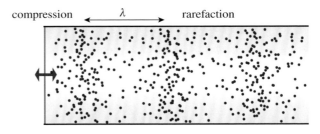

compression λ rarefaction

C.2 Figure 21 A sound wave moves along a pipe.

A sound wave is a propagation of **changing pressure**. This causes layers of gas to oscillate, but remember, the individual molecules of the gas are moving with random motion. Since the disturbance is in the same direction as the transfer of energy, sound is a longitudinal wave.

The loudness of a sound is measured using a sound meter in decibels (dB).

Reflection of sound

If you shout in front of a cliff, the sound reflects back as an echo. In fact, any wall is a good reflector of sound, so when speaking in a room, the sound is reflected off all the surfaces. This is why your voice sounds different in a room and outside.

Refraction of sound

When sound passes from warm air into cold air, it refracts. This is why sounds carry well on a still night.

The sound travels to the listener by two paths: one direct and one by refraction through the layers of air (Figure 22). This results in an extra loud sound.

C.2 Figure 22 Sound refracts through layers of air.

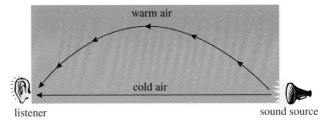

warm air

cold air

listener   sound source

The electromagnetic spectrum

Light is an **electromagnetic wave**, which means that it is a propagation of disturbance in an electric and magnetic field (more about this in D.2). Electromagnetic waves are classified according to their wavelength as represented by the spectrum shown in Figure 23 (on the following page). Unlike the other waves studied so far, electromagnetic waves do not need to have a medium to propagate through.

Electromagnetic waves are transverse waves since the changing electric and magnetic fields are perpendicular to the direction of propagation.

Nature of Science

The first time thin film interference was observed in the laboratory was when Joseph Fraunhofer saw colors appearing while alcohol evaporated from a sheet of glass. Sometimes luck plays a part in scientific discovery, but if Fraunhofer had not realized that the colors were interesting, the discovery would have had to wait for someone else.

How can light be modeled as an electromagnetic wave? (E.2)

Color is perceived but wavelength is measured.

Some light facts

In this chapter, we are particularly interested in visible light, which has wavelengths between 400 and 800 nm. Different wavelengths of visible light have different colors.

However, our perception of color is not so simple. For example, red light mixed with green light gives yellow light.

The velocity of light in a vacuum is approximately $3 \times 10^8 \, \text{m s}^{-1}$.

The brightness or **intensity** of light, measured in W m^{-2}, is proportional to the square of the wave amplitude:

$$I \propto A^2$$

Light intensity is measured with a light meter. As you move away from a point light source, its intensity gets less and less. This is because the light spreads out so the power per unit area is reduced. The light from a point source spreads out in a sphere, so at a distance r, the power will be spread over an area equal to $4\pi r^2$. If the power of the source is P, then the intensity at this distance will be $\frac{P}{4\pi r^2}$.

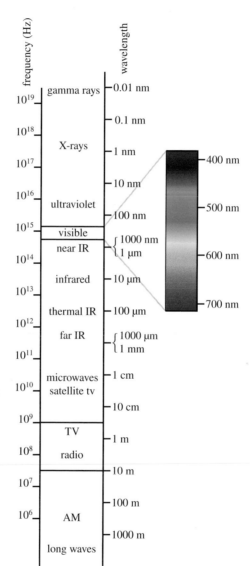

◀ **C.2 Figure 23** The electromagnetic spectrum. Waves can be classified in terms of their wavelength. Each range of wavelengths has a different name, different mode of production, and different uses.

Why does the intensity of an electromagnetic wave decrease with distance according to the inverse square law? (B.1)

Challenge yourself

1. Conduct research into the following Nature of Science linking questions and consider the connections to Theory of Knowledge:

   How were X-rays discovered? (NOS)

   How are waves used in technology to improve society? (NOS)

Guiding Questions revisited

What are the similarities and differences between different types of wave?

How can the wave model describe the transmission of energy as a result of local disturbances in a medium?

What effect does a change in the frequency of oscillation or medium through which the wave is traveling have on the wavelength of a traveling wave?

In this chapter, we have looked at one particular means of categorizing waves, as well as key examples, to examine how:

- All traveling waves transfer energy.
- This energy is transmitted through local disturbances in a medium or field.
- Transverse waves have oscillations that are perpendicular to the direction of energy transfer. An example is electromagnetic radiation, which requires no medium for propagation, and which travels at the speed of light in a vacuum.
- Longitudinal waves have oscillations that are perpendicular to the direction of energy transfer. An example is sound waves, which require matter for propagation, and which are fastest in high-density media.
- The speed of any wave can be calculated as the product of wavelength (the distance between in-phase points on consecutive oscillations) and frequency.
- For a given wave speed (usually defined by the medium), an increase in the frequency of oscillations results in a decrease in the wavelength.

Practice questions

1. A sound wave of frequency 660 Hz passes through air. The variation of particle displacement with distance along the wave at one instant of time is shown in the diagram.

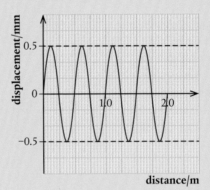

(a) State whether this wave is an example of a longitudinal or a transverse wave. (1)

(b) Using data from the diagram, deduce for this sound wave:

   (i) the wavelength (1)

   (ii) the amplitude (1)

   (iii) the speed. (2)

(Total 5 marks)

2. The speed of a wave in the ocean is greater in deep water than shallow water. The diagram represents the waves approaching the deep water channel at Nazaré, a famous big-wave surfing location.

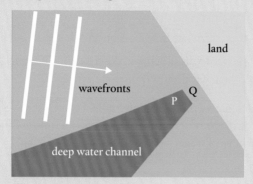

(a) Copy the diagram and sketch the wavefronts as they pass into the deep water channel. (1)

(b) Explain why the waves are so big at point P. (1)

A water wave transfers energy from a storm at sea to the land. The energy is the potential energy of the water lifted up in the peaks.

(c) Explain the change of wavelength when the wave travels from P to Q. (1)

(d) Explain using a diagram why the wave gets steeper after passing from P to Q. (1)

(Total 4 marks)

3. The diagram shows a wave simulation made of eight pendulums 2 cm apart, each with time period 2 s. The end pendulums are in phase.

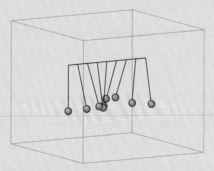

(a) (i) Calculate the wave speed. (3)

(ii) Calculate the phase difference between consecutive pendulums. (2)

(b) The length of the pendulum strings is halved but the phase difference stays the same. Determine :

(i) the wavelength (1)

(ii) the frequency (1)

(iii) the wave speed. (1)

(Total 8 marks)

4. **(a)** **(i)** Define what is meant by the *speed of a wave*. (2)

(ii) Light is emitted from a candle flame. Explain why, in this situation, it is correct to refer to the 'speed of the emitted light', rather than its velocity. (2)

(b) **(i)** Define, by reference to wave motion, what is meant by *displacement*. (2)

(ii) By reference to displacement, describe the difference between a longitudinal wave and a transverse wave. (3)

The center of an earthquake produces longitudinal waves (P waves) and transverse waves (S waves). The diagram shows the variation with time t of the distance d moved by the two types of wave.

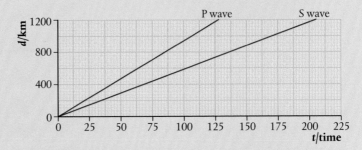

(c) Use the diagram to determine the speed of:

(i) the P waves (1)

(ii) the S waves. (1)

The waves from an earthquake close to the Earth's surface are detected at three laboratories, L_1, L_2, and L_3. The laboratories are at the corners of a triangle so that each is separated from the others by a distance of 900 km, as shown in the diagram.

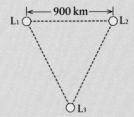

The records of the variation with time of the vibrations produced by the earthquake as detected at the three laboratories are shown in the diagram below. All three records were started at the same time. On each record, one pulse is made by the S wave and the other by the P wave. The separation of the two pulses is referred to as the S–P interval.

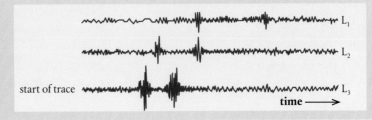

(d) (i) Copy the diagram above and on the trace produced by laboratory L_2 identify, by reference to your answers in (c), the pulse due to the P wave (label the pulse P). (1)

(ii) Using evidence from the records of the earthquake, state which laboratory was closest to the site of the earthquake. (1)

(iii) State **three** separate pieces of evidence for your statement in (d)(ii). (3)

(iv) The S–P intervals are 68 s, 42 s and 27 s for laboratories L_1, L_2 and L_3, respectively. Use the diagrams, or otherwise, to determine the distance of the earthquake from each laboratory. Explain your working. (4)

(v) Copy the triangle diagram and mark on it a possible site of the earthquake. (1)

(Total 21 marks)

5. A longitudinal wave moves through a medium. Relative to the direction of energy transfer through the medium, what are the displacement of the medium and the direction of propagation of the wave?

	Displacement of medium	Direction of propagation of wave
A	parallel	perpendicular
B	parallel	parallel
C	perpendicular	parallel
D	perpendicular	perpendicular

(Total 1 mark)

6. A girl in a stationary boat observes that 10 wave crests pass the boat every minute. What is the period of the water waves?

A $\frac{1}{10}$ min   **C** 10 min

B $\frac{1}{10}$ min^{-1} **D** 10 min^{-1} *(Total 1 mark)*

7. A traveling wave has a frequency of 500 Hz. The closest distance between two points on the wave that have a phase difference of 60° ($\frac{\pi}{3}$ rad) is 0.050 m. What is the speed of the wave?

A 25 m s^{-1} **C** 150 m s^{-1}

B 75 m s^{-1} **D** 300 m s^{-1} *(Total 1 mark)*

8. A sound wave has a frequency of 1.0 kHz and a wavelength of 0.33 m. What is the distance traveled by the wave in 2.0 ms and the nature of the wave?

	Distance traveled in 2.0 ms	Nature of the wave
A	0.17 m	longitudinal
B	0.17 m	transverse
C	0.66 m	longitudinal
D	0.66 m	transverse

(Total 1 mark)

9. A transverse traveling wave is moving through a medium. The graph shows, for one instant, the variation with distance of the displacement of particles in the medium. The frequency of the wave is 25 Hz and the speed of the wave is 100 m s^{-1}. What is correct for this wave?

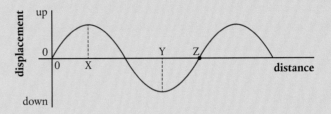

A   The particles at X and Y are in phase.

B   The velocity of the particle at X is a maximum.

C   The horizontal distance between X and Z is 3.0 m.

D   The velocity of the particle at Y is 100 m s^{-1}.

(Total 1 mark)

10. The graph shows the variation with time for the displacement of a particle in a traveling wave. What are the frequency and amplitude for the oscillation of the particle?

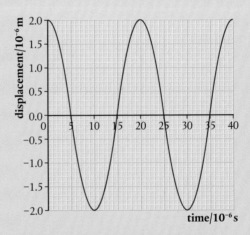

	Frequency / kHz	Amplitude / μm
A	20	2
B	20	4
C	50	2
D	50	4

(Total 1 mark)

11. The graphs show the variation of the displacement y of a medium with distance x and with time t for a traveling wave. What is the speed of the wave?

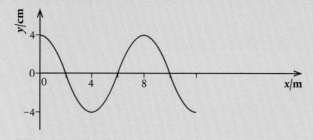

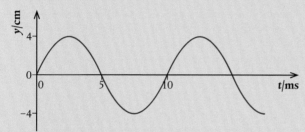

A	$0.6\,\text{m s}^{-1}$	**B**	$0.8\,\text{m s}^{-1}$
C	$600\,\text{m s}^{-1}$	**D**	$800\,\text{m s}^{-1}$

(Total 1 mark)

12. Which statement about X-rays and ultraviolet radiation is correct?

A X-rays travel faster in a vacuum than ultraviolet waves.

B X-rays have a higher frequency than ultraviolet waves.

C X-rays cannot be diffracted unlike ultraviolet waves.

D Microwaves lie between X-rays and ultraviolet in the electromagnetic spectrum.

(Total 1 mark)

13. A student stands a distance L from a wall and claps her hands. Immediately on hearing the reflection from the wall, she claps her hands again. She continues to do this, so that successive claps and the sound of reflected claps coincide. The frequency at which she claps her hands is f. What is the speed of sound in air?

A	$\dfrac{L}{2f}$	**B**	$\dfrac{L}{f}$
C	Lf	**D**	$2Lf$

(Total 1 mark)

14. A water wave moves on the surface of a lake. P and Q are two points on the water surface. The wave is traveling toward the right. The diagram on the left shows the wave at time $t = 0$. Which graph shows how the displacements of P and Q vary with t?

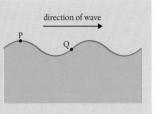

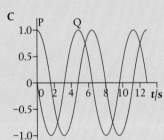

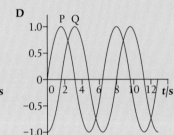

(Total 1 mark)

15. Which graph shows the variation of amplitude with intensity for a wave?

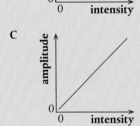

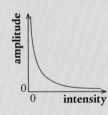

(Total 1 mark)

16. A longitudinal wave is traveling in a medium from left to right. The graph shows the variation with distance x of the displacement y of the particles in the medium. The solid line and the dotted line show the displacement at $t = 0$ and $t = 0.882$ ms, respectively. The period of the wave is greater than 0.882 ms. A displacement to the right of the equilibrium position is positive.

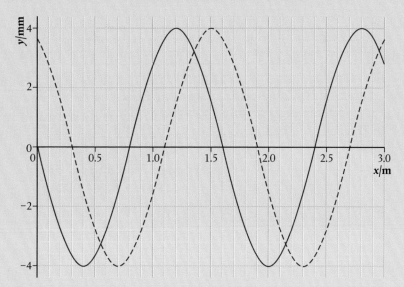

(a) State what is meant by a longitudinal traveling wave. (1)

(b) Calculate, for this wave:

   (i) the speed (2)

   (ii) the frequency. (2)

(c) The equilibrium position of a particle in the medium is at $x = 0.80$ m. For this particle at $t = 0$, state and explain:

   (i) the direction of motion (2)

   (ii) whether the particle is at the center of a compression or a rarefaction. (2)

(Total 9 marks)

17. An earthquake off the coast of Sumatra, Indonesia, at A produces mechanical P waves and S waves that travel through the mantle of the Earth at speeds of 5.50 km s^{-1} and 3.00 km s^{-1}, respectively.

(a) If the S wave arrives at the coastal station B, across the Indian Ocean near Mombasa, Kenya, 15 mins 17 s after the P wave, determine:

   (i) the distance, D, of B from A (2)

   (ii) the time, T, taken by a tsunami, produced by the earthquake, to arrive at B if it travels at 800 km h^{-1}. (2)

(b) On what principle could a tsunami warning system be established? (2)

(Total 6 marks)

C.3 Wave phenomena

A bubble is a thin, spherical soap and water film. When white light from the Sun shines on a bubble, some is reflected at the uppermost air–soap surface and some refracts inside the film, with additional reflection taking place at the inner soap–air surface. The distance traveled by the light within the film is greater than the distance traveled by the light reflecting on the outside, which means that interference occurs. The colors that are most prominent from a given viewpoint are the result of constructive interference. Those that are not seen have undergone destructive interference. These change with the observer's position in an effect called iridescence.

Guiding Questions

How are observations of wave behaviors at a boundary between different media explained?

How is the behavior of waves passing through apertures represented?

What happens when two waves meet at a point in space?

If C.1 is for the study of oscillations of individual particles and C.2 is for the study of how the particles in a medium can be disturbed to transmit energy, C.3 comprises what traveling waves *do*.

They reflect, refract, transmit, diffract and interfere, and we'll make use of ray and wavefront diagrams and graphs of displacements to visualize these behaviors.

As you read, try to notice not only *what* these behaviors consist of and emerge from but also *how* you as a physicist are being asked to understand them: qualitatively and/or quantitatively?

Students should understand:

waves traveling in two and three dimensions can be described through the concepts of wavefronts and rays
wave behavior at boundaries in terms of reflection, refraction and transmission
wave diffraction around a body and through an aperture
wavefront-ray diagrams showing refraction and diffraction
Snell's law, critical angle and total internal reflection
Snell's law as given by $\frac{n_1}{n_2} = \frac{\sin\theta_1}{\sin\theta_2} = \frac{v_2}{v_1}$ where n is the refractive index and θ is the angle between the normal and the ray
superposition of waves and wave pulses
double-source interference requires coherent sources
the condition for constructive interference as given by path difference $= n\lambda$
the condition for destructive interference as given by path difference $= \left(n + \frac{1}{2}\right)\lambda$
Young's double-slit interference as given by $s = \frac{\lambda D}{d}$ where s is the separation of fringes, d is the separation of the slits, and D is the distance from the slits to the screen.
HL   single-slit diffraction including intensity patterns as given by $\theta = \frac{\lambda}{b}$ where b is the slit width
HL   the single-slit pattern modulates the double-slit interference pattern
HL   interference patterns from multiple slits and diffraction gratings as given by $n\lambda = d\sin\theta$

▲
These narrow beams of light at a concert are like rays; they are parallel to the direction of the energy transfer.

▲
These 'ripples' in the sand are like wavefronts; they are perpendicular to the direction of energy transfer.

Wavefronts

In reality, not all waves are sinusoidal. The troughs between the peaks of water waves traveling toward a beach are much longer than the peaks

If a stone is thrown into a pond, then a pulse will be seen to spread out across the surface in two dimensions: energy has been transferred from the stone to the surface of the water. If the surface is disturbed continuously by an oscillating object (or the wind), a continuous wave will be formed whose profile resembles a sine wave. Viewed from directly above, the wave spreads out in circles. The circles that we see are actually the peaks and troughs of the wave; we call these lines **wavefronts**. A wavefront is any line joining points that are in phase. Wavefronts are perpendicular to the direction of energy transfer, which can be represented by an arrow called a **ray**.

Ripples spreading out in circles after the surface is disturbed.

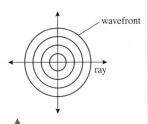

C.3 Figure 1 A circular wavefront spreading out from a point.

Point sources produce circular wavefronts, but if the source is far away, the waves will appear plane.

A plane wavefront moves toward the beach.

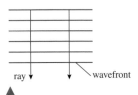

C.3 Figure 2 Parallel plane wavefronts.

Tidal bores, beloved of surfers, occur in rivers with large tidal ranges. These surfers take the path of rays respective to the wavefront crests.

Wave propagation (Huygens' construction)

We can think of a wavefront as being made up of an infinite number of new centers of disturbance. Each disturbance creates its own wavelet that progresses in the direction of the wave. The wavefront is made up of the sum of all these wavelets. Figure 3 shows how circular and plane wavefronts propagate according to this construction.

Nature of Science

The Huygens' construction treats a wavefront as if it is made of an infinite number of small point sources that only propagate forward. Huygens gave no explanation for the fact that propagation is only forward but the model correctly predicts the laws of reflection and Snell's law of refraction. Snell's law was the result of many experiments measuring the angles of light rays passing from one medium to another. The result gives the path with the shortest time, a result that is in agreement with Einstein's theory of relativity. There can be more than one theory to explain a phenomenon but they must give consistent predictions.

Reflection of water waves

When a wavefront hits a barrier, the barrier now behaves as a series of wavelet sources sending wavelets in the opposite direction. In this way, a circular wavefront is reflected as a circular wavefront that appears to originate from a point behind the barrier as in Figure 4.

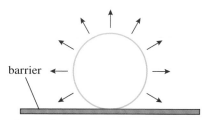
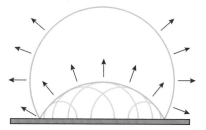

barrier becomes source of disturbance   wavelets add to give reflected wavefront

A plane wavefront reflects as a plane wavefront, making the same angle to the barrier as the incident wave, as shown in Figure 5.

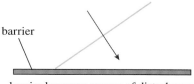
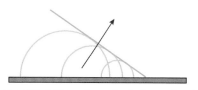

barrier becomes source of disturbance   wavelets add to give reflected wavefront

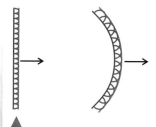

C.3 Figure 3 Huygens' construction used to find the new position of plane and circular wavefronts.

C.3 Figure 4 Reflection of a circular wavefront.

C.3 Figure 5 Reflection of a plane wavefront.

Refraction of water waves

Refraction is the change of direction of propagation when a wave passes from one medium to another. In the case of water waves, it is difficult to change the medium but we can change the depth. This changes the speed of the wave and causes the ray to change direction. This can again be explained using Huygens' construction as shown in Figure 6, where the wave is passing into shallower water, where it travels more slowly.

C.3 Figure 6 The wavefront and ray directions change.

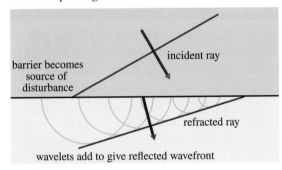

C.3 Figure 7 The optical density need not change for a change in direction to emerge.

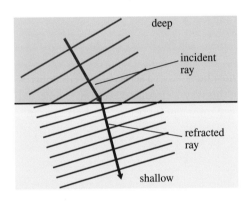

The frequency of the wave does not change when the wave slows down so the wavelength must be shorter ($v = f\lambda$). Note that, although not drawn in Figure 7, when a wave meets a boundary such as this, it will be reflected as well as refracted.

Snell's law

Snell's law relates the angles of incidence and refraction to the ratio of the velocity of the wave in the different media. The ratio of the sine of the angle of incidence to the sine of the angle of refraction is equal to the ratio of the velocities of the wave in the different media:

$$\frac{\sin i}{\sin r} = \frac{v_1}{v_2}$$

Note that the angles are measured between the ray and the **normal**, or between the wavefront and the boundary.

C.3 Figure 8 Angles of incidence and refraction.

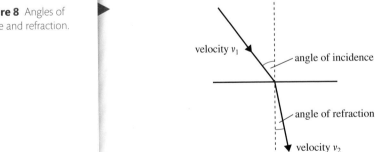

We refer to the proportion of the energy of the wave that refracts at a boundary as being 'transmitted' to the new medium. The remainder is reflected.

Worked example

A water wave traveling at $20\,\mathrm{m\,s^{-1}}$ in deep water enters a shallow region where its velocity is $15\,\mathrm{m\,s^{-1}}$ (Figure 9). If the angle of incidence between the water wave and the normal of the boundary between regions is 50°, what is the angle of refraction?

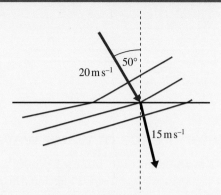

Solution

Applying Snell's law:

$$\frac{\sin i}{\sin r} = \frac{v_1}{v_2} = \frac{20}{15}$$

$$\sin r = \frac{\sin 50°}{1.33} = 0.576$$

$$r = 35°$$

Exercise

Q1. A water wave with wavelength 30 cm traveling with velocity $0.50\,\mathrm{m\,s^{-1}}$ meets the straight boundary to a shallower region at an angle of incidence 30°. If the velocity in the shallow region is $0.40\,\mathrm{m\,s^{-1}}$, calculate:

   (a) the frequency of the wave

   (b) the wavelength of the wave in the shallow region

   (c) the angle of refraction.

Q2. A water wave traveling in a shallow region at a velocity of $0.30\,\mathrm{m\,s^{-1}}$ meets the straight boundary to a deep region at angle of incidence 20°. If the velocity in the deep region is $0.50\,\mathrm{m\,s^{-1}}$, calculate the angle of refraction.

Diffraction of water waves

Diffraction takes place when a wave passes through a small opening. If the opening is very small, then the wave behaves just like a point source as shown in Figure 9.

Using Huygens' construction, we can explain why this happens. In the case of the very narrow slit, the wavefront is reduced to one wavelet that propagates as a circle.

Water waves diffracting through two different sized openings. The waves are diffracted more through the narrower opening.

C.3 Figure 9 If the opening is a bit bigger then the effect is not so great.

Waves are also diffracted by objects and edges as shown in Figure 10. Notice how the wave seems to pass round the very small object.

C.3 Figure 10 Diffraction around obstacles.

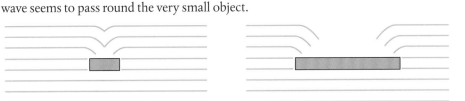

What evidence is there that particles possess wave-like properties such as wavelength? (NOS)

Interference of water waves

If two disturbances are made in a pool of water, two different waves will be formed. When these waves meet, the individual displacements will add vectorially. This is called **superposition**. If the frequency of the individual waves is equal, then the resulting amplitude will be constant and related to the **phase difference** between the two waves.

C.3 Figure 11 Constructive and destructive interference.

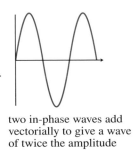
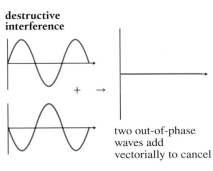

constructive interference

two in-phase waves add vectorially to give a wave of twice the amplitude

destructive interference

two out-of-phase waves add vectorially to cancel

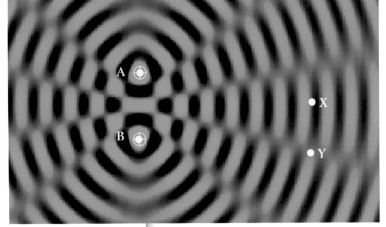

When two identical point sources produce waves on the surface of a pool of water, a pattern like the one in Figure 12 is produced.

We can see that there are regions where the waves are interfering constructively (X) and regions where they are interfering destructively (Y). If we look carefully at the waves arriving at X and Y from A and B, we see that at X they are in phase and at Y they are out of phase (Figure 13). This is because the waves have traveled the same distance to get to X, but the wave from A has traveled $\frac{1}{2}\lambda$ extra to get to Y.

C.3 Figure 12 Ripple tank www.falstad.com

In general, constructive interference occurs if:

path difference = $n\lambda$

or:

phase difference = $2n\pi$

where n is a whole number.

Destructive interference occurs if:

path difference = $\left(n + \frac{1}{2}\right)\lambda$

or:

phase difference = $(2n + 1)\pi$

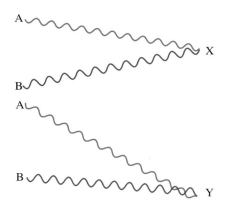

◀ **C.3 Figure 13** Path difference leads to phase difference.

Path difference and phase difference

We can see from the previous example that a path difference of $\frac{1}{2}\lambda$ introduces a phase difference of π, so if the path difference is d, then the phase difference, $\theta = \frac{2\pi d}{\lambda}$.

Worked example

Two boys playing in a pool make identical waves that travel toward each other. The boys are 10 m apart and the waves have a wavelength 2 m. Their little sister is swimming from one boy to the other. When she is 4 m from the first boy, will she be in a big wave or a small wave?

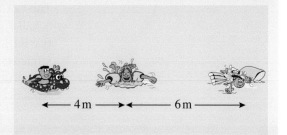

A diagram always helps, no matter how simple it is.

Solution

The waves from the boys will interfere when they meet. If the girl is 4 m from the first boy, then she must be 6 m from the other. This is a path difference of 2 m, one whole wavelength. The waves are therefore in phase and will interfere constructively.

Exercise

Q3. Two wave sources, A and B, produce waves of wavelength 2.0 cm. What is the phase angle between the waves at:

   (a) a point C, distance 6.0 cm from A and 6.2 cm from B?

   (b) a point D, distance 8.0 cm from A and 7.0 cm from B?

   (c) a point E, distance 10.0 cm from A and 11.5 cm from B?

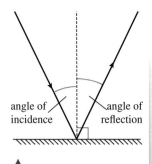

C.3 Figure 14 The angles of incidence and reflection are the same.

Reflection of light

When light hits an object, part of it is absorbed and part of it is reflected. It is the reflected light that enables us to see things. If the reflecting surface is uneven, the light is reflected in all directions, but if it is flat, the light is reflected uniformly so we can see that the angle of reflection equals the angle of incidence (Figure 14).

Refraction of light

The velocity of light is different for different transparent media, so when light passes from one medium to another, it changes direction. For example, the velocity of light is greater in air than it is in glass, so when light passes from air to glass, it refracts as in Figure 15.

The refractive index of a medium, n, is the ratio of the speed of light in a vacuum to the speed of light in the medium. This means that: $n_1 = \dfrac{c}{c_1}$ and $n_2 = \dfrac{c}{c_2}$

C.3 Figure 15 Light refracts from air to glass.

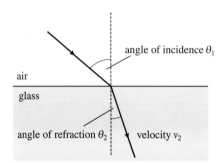

Rearranging: $c_1 = \dfrac{c}{n_1}$ and $c_2 = \dfrac{c}{n_2}$

Combining: $\dfrac{c_1}{c_2} = (\dfrac{c}{n_1})/(\dfrac{c}{n_2})) = \dfrac{n_2}{n_1}$

Applying Snell's law, we get: $\dfrac{\sin\theta_1}{\sin\theta_2} = \dfrac{v_1}{v_2}$

This can also be written $\dfrac{\sin\theta_1}{\sin\theta_2} = \dfrac{n_2}{n_1}$ where n_1 and n_2 are the **refractive indices** of the two media. The bigger the difference in refractive index, the more the light ray will be deviated. We say a medium with a high refractive index is 'optically dense'.

Light reflected off the straw is refracted as it comes out of the water, causing the straw to appear bent.

SKILLS

To measure the refractive index of a glass block, you can pass a narrow beam of light (e.g. from a laser) through it and measure the angles of incidence and refraction as the light passes from air into the glass. It is not possible to trace the ray as it passes through the block, but if you place the block on a sheet of paper and mark where the ray enters and leaves the block (at B and C in the figure on the right), then you can plot the path of the ray. If you do not have a light source, you can use an alternative method with pins. Place two pins on one side of the block in positions A and B then, looking through the block, place a third pin in line with the other two. Joining the dots will give the path of a ray from A through the block.

The angles can then be measured using a protractor and the refractive index calculated.

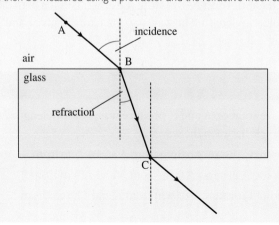

A worksheet with full details of how to carry out this experiment is available in the eBook.

SKILLS

Worked example

A ray of light traveling in air is incident on a glass block at an angle of 56°. Calculate the angle of refraction if the refractive index of glass is 1.5.

Solution

Applying Snell's law:

$$\frac{\sin \theta_1}{\sin \theta_2} = \frac{n_2}{n_1}$$

Where:

$$\theta_1 = 56°$$
$$n_1 = 1 \text{ (air)}$$
$$n_2 = 1.5 \text{ (glass)}$$

$$\frac{\sin 56°}{\sin \theta_2} = \frac{1.5}{1}$$

$$\sin \theta_2 = \sin \frac{\sin 56°}{1.5} = 0.55$$

$$\theta_2 = 34°$$

Exercise

Use the refractive indices in the table on the right to solve the following problems.

Q4. Light traveling through the air is incident on the surface of a pool of water at an angle of 40°. Calculate the angle of refraction.

Q5. Calculate the angle of refraction if a beam of light is incident on the surface of a diamond at an angle of 40°.

Q6. If the velocity of light in air is $3.00 \times 10^8 \,\text{m s}^{-1}$, calculate its velocity in glass.

Q7. A fish tank made of glass contains water (and fish). Light travels from a fish at an angle of 30° to the side of the tank. Calculate the angle between the light and the normal to the glass surface as it emerges into the air.

Q8. Light incident on a block of transparent plastic at an angle of 30° is refracted at an angle of 20°. Calculate the angle of refraction if the block is immersed in water and the ray is incident at the same angle.

Material	Refractive index
Air	1.0003
Water	1.33
Glass	1.50
Diamond	2.42

Some refractive indexes of different media.

Dispersion

The angle of refraction is dependent on the wavelength of the light. If red light and blue light pass into a block of glass, the blue light will be refracted more than the red, causing the colors to **disperse**. This is why you see a spectrum when light passes through a prism as in the photo. It is also the reason why rainbows are formed when light is refracted by raindrops.

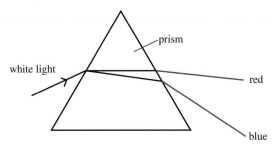

The white light is dispersed into the colors of the spectrum because the different colors of light travel at different speeds in glass.

C.3 Figure 16 When white light is passed through a prism, blue light is refracted more than red.

The critical angle

If light passes into an optically less dense medium, e.g. from glass to air, then the ray will be refracted away from the normal as shown in Figure 17.

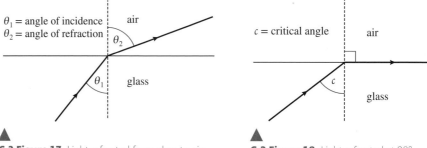

▲ **C.3 Figure 17** Light refracted from glass to air.

▲ **C.3 Figure 18** Light refracted at 90°.

If the angle of incidence increases, a point will be reached where the refracted ray is refracted along the boundary. The angle at which this happens is called the **critical angle**.

Applying Snell's law to this situation:

$$\frac{\sin \theta_1}{\sin \theta_2} = \frac{n_2}{n_1}$$

Where:
$$\theta_1 = c$$
$$\theta_2 = 90°$$
$$n_1 = 1.5 \text{ for glass}$$
$$n_2 = 1$$
$$c = 42°$$

Total internal reflection

If the critical angle is exceeded, all of the light is reflected. This is known as **total internal reflection**. Since all the light is reflected, none is transmitted. This is not the case when light is reflected off a mirror when some is absorbed.

C.3 Figure 19 Light totally internally reflected.

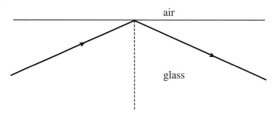

Optical fibers

An optical fiber is a thin strand of glass or clear plastic. If a ray of light enters its end at a small angle, the ray will be totally internally reflected when it meets the side. Since the sides are parallel, the ray will be reflected back and forth until it reaches the other end as in Figure 20. Optical fibers are used extensively in communication.

C.3 Figure 20 Light reflected along a fiber.

light refracted when entering fiber

light reflected at the sides

Exercise

Q9. Light enters a glass block of refractive index 1.5 at an angle 70°.

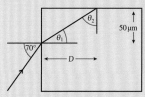

    (a) Use Snell's law to calculate the angle of refraction θ_1.

    (b) Use geometry to find the angle θ_2.

    (c) Calculate the critical angle for glass.

    (d) Will the ray be totally internally reflected?

    (e) Calculate length D.

HL

Diffraction of light at a single slit

When light passes through a narrow slit, it diffracts, forming a series of bright and dark bands, as shown in Figure 21.

We can derive an equation for the first minima in this pattern by applying Huygens' construction.

When a wavefront passes through a narrow slit, it will propagate as if there were a large number of wavelet sources across the slit, as in Figure 22.

The resultant intensity at some point P in front of the slit is found by summing all the wavelets. This is not a simple matter since each wavelet has traveled a different distance so they will be out of phase when they arrive at P. To simplify the problem, we will consider a point Q a long way from the slits. Light traveling through the slit arriving at point Q is almost parallel, and if we say that it is parallel, then the geometry of the problem becomes much simpler.

C.3 Figure 21 Single-slit diffraction.

C.3 Figure 22 Wavelets add vectorially to give resultant intensity.

Nearly parallel wavelets going to Q

The central maximum

The central maximum occurs directly ahead of the slit. If we take a point a long way from the slit, all wavelets will be parallel and will have traveled the same distance, as shown in Figure 23. If all the wavelets have traveled the same distance, they will be in phase, so will interfere constructively to give a region of high intensity (bright).

C.3 Figure 23 Wavelets traveling to the central maximum.

The first minimum

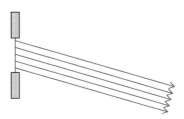

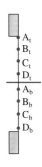

C.3 Figure 25 Wavelets in the top half cancel with wavelets in the bottom.

If we now consider wavelets traveling toward the first minimum, as in Figure 24, they are traveling at an angle so will not all travel the same distance. The wavelet at the top will travel further than the wavelet at the bottom. When these wavelets add together, they interfere destructively to form a region of low intensity (dark).

We can calculate the angle at which the first minimum is formed by splitting the slit into two halves, top and bottom, as in Figure 25. If all the wavelets from the top half cancel out all the wavelets from the bottom, the result will be a dark region. So if we have eight wavelet sources, four in the top half (A_t, B_t, C_t, D_t) and four in the bottom (A_b, B_b, C_b, D_b), and if A_t cancels with A_b and B_t cancels with B_b, etc., then all the wavelets will cancel with each other. For each pair to cancel, the path difference must be $\frac{1}{2}\lambda$. Figure 26 shows the situation for the top wavelet and the one in the middle.

C.3 Figure 26 Geometric construction for the first minimum.

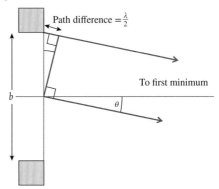

The orange line cuts across the two wavelets at 90° showing that the top one travels further than the bottom one. If the path difference shown is $\frac{\lambda}{2}$, then these wavelets will cancel and so will all the others. If the first minimum occurs at an angle θ as shown, then this will also be the angle of the triangle made by the orange line. We can therefore write:

$$\sin\theta = \frac{\frac{\lambda}{2}}{\frac{b}{2}} = \frac{\lambda}{b}$$

But the angles are very small, so if θ is measured in radians: $\sin\theta = \theta$

So:
$$\theta = \frac{\lambda}{b}$$

Knowing the position of the first minimum tells us how spread out the diffraction pattern is. From the equation, we can see that if b is small, then θ is big, so the pattern is spread out as shown in Figure 27.

If white light is passed through the slit, light of different wavelengths forms peaks at different angles, resulting in colored fringes.

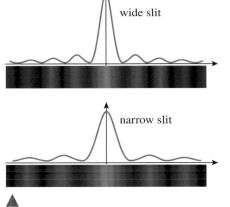

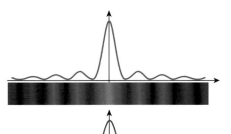

wide slit

narrow slit

What are the similarities and differences between single-slit diffraction and diffraction to study atomic structures? (E.1)

SKILLS

A worksheet linked to this topic is available in your eBook.

▲ **C.3 Figure 27** Notice that with a narrower slit the pattern is wider but less intense.

▲ **C.3 Figure 28** Same size slit but different wavelength light; longer wavelength gives a wider pattern.

Example

A diffraction pattern is formed on a wall 2.0 m from a 0.10 mm slit illuminated with light of wavelength 600 nm. How wide will the central maximum be?

First draw a diagram showing the relative positions of the slit and screen.

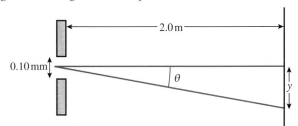

◀ **C.3 Figure 29** The angle between the center and the first minimum.

The angle θ can be calculated from:

$$\theta = \frac{\lambda}{b}$$

$$= \frac{600 \times 10^{-9}}{0.10 \times 10^{-3}}$$

$$= 0.0060 \, \text{rad}$$

Since this angle is small, we can say that:

$$\theta = \frac{y}{2.0} \quad \text{so} \quad y = 0.006 \times 2.0$$

$$= 0.012 \, \text{m}$$

$$= 1.2 \, \text{cm}$$

This is half the width of the maximum, so: width = 2.4 cm

Note that the intensity of the diffraction maxima decreases as you move away from the central maximum. In reality, the first maximum would be smaller than shown in these diagrams (about $\frac{1}{20}$ the height of the central maximum).

Exercise

Q10. Light of wavelength 550 nm is passed through a slit of size 0.050 mm. Calculate the width of the central maximum formed on a screen that is 5.0 m away.

Q11. Calculate the size of the slit that would cause light of wavelength 550 nm to diffract, forming a diffraction pattern with a central maximum 5.0 cm wide on a screen 4.0 m from the slit.

HL end

Interference of light

For the light from two sources to interfere, the light sources must be coherent. This means they have the same frequency, similar amplitude, and a constant phase difference. This can be achieved by taking one source and splitting it in two, using slits or thin films.

Two-slit interference

Light from a single source is split in two by parallel narrow slits. Since the slits are narrow, the light diffracts, creating an overlapping region where interference takes place. This results in a series of bright and dark parallel lines called fringes, as shown in Figure 30. This set-up is called Young's slits.

C.3 Figure 30 Double-slit interference.

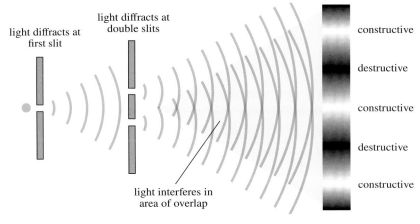

To simplify matters, let us consider rays from each slit arriving at the first maximum as shown in Figure 31.

C.3 Figure 31 Paths taken by the rays from each slit to the first maximum.

C.3 Figure 32 Light rays interfere at the first maximum.

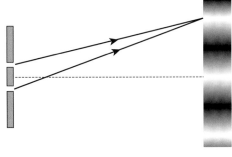

Since this is the first interference maximum, the path difference between the waves must be λ. If we fill in some angles and lengths as in Figure 32 we can use trigonometry to derive an equation for the separation of the fringes.

Since the wavelength of light is very small, the angle θ will also be very small. This means that we can approximate θ in radians to $\frac{\lambda}{d} = \frac{s}{D}$ so the distance from the central bright fringe to the next one, $s = \frac{\lambda D}{d}$.

What can an understanding of the results of Young's double-slit experiment reveal about the nature of light? (C.2)

This equation is of great importance. This is the fringe spacing. We can therefore conclude that if the slits are made closer together, the fringes become more separated.

Exercise

Q12. Two narrow slits, 0.01 mm apart (d), are illuminated by a laser of wavelength 600 nm. Calculate the fringe spacing (y) on a screen 1.5 m (D) from the slits.

Q13. Calculate the fringe spacing if the laser is replaced by one of wavelength 400 nm.

HL

Effect of diffraction

As we have seen, light is diffracted when it passes through each slit. This means that each slit will form a diffraction pattern on the screen, which causes the fringes to vary in brightness as shown in Figure 33. Here it can also be seen how the diffraction pattern has modulated the fringes.

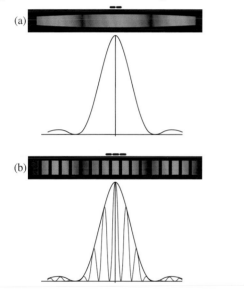
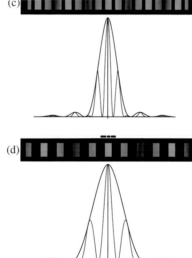

C.3 Figure 33 Fringes modulated by diffraction pattern.
(a) A single slit of width b.
(b) Double slits, each of width b and separation d.
(c) Double slits of width $>b$, giving a less spread diffraction pattern. The separation of the slits is d so the fringes are the same as in (b).
(d) Double slits of width b so the diffraction pattern is the same as (a). Separation of the slits $<d$ so the fringes are further apart.

Multiple-slit diffraction

The intensity of double-slit interference patterns is very low but can be increased by using more than two slits. A diffraction grating is a series of very fine parallel slits mounted on a glass plate.

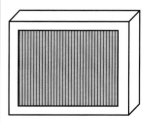

C.3 Figure 34 Diffraction grating (the number of lines per millimeter can be very high: school versions usually have 600 lines per millimeter).

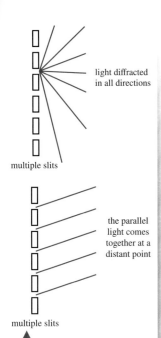

multiple slits

light diffracted
in all directions

multiple slits

the parallel
light comes
together at a
distant point

C.3 Figure 35 Light diffracted at each slit undergoes interference at a distant screen.

C.3 Figure 36 Parallel light travels through the grating and some is diffracted at an angle *θ*. The expansion shows just slits A and B. If the path difference is *nλ*, then constructive interference takes place.

Diffraction at the slits

When light is incident on the grating, it is diffracted at each slit. The slits are very narrow so the diffraction causes the light to propagate as if coming from a point source.

Interference between slits

To make the geometry simpler, we will consider what would happen if the light passing through the grating were observed from a long distance. This means that we can consider the light rays to be almost parallel. So the parallel light rays diffracted through each slit will come together at a distant point. When they come together, they will interfere.

Geometrical model

Let us consider waves that have been diffracted at an angle *θ* as shown in Figure 36 (remember, light is diffracted at all angles – this is just one angle that we have chosen to consider).

We can see that when these rays meet, the ray from A will have traveled a distance *x* further than the ray from B. The ray from D has traveled the same distance further than C, and so on. If the path difference between neighbors is *λ*, then they will interfere constructively; if $\frac{1}{2}\lambda$, then the interference will be destructive.

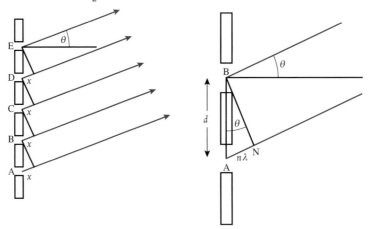

The line BN is drawn perpendicular to both rays so angle N is 90°.

Therefore from triangle ABN, we see that: $\sin\theta = \dfrac{n\lambda}{d}$

Rearranging gives: $d\sin\theta = n\lambda$

If you look at a light source through a diffraction grating and move your head around, bright lines will be seen every time $\sin\theta = \dfrac{n\lambda}{d}$.

Producing spectra

If white light is viewed through a diffraction grating, each wavelength undergoes constructive interference at different angles. This results in a spectrum. The individual wavelengths can be calculated from the angle using the formula $d\sin\theta = n\lambda$.

Worked example

If blue light of wavelength 450 nm and red light of wavelength 700 nm are viewed through a grating with 600 lines mm^{-1}, at what angle will the first bright blue and red lines be seen?

Solution

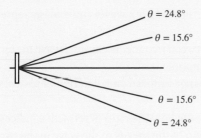

If there are 600 lines/mm:	$d = \dfrac{1}{600}\,\text{mm} = 0.001\,67\,\text{mm}$
For the first line:	$n = 1$
For blue light:	$\sin\theta = \dfrac{450 \times 10^{-9}}{0.001\,67 \times 10^{-3}} = 0.269$
Therefore:	$\theta_{\text{blue}} = 15.6°$
For red light:	$\sin\theta = \dfrac{700 \times 10^{-9}}{0.001\,67 \times 10^{-3}} = 0.419$
Therefore:	$\theta_{\text{red}} = 24.8°$

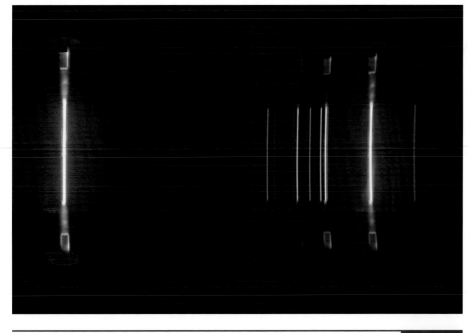

C.3 Figure 37 A hydrogen lamp viewed through a grating.

HL end

Guiding Questions revisited

How are observations of wave behaviors at a boundary between different media explained?

How is the behavior of waves passing through apertures represented?

What happens when two waves meet at a point in space?

In this chapter, we have considered both wavefront and ray representations of the wave model to understand:

- That a boundary between two media causes a wave to separate into reflected and transmitted components.
- Refraction as the change in speed and direction of a wave when transmitted into a new medium.
- Snell's law as describing how the ratio of refractive indices of the two media is the inverse of the ratio of the speeds of the wave in the two media.
- Interference as a property of waves.
- Constructive interference as being associated with path differences of whole wavelengths and destructive interference as resulting from odd numbers of half-wavelengths.
- Superposition as the vector sum of the amplitudes of waves that interfere at a point.
- The characteristic interference pattern of a double slit, with the separation of fringes increasing with wavelength and distance from the slits to the screen and decreasing with the slits' separation.
- Diffraction as the spreading out of a wave's energy when passing around a body or through an aperture, with the effects most noticeable when the barrier or gap is approximately equal to the wavelength.
- **HL** The characteristic diffraction pattern of a single slit as a modulator to any double-slit interference pattern.
- **HL** The characteristic interference pattern of multiple slits and diffraction gratings.

Practice questions

1. The diagram shows an arrangement (not to scale) for observing the interference pattern produced by the superposition of two light waves. S_1 and S_2 are two very narrow slits. The single slit S ensures that the light leaving the slits S_1 and S_2 is coherent.

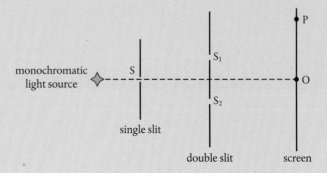

(a) **(i)** Define *coherent*. (1)

(ii) Explain why slits S_1 and S_2 need to be very narrow. (2)

The point O on the diagram is equidistant from S_1 and S_2 and there is maximum constructive interference at point P on the screen. There are no other points of maximum interference between O and P.

(b) **(i)** State the condition necessary for there to be maximum constructive interference at the point P. (1)

(ii) Copy the axes below and draw a graph to show the variation of intensity of light on the screen between the points O and P. (2)

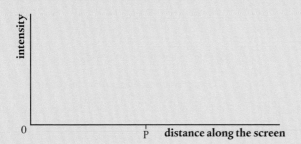

(c) In this particular arrangement, the distance between the double slit and the screen is 1.50 m and the separation of S_1 and S_2 is 3.00×10^{-3} m. The distance OP is 0.25 mm. Determine the wavelength of the light. (2)

(Total 8 marks)

2. **(a)** By making reference to waves, distinguish between a *ray* and a *wavefront*. (3)

The diagram on the right shows three wavefronts incident on a boundary between medium I and medium R. Wavefront CD is shown crossing the boundary. Wavefront EF is incomplete.

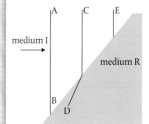

(b) **(i)** Copy the diagram and draw a line to complete the wavefront EF. (1)

(ii) Explain in which medium, I or R, the wave has the higher speed. (3)

(iii) By taking appropriate measurements from the diagram, determine the ratio of the speeds of the wave traveling from medium I to medium R. (2)

The diagram below shows the variation with time t of the velocity v of one particle of the medium through which the wave is traveling.

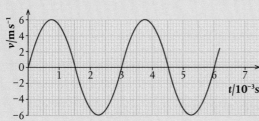

(c) (i) Explain how it can be deduced from the diagram that the particle is oscillating. (2)

    (ii) Determine the frequency of oscillation of the particle. (2)

    (iii) Copy the diagram and mark on the graph, with the letter M, one time at which the particle is at maximum displacement. (1)

    (iv) Estimate the area between the curve and the x-axis from the time $t = 0$ to the time $t = 1.5$ ms. (2)

    (v) Suggest what the area in (c)(iv) represents. (1)

(Total 17 marks)

3. The diagram represents the direction of oscillation of a disturbance that gives rise to a wave.

  (a) Copy the diagram and add arrows to show the direction of wave energy transfer to illustrate the difference between:

    (i) a transverse wave (1)

    (ii) a longitudinal wave. (1)

A wave travels along a stretched string. The diagram below shows the variation with distance along the string of the displacement of the string at a particular instant in time. A small marker is attached to the string at the point labeled M. The undisturbed position of the string is shown as a dotted line.

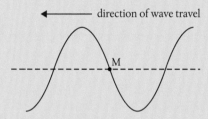

direction of wave travel

M

  (b) Copy the diagram.

    (i) Draw an arrow on the diagram to indicate the direction in which the marker is moving. (1)

    (ii) Indicate, with the letter A, the amplitude of the wave. (1)

    (iii) Indicate, with the letter λ, the wavelength of the wave. (1)

    (iv) Draw the displacement of the string a time $\frac{T}{4}$ later, where T is the period of oscillation of the wave. Indicate, with the letter N, the new position of the marker. (2)

The wavelength of the wave is 5.0 cm and its speed is 10 cm s^{-1}.

  (c) Determine:

    (i) the frequency of the wave (1)

    (ii) how far the wave has moved in $\frac{T}{4}$ s. (2)

  (d) By reference to the principle of superposition, explain what is meant by *constructive interference*. (4)

The diagram below (not drawn to scale) shows an arrangement for observing the interference pattern produced by the light from two narrow slits S_1 and S_2.

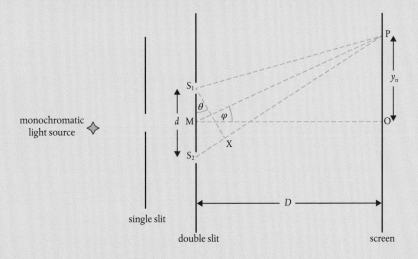

The distance S_1S_2 is d, the distance between the double slit and screen is D and $D \gg d$ such that the angles θ and φ shown on the diagram are small. M is the midpoint of S_1S_2 and it is observed that there is a bright fringe at point P on the screen, a distance y_n from point O on the screen. Light from S_2 travels a distance S_2X further to point P than light from S_1.

(e) **(i)** State the condition in terms of the distance S_2X and the wavelength of the light λ, for there to be a bright fringe at P. (2)

    (ii) Deduce an expression for θ in terms of S_2X and d. (2)

    (iii) Deduce an expression for φ in terms of D and y_n. (1)

For a particular arrangement, the separation of the slits is 1.40 mm and the distance from the slits to the screen is 1.50 m. The distance y_n is the distance of the eighth bright fringe from O and the angle $\theta = 2.70 \times 10^{-3}$ rad.

(f) Using your answers to (e) to determine:

    (i) the wavelength of the light (2)

    (ii) the separation of the fringes on the screen. (3)

(Total 24 marks)

4. Three quantities used to describe a light wave are:

I. frequency   **II.** wavelength   **III.** speed

Which quantities increase when the light wave passes from water to air?

A I and II only   **C** II and III only

B I and III only   **D** I, II and III

(Total 1 mark)

5. A glass block of refractive index 1.5 is immersed in a tank filled with a liquid of higher refractive index. Light is incident on the base of the glass block. Which is the correct diagram for rays incident on the glass block at an angle greater than the critical angle?

A
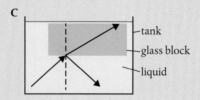

B
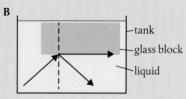

C
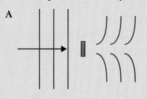

D
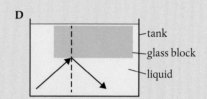

(Total 1 mark)

6. A glass block has a refractive index in air of n_g. The glass block is placed in two different liquids: liquid X with a refractive index of n_X and liquid Y with a refractive index of n_Y. In liquid X, $\frac{n_g}{n_X} = 2$, and in liquid Y, $\frac{n_g}{n_Y} = 1.5$. What is $\frac{\text{speed of light in liquid X}}{\text{speed of light in liquid Y}}$?

A $\frac{2}{4}$ **B** $\frac{3}{4}$ **C** $\frac{4}{3}$ **D** 3   *(Total 1 mark)*

7. Which diagram shows the shape of the wavefront as a result of the diffraction of plane waves by an object?

A

B

C
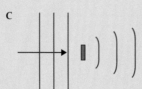

D
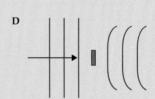

(Total 1 mark)

8. Two identical waves, each with amplitude X_0 and intensity I, interfere constructively. What are the amplitude and intensity of the resultant wave?

	Amplitude of the resultant wave	Intensity of the resultant wave
A	X_0	$2I$
B	$2X_0$	$2I$
C	X_0	$4I$
D	$2X_0$	$4I$

(Total 1 mark)

9. X and Y are two coherent sources of waves. The phase difference between X and Y is zero. The intensity at P due to X and Y separately is I. The wavelength of each wave is 0.20 m. What is the resultant intensity at P?

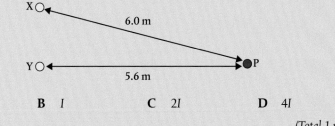

| **A** | 0 | **B** | I | **C** | $2I$ | **D** | $4I$ |

(Total 1 mark)

10. In a Young's double-slit experiment, the distance between fringes is too small to be observed. What change would increase the distance between fringes?

A   Increasing the frequency of light

B   Increasing the distance between slits

C   Increasing the distance from the slits to the screen

D   Increasing the distance between light source and slits

(Total 1 mark)

11. Monochromatic light of wavelength λ is incident on a double slit. The resulting interference pattern is observed on a screen a distance y from the slits. The distance between consecutive fringes in the pattern is 55 mm when the slit separation is a. λ, y and a are all doubled. What is the new distance between consecutive fringes?

| **A** | 55 mm | **B** | 110 mm | **C** | 220 mm | **D** | 440 mm |

(Total 1 mark)

12. ⬛ H L   The diagram shows the diffraction pattern for light passing through a single slit. What is $\frac{\text{wavelength of light}}{\text{width of slit}}$?

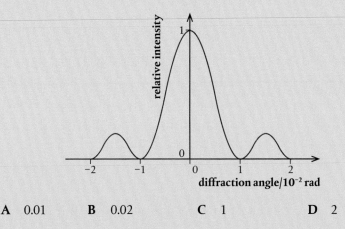

| **A** | 0.01 | **B** | 0.02 | **C** | 1 | **D** | 2 |

(Total 1 mark)

13.   **HL** White light is incident normally on separate diffraction gratings, X and Y. Y has a greater number of lines per meter than X. Three statements about differences between X and Y are:

    I.   Adjacent slits in the gratings are further apart for X than for Y.

    II. The angle between red and blue light in a spectral order is greater in X than in Y.

    III. The total number of visible orders is greater for X than for Y.

    Which statements are correct?

    A I and II only   **C** II and III only

    B I and III only   **D** I, II and III

(Total 1 mark)

14. A large cube is formed from ice. A light ray is incident from a vacuum at an angle of 46° to the normal on one surface of the cube. The light ray is parallel to the plane of one of the sides of the cube. The angle of refraction inside the cube is 33°.

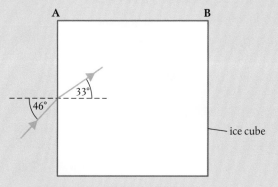

    (a) Calculate the speed of light inside the ice cube.   (2)

    (b) Show that no light emerges from side AB.   (3)

    (c) Copy the diagram and sketch the subsequent path of the light ray.   (2)

(Total 7 marks)

15. Monochromatic light from two identical lamps arrives on a screen. The intensity of light on the screen from each lamp separately is I_0.

(a) Copy the axes below and sketch a graph to show the variation with distance x on the screen of the intensity I of light on the screen. (1)

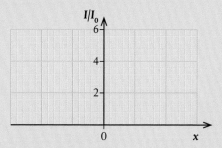

Monochromatic light from a single source is incident on two thin, parallel slits. The following data are available:

Slit separation = 0.12 mm   Wavelength = 680 nm   Distance to screen = 3.5 m

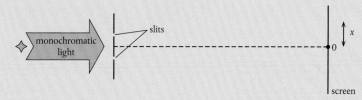

(b) The intensity I of light at the screen from each slit separately is I_0. Copy the axes below and sketch a graph to show the variation with distance x on the screen of the intensity of light on the screen for this arrangement. (3)

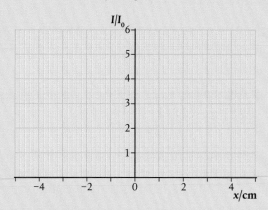

(c) The slit separation is increased. Outline **one** change observed on the screen. (1)

(Total 5 marks)

16. **HL** Monochromatic coherent light is incident on two parallel slits of negligible width, a distance d apart. A screen is placed a distance D from the slits. Point M is directly opposite the midpoint of the slits. Initially, the lower slit is covered and the intensity of light at M due to the upper slit alone is $22\,\mathrm{W\,m^{-2}}$. The lower slit is now uncovered.

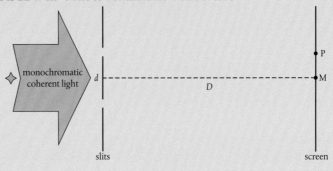

diagram not scale

(a) Deduce, in $\mathrm{W\,m^{-2}}$, the intensity at M. (3)

(b) P is the first maximum of intensity on one side of M. The following data are available:
$d = 0.12$ mm
$D = 1.5$ m
Distance MP $= 7.0$ mm

Calculate, in nm, the wavelength λ of the light. (2)

The width of each slit is increased to 0.030 mm. D, d and λ remain the same.

(c) Suggest why, after this change, the intensity at P will be less than that at M. (1)

(d) Show that, due to single slit diffraction, the intensity at a point on the screen a distance of 28 mm from M is zero. (2)

(Total 8 marks)

17. **HL** Monochromatic light of wavelength λ is normally incident on a diffraction grating. The diagram shows adjacent slits of the diffraction grating, labeled V, W and X. Light waves are diffracted through an angle θ to form a second-order diffraction maximum. Points Z and Y are labeled.

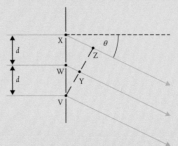

(a) State the phase difference between the waves at V and Y. (1)

(b) State, in terms of λ, the path length between points X and Z. (1)

(c) The separation of adjacent slits is d. Show that for the second-order diffraction maximum, $2\lambda = d\sin\theta$. (1)

Monochromatic light of wavelength 633 nm is normally incident on a diffraction grating. The diffraction maxima incident on a screen are detected and their angle θ to the central beam is determined. The graph shows the variation of sin θ with the order n of the maximum. The central order corresponds to $n = 0$.

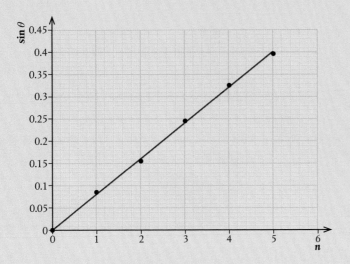

(d) Determine a mean value for the number of slits per millimeter of the grating. (4)

(e) State the effect on the graph of the variation of sin θ with n of:

   (i) using a light source with a smaller wavelength (1)

   (ii) increasing the distance between the diffraction grating and the screen. (1)

(Total 9 marks)

18. A lifeguard at L on the beach must rescue a person in need in the water at P. Time is of the essence. Which path from L to P in the figure on the right will take the least time and why? (2)

(Total 2 marks)

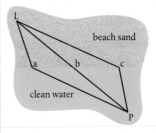

19. A coin is underwater. What does it appear to be? (1)

  A   Nearer the surface than it really is

  B   Farther from the surface than it really is

  C   As deep as it really is

20. The two wave forms in the diagram are traveling in opposite directions. Draw three diagrams to show the resultant wave forms when point O reaches A, B and C.

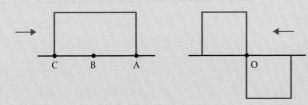

(Total 4 marks)

C.4

Standing waves and resonance

Music, no matter whether classical, pop, reggae or jazz, is all about standing waves. The notes that an instrument can produce depend on its length and boundary conditions, and a musician might use skillful techniques (like blowing harder or touching a string) to play different harmonics.

Guiding Questions

What distinguishes standing waves from traveling waves?

How does the form of standing waves depend on the boundary conditions?

How can the application of force result in resonance within a system?

The difference between a standing wave and a traveling wave can be illustrated by looking at the difference between a skipping rope and a stadium wave. (Note that neither are really waves.)

Viewed from the front, a skipping rope has a sinusoidal shape (a quarter of a wavelength) so it looks like a wave but the wave profile does not move; the rope simply moves up and down. All the parts of the rope are in phase but do not have the same amplitude.

With a stadium wave, each person gets up at a different time. But if we assume they are all the same height, the amplitude of each is the same.

A child's swing can be used to illustrate the concept of resonance. If displaced and released, a swing will oscillate with a frequency that depends on its length. If you push the swing, you have to push it at this frequency. If you push at any other frequency, the amplitude will never be large enough to be fun.

Students should understand:

the nature and formation of standing waves in terms of superposition of two identical waves traveling in opposite directions
nodes and antinodes, relative amplitude and phase difference of points along a standing wave
standing wave patterns in strings and pipes
the nature of resonance, including natural frequency and amplitude of oscillation based on driver frequency
the effect of damping on the maximum amplitude and resonant frequency of oscillation
the effects of light, critical and heavy damping on the system.

Standing waves in strings

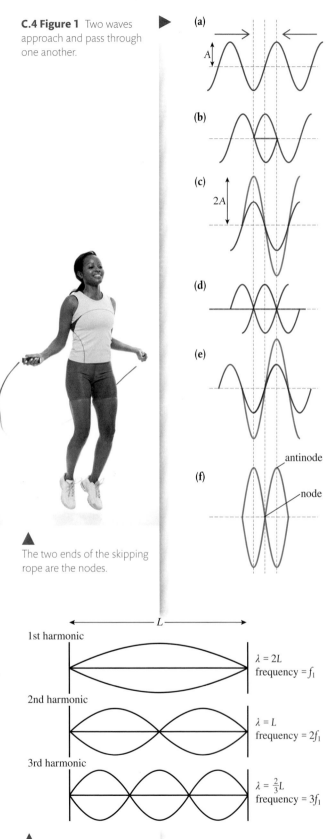

C.4 Figure 1 Two waves approach and pass through one another.

The two ends of the skipping rope are the nodes.

C.4 Figure 2 Standing waves in a string.

Consider two identical waves traveling along a string in opposite directions as shown in Figure 1(a). As the waves progress, they cross over each other and will superpose. In Figure 1(b), the waves have each progressed $\frac{1}{4}\lambda$, and are out of phase so they cancel out. The green line shows the resultant wave. After a further $\frac{1}{4}\lambda$, the waves will be in phase as shown in Figure 1(c). Notice that even though the waves are adding, the displacement of the midpoint remains zero. Figures 1(a) to (e) show the waves in $\frac{1}{4}\lambda$ steps, illustrating how they alternately add and cancel in such a way that the midpoint never moves: this is called a **node**. Either side of the midpoint, the waves sometimes add to give a peak and sometimes a trough: these points are called **antinodes**. Figure 1(f) shows the two extreme positions of the resultant wave. Notice that the nodes are separated by a distance $\frac{1}{2}\lambda$.

Differences between progressive waves and standing waves

The most obvious difference between a wave that travels along the rope and the standing wave is that the wave profile of a standing wave does not progress and nor does the energy associated with it. A second difference is that all points between two nodes on a standing wave are in phase (think of the skipping rope), whereas points on a progressive wave that are closer than one wavelength are all out of phase (think of a stadium wave). The third difference is related to the amplitude. All points on a progressive wave have the same amplitude, but on a standing wave, some points have zero amplitude (nodes) and some points have large amplitude (antinodes).

Stringed instruments

Many musical instruments (guitar, violin, piano) use stretched strings to produce sound waves. When the string is plucked, a wave travels along the string to one of the fixed ends. When it gets to the end, it reflects back along the string, superposing with the original wave. The result is a standing wave. The important thing to realize about the standing wave in a stretched string is that since the ends cannot move, they must become nodes, so only standing waves with nodes at the ends can be produced. Figure 2 shows some of the possible standing waves that can be formed in a string of length L.

As shown in the diagram, each of the possible waves is called a harmonic. The first harmonic (sometimes called the fundamental) is the wave with the lowest possible frequency. To calculate the frequency, we can use the formula $f = \frac{v}{\lambda}$ so for the first harmonic:

$$f_1 = \frac{v}{2L}$$

For the second harmonic:

$$f_2 = \frac{v}{L}$$

The wave velocity is the same for all harmonics, so we can deduce that: $f_2 = 2f_1$

Playing the guitar

When the guitar string is plucked, it does not just vibrate with one frequency but with many frequencies. However, the only ones that can create standing waves are the ones with nodes at the ends (as shown in Figure 2). You can try this with a length of rope. Get a friend to hold one end and shake the other. When you shake the end at certain specific frequencies, you get a standing wave, but if the frequency is not right, you do not. This is an example of resonance: hit the right frequency and the amplitude is big. So when the guitar string is plucked, all the possible standing waves are produced. If the signal from an electric guitar pickup is fed into a computer, the frequencies can be analyzed to get a frequency spectrum. Figure 3 shows an example.

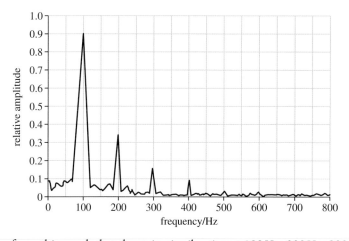

We can see from this graph that the string is vibrating at 100 Hz, 200 Hz, 300 Hz and so on. However, the largest amplitude note is the first harmonic (100 Hz) so this is the frequency of the note you hear.

Playing different notes

A guitar has six strings. Each string is the same length but has a different diameter and therefore different mass per unit length. The wave in the thicker strings is slower than in the thin strings so, since $f = \frac{v}{\lambda}$, the thick strings will be lower notes.

To play different notes on one string, the string can be shortened by placing a finger on one of frets on the neck of the guitar. Since $f = \frac{v}{\lambda}$, the shorter string will be a higher note.

The speed of a wave in a string is given by the formula:

$$v = \sqrt{\frac{F_t}{\mu}}$$

where F_t = tension and μ = mass per unit length.

This is why a thick guitar string is a lower note than a thin one, and why the note gets higher when you increase the tension.

Here we are focusing on the vibrating string. The string will cause the body of the guitar to vibrate, which in turn causes the pressure of the air to vary. It is the pressure changes in the air that cause the sound wave that we hear.

C.4 Figure 3 Frequency spectrum for a string.

A guitarist's pressure on the strings brings them tight to the frets, which defines their effective lengths.

An alternative way to play a higher note is to play a harmonic. This is done by placing a finger on the node of a harmonic (e.g. in the middle for the second harmonic) then plucking the string. Immediately after the string is plucked, the finger is removed to allow the string to vibrate freely.

Exercise

Q1. The mass per unit length of a guitar string is 1.2×10^{-3} kg m^{-1}. If the tension in the wire is 40 N, calculate:

(a) the velocity of the wave

(b) the frequency of the first harmonic if the vibrating length of the guitar string is 63.5 cm.

Q2. A 30 cm long string of mass per unit length 1.2×10^{-3} kg m^{-1} is tensioned so that its first harmonic is 500 Hz. Calculate the tension of a second string with half the mass per unit length but the same length that has the same first harmonic.

Q3. The first harmonic of a 1.0 m long stretched string is 650 Hz. What will its first harmonic be if its length is shortened to 80 cm, keeping the tension constant?

Challenge yourself

1. By adding the wave equations for two waves traveling in opposite directions, show that the distance between nodes on a standing wave is $\frac{1}{2}\lambda$.

Standing waves in closed pipes

When a sound wave travels along a pipe with closed boundaries, it will reflect off the closed end. The reflected wave and original wave then superpose to give a standing wave. A sound wave is a propagation of disturbance in air pressure. The change in air pressure causes the air to move backward and forward in the direction of the propagation. If the end of the pipe is closed, then the air cannot move back and forth so a node must be formed. This limits the possible standing waves to the ones shown in Figure 4.

In these diagrams, the sound waves are represented by displacement–position graphs. These make the sound look like a transverse wave but remember the displacement is in the direction of the disturbance.

C.4 Figure 4 Standing waves in a closed pipe.

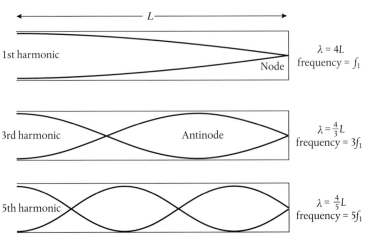

1st harmonic — Node — $\lambda = 4L$ frequency $= f_1$

3rd harmonic — Antinode — $\lambda = \frac{4}{3}L$ frequency $= 3f_1$

5th harmonic — $\lambda = \frac{4}{5}L$ frequency $= 5f_1$

These diagrams show how the displacement of the air varies along the length of the pipe, but remember, the displacement is actually along the pipe not perpendicular to it as shown. The frequency of each harmonic can be calculated using $f = \frac{v}{\lambda}$.

For the first harmonic:

$$f_1 = \frac{v}{4L}$$

For the next harmonic:

$$f_3 = \frac{v}{\frac{4}{3}L} = \frac{3v}{4L} = 3f_1 \text{ so this is the third harmonic}$$

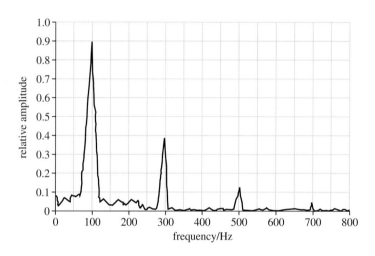

C.4 Figure 6 Frequency spectrum for a closed pipe.

So when a standing wave is formed in a closed pipe, only odd harmonics are formed, resulting in the frequency spectrum shown in Figure 6.

Standing waves in open pipes

If a wave is sent along an open-ended pipe, a wave is also reflected. The resulting superposition of reflected and original waves again leads to the formation of a standing wave. This time, there will be an antinode at both ends, leading to the possible waves shown in Figure 7.

This time all the harmonics are formed.

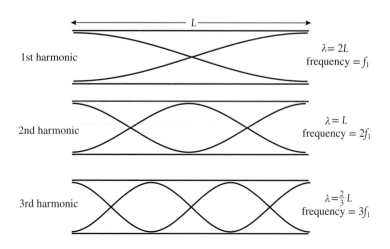

1st harmonic $\lambda = 2L$ frequency $= f_1$

2nd harmonic $\lambda = L$ frequency $= 2f_1$

3rd harmonic $\lambda = \frac{2}{3}L$ frequency $= 3f_1$

C.4 Figure 5 Remember, this is one quarter of a wave. It can be useful to split the harmonics into quarters when determining the frequency.

C.4 Figure 7 Standing waves in an open pipe.

341

In a clarinet, the reed is made to vibrate.

Wind instruments

All wind instruments (e.g. flute, clarinet, trumpet, and church organ) make use of the standing waves set up in pipes. The main difference between the different instruments is the way that the air is made to vibrate. In a clarinet, a thin piece of wood (a reed) is made to vibrate, in a trumpet, the lips vibrate, and in a flute, air is made to vibrate as it is blown over a sharp edge. Different notes are played by opening and closing holes. This has the effect of changing the length of the pipe. You can also play higher notes by blowing harder. This causes the higher harmonics to sound louder, resulting in a higher frequency note. If you have ever played the recorder, you might have had problems with this – if you blow too hard, you get a high-pitched noise that does not sound so good.

Exercise

The speed of sound in air is $340 \, \text{m s}^{-1}$.

Q4. Calculate the first harmonic produced when a standing wave is formed in a closed pipe of length 50 cm.

Q5. The air in a closed pipe in Figure 8 is made to vibrate by holding a tuning fork of frequency 256 Hz over its open end. As the length of the pipe is increased, loud notes are heard as the standing wave in the pipe resonates with the tuning fork.

(**a**) What is the shortest length that will cause a loud note?

(**b**) If the pipe is 1.5 m long, how many loud notes will you hear as the plunger is withdrawn?

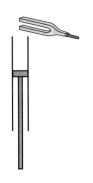

SKILLS

The speed of sound can be measured by sampling the sound made by a drinking straw whistle with your computer. In this example, Audacity®, free open source software, was used. The drinking straw is turned into a whistle by cutting a notch close to one end. When blown, the noise is not very loud (more of a rush of air than a whistle) but it is enough to analyze.

C.4 Figure 8 A drinking straw whistle.

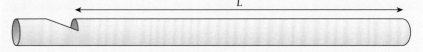

The whistle was blown into the microphone of the computer and the sound recorded using Audacity® (Figure 9). The frequency spectrum was then analyzed to find the harmonics of the sound.

C.4 Figure 9 Screenshot from Audacity® showing sampled sound.

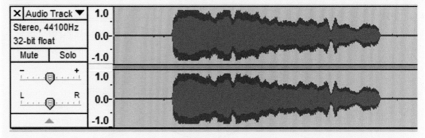

A worksheet with full details of how to carry out this experiment is available in your eBook.

SKILLS

By measuring the sound from pipes of different lengths, it is possible to plot a graph of f vs $\frac{1}{\lambda}$ (Figure 10). This should be a straight line with gradient = velocity of sound.

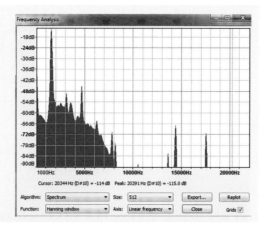

When we say the source is moving, we mean that it moves relative to the medium (air). The observer is at rest relative to the medium.

C.4 Figure 10 The frequency spectrum from Audacity®.

Nature of Science

Damping occurs when a pendulum moves through air. By combining what we know from the study of simple harmonic motion and motion through fluids, we can derive a model for damped harmonic motion.

Resonance is a phenomenon related to mechanical vibrations. However, the same model can be applied to electrical circuits, molecules, atoms, and nuclei.

Damped harmonic motion

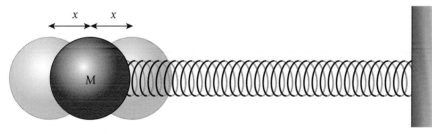

C.4 Figure 11 A mass on a spring (in space).

When dealing with oscillations, we only considered simple harmonic motion. That is, motion where the acceleration of a body is proportional to the displacement from a fixed point and always directed toward that point; for example, a mass on the end of a stiff spring attached to the side of a spaceship (no gravity or air) as in Figure 11. When the mass moves to the left, the spring gets stretched and, according to Hooke's law, the force will be proportional to the extension. When the mass moves to the right, the spring is compressed and the force pushing back is proportional to compression.

This motion can be represented by the equation $ma = -kx$.

This implies that the displacement, $x = a \cos \omega t$. In other words, the motion is sinusoidal as represented by Figure 13: ω is the angular frequency, $2\pi f$, where f is the **natural frequency** of the oscillation.

343

C.4 Figure 12 Free oscillation of a mass on a spring.

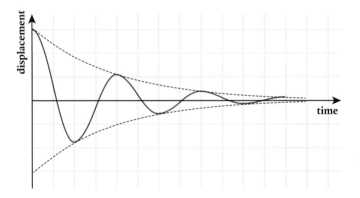

If we now add a fluid around the ball, then it will experience a viscous force opposing the motion. This force is proportional to the velocity of the ball so the equation for the force becomes:

$$ma = -kx - bv \text{ where } b \text{ is some constant}$$

Solving this equation reveals that: $x = ae^{-\frac{bt}{m}} \cos \omega t$

So the amplitude decreases exponentially with time as shown in Figure 13.

C.4 Figure 13 Lightly damped (or **under-damped**) harmonic motion; the dashed lines show the exponential change of amplitude.

The value of bm in the equation is equivalent to the decay constant in nuclear decay. If it is large, then the amplitude will decrease in a short time. b is the constant of proportionality relating the drag force with the velocity ($F_d = bv$), so if the mass was surrounded by a much more viscous fluid, then the drag force would be greater, resulting in heavier damping. If the fluid is very viscous, the damping could be so heavy that the spring would not oscillate at all. The damping is said to be **critical** if the mass returns to the equilibrium position as quickly as possible but without crossing it, and **over-damped** if the system tends toward the equilibrium position more slowly, as in Figure 14.

A car suspension is critically damped to prevent the car oscillating after it goes over a bump in the road.

C.4 Figure 14 Critically damped and over-damped harmonic motion.

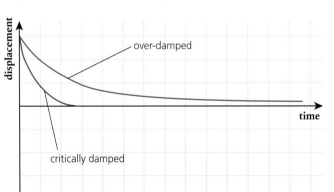

Forced vibration

When an oscillating system such as a mass on a spring is disturbed, it will oscillate at its own natural frequency $= \frac{1}{2\pi}\sqrt{\frac{k}{m}}$. The mass can also be made to vibrate at other frequencies by applying a sinusoidally varying force as in Figure 16. Here energy is being transferred from the motor driving the oscillating platform to the ball and spring.

When driven in this way, the mass and spring will oscillate with the frequency of the driver.

Simulation of forced harmonic motion

Using a simulation program like Algodoo® or Interactive physics®, it is possible to make a simulation just like the one in Figure 15. By varying the speed of the wheel, we can investigate how the amplitude depends on the driving frequency. Figure 16 shows the results from such a simulated experiment where the natural frequency of the mass–spring system was 2 Hz.

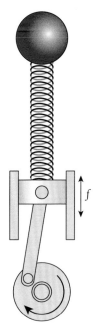

C.4 Figure 15 A driven oscillator.

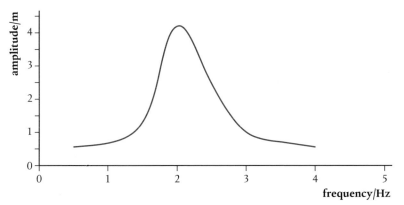

C.4 Figure 16 Graph of amplitude vs frequency for a driven oscillator.

From this graph, we can see that the amplitude is biggest when the driving frequency is equal to the natural frequency of the mass and spring. This is called **resonance**. You will have experienced the same phenomenon if you have ever been pushed on a swing by someone with no sense of rhythm. If the pushes are not at the natural frequency of the swing, the amplitude is small. However, if the pushes are at the same frequency, the amplitude becomes large.

Resonance

Resonance is an important phenomenon, both in mechanics and in other areas of physics, so is worth studying in detail.

Effect of damping

In a lightly damped system, the resonant frequency is the same as the natural frequency, but if the damping is heavy, the resonant frequency is reduced. Figure 18 shows two resonance curves, one with light damping and one with heavy damping. The resonance peak for the lightly damped situation is also sharper than the heavily damped example.

You can see how the different constants affect the oscillation by simulating damping with Excel®.

What is the relationship between resonance and simple harmonic motion? (C.1)

How can resonance be explained in terms of conservation of energy? (A.3)

It is possible to shatter a wine glass if it is made to vibrate at its own natural frequency.

C.4 Figure 17 Resonance curves with light and heavy damping. Notice that the resonant frequency is slightly less when the damping is heavy.

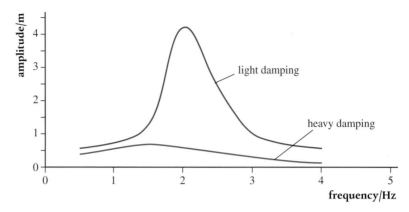

How does the amplitude of vibration at resonance depend on the dissipation of energy in the driven system? (A.3)

When a damped oscillator is driven, the energy supplied by the driver is providing energy to drive the system plus doing work against the damping force. Damping reduces the maximum amplitude and resonant frequency of the resonating system.

Exercise

Q6. A series of pendula is suspended on one string. When pendulum A starts to swing, it disturbs the suspension, forcing pendulums B, C, D, E, and F to oscillate.

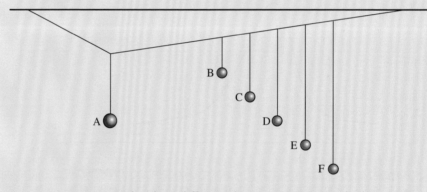

(a) Comment on the phase difference between B, C, D, E, F, and A.

(b) Which pendulum will have the largest amplitude?

Examples of resonance

- A truck drives past a room and makes plates rattle on a shelf.
- An opera singer shatters a wine glass by singing a note with the same frequency as its resonant frequency.
- Infrared radiation resonates with a CO_2 molecule.
- The air in a tube resonates to produce a sound when a wind instrument is blown.
- A resonating quartz crystal is the basis of many clocks.
- An engine part may break if it resonates with the frequency of the engine.
- A tall building can collapse if its natural frequency is the same as the frequency of an earthquake.

How can the idea of resonance of gas molecules be used to model the greenhouse effect? (NOS)

Electrical resonance – an interesting link

Consider a capacitor C, resistor R, and coil of inductance L connected as shown in Figure 18. (An understanding of capacitors is not required for your course but, in short, they store charge so that a potential difference exists between the two plates. Awareness of inductance is also not required.)

At the present time, the capacitor is charged, creating a potential difference across the resistance, resulting in a flow of current through the coil. The current in the coil will induce a magnetic field, but the changing magnetic field in the coil induces a current in the coil that opposes the change producing it, causing current to flow back onto the capacitor. The charge oscillates back and forth, energy transferring from electric to magnetic. The frequency of the oscillation depends on the size of the capacitor and coil and the resistance causes damping. If this circuit is connected to a variable frequency AC supply, the current will be made to oscillate in the circuit. If the frequency is the same as the natural frequency, then resonance occurs and the current flowing is a maximum. Figure 19 shows the resonance curves for different values of R.

This circuit is used in a radio receiver. The values of L (this is related to the coil) and C are varied to match the frequency of a radio station. An aerial picks up all radio stations but the circuit resonates with the one that has the same frequency as the natural frequency. By reducing the resistance of the circuit, the peak can be made narrower (reduced bandwidth), making the circuit more selective.

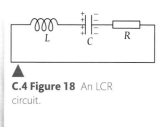

C.4 Figure 18 An LCR circuit.

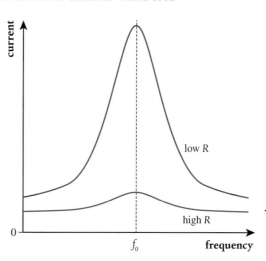

C.4 Figure 19 Resonance curves for different values of R.

Guiding Questions revisited

What distinguishes standing waves from traveling waves?

How does the form of standing waves depend on the boundary conditions?

How can the application of force result in resonance within a system?

In this chapter, we have found that waves of equal amplitude and frequency traveling in opposite directions (or a wave reflected on itself) can result in no net energy propagation. In doing so, we now understand:

- Standing waves are common, featuring in musical instruments, but sometimes have undesirable effects (i.e. failure or harm to the user) when a system's driving frequency matches its natural frequency.
- A node is a position of zero displacement.
- An antinode is a position of maximum displacement.
- All particles between two consecutive nodes oscillate in phase (and in antiphase with the particles on adjacent loops).
- A standing wave on a string with two clasped ends will have nodes at the boundaries, and its first harmonic as a single loop.

- A closed end on a pipe will mean that a node forms at this boundary.
- An open end on a pipe will mean that an antinode forms at this boundary.
- When a driver forces oscillations of a system with increasing frequency, the amplitude increases to a maximum (resonance) at the natural frequency and then decreases.
- Damping is the interaction of an oscillating body and the fluid that surrounds it. Heavy damping prevents even a quarter of a cycle from being completed, critical damping results in a return to the equilibrium position as quickly as possible without overshoot, and light damping reduces the amplitude of oscillations.
- Damping as reducing the natural frequency of a system.

Practice questions

1. (a) State the principle of superposition. (2)

   A wire is stretched between two points A and B.

   A ——————————————— B

   A standing wave is set up in the wire. This wave can be thought of as being made up from the superposition of two waves: wave X traveling from A to B, and wave Y traveling from B to A. At one particular instant in time, the displacement of the wire is as shown in the diagram below. A background grid is given for reference and the equilibrium position of the wire is shown as a dotted line.

   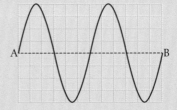

   (b) Copy the grids below and draw the displacements of the wire due to wave X and wave Y. (4)

   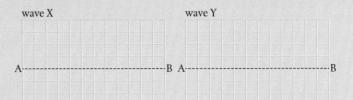

   wave X   wave Y

   (Total 6 marks)

2. There is a tall building near to the site of the earthquake. The base of the building vibrates horizontally due to the earthquake.

   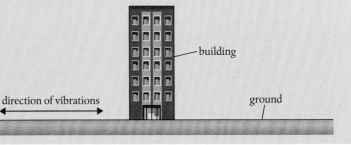

   direction of vibrations   building   ground

(a) Copy the diagram above and draw the fundamental mode of vibration of the building caused by these vibrations. (1)

The building is of height 280 m and the mean speed of waves in the structure of the building is $3.4 \times 10^3 \, ms^{-1}$.

(b) Explain quantitatively why earthquake waves of frequency about 6 Hz are likely to be very destructive. (3)

(Total 4 marks)

3. The diagram shows two pipes of the same length. Pipe A is open at both ends and pipe B is closed at one end.

pipe A   pipe B

(a) (i) Copy the diagram and draw lines to represent the waveforms of the fundamental (first harmonic) resonant note for each pipe. (2)

(ii) On each pipe, label the position of the nodes with the letter N and the position of the antinodes with the letter A. (2)

The frequency of the fundamental note for pipe A is 512 Hz.

(b) (i) Calculate the length of pipe A (speed of sound in air = 340 m s^{-1}). (3)

(ii) Suggest why organ pipes designed to emit low frequency fundamental notes (e.g. frequency ≈ 32 Hz) are often closed at one end. (2)

(Total 9 marks)

4. The diagram shows two images of a machine designed to make a piston oscillate. The bottom image was taken 0.125 s after the top image. The wheel turns anticlockwise at a constant speed. The long rod is attached 1.0 cm from the center of the wheel.

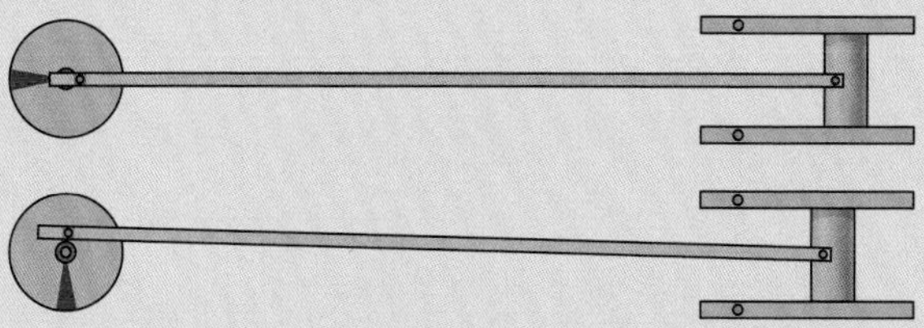

(a) (i) Explain why the motion of the piston is approximately simple harmonic. (2)

(ii) The motion is not exactly simple harmonic. Explain why and state which feature reduces this effect. (2)

(iii) State the amplitude of the oscillation. (1)

(iv) Calculate the frequency of the oscillation. (1)

The machine is connected to a 2.0 m long slinky spring suspended on strings. The diagrams show before switching on and one revolution after switching on.

(b) **(i)** Determine the wavelength of the wave. (1)

(ii) Calculate the velocity of the wave. (1)

(iii) The right end of the spring is free to move. Describe what happens when the wave reaches this end. (2)

(iv) Deduce whether or not a standing wave will be formed. (3)

(c) The speed of the wave is given by the same equation as the speed of a wave in a string:

$$v = \sqrt{\frac{F_t}{\mu}}$$

where F_t = tension and μ = mass per unit length.

Assuming that the unstretched length is negligible in comparison with the stretched length, show that doubling the length will double the velocity. (3)

(d) The machine is now turned so that the piston is horizontal. A small ball is placed on the piston so that it goes up and down with the piston. The speed of the wheel is increased until the ball starts to rattle as it loses contact with the piston. Determine the frequency at which this takes place. (3)

(Total 19 marks)

5. The circuit represents a simple electrical oscillator. When the switch is closed, charge flows from the capacitor to the coil. The flow of current causes a changing magnetic field that induces an emf, sending the charge back to the capacitor, and the cycle repeats. The frequency of the cycle depends on the physical properties of the coil and capacitor:

$$\omega = \frac{1}{\sqrt{LC}}$$

where L = inductance and C = capacitance.

Radio waves cause electrons to oscillate in the wire.

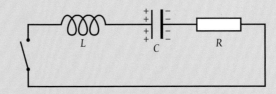

(a) Explain how this circuit could be used to tune a radio to one particular frequency. (4)

(b) If L = 4.00 mH and C = 2.00 pF, calculate the frequency of the radio signal that will cause the most current to flow in the circuit. (2)

The frequency of the radio signal is varied from below the frequency calculated in (ii) to above this frequency.

(c) Sketch a graph of the variation of current (y-axis) with frequency (x-axis). (2)

(*Total 8 marks*)

6. A pipe is open at both ends. Which is correct about a standing wave formed in the air of the pipe?

A The sum of the number of nodes plus the number of antinodes is an odd number.

B The sum of the number of nodes plus the number of antinodes is an even number.

C There is always a central node.

D There is always a central antinode.

(*Total 1 mark*)

7. A pipe of fixed length is closed at one end.
What is $\dfrac{\text{third harmonic frequency of pipe}}{\text{first harmonic frequency of pipe}}$?

   A $\frac{1}{5}$ B $\frac{1}{3}$ C 3   D 5

(*Total 1 mark*)

8. On a guitar, the strings played vibrate between two fixed points. The frequency of vibration is modified by changing the string length using a finger. The different strings have different wave speeds. When a string is plucked, a standing wave forms between the bridge and the finger.

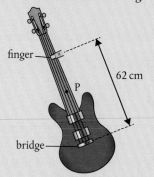

(a) Outline how a standing wave is produced on the string. (2)

The string is displaced 0.4 cm at point P to sound the guitar. Point P on the string vibrates with simple harmonic motion (SHM) in its first harmonic with a frequency of 195 Hz. The sounding length of the string is 62 cm.

(b) Show that the speed of the wave on the string is about 240 m s^{-1}. (2)

(c) Copy the axes below and sketch a graph to show how the acceleration of point P varies with its displacement from the rest position. (1)

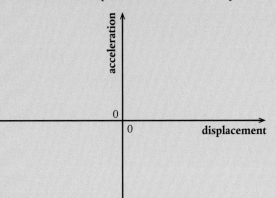

(d) The string is made to vibrate in its third harmonic. State the distance between consecutive nodes. (1)

(Total 6 marks)

9. A pipe is open at both ends. A first-harmonic standing wave is set up in the pipe. The diagram shows the variation of displacement of air molecules in the pipe with distance along the pipe at time $t = 0$. The frequency of the first harmonic is f.

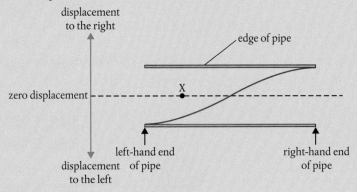

(a) Copy the diagram and sketch the variation of displacement of the air molecules with distance along the pipe when $t = \frac{3}{4f}$. (1)

(b) An air molecule is situated at point X in the pipe at $t = 0$. Describe the motion of this air molecule during one complete cycle of the standing wave beginning from $t = 0$. (2)

The speed of sound c for longitudinal waves in air is given by:

$$c = \sqrt{\frac{K}{\rho}}$$

where ρ is the density of the air and K is a constant.

(c) A student measures f to be 120 Hz when the length of the pipe is 1.4 m. The density of the air in the pipe is 1.3 kg m⁻³. Determine the value of K for air. Give your answer with the appropriate unit. (4)

(Total 7 marks)

10. (a) Describe **two** ways in which standing waves differ from traveling waves. (2)

A vertical tube, open at both ends, is completely immersed in a container of water. A loudspeaker above the container connected to a signal generator emits sound. As the tube is raised, the loudness of the sound heard reaches a maximum because a standing wave has formed in the tube.

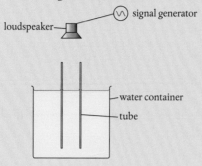

(b) Outline how a standing wave forms in the tube. (2)

The tube is raised until the loudness of the sound reaches a maximum for a second time.

(c) Copy the diagram below and draw the position of the nodes in the tube when the second maximum is heard. (1)

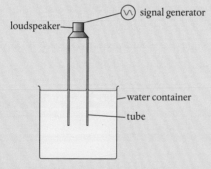

(d) Between the first and second positions of maximum loudness, the tube is raised through 0.37 m. The speed of sound in the air in the tube is 320 m s⁻¹. Determine the frequency of the sound emitted by the loudspeaker. (2)

(Total 7 marks)

11. A mass–spring system is forced to vibrate vertically at the resonant frequency of the system. The motion of the system is damped using a liquid.

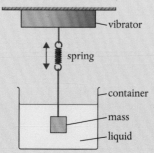

At time $t = 0$, the vibrator is switched on. At time t_A, the vibrator is switched off and the system comes to rest. The graph shows the variation of the vertical displacement of the system with time until t_B.

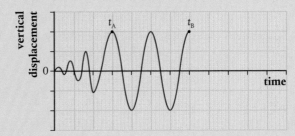

(a) Explain, with reference to energy in the system, the amplitude of oscillation between:

   (i) $t = 0$ and t_A (1)

   (ii) t_A and t_B. (1)

(b) The system is critically damped. Copy the graph above and draw the variation of the displacement with time from t_B until the system comes to rest. (2)

(Total 4 marks)

12. A railway track passes over a bridge that has a span of 20 m. The bridge is subject to a periodic force as a train crosses. This is caused by the weight of the train acting through the wheels as they pass the center of the bridge. The wheels of the train are separated by 25 m.

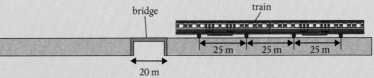

(a) Show that, when the speed of the train is 10 m s^{-1}, the frequency of the periodic force is 0.4 Hz. (1)

The graph below shows the variation of the amplitude of vibration A of the bridge with driving frequency f_D when the damping of the bridge system is small.

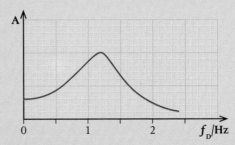

(b) Outline, with reference to the curve, why it is unsafe to drive a train across the bridge at 30 m s^{-1} for this amount of damping. (2)

(c) The damping of the bridge system can be varied. Copy the graph and draw a second curve when the damping is larger. (2)

(Total 5 marks)

13. You are emptying a large bottle of water. As the liquid runs out it makes a 'gluug gluug gluug' sound. As the bottle becomes empty, what happens to the frequency of the sound?

A It gets lower.

B It does not change.

C It gets higher. *(Total 1 mark)*

14. A guitar string is stretched from point A to point G. Equal intervals (A, B, C, D, E, F, G) are marked and paper 'riders' are placed on the string at D, E and F. The string is pinched at C and twanged at B. What happens?

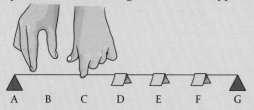

A All the riders jump off.

B None of the riders jump off.

C The rider at E jumps off.

D The riders at D and F jump off.

E The riders at E and F jump off. *(Total 1 mark)*

15. A mouse wants to get a ball bearing up and out of a bowl, but the ball is too heavy and the sides of the bowl too steep for the mouse to support the ball's weight. Using only its own strength, without the help of levers and such, what will the mouse be able to do?

A Not be able to get the ball bearing up and out of the bowl

B Be able to get the ball bearing up and out of the bowl (how?)

(Total 2 marks)

16. A body, mass m, rests on a scale pan which is supported by a spring. The period of oscillation of the scale pan is 0.50 s.

(a) It is observed that when the amplitude of the oscillations is increased, and exceeds a certain value, the mass leaves the pan. Explain why the mass leaves the pan. (3)

(b) At what point in the motion does the mass initially leave the pan? (3)

(Total 6 marks)

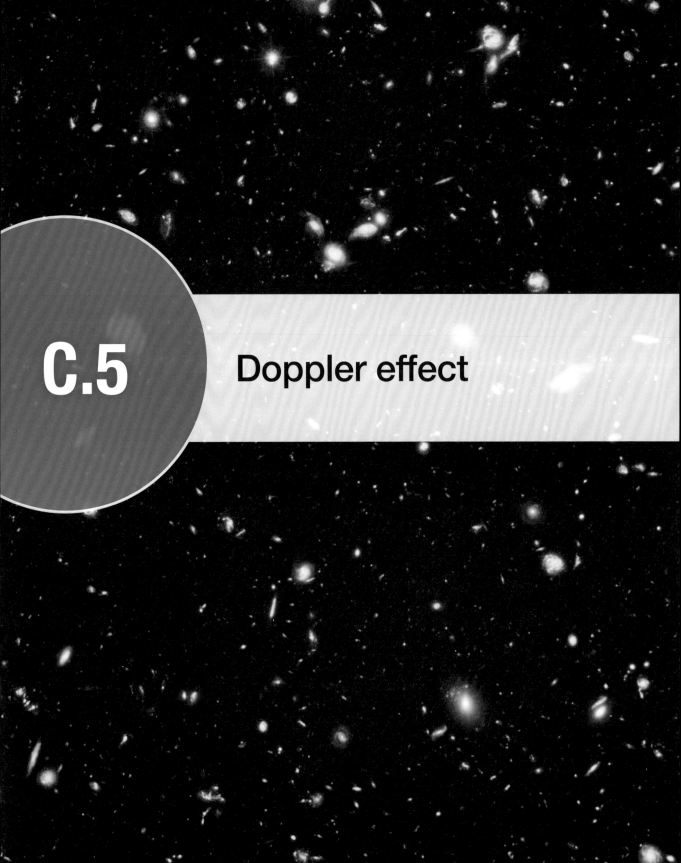

C.5

Doppler effect

◄ The Hubble Deep Field is an image of a small section of the night sky captured by the Hubble Space Telescope over 100 hours in 1995. Most of the bright objects captured are galaxies (some of them are among the most distant ever known) and the light from many of these galaxies is red-shifted. A red shift of electromagnetic radiation is an increase in the wavelength of the light and is evidence for the expansion of the Universe. For sound waves, we refer to this change in the observed wavelength as the Doppler effect.

Guiding Questions

How can the Doppler effect be explained qualitatively and quantitatively?

What are some practical applications of the Doppler effect?

Why are there differences when applying the Doppler effect to different types of wave?

The Doppler effect is the change in frequency of a wave due to the relative motion of source and observer. It can be observed in both light and sound but the reason for the shift is quite different.

When sound is involved, there is a reference frame that we can measure the velocity of sound relative to – the air. If the source or observer are not moving relative to the air, then they are not moving.

Light does not need a medium so there is no difference between moving source or observer. The shift in frequency is due to relativistic effects rather than a change in wave velocity. For light, it makes no difference if you take the source or the observer to be the one that is moving; it is their velocity relative to each other that is important.

Students should understand:

the nature of the Doppler effect for sound waves and electromagnetic waves
the representation of the Doppler effect in terms of wavefront diagrams when either the source or the observer is moving
the relative change in frequency or wavelength observed for a light wave due to the Doppler effect where the speed of light is much larger than the relative speed between the source and the observer as given by $\frac{\Delta f}{f} = \frac{\Delta \lambda}{\lambda} \approx \frac{v}{c}$
shifts in spectral lines provide information about the motion of bodies like stars and galaxies in space
HL the observed frequency for sound waves and mechanical waves due to the Doppler effect as given by: moving source $f' = f\left(\frac{v}{v \pm u_s}\right)$, where u_s is the velocity of the source   moving observer $f' = f\left(\frac{v \pm u_o}{v}\right)$ where u_o is the velocity of the observer.

The Doppler effect

If you have ever stood next to a busy road, or even better a race track, you might have noticed that the cars sound different when they come toward you and when they go away. It is difficult to put this into words but the sound is something like this:

What are the similarities and differences between light and sound waves? (C.2)

'eeeeeeeeeeeoowwwwwwww'. The sound on approach is a higher frequency than on retreat. This effect is called the Doppler effect and can occur when the source of the sound is moving or when the observer is moving, or when both the source and observer are moving.

C.5 Figure 1 The car starts from the red spot and moves forward. The largest circle is the wavefront formed when the car began.

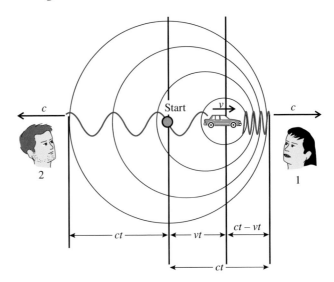

Moving source

Standard Level students will notice that the wavelength and speed effects for moving sources and moving observers are different. Higher Level students are required to calculate the frequencies in each case by carefully selecting the appropriate equation.

The change in frequency caused when a source moves is due to the change in wavelength in front of and behind the source. This is illustrated in Figure 1. You can see that the waves ahead of the source have been squashed as the source 'catches up' with them. The velocity of sound is not affected by the movement of the source, so the reduction in wavelength results in an increased frequency ($v = f\lambda$).

The car in Figure 1 starts at the red spot and drives forward at speed v for t seconds, producing a sound of frequency f_0. In this time t, the source produces $f_0 t$ complete waves ($3\frac{1}{2}$ in this example).

TOK

In this course, we often explain phenomena that you have observed such as the change in frequency as an ambulance drives past. What if you have never observed such a thing?

Ahead of the car, these waves have been squashed into a distance $ct - vt$ so the wavelength must be:

$$\lambda_1 = \frac{ct - vt}{f_0 t} = \frac{c - v}{f_0}$$

The velocity of these waves is c so the observed frequency, $f_1 = \frac{c}{\lambda_1}$.

$$f_1 = \frac{c}{\frac{c - v}{f_0}} = \frac{cf_0}{c - v}$$

We could do a similar derivation for the waves behind the source. This time, the waves are stretched out to fit into a distance $ct + vt$. The observed frequency is then:

$$f_2 = \frac{cf_0}{c + v}$$

Moving observer

If the observer moves relative to a stationary source, the change in frequency observed is due to the fact that the velocity of the sound changes relative to the observer. This is because the velocity of sound is relative to the air, so if you travel through the air toward a sound, the velocity of the sound will increase. Figure 2 illustrates the effect on frequency of a moving observer.

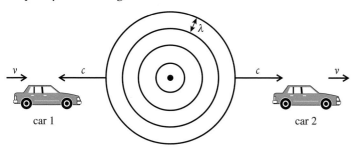

If the source moves at the speed of sound, the sound in front bunches up to form a shock wave. This causes the bang (sonic boom) you hear when a plane breaks the sound barrier.

C.5 Figure 2 The wavelength of the sound is constant because the source is stationary.

The relative velocity of the sound coming toward car 1 approaching the source is $(c + v)$ so the frequency, $f_1 = \frac{c + v}{\lambda}$.

But:
$$\lambda = \frac{c}{f_0}$$

So:
$$f_1 = \frac{(c + v)f_0}{c} \quad \text{(approaching)}$$

For car 2, receding from the source:
$$f_2 = \frac{(c - v)f_0}{c} \quad \text{(receding)}$$

Worked example   HL

A car traveling at $30\,\mathrm{m\,s^{-1}}$ emits a sound of frequency $500\,\mathrm{Hz}$. Calculate the frequency of the sound measured by an observer in front of the car.

Solution

This is an example where the source is moving relative to the medium and the observer is stationary relative to the medium. To calculate the observed frequency, we use the equation:

$$f_1 = \frac{c f_0}{c - v}$$

Where:
$$f_0 = 500\,\mathrm{Hz}$$
$$c = 340\,\mathrm{m\,s^{-1}}$$
$$v = 30\,\mathrm{m\,s^{-1}}$$

So:
$$f_1 = \frac{340 \times 500}{340 - 30} = 548\,\mathrm{Hz}$$

The rate at which blood flows can be found by measuring the Doppler shift of ultrasound waves reflected off blood cells. In this case, the blood cell is moving relative to the source, and when it reflects the wave, the reflected wave is moving relative to the receiver. This causes a double shift in frequency.

Exercise · HL

Q1. A person jumps off a high bridge attached to a long elastic rope (a bungee jump). As they begin to fall, they start to scream at a frequency of 1000 Hz. They reach a terminal velocity of 40 m s⁻¹ on the way down and 30 m s⁻¹ on the way back up.

    (a) Describe what you would hear if you were standing on the bridge.

    (b) Calculate the maximum frequency you would hear from the bridge.

    (c) Calculate the minimum frequency you would hear from the bridge.

    (d) What would you hear if you were standing directly below the bungee jumper?

Q2. The highest frequency you can hear is 20 000 Hz. If a plane making a sound of frequency 500 Hz went fast enough, you would not be able to hear it. How fast would the plane have to go?

Q3. Calculate the frequency of sound you would hear as you drove at 20 m s⁻¹ toward a sound source emitting a sound of frequency 300 Hz.

The Doppler effect and electromagnetic radiation

How can the use of the Doppler effect for light be used to calculate speed? (NOS)

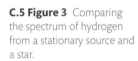

Unlike the Doppler effect for sound, it does not matter whether it is the source or observer that moves.

The Doppler effect also applies to electromagnetic radiation (radio waves, microwaves and light). The derivation of the formula is rather more complicated since the velocity of light is not changed by the relative movement of the observer. However, if the relative velocities are much smaller than the speed of light, we can use the following approximation:

$$\Delta f = \frac{v}{c} f_0$$

Where:

Δf = the change in frequency

v = the relative speed of the source and observer

c = the speed of light in a vacuum

f_0 = the original frequency.

Red shift

C.5 Figure 3 Comparing the spectrum of hydrogen from a stationary source and a star.

Short wavelength   Long wavelength

Full visible light spectrum

Hydrogen spectrum

Hydrogen spectrum from a receding star (red-shifted)

If a source of light is moving away from an observer, the light received by the observer will have a longer wavelength than when it was emitted. If we look at the spectrum as shown in Figure 3, we see that the red end of the spectrum is long wavelength and the blue end is short wavelength. The change in wavelength will therefore cause the light to shift toward the red end of the spectrum – this is called **red shift**. This effect is very useful to astronomers as they can use it to calculate how fast stars are moving away from us. This is made possible by the fact that the spectrum of light from stars contains characteristic absorption lines from elements such as hydrogen. The wavelength of these lines is known so their shifted position in the spectrum can be used to calculate velocity.

Example: Radars

Police speed traps use the Doppler effect to measure the speed of passing cars. When they aim the device at the car, a beam of electromagnetic radiation (radio waves, microwaves or infrared) is reflected off the car. The reflected beam undergoes a double Doppler shift. First the car is approaching the beam so there will be a shift due to the moving observer, and second the reflected beam is emitted from the moving source. When the device receives the higher frequency reflected beam, the speed of the car can be calculated from the change in frequency.

Speed cameras use the Doppler effect.

How can the Doppler effect be utilized to measure the rotational speed of extended bodies? (A.4)

Worked example

A speed trap uses a beam with a wavelength 1.0 cm. What is the change in frequency received by the detector if the beam reflects off a car traveling at 150 km h^{-1}?

Solution

First convert the car's speed to m s^{-1}:

$$150\,\text{km h}^{-1} = 42\,\text{m s}^{-1}$$

$$\text{frequency of the 1.0 cm wave} = \frac{c}{\lambda} = \frac{3.00 \times 10^8}{1.0 \times 10^{-2}} = 3.0 \times 10^{10}\,\text{Hz}$$

Change in f of signal received by car is given by:

$$\Delta f = \frac{v}{c} f_0$$

$$= \frac{42}{3.00 \times 10^8} \times 3.0 \times 10^{10} = 4.2 \times 10^3\,\text{Hz}$$

This shift is then doubled since the car, now the source of the reflected wave, is traveling toward the detector, so the change in frequency = 8.4×10^3 Hz.

Exercise

Q4. A star emits light of wavelength 650 nm. If the light received at the Earth from this star has a wavelength of 690 nm, how fast is the star moving away from the Earth?

Q5. An atom of hydrogen traveling toward the Earth at 2×10^6 m s^{-1} emits light of wavelength 658 nm. What is the change in wavelength experienced by an observer on the Earth?

The expanding Universe

In the early 1920s, Albert Einstein developed his theory of general relativity, and while Einstein and Friedmann were developing their theories, Vesto Slipher and Edwin Hubble were taking measurements. Slipher was measuring the line spectra from distant galaxies and Hubble was measuring how far away they were. Slipher discovered that the spectral lines from *all* the galaxies were shifted toward the red end of the spectrum. If this was due to the Doppler shift, it would imply that *all* of the galaxies were moving away from the Earth. Given the change in wavelength, Hubble then calculated the velocity of the galaxies using the Doppler formula:

$$\frac{v}{c} = \frac{\Delta\lambda}{\lambda_{em}} = z$$

Where:

v = the recessional velocity
c = the speed of light
$\Delta\lambda$ = the change in wavelength
λ_{em} = the wavelength originally emitted from the galaxy
z = the fractional increase or z parameter (no units).

What gives rise to emission spectra and how can they be used to determine astronomical distances? (E.1)

Exercise

Q6. A spectral line from a distant galaxy of wavelength 434.0 nm is red-shifted to 479.8 nm. Calculate the recession speed of the galaxy.

Q7. The same line from a second galaxy is shifted to 481.0 nm. Calculate its recession speed. Is this galaxy closer or further away?

C.5 Figure 4 Graph of recession speed of galaxies against their distance from the Earth.

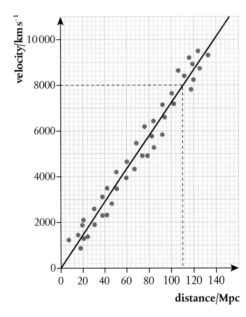

Hubble's law

In 1929, Hubble published his discovery that there appeared to be a linear relationship between the recessional velocity and distance to the galaxy. This can be illustrated by plotting the data on a graph as shown in Figure 4.

The recessional velocity of a distant galaxy is directly proportional to its distance.

In other words, the further away a galaxy is, the faster it moves away from us.

This can be expressed in terms of the formula:

$$v \propto d$$

Or:

$$v = H_0 d$$

where H_0 is the **Hubble constant**. This is the gradient of the line and has the value 72 km s^{-1} Mpc^{-1}. This gives a measure of the rate of expansion of the Universe. This is probably not a constant rate since the effect of gravity might slow down the rate over time. H_0 is the current value.

What happens if the speed of light is not much larger than the relative speed between the source and the observer? (A.5)

Exercise

Q8. Use Hubble's law to estimate the distance from the Earth to a galaxy with a recessional velocity of $150\,\text{km s}^{-1}$.

Q9. If a galaxy is 20 Mpc from Earth, how fast will it be receding?

Guiding Questions revisited

How can the Doppler effect be explained qualitatively and quantitatively?

What are some practical applications of the Doppler effect?

Why are there differences when applying the Doppler effect to different types of wave?

In this chapter, we have studied the interactions between the velocity of a wave and the velocities of its source and observer to appreciate how:

- The motion of a source of a wave causes the wavelength of the wave to decrease in front and increase behind, which means that an increased frequency is observed for an approaching source and a decreased frequency is observed for a receding source.
- The same qualitative effects are observed when an observer approaches a wave source (increased observed frequency) and recedes from the source (decreased observed frequency), although, in this case, the emitted wavelengths are constant throughout.
- This 'Doppler effect' is useful in medicine (to examine circulation), transport (radar for speed measurements) and cosmology (motion of galaxies relative to the Earth).
- When the wave speed is much higher than the relative speed of the source and observer (e.g. light), the ratio of the change in the observed frequency to the frequency at source is approximately equal to the ratio of the velocity of the source to the speed of the wave.
- **HL** The same extreme difference between the speed of the wave and the relative speed of its observers and sources is less likely for sound or mechanical waves, so alternative equations are needed for quantitative evaluation.

Practice questions

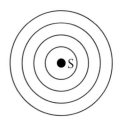

1. The diagram on the left shows wavefronts produced by a stationary wave source S. The spacing of the wavefronts is equal to the wavelength of the waves. The wavefronts travel with speed V.

   (a) The source S now moves to the right with speed v. Copy the diagram and draw four successive wavefronts to show the pattern of waves produced by the moving source. (3)

   (b) [HL] Derive the Doppler equation for the observed frequency f_0 of a sound source, as heard by a stationary observer, when the source approaches the stationary observer with speed v. The speed of sound is V and the frequency of the sound emitted by the source is f. (3)

   The Sun rotates about its center. The light from one edge of the Sun, as seen by a stationary observer, shows a Doppler shift of 0.004 nm for light of wavelength 600.000 nm.

   (c) Assuming that the Doppler equation for sound may be used for light, estimate the linear speed of a point on the surface of the Sun due to its rotation. (3)

   (Total 9 marks)

2. A source emits sound of wavelength λ_0 and wave speed v_0. A stationary observer hears the sound as the source moves away. What are the wavelength of the sound and the wave speed of the sound as measured by the stationary observer?

	Wavelength	Wave speed
A	less than λ_0	equal to v_0
B	greater than λ_0	equal to v_0
C	less than λ_0	less than v_0
D	greater than λ_0	less than v_0

   (Total 1 mark)

3. A train approaches a station and sounds a horn of constant frequency and constant intensity. An observer waiting at the station detects a frequency f_{obs} and an intensity I_{obs}. What are the changes, if any, in I_{obs} and f_{obs} as the train slows down?

	I_{obs}	f_{obs}
A	no change	decreases
B	increases	increases
C	no change	increases
D	increases	decreases

(Total 1 mark)

4. On approaching a stationary observer, a train sounds its horn and decelerates at a constant rate. At time t, the train passes by the observer and continues to decelerate at the same rate. Which diagram shows the variation with time of the frequency of the sound measured by the observer?)

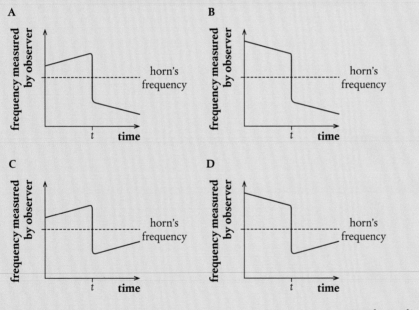

(Total 1 mark)

5. **HL** A train is moving in a straight line away from a stationary observer when the train horn emits a sound of frequency f_0. The speed of the train is $0.10v$ where v is the speed of sound. What is the frequency of the horn as heard by the observer?

A $\quad \frac{0.9}{1}f_0$ B $\quad \frac{1}{1.1}f_0$ C $\quad \frac{1.1}{1}f_0$ D $\quad \frac{1}{0.9}f_0$

(Total 1 mark)

6. **HL** A train is approaching an observer with constant speed: $\frac{c}{34}$ where c is the speed of sound in still air. The train emits sound of wavelength λ. What is the observed speed of the sound and observed wavelength as the train approaches?

	Speed of sound	Wavelength
A	c	$\frac{33\lambda}{34}$
B	$\frac{35c}{34}$	$\frac{33\lambda}{34}$
C	c	λ
D	$\frac{35c}{34}$	λ

(Total 1 mark)

7. A stationary sound source emits waves of wavelength λ and speed v. The source now moves away from a stationary observer. What are the wavelength and speed of the sound as measured by the observer?

	Wavelength	Speed
A	longer than λ	equal to v
B	longer than λ	less than v
C	shorter than λ	equal to v
D	shorter than λ	less than v

(Total 1 mark)

8. **HL** Sea waves move toward a beach at a constant speed of 2.0 m s^{-1}. They arrive at the beach with a frequency of 0.10 Hz. A girl on a surfboard is moving in the sea at right angles to the wavefronts. She observes that the surfboard crosses the wavefronts with a frequency of 0.40 Hz. What is the speed of the surfboard and what is the direction of motion of the surfboard relative to the beach?

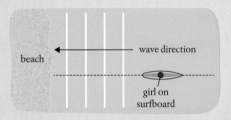

	Speed of surfboard relative to the beach / m s⁻¹	Direction of motion of surfboard relative to the beach
A	6.0	toward beach
B	6.0	away from beach
C	1.5	toward beach
D	1.5	away from beach

(Total 1 mark)

9. Two lines, X and Y, in the emission spectrum of hydrogen gas are measured by an observer stationary with respect to the gas sample. The emission spectrum is then measured by an observer moving away from the gas sample. What are the correct shifts, X* and Y*, for spectral lines X and Y?

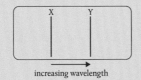

increasing wavelength

A
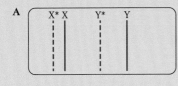

B
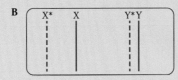

C
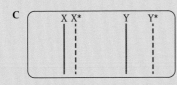

D
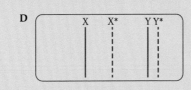

(Total 1 mark)

10. Police use radar to detect speeding cars. A police officer stands at the side of the road and points a radar device at an approaching car. The device emits microwaves that reflect off the car and return to the device. A change in frequency between the emitted and received microwaves is measured at the radar device.

The frequency change Δf is given by:

$$\Delta f = \frac{2fv}{c}$$

where f is the transmitter frequency, v is the speed of the car and c is the wave speed.

The following data are available:

Transmitter frequency, f = 40 GHz

Δf = 9.5 kHz

Maximum speed allowed = 28 m s^{-1}

(a) Explain the reason for the frequency change. (2)

(b) Suggest why there is a factor of 2 in the frequency change equation. (2)

(c) Determine whether the speed of the car is below the maximum speed allowed. (2)

(Total 6 marks)

11. **HL** Two loudspeakers, A and B, are driven in phase and with the same amplitude at a frequency of 850 Hz. Point P is located 22.5 m from A and 24.3 m from B. The speed of sound is 340 m s⁻¹.

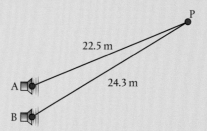

(**a**) Deduce that a minimum intensity of sound is heard at P. (4)

A microphone moves along the line from P to Q. PQ is normal to the line midway between the loudspeakers.

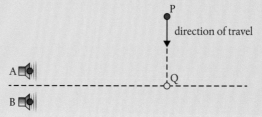

(**b**) The intensity of sound is detected by the microphone. Predict the variation of detected intensity as the microphone moves from P to Q. (2)

(**c**) When both loudspeakers are operating, the intensity of sound recorded at Q is I_0. Loudspeaker B is now disconnected. Loudspeaker A continues to emit sound with unchanged amplitude and frequency. The intensity of sound recorded at Q changes to I_A.

Estimate $\dfrac{I_A}{I_0}$. (2)

(**d**) In another experiment, loudspeaker A is stationary and emits sound with a frequency of 850 Hz. The microphone is moving directly away from the loudspeaker with a constant speed v. The frequency of sound recorded by the microphone is 845 Hz. Explain why the frequency recorded by the microphone is lower than the frequency emitted by the loudspeaker. (2)

(**e**) Calculate v. (2)

(*Total 12 marks*)

12. Objects are detected by radar; that is, by sending out radio signals and then receiving the radio signals reflected back from the objects. In the diagram, a radar signal is sent from Earth to a planet.

 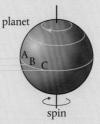

(a) The signal from which part of the planet returns to the Earth first? (2)

(b) The signal from which part of the planet returns with the highest frequency? (2)

(Total 4 marks)

13. | HL | A source of sound, emitting a note of frequency 500 Hz, starts from a stationary observer and travels directly toward a wall at speed v. The speed of sound c_s is 340 m s^{-1}. v is much less than c_s.

(a) Derive an expression for the frequency received by the stationary observer:

(i) directly from the source (3)

(ii) after reflection from the wall. (3)

In acoustics, a 'beat' is an interference pattern between two sounds of similar frequencies. The 'beat frequency' is the difference in the two frequencies that interfere to produce the beats.

(b) Determine the value of v if the observer detects a beat frequency of 30 Hz. (4)

(Total 10 marks)

THEME D Fields

D Fields

The aurora borealis ('northern lights') and aurora australis ('southern lights') can sometimes be seen on clear, dark nights from spots near to the poles. They are caused by interactions between the solar wind and the Earth's magnetic field.

A field is a region of space in which a force is experienced. Particles with mass experience forces when in gravitational fields, charged particles experience forces in electric fields and magnetic domains and moving charges experience forces in magnetic fields. The direction of force can be represented by field lines and the strength of field by the proximity at which the lines are drawn. While gravitational fields are always attractive, electric fields can be either attractive (for unlike charges) or repulsive (for like charges).

A field can be uniform, with constant field strength throughout. In these instances, potential energy changes in equal steps with equal distances moved. Alternatively, fields can be radial, for example when a result of a point or spherical mass or charge. In gravitational radial fields, the force is proportional to the product of the masses and inversely proportional to the square of the distance between them. In electric fields, the force is proportional to the product of the charges and inversely proportional to the square of the distance between them. You will also learn about Kepler's laws, potential, equipotential surfaces, potential gradient and escape speed.

Electromagnetism emerges from electric currents because moving charges have associated magnetic fields. When placed within the field of an existing magnet, a perpendicular force is exerted on the charges or the conductor that contains them. The size of force depends on the charge (or length of conductor), its speed (or current) and the magnetic field strength, and you will also learn how to determine the force per unit length between parallel wires. Electromagnetism can be put to use in spinning motors or oscillating loudspeaker cones.

Induction is the other way round; an emf is induced in a conductor that moves relative to or lies within a changing magnetic field. According to Faraday's law, the emf depends on the rate of change of magnetic flux and the number of turns or repeats in the conductor. Lenz's law, on the other hand, dictates that the direction of the emf must oppose the change that induces it and is an implication of the conservation of energy.

D.1

Gravitational fields

◀ Gravitational field strength, the gravitational force acting per unit mass, is often assumed to be constant worldwide. However, when Hirt *et al.* (*Geophysical Research Letters*, 2013) used ultrahigh-resolution pictures to map the Earth's surface, gravitational field strength was found to vary from 9.83 N kg⁻¹ in the Arctic Ocean to 9.76 N kg⁻¹ at the summit of Huascarán, a mountain in Peru (shown here). As you study the content in this chapter, think about why this variation exists and what might be responsible.

Guiding Questions

How are the properties of a gravitational field quantified?

How does an understanding of gravitational fields allow humans to explore the solar system?

Nature of Science

Newton's universal law of gravitation suggests a very simple relationship between the mass of a body and the force between it and every other particle of mass in the Universe. However, it does not explain why matter behaves in this way.

To help us understand the way a gravitational field varies throughout space, we will use the visual models of field lines and potential surfaces.

If we want to make a body move, we have to apply a force. We could push or pull it with our hands. If the body moved without us touching it, we would call it magic.

But when a body is released from a height, it falls without being in contact with anything. How? The gravitational force.

What is gravity? It is the thing that makes a ball fall toward the Earth. Why does a ball fall to the Earth? Because of gravity. We made up gravity to answer the question. The gravitational force on a body is proportional to its mass. We have been using weight = mg since it was introduced as a force. But mass is not just related to weight. From Newton's second law, we know that mass is the ratio of force to acceleration, which does not refer to gravity.

We end up with two definitions of mass: inertial mass and gravitational mass. Newton did not manage to explain the connection between these two definitions. Einstein did, however, by realizing that if you were in a box accelerating upward at 9.8 m s⁻², it would feel the same as if you were in a box that was stationary on the Earth. This principle of equivalence is the basis of Einstein's general theory of relativity, the modern gravitational model, which explains gravitation in terms of the curvature of space–time.

Students should understand:

Kepler's three laws of orbital motion
Newton's universal law of gravitation as given by $F = G\frac{m_1 m_2}{r^2}$ for bodies treated as point masses
conditions under which extended bodies can be treated as point masses
gravitational field strength g at a point is the force per unit mass experienced by a small point mass at that point as given by $g = \frac{F}{m} = G\frac{M}{r^2}$

gravitational field lines.
HL the gravitational potential energy E_p of a system is the work done to assemble the system from infinite separation of the components of the system
HL the gravitational potential energy for a two-body system as given by $E_p = -G\frac{m_1 m_2}{r}$ where r is the separation between the center of mass of the two bodies
HL the gravitational potential V_g at a point is the work done per unit mass in bringing a mass from infinity to that point as given by $V_g = -G\frac{M}{r}$
HL the gravitational field strength g as the gravitational potential gradient as given by $g = -\frac{\Delta V_g}{\Delta r}$
HL the work done in moving a mass m in a gravitational field as given by $W = m\Delta V_g$
HL equipotential surfaces for gravitational fields
HL the relationship between equipotential surfaces and gravitational field lines
HL the escape speed v_{esc} at any point in a gravitational field as given by $v_{esc} = \sqrt{\frac{2GM}{r}}$
HL the orbital speed $v_{orbital}$ of a body orbiting a large mass as given by $v_{orbital} = \sqrt{\frac{GM}{r}}$
HL the qualitative effect of a small viscous drag due to the atmosphere on the height and speed of an orbiting body.

Gravitational force and field

We have all seen how an object falls to the ground when released. Newton was not the first person to realize that an apple falls to the ground when dropped from a tree. However, he did recognize that the force that pulls the apple to the ground is the same as the force that holds the Earth in its orbit around the Sun. This was not obvious – after all, the apple moves in a straight line and the Earth moves in a circle. In this chapter, we will see how these forces are connected.

Newton's universal law of gravitation

Newton extended his ideas further to say that every single particle of mass in the Universe 'gravitates toward' every other particle of mass. In other words, everything in the Universe is attracted to everything else. So there is a force between the end of your nose and a lump of rock on the Moon.

Newton's universal law of gravitation is often phrased as:

> *Every single point mass attracts every other point mass with a force that is directly proportional to the product of their masses and inversely proportional to the square of their separation.*

If two point masses with mass m_1 and m_2 are separated by a distance r, then the force, F,

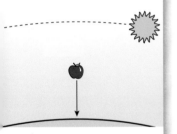

▲ **D.1 Figure 1** The apple drops and the Sun seems to move in a circle, but it is gravity that makes both things happen.

Was it reasonable for Newton to think that his law applied to the whole Universe?

TOK

experienced by each will be given by:

$$F \propto \frac{m_1 m_2}{r^2}$$

◀ **D.1 Figure 2** The gravitational force F between two point masses.

The constant of proportionality is the universal gravitational constant G.

$$G = 6.67 \times 10^{-11}\,\mathrm{m^3\,kg^{-1}\,s^{-2}}$$

Therefore the equation is:

$$F = G\frac{m_1 m_2}{r^2}$$

Extended bodies

An extended body is any object at which the mass is not confined to a point.

Every extended body has a center of mass around which mass is uniformly distributed. In regularly shaped bodies with uniform distribution of mass, this is the center of the object.

The center of gravity of a body is the point at which the entirety of its weight can be taken to act.

In a uniform field, the center of mass and the center of gravity are identical. The center of gravity of a sphere is at its center and this particular example of an extended body can be treated as a point mass.

In a non-uniform field, the gravitational force acting on the mass is greater where field strength is greater; the center of gravity of the Moon (for example) is not in the center but instead slightly toward the Earth.

Spheres of mass

By working out the total force between every particle of one sphere and every particle of another sphere, Newton deduced that spheres of mass follow the same law, where r is the separation between their centers. This is because spheres are symmetrically spherical.

How fast does the apple drop?

If we apply Newton's universal law to the apple on the surface of the Earth, we find that it will experience a force given by:

$$F = G\frac{m_1 m_2}{r^2}$$

where:

m_1 = mass of the Earth = $5.97 \times 10^{24}\,\mathrm{kg}$

m_2 = mass of the apple = $250\,\mathrm{g}$

r = radius of the Earth = $6378\,\mathrm{km}$ (at the equator)

So: $F = 2.43\,\mathrm{N}$

We often look at the applications of physics with an international perspective but perhaps we should adopt a universal perspective.

The modern equivalent of the apparatus used by Cavendish to measure G in 1798.

Physics utilizes a number of constants such as G. What is the purpose of these constants and how are they determined? (NOS)

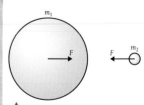

▲ **D.1 Figure 3** Forces between two spheres. Even though these bodies do not have the same mass, the force on them is the same size. This is due to Newton's third law: if mass m_1 exerts a force on mass m_2, then m_2 will exert an equal and opposite force on m_1.

From Newton's second law, we know that $F = ma$.

So: acceleration a of the apple $= \dfrac{2.43}{0.25}$ m s^{-2}

$$a = 9.79 \,\text{m s}^{-2}$$

This is very close to the average value for the acceleration of free fall on the Earth's surface. It need not be exactly the same since 9.81 m s^{-2} is an average for the whole Earth, the radius of the Earth being maximum at the equator.

Exercise

Q1. The mass of the Moon is 7.35×10^{22} kg and its radius is 1.74×10^{3} km. What is the acceleration due to gravity on the Moon's surface?

Gravitational field

The fact that both the apple and the Earth experience a force without being in contact makes gravity different from the other forces we have come across. To model this situation, we introduce the idea of a **field**. A field (in the physical sense) is a region of space where forces are experienced by bodies with the appropriate property. A gravitational field is a region where you find gravity. More precisely, a gravitational field is defined as a region of space where a mass experiences a force because of its mass.

So there is a gravitational field in your classroom since masses experience a force in it.

Field strength on the Earth's surface:
Substituting:
M = mass of the Earth = 5.97×10^{24} kg
r = average radius of the Earth = 6371 km
gives
$$g = G\frac{M}{r^2} = 9.81 \text{ N kg}^{-1}$$
This is the same as the acceleration due to gravity, which is what you might expect, since Newton's second law says that $a = \frac{F}{m}$.

Gravitational field strength, g

This gives a measure of how much force a body will experience in the field. It is defined as the force per unit mass experienced by a small point mass placed in the field.

So if a test mass, m, experiences a force F at some point in space, then the field strength, g, at that point is given by: $g = \dfrac{F}{m}$

Gravitational field strength g is measured in N kg^{-1} and is a vector quantity.

Gravitational field around a spherical object

The force experienced by a mass m in the field produced by a mass M is given by:

$$F = G\frac{Mm}{r^2}$$

So: field strength at this point in space, $g = \dfrac{F}{m}$

So:
$$g = G\frac{M}{r^2}$$

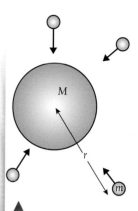

D.1 Figure 4 The region surrounding M is a gravitational field since all the test masses experience a force.

Exercise

Q2. The mass of Jupiter is 1.89×10^{27} kg and its radius is 71 492 km. What is the gravitational field strength on the surface of Jupiter?

Q3. What is the gravitational field strength at a distance of 1000 km from the surface of the Earth?

Field lines

Field lines are drawn in the direction that a mass would accelerate if placed in the field – they are used to help us visualize the field.

The field lines for a spherical mass are shown in Figure 5:

- The arrows give the direction of the field.

- The field strength, g, is given by the density of the lines.

What measurements of a binary star system need to be made in order to determine the nature of the two stars? (C.5, E.5)

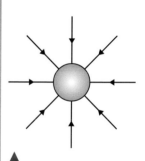

D.1 Figure 5 Field lines for a sphere of mass.

Gravitational field close to the Earth

When we are doing experiments close to the Earth, in the classroom, for example, we assume that the gravitational field strength is constant throughout. This means that wherever you put a mass in the classroom it is always pulled downward with the same force. We say that the field is **uniform**.

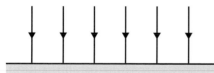

 D.1 Figure 6 Close to the Earth the field is uniform.

Addition of field

Since field strength is a vector, when we add field strengths caused by several bodies, we must remember to add them vectorially.

In this example, the angle between the vectors is 90°. This means that we can use Pythagoras to find the resultant.

$$g = \sqrt{g_1^2 + g_2^2}$$

The value of g is different in different places on the Earth. Do you know what it is where you live?

What are the benefits of using consistent terminology to describe different types of fields? (NOS)

 **D.1 Figure 7** Vector addition of field strength.

Worked example

Calculate the gravitational field strength at point A.

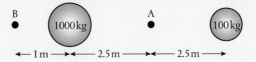

Solution

The gravitational field strength at A is equal to the sum of the fields due to the two masses.

$$\text{field strength due to large mass} = \frac{G \times 1000}{2.5^2} = 1.07 \times 10^{-8}\,\text{N kg}^{-1}$$

$$\text{field strength due to small mass} = \frac{G \times 100}{2.5^2} = 1.07 \times 10^{-9}\,\text{N kg}^{-1}$$

$$\text{field strength} = 1.07 \times 10^{-8} - 1.07 \times 10^{-9}$$
$$= 9.63 \times 10^{-9}\,\text{N kg}^{-1}$$

Since field strength g is a vector, the resultant field strength equals the vector sum.

Exercise

Q4. Calculate the gravitational field strength at point B in the figure in the worked example above.

Q5. Calculate the gravitational field strength at point A if the big mass were changed for a 100 kg mass.

HL

Gravitational potential in a uniform field

As you lift a mass *m* from the ground, you do work. This increases the potential energy of the object. As potential energy = *mgh* (as we learned in A.3), we know the potential energy gained by the mass depends partly on the size of the mass, *m*, and partly on where it is, *gh*. The 'where it is' part is associated with a new quantity: 'gravitational potential (V_g). This is a useful quantity because, if we know it, we can calculate how much potential energy a given mass would have if placed there.

Dimensionally, potential is potential energy per unit mass. For calculations performed in a uniform field, it is convenient to assume a given surface (e.g. the surface of the Earth) as having zero potential. Gravitational potential can therefore be calculated in uniform fields as work done per unit mass in taking a mass from zero potential to the point in question.

In the simple example of masses in a room in which we define the floor as having zero potential, the potential is proportional to height, so a mass *m* placed at the same height in the room will have the same potential energy. By joining all positions of the same potential, we get a **surface of equal potential** (also known as an equipotential), and these are useful for visualizing the changes in potential energy as an object moves around the room.

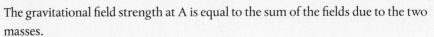

Worked example

What is the potential at A? Assume that gravitational field strength, $g = 10\,N\,kg^{-1}$.

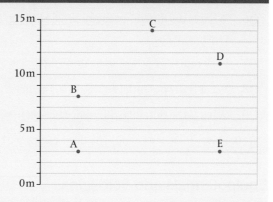

Close to the Earth, lines of equipotential join points that are the same height above the ground. These are the same as contours on a map.

Solution

$V_A = gh$

So:

potential at A = $10 \times 3 = 30\,J\,kg^{-1}$

Worked example

Referring to the previous worked example, what is the difference in potential between points A and B?

Solution

$V_A = 30\,J\,kg^{-1}$

$V_B = 80\,J\,kg^{-1}$

change in potential = $80 - 30 = 50\,J\,kg^{-1}$

Worked example

Referring to the figure on the previous page, how much work is done moving a 2 kg mass from A to B?

Solution

work done moving from A to B is equal to the change in potential × mass = $50 \times 2 = 100\,J$

Exercise

Q6. What is the difference in potential between C and D in the figure on the previous page?

Q7. How much work would be done moving a 3 kg mass from D to C?

Q8. What is the potential energy of a 3 kg mass placed at B?

Q9. What is the potential difference between A and E?

Q10. How much work would be done taking a 2 kg mass from A to E?

Equipotentials and field lines

If we draw the field lines in a 15 m room, they will look like Figure 8. The field is uniform so they are parallel and equally spaced. If you were to move upward along a field line (A–B), you would have to do work and therefore your potential energy would increase. On the other hand, if you traveled perpendicular to the field lines (A–E), no work would be done, in which case you must be traveling along a line of equipotential. For this reason, field lines and equipotentials are perpendicular.

D.1 Figure 8
Equipotentials and field lines.

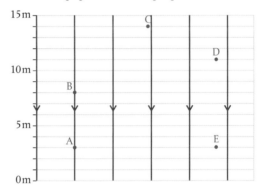

The amount of work done as you move up is equal to the change in potential × mass.

$$\text{work done} = \Delta V_g m$$

But the work done is also equal to:

$$\text{force} \times \text{distance} = mg\Delta h$$

So:

$$\Delta V_g m = mg\Delta h$$

Rearranging gives:

$$\frac{\Delta V_g}{\Delta h} = g$$

Or: potential gradient = field strength

So lines of equipotential that are close together represent a strong field.

This is similar to the situation with contours as shown in Figure 9. Contours that are close together mean that the gradient is steep, and where the gradient is steep, there will be a large force pulling you down the slope.

Take care to notice when you are working with *height* and *potential* and when you are working with *change in height* and *change in potential*. In a uniform field where zero potential is defined arbitrarily, V_g and ΔV_g have the same value if ΔV_g is calculated from an initial value of zero.

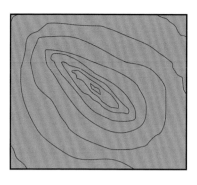

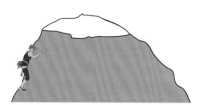

D.1 **Figure 9** Close contours mean a steep mountain.

In this section, we have been dealing with the simplified situation. Firstly, we have only been dealing with bodies close to the Earth, where the field is uniform. Secondly, we have been assuming that the ground is a position of zero potential. A more general situation would be to consider large distances away from a sphere of mass. This is rather more difficult but the principle is the same, as are the relationships between field lines and equipotentials.

Gravitational potential due to a massive sphere

The gravitational potential at point P is defined as:

The work done per unit mass taking a mass from a position of zero potential to the point P.

In the previous example, we took the Earth's surface to represent zero potential but a better choice would be somewhere where the mass is not affected by the field at all. Since $g = G\frac{M}{r^2}$, the only place completely out of the field is at an infinite distance from the mass, so let us start there.

Figure 10 represents the journey from infinity to point P, a distance r from the center of a large mass, M. The work done making this journey = $-W$ so the potential, $V = -\frac{W}{m}$.

TOK

We cannot really take a mass from infinity and bring it to the point in question, but we can calculate how much work would be required if we did. Is it OK to calculate something we can never do?

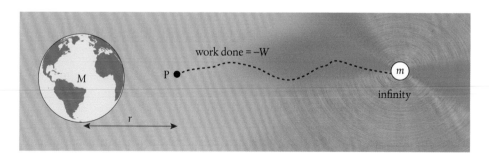

work done = $-W$

D.1 **Figure 10** The journey from infinity to point P.

The negative sign is because the mass is being pulled to the Earth by the attractive force of gravity, so you would not have to pull it; it would pull you. The direction of force applied by you, holding the mass, is opposite to the direction of motion, so the work done by you would be negative.

Calculating the work done

For a system composed of many bodies, the gravitational potential energy is the total work that would be done in assembling that system, assuming that all the bodies started with infinite separation from one another. There are two problems when you try to calculate the work done from infinity to P: the distance is infinite (obviously) and the force gets bigger as you get closer. To solve this problem, we use the area under the force–distance graph (remember the work done stretching a spring). From Newton's universal law of gravitation, we know that the force varies according to the equation $F = G\frac{Mm}{x^2}$ so the graph will be as shown in Figure 11.

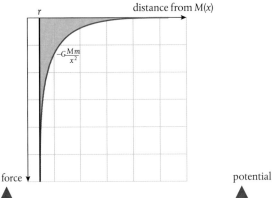

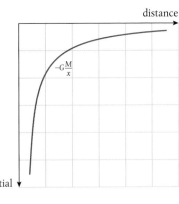

▲ **D.1 Figure 11** Graph of force against distance as the test mass is moved toward M.

▲ **D.1 Figure 12** Graph of potential against distance.

The area under this graph can be found by integrating the function $-G\frac{Mm}{x^2}$ from infinity to r (you will do this in math). This gives the result:

$$W = -G\frac{Mm}{r}$$

So: potential, $V_g = \frac{W}{m} = -G\frac{M}{r}$

The graph of potential against distance is shown in Figure 12. The gradient of this line gives the field strength, but notice that the gradient is positive and the field strength is negative so we get the formula:

$$g = -\frac{\Delta V_g}{\Delta x}$$

The integral mentioned here is:
$$V = \int_{\infty}^{r} G\frac{M}{x^2}\,dx$$

Equipotentials and potential wells

If we draw the lines of equipotential for the field around a sphere on paper, we get concentric circles, as in Figure 13. An equipotential can be either a line or a surface.

D.1 Figure 13 The lines of equipotential and potential well for a sphere.

An alternative way of representing this field is to draw the hole or well that these contours represent. This is a very useful visualization, since it not only represents the change in potential but, by looking at the gradient, we can also see where the force is biggest. If you imagine a ball rolling into this well, you can visualize the field.

Relationship between field lines and potential

If we draw the field lines and the equipotential lines as in Figure 14, we see that, as before, they are perpendicular. We can also see that the lines of equipotential are closest together where the field is strongest (where the field lines are most dense). This agrees with our earlier finding that $g = -\dfrac{\Delta V_g}{\Delta x}$.

Addition of potential

Potential is a scalar quantity, so adding potentials is just a matter of adding the magnitudes. If we take the example shown in Figure 15, to find the potential at point P, we calculate the potential due to the masses at A and B then add them together.

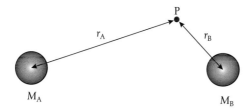

total potential at P $= -G\dfrac{M_A}{r_A} - G\dfrac{M_B}{r_B}$

The lines of equipotential for this example are shown in Figure 16.

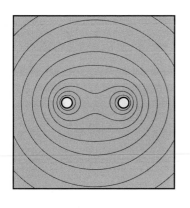
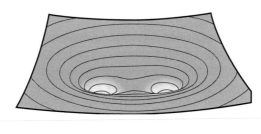

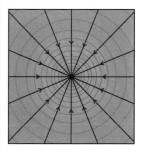

▲
D.1 Figure 14
Equipotentials and field lines.

◀ **D.1 Figure 15** Two masses.

◀ **D.1 Figure 16**
Equipotentials and potential wells for two equal masses. If you look at the potential well, you can imagine a ball could sit on the hump between the two holes. This is where the field strength is zero.

Exercise

Q11. The Moon has a mass of 7.4×10^{22} kg and the Earth has mass of 6.0×10^{24} kg. The average distance between the Earth and the Moon is 3.8×10^5 km. You travel directly between the Earth and the Moon in a rocket of mass 2000 kg.

   (a) Calculate the gravitational potential when you are 1.0×10^4 km from the Moon.

   (b) Calculate the rocket's potential energy at the point in part (a).

   (c) Draw a sketch graph showing the change in potential.

   (d) Mark the point where the gravitational field strength is zero.

Escape speed

If a body is thrown straight up, its kinetic energy decreases as it rises. If we ignore air resistance, this kinetic energy is transferred to potential energy. When it gets to the top, the final potential energy will equal the initial kinetic energy, so $\frac{1}{2}mv^2 = mgh$.

If we throw a body up really fast, it might get so high that the gravitational field strength would start to meaningfully decrease. In this case, we would have to use the formula for the potential energy around a sphere:

$$E_p = -G\frac{Mm}{r}$$

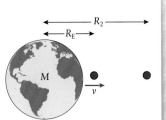

D.1 Figure 17 A mass m thrown away from the Earth.

So when it gets to its furthest point as shown in Figure 17:

$$\text{loss of } E_k = \text{gain in } E_p$$

$$\frac{1}{2}mv^2 - 0 = -G\frac{Mm}{R_2} - \left(-G\frac{Mm}{R_E}\right)$$

If we throw the ball fast enough, it will never come back. This means that it has reached a place where it is no longer attracted back to the Earth, which we call infinity. Of course it cannot actually reach infinity but we can substitute $R_2 = \infty$ into our equation to find out how fast that would be.

$$\frac{1}{2}mv^2 = -G\frac{Mm}{\infty} - \left(-G\frac{Mm}{R_E}\right)$$

$$\frac{1}{2}mv^2 = G\frac{Mm}{R_E}$$

Rearranging gives:
$$v_{escape} = \sqrt{\frac{2GM}{R_E}}$$

If we calculate the escape velocity for the Earth, it is about $11\,\text{km s}^{-1}$.

Why the Earth has an atmosphere but the Moon does not

The average velocity of an air molecule at the surface of the Earth is about $500\,\text{m s}^{-1}$. This is much less than the velocity needed to escape from the Earth, and for that reason, the atmosphere does not escape.

The escape velocity on the Moon is $2.4\,\text{km s}^{-1}$ so you might expect the Moon to have an atmosphere. However, $500\,\text{m s}^{-1}$ is the *average* speed. A lot of the molecules would be traveling faster than this, leading to a significant number escaping. Over time, all would escape.

Black holes

A star is a big ball of gas held together by the gravitational force. The reason this force does not cause the star to collapse is that the particles are continuously given kinetic energy from the nuclear reactions taking place (fusion). As time progresses, the nuclear fuel gets used up, so the star starts to collapse. As this happens, the escape velocity increases until it is greater than the speed of light. At this point, not even light can escape and the star has formed a black hole.

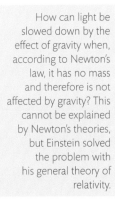

If you threw something up with a velocity of $11\,\text{km s}^{-1}$, it would be rapidly slowed by air resistance. The work done against this force would be transferred to thermal energy, causing the body to vaporize. Rockets leaving the Earth do not have to travel anywhere near this fast, as they are not thrown upward, but instead have a rocket engine that provides a continual force.

How is the amount of fuel required to launch rockets into space determined by considering energy? (A.3)

How can light be slowed down by the effect of gravity when, according to Newton's law, it has no mass and therefore is not affected by gravity? This cannot be explained by Newton's theories, but Einstein solved the problem with his general theory of relativity.

Q12. The mass of the Moon is 7.4×10^{22} kg and its radius is 1738 km. Show that its escape speed is 2.4 km s^{-1}.

Q13. Why does the Earth's atmosphere not contain hydrogen?

Q14. The mass of the Sun is 2.0×10^{30} kg. Calculate how small its radius would have to be for it to become a black hole.

Q15. When traveling away from the Earth, a rocket runs out of fuel at a distance of 1.0×10^5 km. How fast would the rocket have to be traveling for it to escape from the Earth? (Mass of the Earth = 6.0×10^{24} kg, radius = 6400 km.)

`HL end`

The solar system

The solar system consists of the Sun at the center, surrounded by eight orbiting planets. The shape of the orbits is slightly elliptical but, to make things simpler, we will assume them to be circular. We know that for a body to travel in a circle, there must be an unbalanced force (called the centripetal force, $m\omega^2 r$) acting toward the center. The force that holds the planets in orbit around the Sun is the gravitational force, $G\frac{Mm}{r^2}$. Equating these two expressions gives us an equation for orbital motion:

$$m\omega^2 r = G\frac{Mm}{r^2}$$

Now ω is the angular velocity of the planet; that is, the angle swept out by a radius per unit time. If the time taken for one revolution (2π radians) is T, then $\omega = \frac{2\pi}{T}$.

Substituting into the equation above gives:

$$m\left(\frac{2\pi}{T}\right)^2 r = G\frac{Mm}{r^2}$$

Rearranging gives:

$$\frac{T^2}{r^3} = \frac{4\pi^2}{GM}$$

where M is the mass of the Sun.

How can the motion of electrons in the atom be modeled on planetary motion and in what ways does this model fail? (NOS)

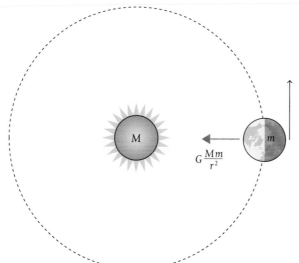

D.1 Figure 18 The Earth orbiting the Sun.

How is uniform circular motion like, and unlike, real-life orbits? (A.2)

So for planets orbiting the Sun, $\frac{T^2}{r^3}$ is a constant, or T^2 is proportional to r^3.

This is **Kepler's third law**.

From this, we can deduce that the planet closest to the Sun (Mercury) has a shorter time period than the planet furthest away. This is supported by measurements:

time period of Mercury = 0.24 years

time period of Neptune = 165 years.

Exercise

Q16. Use a database of planetary information to make a table of the values of time period and radius for all the planets. Plot a graph to show that T^2 is proportional to r^3.

Kepler's laws

Kepler is associated with three laws for planetary motion. All three can be derived from Newton's laws, but Kepler came up with them empirically by measuring orbital times and radii of the planets and looking for relationships among the numbers.

First law

Planetary orbits are elliptical with a star at a focus.

In this course, we deal only with circular orbits (a special example of an ellipse), but Newton's law also predicts elliptical orbits.

Second law

The radius vector from the star to the orbiting body sweeps equal areas in equal times.

This is obvious for circular orbits but not so obvious for elliptical ones.

D.1 Figure 19 Kepler's first and second laws.

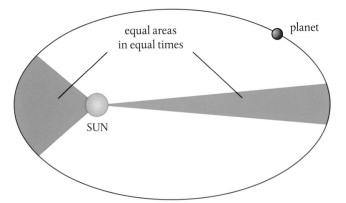

Qualitatively, we can understand the second law as being because the orbiting body (for example, a planet) gets faster as it gets closer to the star. This is because its kinetic energy must increase as its potential energy decreases, which in turn is because its energy is conserved.

Third law

The square of the orbital period is proportional to the cube of the semi-major axis of the ellipse (the distance from the center of the ellipse to the farthest point on the perimeter). For a circular orbit, $T^2 \propto r^3$, where T is the orbital period and r is the radius of the orbit.

This means that the ratio of T^2 to r^3 is constant for all planets in the same solar system. This law comes from equating the gravitational force according to Newton's law with the equation for centripetal force.

HL

Energy of an orbiting body

As planets orbit the Sun, they have kinetic energy due to their movement and potential energy due to their position. We know that their potential energy is given by the equation:

$$E_p = -G\frac{Mm}{r}$$

and

$$E_k = \frac{1}{2}mv^2$$

We also know that if we approximate the orbits to be circular, then equating the centripetal force with gravity gives:

$$G\frac{Mm}{r^2} = \frac{mv^2}{r}$$

> ### Challenge yourself
>
> 1. Using the kinetic energy equation, can you determine an equation for the speed of an orbiting body? How does this compare with the equation for escape speed?

Rearranging and multiplying by $\frac{1}{2}$ gives: $\quad \frac{1}{2}mv^2 = G\frac{Mm}{2r}$

$$E_k = G\frac{Mm}{2r}$$

$$\text{total energy} = E_p + E_k = -G\frac{Mm}{r} + G\frac{Mm}{2r}$$

$$\text{total energy} = -G\frac{Mm}{2r}$$

HL end

Earth satellites

The equations we have derived for the orbits of the planets also apply to the satellites that humans have put into orbits around the Earth. This means that the satellites closer to the Earth have a time period much shorter than the distant ones. For example, a low-orbit spy satellite could orbit the Earth once every two hours while a much higher TV satellite orbits only once a day.

HL The total energy of an orbiting satellite $= -G\frac{Mm}{2r}$ so the energy of a high satellite (big r) is less negative and hence bigger than for a low-orbiting satellite. To move from a low orbit to a high one therefore requires energy to be added (work done).

TOK

It is strange how everyone can describe the motion of the planets as viewed from a long way outside the solar system but few can describe the motion as seen from the Earth.

There are two versions of the equation for centripetal force.
Speed version:

$$F = \frac{mv^2}{r}$$

Angular speed version:

$$F = m\omega^2 r$$

Imagine you are in a spaceship orbiting the Earth in a low orbit. To move into a higher orbit, you would have to use your rocket motor to increase your energy. If you kept doing this, you could move from orbit to orbit, getting further and further from the Earth.

HL

The energy of the spaceship in each orbit can be displayed as a graph as in Figure 20.

D.1 Figure 20
Graph of E_k, E_p, and total energy for a satellite with different orbital radii.

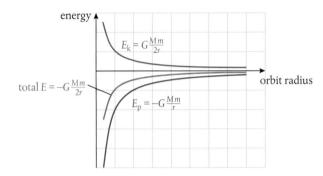

From the graph, we can see that low satellites have greater kinetic energy but less total energy than distant satellites, so although the distant ones move with slower speed, we have to do work to increase the orbital radius. To move from a distant orbit to a close orbit, the spaceship needs to lose energy. Satellites in low-Earth orbit are not completely out of the atmosphere, so lose energy due to air resistance. As they lose energy, they spiral in toward the Earth.

How can air resistance be used to alter the motion of a satellite orbiting Earth? (A.2)

A low-orbit spy satellite.

The lowest satellites orbit the Earth at a height of around 150 km. However, they are not entirely out of the atmosphere so need the occasional boost of power to keep them traveling fast enough or they would move to an even lower orbit. These satellites are mainly used for spying.

TV satellites are geostationary so must be placed about 6 Earth radii from the Earth. With the thousands of TV channels available, you might expect there to be thousands of satellites but there are only about 300.

HL end

Exercise

Q17. So that they can stay above the same point on the Earth, TV satellites have a time period equal to one day. Calculate the radius of their orbit.

Q18. A spy satellite orbits 400 km above the Earth. If the radius of the Earth is 6400 km, what is the time period of the orbit?

Q19.  HL If the satellite in Q18 has a mass of 2000 kg, calculate its:

(**a**) kinetic energy

(**b**) potential energy

(**c**) total energy.

The orbits of spy satellites are set so that they pass over places of interest. Which countries have most spy satellites passing overhead?

Guiding Questions revisited

How are the properties of a gravitational field quantified?

How does an understanding of gravitational fields allow humans to explore the solar system?

In this chapter, we have considered gravitational fields through the models of Kepler and Newton to develop an understanding of how:

- Kepler's three laws state collectively that bodies' orbits of a celestial mass are elliptical with the mass at a focus, that equal areas are swept out by these bodies in equal times, and that the square of a body's orbital period is proportional to the cube of its orbital radius (for circular orbits).
- Newton's universal law of gravitation states that the gravitational force acting between any two masses in the Universe is proportional to the product of the masses and inversely proportional to the square of the distance between their centers of masses.
- Gravitational field strength is defined as the gravitational force acting per unit mass.
- Field lines represent the direction of the force that would act on a mass placed in the field, with the closeness of the lines indicating the strength of the field.
- Knowledge of forces, energy and fields can be used to estimate the escape speed that a solar system probe or telescope would need to be launched with from the Earth, assuming the atmosphere causes no drag.

HL

- Gravitational potential energy is defined as the total work done when bringing a test mass from infinity to a point in space at a low constant speed.
- Gravitational potential is defined as the potential energy per unit mass. The magnitude of potential gradient is equal to the field strength.
- Equipotential surfaces represent locations with the same potential. They are perpendicular to field lines.

HL end

Practice questions

1. An object of mass m released from rest near the surface of a planet has an initial acceleration z. What is the gravitational field strength near the surface of the planet?

    A z

    B $\frac{z}{m}$

    C mz

    D $\frac{m}{z}$

    (Total 1 mark)

2. Which graph shows the relationship between gravitational force F between two point masses and their separation r?

    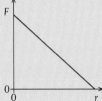

    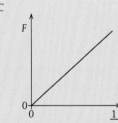
    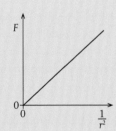

    (Total 1 mark)

3. Two isolated point particles of mass $4M$ and $9M$ are separated by a distance 1 m. A point particle of mass M is placed a distance x from the particle of mass $9M$. The net gravitational force on M is zero. What is x?

    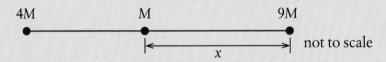

    A $\frac{4}{13}$ m

    B $\frac{2}{5}$ m

    C $\frac{3}{5}$ m

    D $\frac{9}{13}$ m

    (Total 1 mark)

4. A planet is in a circular orbit around a star. The speed of the planet is constant.

   (a) Explain why a centripetal force is needed for the planet to be in a circular orbit. (2)

   (b) State the nature of this centripetal force. (1)

   The following data are given:

   Mass of planet = 8.0×10^{24} kg

   Radius of planet = 9.1×10^6 m

   (c) Determine the gravitational field strength at the surface of the planet. (2)

   (Total 5 marks)

5. The moon Phobos moves around the planet Mars in a circular orbit.

   (a) Outline why the gravitational force does no work on Phobos. (1)

   The orbital period T of a moon orbiting a planet of mass M is given by:

   $$\frac{R^3}{T^2} = kM$$

   where R is the average distance between the center of the planet and the center of the moon.

   (b) Show that $k = \frac{G}{4\pi^2}$. (3)

   The following data for the Mars–Phobos system and the Earth–Moon system are available:

   Mass of Earth = 5.97×10^{24} kg

   The Earth–Moon distance is 41 times the Mars–Phobos distance.

   The orbital period of the Moon is 86 times the orbital period of Phobos.

   (c) Calculate, in kg, the mass of Mars. (2)

   HL The graph shows the variation of the gravitational potential between the Earth and Moon with distance from the center of the Earth. The distance from the Earth is expressed as a fraction of the total distance between the center of the Earth and the center of the Moon.

   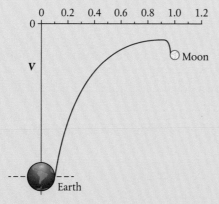

   (d) Determine, using the graph, the mass of the Moon. (3)

   (Total 9 marks)

6. (a) In a cave below the surface of the Earth, which is correct about the gravity? (1)

    A   There is more gravity than at the Earth's surface.

    B   There is less gravity than at the Earth's surface.

    C   There is the same gravity as at the Earth's surface.

  (b) Explain your answer. (1)

(Total 2 marks)

7. (a) As you move away from the Earth, its gravity gets weaker. But suppose it got stronger. If that were so, would it be possible for objects, like the Moon, to orbit the Earth? (1)

    A   Yes, just as they do presently.

    B   Yes, but unlike they do presently.

    C   No, orbital motion could not occur.

  (b) Explain your answer. (1)

(Total 2 marks)

8. The Sun orbits the center of the Milky Way galaxy at a radius of 30 000 light years and with a period of orbit of 200 million years. Assume that the mass distribution of matter in the galaxy is concentrated mainly in a central uniform sphere and the Sun lies in one of the arms.

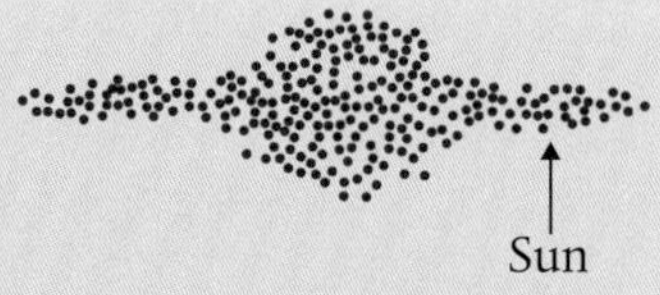

  (a) Obtain an expression for the period of orbit T of the Sun in terms of its radius of orbit, r, G and M_g, the mass of the galaxy. (1)

Sun

  (b) Determine a value of the mass of the galaxy. (1)

  (c) If the Sun represents the mass of an average star, estimate the number of stars in the Milky Way. (1)

(Total 3 marks)

HL

9. (a) Define *gravitational potential* at a point. (2)

  (b) The graph shows the variation of gravitational potential V of a planet and its moon with distance r from the center of the planet. The unit of separation is arbitrary. The center of the planet corresponds to $r = 0$ and the center of the moon to $r = 1$. The curve starts at the surface of the planet and ends at the surface of the moon.

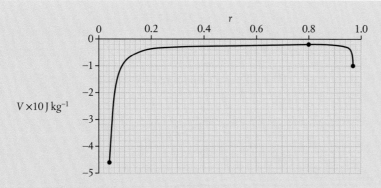

(i) At the position where $r = 0.8$, the gravitational field strength is zero. Determine the ratio:

$$\frac{\text{mass of planet}}{\text{mass of moon}}$$ (3)

(ii) A satellite of mass 1500 kg is launched from the surface of the planet. Determine the **minimum** kinetic energy at launch the satellite must have so that it can reach the surface of the moon. (3)

(Total 8 marks)

10. The graph shows the variation of gravitational potential V due to the Earth with distance R from the center of the Earth. The radius of the Earth is 6.4×10^6 m. The graph does not show the variation of potential V within the Earth.

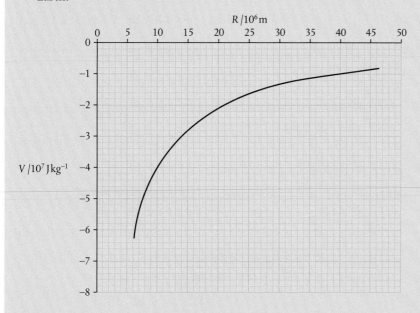

(a) Use the graph to find the gravitational potential:

(i) at the surface of the Earth (1)

(ii) at a height of 3.6×10^7 m above the surface of the Earth. (2)

(b) Use the values you have found in part (a) to determine the minimum energy required to put a satellite of mass 1.0×10^4 kg into an orbit at a height of 3.6×10^7 m above the surface of the Earth. (3)

(c) Give **two** reasons why more energy is required to put this satellite into orbit than that calculated in (b). (2)

(Total 8 marks)

11. A probe of mass m is in a circular orbit of radius r around a spherical planet of mass M.

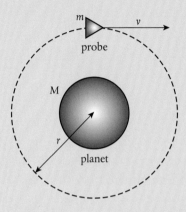

(diagram not to scale)

(a) State why the work done by the gravitational force during one full revolution of the probe is zero. (1)

(b) Deduce for the probe in orbit that its:

(i) speed is $v = \sqrt{\dfrac{GM}{r}}$ (2)

(ii) total energy is $E = -G\dfrac{Mm}{2r}$ (2)

(c) It is now required to place the probe in another circular orbit further away from the planet.

To do this, the probe's engines will be fired for a very short time.

State and explain whether the work done on the probe by the engines is positive, negative **or** zero. (2)

(Total 7 marks)

12. Which is a correct unit for gravitational potential?

A $m^2 s^{-2}$ B $J\,kg$

C $m\,s^{-2}$ D $N\,m^{-1} kg^{-1}$

(Total 1 mark)

13. A planet has radius R. The escape speed from the surface of the planet is v. At what distance from the surface of the planet is the orbital speed $0.5v$?

A 0.5R B R C 2R D 4R

(Total 1 mark)

14. A satellite orbits a planet. Which graph shows how the kinetic energy E_k, the potential energy E_p and the total energy E of the satellite vary with distance x from the center of the planet?

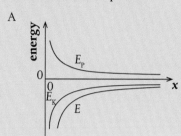

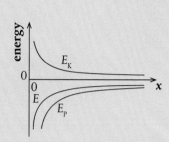

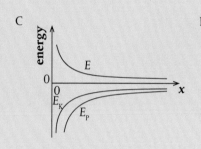

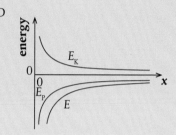

(Total 1 mark)

15. A planet of mass m is in a circular orbit around a star. The gravitational potential due to the star at the position of the planet is V.

(a) Show that the total energy of the planet is given by the equation
$$E = \tfrac{1}{2}mV.$$
(2)

(b) Suppose the star could contract to half its original radius without any loss of mass. Discuss the effect, if any, this has on the total energy of the planet.
(2)

(Total 4 marks)

16. (a) Outline what is meant by *escape speed*. (1)

A probe is launched vertically upward from the surface of a planet with a speed:
$$v = \tfrac{3}{4}v_{esc}$$
where v_{esc} is the escape speed from the planet. The planet has no atmosphere.

(b) Determine, in terms of the radius of the planet R, the maximum height from the surface of the planet reached by the probe. (3)

(c) The total energy of a probe in orbit around a planet of mass M is:
$$E = -G\frac{Mm}{2r}$$
where m is the mass of the probe and r is the orbit radius. A probe in low orbit experiences a small frictional force. Suggest the effect of this force on the speed of the probe. (3)

(Total 7 marks)

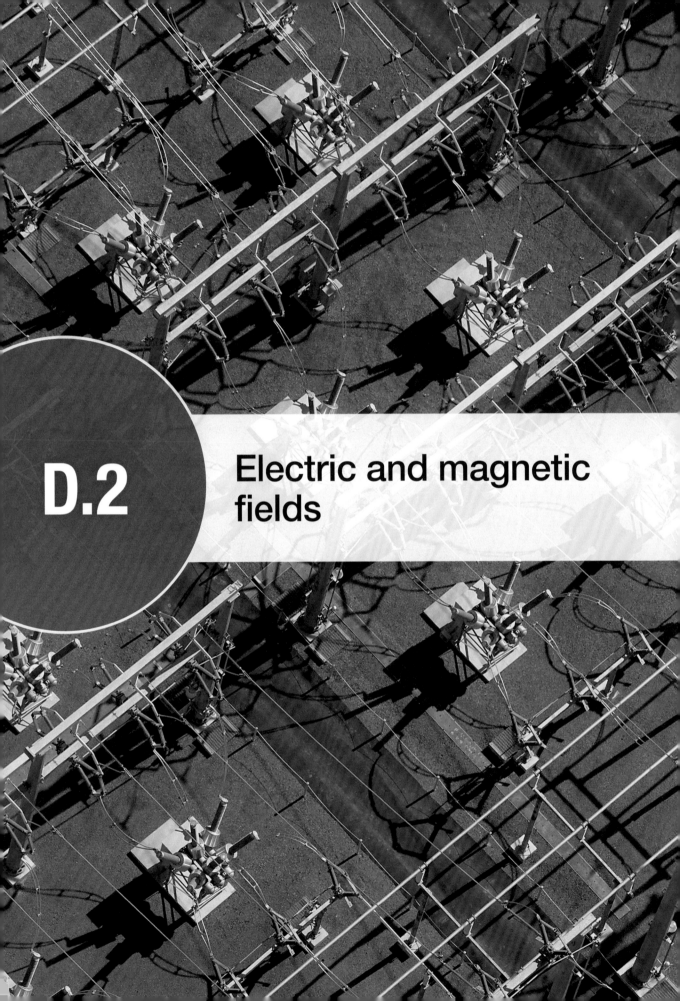

D.2

Electric and magnetic fields

◀ Strong electric fields can break down the air, resulting in a spark. To prevent sparking between high-voltage cables, they must be kept far apart from each other and the Earth.

Guiding Questions

Which experiments provided evidence to determine the nature of the electron?

How can the properties of fields be understood using an algebraic approach and a visual representation?

What are the consequences of interactions between electric and magnetic fields?

Nature of Science

The electric force is very similar to gravitation in that it acts over a distance and its strength is inversely proportional to the distance between affected bodies. This means that we can use the concept of field we developed for gravity to model the electric effect.

Electric fields and gravitational fields can be modeled in similar ways. By changing mass to charge, we go from gravity to electricity. However, they could not be more different and they represent two of the four fundamental forces.

Electricity and magnetism may appear to be quite distinct. There is no magnetic equivalent of charge or mass and no replacement for the laws of Newton and Coulomb. Saying this, they are fundamentally the same. The magnetic force is the electric force between moving charges.

Students should understand:

the direction of forces between the two types of electric charge
Coulomb's law as given by $F = k\dfrac{q_1 q_2}{r^2}$ for charged bodies treated as point charges where $k = \dfrac{1}{4\pi\varepsilon_0}$
the conservation of electric charge
Millikan's experiment as evidence for quantization of electric charge
the electric charge can be transferred between bodies using friction, electrostatic induction and by contact, including the role of grounding (earthing)
the electric field strength as given by $E = \dfrac{F}{q}$
electric field lines
the relationship between field line density and field strength
the electric field strength for parallel plates as given by $E = \dfrac{V}{d}$
magnetic field lines.

HL the electric potential energy E_p in terms of work done to assemble the system from infinite separation
HL the electric potential energy for a system of two charged bodies as given by $E_p = k\dfrac{q_1 q_2}{r}$
HL the electric potential is a scalar quantity with zero defined at infinity
HL the electric potential V_e at a point is the work done per unit charge to bring a test charge from infinity to that point as given by $V_e = \dfrac{kQ}{r}$
HL the electric field strength E as the electric potential gradient as given by $E = -\dfrac{\Delta V_e}{\Delta r}$
HL the work done in moving a charge q in an electric field as given by $W = q\Delta V_e$
HL equipotential surfaces for electric fields
HL the relationship between equipotential surfaces and electric field lines.

Electric force

If you rub a plastic ruler on a woolen sweater (or synthetic fleece) and hold it above some small bits of paper, something interesting happens. The paper jumps up and sticks to the ruler. We know that to make the paper move there must have been an unbalanced force acting on it; we call this the **electric force**. To find out more about the nature of this force, we need to do some more experiments. This time we will use a balloon.

If we rub a balloon on a sweater, we find that it will experience a force in the direction of the sweater as shown in Figure 1. The sweater is causing this force, so according to Newton's third law, the sweater must experience an equal and opposite force toward the balloon: the two objects are said to **attract** each other.

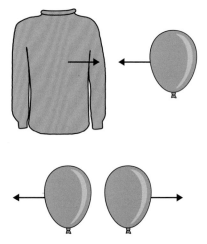

If two balloons are rubbed together against friction and held close to each other, they will experience a force pushing them away from each other as shown in Figure 2. They are said to **repel** each other.

Why does a charged balloon 'stick' to a wall? Let us assume that the balloon is negatively charged. When the balloon is brought close to the wall, the negative electrons in the wall are repelled and move away from the wall's surface. The net migration of electrons within the wall means that the surface has a temporary positive charge, to which the negative balloon is attracted. We call this process 'induction' of charge.

D.2 Figure 1 The balloon is attracted to the wool.

D.2 Figure 2 Balloons repel each other.

In some ways, this effect is similar to gravity, as the force acts over a distance and gets bigger as the bodies are brought closer together. In other ways, it is very different to gravity. The gravitational force is experienced by bodies with mass and always attractive, whereas the electric force can be either attractive or repulsive. Whatever property is responsible for the electric force must exist in two types. We call this property **charge**. Experimenting further, we find that the more balloons we have, the greater the force, as in Figure 3. Also, a balloon attached to the sweater on which it was charged exerts no force on a nearby uncharged balloon as in Figure 4.

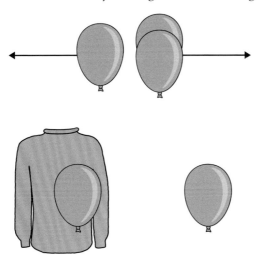

D.2 **Figure 3** More balloons, bigger force.

D.2 **Figure 4** The charge on the sweater cancels the charge on the balloon, which means no force is exerted on a nearby uncharged balloon.

So charges add and charges cancel. For this reason, we can use positive and negative numbers to represent charge but we have to decide which will be positive and which negative. Let us say the balloon is negative and the sweater is positive.

TOK

Historically, the decision was related to silk and glass rather than sweaters and balloons but it was just as arbitrary. In some ways it was a pity this way was chosen.

Charge transfer by contact

Have you ever experienced an electric shock?

As with friction and charge induction, contact is another mechanism through which charge can be transferred, sometimes with painful consequences. For a noticeable shock to occur, one of the two bodies must have a considerable net positive or negative charge already built up, with an insulator preventing this charge from flowing out into neighboring bodies.

The touch of the insulated body and the conducting body means that electrons will flow from one to the other to reduce the difference in their two charges. Humans have quite high conductivity (considerably more than air).

To prevent this transfer of charge from causing serious harm (for example, in the presence of flammable materials), both bodies can be grounded, which means they are connected electrically to the Earth. The Earth's size and abundance of charge means that both bodies are brought to potentials of 0 V (more on this later), which in turn ensures that no electrons flow between them, preventing the formation of sparks. The electrical symbol for the ground is shown in Figure 5.

D.2 **Figure 5** The electrical symbol for Earth (0 V).

Charge is quantized.
Which other physical
quantities are
quantized? (NOS)

D.2 Figure 6 A simple model of an atom (not to scale).

D.2 Figure 7 The charges on the balloon and sweater.

Charge, Q

The property of matter that causes the electrical force is charge. All bodies are positive, negative or neutral (zero charge). Bodies with the same charge repel each other and oppositely charged bodies attract. Charge cannot be created or destroyed so charges add or subtract.

Charge is a **scalar** quantity.

The unit of charge is the coulomb (C).

Fundamental charge, e

Matter is made of atoms so it makes sense to suppose that the atoms are also charged. All atoms are made of three particles: **protons**, which have positive charge, **electrons**, which are negative, and the **neutrons**, which are neutral. Each atom is made of a positive heavy nucleus consisting of protons and neutrons, surrounded by much lighter negative electrons as in Figure 6. The proton and the electron have the same amount of charge $(1.60 \times 10^{-19}\,C)$ but the opposite sign. This is the smallest amount of charge that exists in ordinary matter and is therefore called the **fundamental charge**, e.

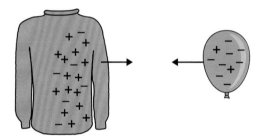

Electrons, being lighter and on the outside of the atom, are easy to move around and provide the answer to what was going on in the balloon experiment. At the start, the balloon and sweater had equal numbers of positive and negative charges so were both neutral. When the balloon is rubbed on the sweater, electrons (negative charges) are rubbed from the sweater onto the balloon so the balloon becomes negative and the sweater positive as shown in Figure 7. We say that charge is conserved; it can be transferred but not created or destroyed.

Millikan's oil drop experiment

Millikan and Fletcher
used X-rays to ionize
the oil drops. Some
modern-day lab
experiments use
friction.

D.2 Figure 8 Millikan's oil-drop apparatus.

Our knowledge of the fundamental charge comes from an experiment conducted over one hundred years ago in Chicago. Millikan and Fletcher sprayed a fine mist of oil drops into a chamber, where they were ionized (adding or removing electrons) by X-rays as they passed through the nozzle.

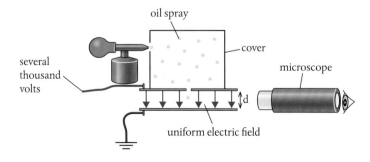

The charged oil drops were released into a region of uniform electric field. When no potential difference was applied across the plates, the oil would fall under gravity and reach terminal velocity when drag was the same as the gravitational force. When an upward electric force was applied, some drops (depending on their mass) would become stationary or rise upward with a new terminal velocity. The balance of forces enabled the charge on individual oil drops to be determined.

Although no single oil drop had the fundamental charge exactly, all charges found through this method shared a common factor: 1.60×10^{-19} C. This quantity of charge remains the smallest that has been found by any experimental procedure.

Electric field

A region of space where an electric force is experienced is called an **electric field**. The region around the sphere of charge in Figure 9 is an electric field since a small positive charge placed in that region experiences a force.

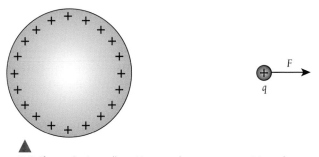

D.2 Figure 9   A small positive test charge near a positive sphere.

Field strength, *E*

The size of the field at a given point is given by the electric field strength. This is defined as the force per unit charge experienced by a small positive test charge placed at that point. So if we consider a point some distance from a sphere of positive charge as shown in Figure 9, the field strength E would be $\frac{F}{q}$. Since force is a vector, field strength will also be a vector with the same direction as the force. The unit of field strength is N C^{-1}.

If the field strength at a point is E, a charge $+q$ placed at that point will experience a force Eq in the direction of the field. A charge $-q$ will experience the same size force but in the opposite direction.

When you drive around in a car on a dry day, the friction between the tires and the road causes the body of the car to become charged. The electric field inside the car is zero so you will not feel anything as long as you stay inside. When you get out of the car, there will be a potential difference between you and the car, resulting in a small discharge, which can be unpleasant. One way to avoid this is to ask a passer-by to touch the car before you get out.

Exercise

Q1. Calculate the force experienced by a charge of $+5 \times 10^{-6}$ C placed at a point where the field strength is 40 N C^{-1}.

Q2. A charge of -1.5×10^{-6} C experiences a force of 3×10^{-5} N toward the north. Calculate the magnitude and direction of the field strength at that point.

Q3. An electron has an instantaneous acceleration of 100 m s^{-2} due to an electric field. Calculate the field strength of the field.

electron charge $= -1.6 \times 10^{-19}$ C

electron mass $= 9.1 \times 10^{-31}$ kg

Field lines

Field lines are drawn to show the direction and magnitude of the field. So for a point positive charge, the field lines would be radial, showing that the force is always away from the charge as in Figure 10. The fact that the force gets stronger as the distance from the charge decreases can be seen from the density of the lines.

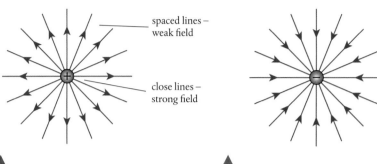

spaced lines – weak field

close lines – strong field

▲ **D.2 Figure 10** Field lines for a negative point charge.

▲ **D.2 Figure 11** Field lines for a positive point charge.

The field around a hollow sphere of charge is the same as it would be if all the charge was placed at the center of the sphere (Figure 12). This is similar to the gravitational field around a spherical mass. A point charge placed inside the sphere will experience forces in all directions due to each of the charges on the surface. The resultant of these forces is zero no matter where the point charge is placed. The electric field inside a charged sphere is therefore zero.

The field between positively and negatively charged parallel plates is uniform. As can be seen in Figure 13, the lines are parallel and equally spaced, meaning that the force experienced by a small test charge placed between the plates will have the same magnitude and direction wherever it is placed.

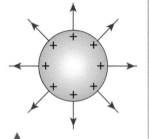

▲ **D.2 Figure 12** A hollow sphere.

D.2 Figure 13 A uniform field. Note that the field is not uniform at the edges.

The constant
$k = 8.99 \times 10^9 \, \text{N m}^2 \, \text{C}^{-2}$

This can also be expressed in terms of the permittivity of a vacuum, ε_0:

$$k = \frac{1}{4\pi\varepsilon_0}$$

$\varepsilon_0 = 8.85 \times 10^{-12} \, \text{C}^2 \text{N}^{-1} \text{m}^{-2}$

The permittivity is different for different media but we will usually be concerned with fields in a vacuum.

The relative permittivity, ε_r, of a material is the ratio of its permittivity to the permittivity of a vacuum.

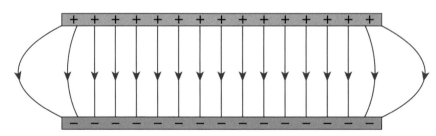

Coulomb's law

In a gravitational field, the force between masses is given by Newton's law, and the equivalent for an electric field is Coulomb's law.

The force experienced by two point charges is directly proportional to the product of their charge and inversely proportional to the square of their separation.

The force experienced by two point charges Q_1 and Q_2 separated by a distance r in a vacuum is given by the formula:

$$F = k\frac{Q_1 Q_2}{r^2}$$

The constant of proportionality, $k = 8.99 \times 10^9 \, \text{N m}^2 \, \text{C}^{-2}$.

Note: Coulomb's law also applies to spheres of charge; the separation being the distance between the centers of the spheres.

This means that we can now calculate the field strength at a distance from a sphere of charge.

Worked example

A 5 μC point charge is placed 20 cm from a 10 μC point charge.

(a) Calculate the force experienced by the 5 μC charge.

(b) What is the force on the 10 μC charge?

(c) What is the field strength 20 cm from the 10 μC charge?

Solution

(a) Using the equation: $F = k\dfrac{Q_1Q_2}{r^2}$

$$Q_1 = 5 \times 10^{-6}\,C, Q_2 = 10 \times 10^{-6}\,C \text{ and } r = 0.20\,m$$

$$F = \frac{8.99 \times 10^9 \times 5 \times 10^{-6} \times 10 \times 10^{-6}}{0.20^2}$$

$$= 11.3\,N$$

(b) According to Newton's third law, the force on the 10 μC charge is the same as on the 5 μC.

(c) force per unit charge $= \dfrac{11.3}{5 \times 10^{-6}}$

$$E = 2.26 \times 10^6\,N\,C^{-1}$$

Challenge yourself

1. Two small conducting spheres, each of mass 10 mg, are attached to two 50 cm long pieces of thread connected to the same point. This causes them to hang next to each other. A charge of 11×10^{-10} C is shared evenly between the spheres, causing them to repel each other. Show that when they stabilize, the spheres will be 3 cm apart.

What are the relative strengths of the four fundamental forces? (D.1, E.3, E.4)

Exercise

Q4. If the charge on a 10 cm radius metal sphere is 2 μC, calculate:
   (a) the field strength on the surface of the sphere
   (b) the field strength 10 cm from the surface of the sphere
   (c) the force experienced by a 0.1 μC charge placed 10 cm from the surface of the sphere.
   (d) Calculate the force if the space between the small sphere and the surface of the big sphere were filled with concrete of relative permittivity 4.5.

Q5. A small sphere of mass 0.01 kg and charge 0.2 μC is placed at a point in an electric field where the field strength is 0.5 N C⁻¹.
   (a) What force will the small sphere experience?
   (b) If no other forces act, what is the acceleration of the sphere?

D.2 Figure 14 Charging a sphere.

Electric field and energy

Imagine taking an uncharged sphere and charging it by taking, one at a time, some small positive charges from a great distance and putting them on the sphere as shown in Figure 14. As the charge on the sphere gets bigger and bigger, it would be more and more difficult to add charges as the small positive charges would be repelled from the now positive sphere. As we pushed the charges onto the sphere, we would do work transferring energy to the charge. The charges on the sphere now have energy due to their position. This is **electrical potential energy** and it is equal to the amount of work done putting the charges on the sphere from a place where they had no potential energy (at an infinite distance from the sphere).

Potential, *V*

If the same amount of charge was put onto a smaller sphere, the charges would be closer together. This would make it more difficult to put more charge onto it. The potential energy of one of the small charges on the sphere depends on the magnitude of its charge, how much charge is on the sphere, and how big the sphere is. To quantify how much work is required per unit charge, we define **potential**.

The potential of the charged sphere is defined as: *the amount of work done per unit charge in taking a small positive test charge, q, from infinity and putting it onto the sphere.*

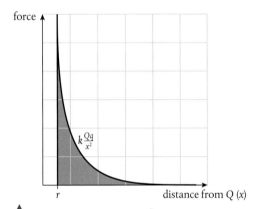

D.2 Figure 15 Graph of force against distance as charge +*q* approaches +*Q*.

Because the force is not constant, we will use the area under the force–distance curve (Figure 15) to find the work done. This is the same method we used to find the amount of potential energy stored in a stretched spring.

Since this is not linear, it is not so simple to find the area under the curve. You will probably do it in math one day but until then you just have to know that the solution is:

$$\text{work done} = k\frac{Qq}{r}$$

where *k* is a constant.

Since energy is conserved, the electrical potential energy of the particle will be the same as the work done. The potential at the surface of the charged sphere = work done per unit charge:

$$V_e = k\frac{Q}{r}$$

The unit of potential is J C^{-1}, which is the same as the volt (V).

The field inside a hollow sphere is zero, so if you were to move a charge around inside the sphere, no work would need to be done. Taking a point charge from infinity to the inside of a sphere, would therefore require the same amount of work as it would take to bring the charge to the surface. We can therefore deduce that the potential inside the sphere equals the potential at the surface. This leads to the horizontal part of the graph in Figure 16.

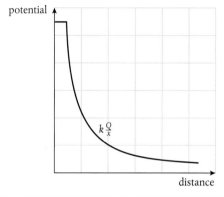

D.2 Figure 16 Graph of potential against distance for a positive charge.

Exercise

Q6. Calculate the electric potential at a distance of 20 cm from the center of a small sphere of charge +50 µC.

Q7. Calculate the potential difference between the point in Q6 and a second point 40 cm from the center of the sphere.

Figure 16 represents the change in potential along one dimension. We can think of this as a hill that gets steeper and steeper as we approach the top. We can extend this to two dimensions as shown in Figure 17. This gives a useful visual representation of the field. Moving a charge toward the sphere would be like pushing a ball up the hill: the steeper the hill becomes, the more difficult it will be to move the ball. If we draw contour lines for the hill, we get a set of equipotential concentric circles. Alternatively we can draw equipotential spheres around a point charge.

You may remember that you can find displacement from the area under a velocity–time graph and the velocity from the gradient of a displacement–time graph. The same holds here: the potential is the area under the field strength–distance graph so the field strength is the gradient of the potential–distance graph. In other words:

$$E = -\frac{dV}{dx}$$

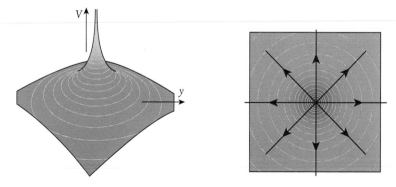

D.2 Figure 17 Field lines and equipotentials for a positive point charge.

Comparing the field lines and lines of equipotential, we see that the field is always perpendicular to the lines of equipotential. So if a charge moves along a line of equipotential, the force acting on it will be perpendicular to the direction of motion. This means no work is done so the potential energy of the particle does not change. This is what would be expected moving along a line of equipotential.

405

D.2 Figure 18 Field lines
and equipotentials for a
dipole.

Dipoles

A dipole is a pair of opposite charges. The field lines and lines of equipotential for a dipole, along with the associated potential well and hill, are shown in Figure 18.

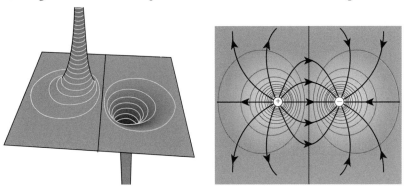

Drawing potential
hills and wells gives
an easy-to-interpret
representation of the
electric field even
though there are no
changes in height
involved.

TOK

Potential difference

The potential difference is the difference in potential between two points. This is the difference in work done per unit charge bringing a small test charge from infinity to each of the two points. This is the same as the amount of work done per unit charge taking a test charge from one point to the other point.

D.2 Figure 19 Notice how
the potential difference
is indicated with arrows,
showing that it is the
difference between two
points.

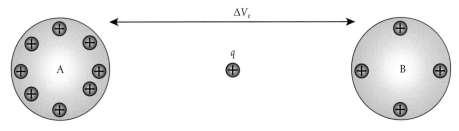

Referring to Figure 19, it takes more work to move q from infinity to A than from infinity to B, so A is at a higher potential than B. The potential difference between A and B, V_{AB}, is the work done per unit charge moving from B to A. So if a charge $+q$ moves from B to A, it will gain an amount of potential energy, $E_p = qV_{AB}$.

Exercise

Refer to the figure below for the questions on the following page.

Q8. (a) One of the charges is positive and the other is negative. Which is which?

    (b) If a positive charge was placed at A, would it move, and if so, in which direction?

Q9. At which point, A, B, C, D, or F, is the field strength greatest?

Q10. What is the potential difference between the following pairs of points?

    (a) A and C

    (b) C and E

    (c) B and E

Q11. How much work would be done taking a +2 C charge between the following points?

    (a) C to A   (b) E to C   (c) B to E

Q12. Using the scale on the diagram, estimate the field strength at point D. Why is this an estimate?

Q13. Write an equation for the potential at point A due to Q_1 and Q_2. If the charge Q_1 is 1 nC, find the value Q_2.

Q14. If an electron is moved between the following points, calculate the work done in eV (remember an electron is negative).

    (a) E to A   (b) C to F   (c) A to C

HL end

Potential in a uniform field

As we have already seen, a uniform field can be produced between two parallel plates as shown in Figure 20.

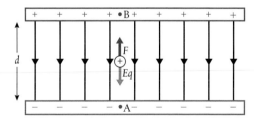

◀ **D.2 Figure 20** Uniform electric field (shown without edge effects).

Because this is a uniform field, the force is constant and equal to Eq, so the work done in moving a charge from A to B is force × distance moved in direction of force = Eqd. The potential difference is the work done per unit charge so: $V_{AB} = \dfrac{Eqd}{q} = Ed$

This result could also be reached by using the fact that E is the potential gradient:

$$E = \frac{V}{d}$$

The electronvolt (eV) is a unit of energy used in atomic physics. 1 eV = energy gained by an electron accelerated through a potential difference of 1 V.

Exercise

Refer to the figure below for these questions.

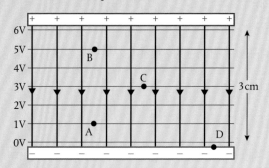

Q15. What is the potential difference between A and C?

Q16. What is the potential difference between B and D?

Q17. If a charge of +3 C was placed at B, how much potential energy would it have?

Q18. If a charge of +2 C was moved from C to B, how much work would be done?

Q19. If a charge of −2 C moved from A to B, how much work would be done?

Q20. If a charge of +3 C was placed at B and released:

  (**a**) what would happen to it?

  (**b**) how much kinetic energy would it gain when it reached A?

Q21. If an electron was released at A and accelerated to B, how much kinetic energy would it gain in eV?

Q22. If an electron was taken from C to D, how much work would be done in eV?

Earthing

Rather like we did in gravitational fields we can take the Earth to represent zero potential. We take the Earth as having a potential of 0 V because all the bodies we will consider are close to the Earth, and therefore any change in the potential of the Earth changes the potential of all nearby bodies equally.

If a charged body is grounded (earthed), current will flow until the potential of the body is zero.

D.2 Figure 21 Before and after a charged body is earthed.

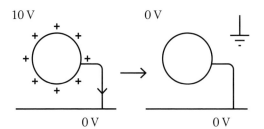

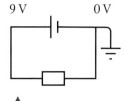

D.2 Figure 22 The Earth symbol within a circuit diagram.

Adding an 'Earth' symbol to a point in a circuit diagram indicates that it is at zero potential.

Charging by induction

The potential of a Body A not only depends on the charge of a body but also on charges nearby.

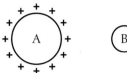

D.2 Figure 23 B is uncharged but experiences a positive potential due to A.

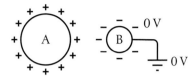

D.2 Figure 24 B gains negative charges from the Earth.

If Body B is connected to Earth, a current will flow until B is at zero potential.

B now has a negative charge. No charge was exchanged between A and B. This is called electrostatic induction.

Conductors and current

A conductor contains free charge carriers. In metals, these are electrons: negatively charged particles that are able to move around.

When electrons are added to a conducting sphere, they spread over the surface.

If this body is connected to an uncharged body, electrons flow from one to the other. This happens because the bodies are at different potentials. Electrons (being negative) flow from low to high potential.

D.2 Figure 25 Electrons repel one another until they are as spread as possible.

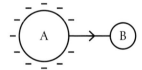

D.2 Figure 26 A and B combine to form one overall conductor with a net movement of electrons from A to B.

We know from B.5 that this is analogous to the electrons flowing uphill. Instead, let us ponder the equivalent transition in terms of conventional current.

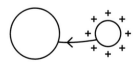

D.2 Figure 27 An equivalent model involving high to low potential would be net movement of positive charges from B to A.

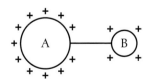

D.2 Figure 28 There is no current flow when there is no potential difference.

A current flows until there is no potential difference across the conductor as a whole. Note that A has more charge than B but the same potential.

 Models do not have to be true.

Not all magnets are human-made; some rocks (for example, this piece of magnetite) are naturally magnetic.

Nature of Science

The first recorded observation of the magnetic effect was 2500 years ago in the ancient city of Magnesia (Manisa) in what is now western Turkey. Experiments showed that one end of a magnet always pointed toward the North Pole so a theory was developed related to the attraction of opposite poles. This theory gave correct predictions, enabling magnets to be used in navigation, even though it was not entirely correct. The connection between magnetism and electricity was not made until a chance observation by Hans Christian Oersted in 1819.

Magnetic fields

It is not obvious that there is a connection between the magnetic force and the electric force. However, we will discover that they are the same thing. First, let us investigate the nature of magnetism.

Magnets are not all human-made; some stones are magnetic. If we place a small magnetic stone next to a big one, it experiences forces that make it rotate. We can define a magnetic field as a region of space where a small magnet would experience a turning force. Magnets are dipoles so one end attracts and the other repels.

If one of these small magnets is placed close to the Earth, we will find that it rotates so that one end always points north. Because of this, we call that end the **north-seeking pole**. The other end is called the south-seeking pole. The direction of the field is defined as the direction that a north-seeking pole points.

Magnetic field lines

In practice, a small compass can be used as our test magnet. Magnetic field lines are drawn to show the direction that the north (N) pole of a small compass would point if placed in the field.

If we join the directions pointed by the compass, we get the field lines shown. These not only show the direction of the field but their density shows us where the field is strongest. We call the density of lines the **flux density**.

D.2 Figure 29 The small magnet is caused to turn, so must be in a magnetic field.

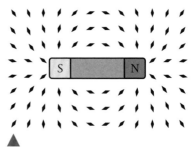

D.2 Figure 30 If the whole field were covered in small magnets, then they would show the direction of the field lines.

Notice that since unlike poles attract, the north-seeking pole of the small compass points to the *south* magnetic pole of the big magnet. So, if we treat the Earth like a big magnet, the north-seeking end of the compass points north because there is a *south* magnetic pole there.

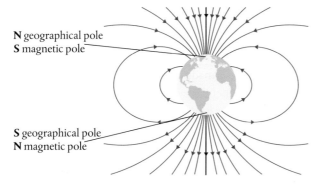

D.2 Figure 31 The Earth's magnetic field.

A uniform field can be created between two flat magnets as in Figure 32. As with uniform gravitational and electric fields, the field lines are parallel and uniformly spaced, apart from at the ends of the magnets.

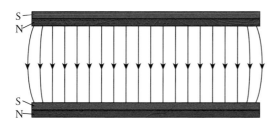

D.2 Figure 32 Uniform magnetic field.

Field caused by currents

If a small compass is placed close to a straight wire carrying an electric current, then it experiences a turning force that makes it always point around the wire. The region around the wire is therefore a magnetic field. This leads us to believe that magnetic fields are caused by moving charges.

We can work out the direction of the field by pretending to grip the wire with our right hand. If the thumb points in the direction of the current, then the fingers will curl in the direction of the field.

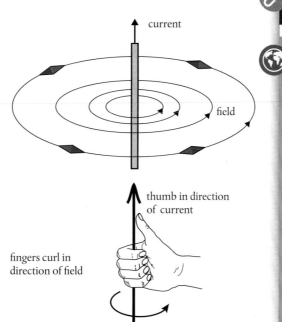

D.2 Figure 33 The field due to a long straight wire carrying a current is in the form of concentric circles. Notice that the field is strongest close to the wire.

How are electric fields and magnetic fields like gravitational fields? (D.1)

When charged particles ejected from the Sun meet the Earth's magnetic field, they are made to follow a helical path. As these particles move toward the poles, the field becomes stronger so the helix becomes tighter until the particles turn back on themselves and head for the other pole. If the particles have enough energy, they sometimes get close enough to the Earth to reach the upper atmosphere. If this happens, light is emitted as the particles ionize the air, forming the northern/southern lights.

Magnetic flux density, *B*

From what we know about fields, the strength of a field is related to the density of field lines. This tells us that the magnetic field is strongest close to the poles. The magnetic flux density is the quantity that is used to measure how strong the field is. However it is not quite the same as field strength as used in gravitational and electric fields.

The field inside a coil

When a current-carrying wire is made into a circular loop, the field inside is due to the addition of all the field components around the loop, making the field at the center greater (Figure 34). Adding more loops to form a coil will increase the field.

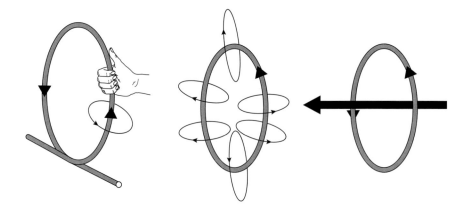

Since the letter *B* is used to denote flux density, the magnetic field is often called a *B* field.

How can moving charges in magnetic fields help probe the fundamental nature of matter? (B.1, C.3)

D.2 Figure 34 The direction of the field can be found by applying the right-hand grip rule to the wire. The circles formed by each bit of the loop add together in the middle to give a stronger field.

The field inside a solenoid

A solenoid is a special type of coil where the loops are wound next to each other along a cylinder to form a helix as shown in Figure 35.

D.2 Figure 35 The direction of the field in a solenoid can be found using the grip rule on one coil.

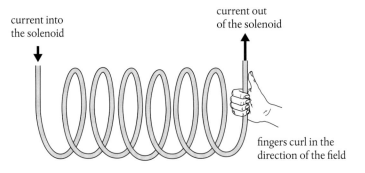

current into the solenoid

current out of the solenoid

fingers curl in the direction of the field

The magnetic field caused by each loop of the solenoid adds to give a field pattern similar to a bar magnet, as shown in Figure 36.

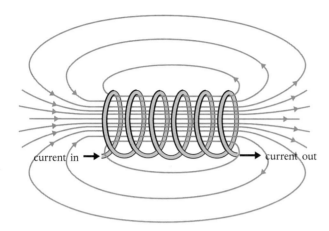

Guiding Questions revisited

Which experiments provided evidence to determine the nature of the electron?

How can the properties of fields be understood using an algebraic approach and a visual representation?

What are the consequences of interactions between electric and magnetic fields?

In this chapter, we have examined electric fields (in a similar way to gravitational fields) as well as magnetic fields, to find out how:

- The combination of ionization of oil droplets and an understanding of gravitational fields enabled Millikan to recognize that all charges measured shared a common factor, which we now know to be the magnitude of the charge on an electron or proton.
- Charge can be transferred from one body to another and is conserved in the process.
- Electric forces are attractive for unlike charges and repulsive for like charges. Magnetic forces are attractive for opposite poles and repulsive for like poles.
- Electric field lines represent the direction of force that would act on a small positive test charge placed in the field. Magnetic field lines represent the direction of force that would act on a test north pole.
- Coulomb's law states that the force between two charges is proportional to the product of the charges and inversely proportional to the square of the distance between them.
- Electric field strength is the force acting per unit charge.
- Just as we can approximate the region close to the surface of the Earth as having a uniform gravitational field, we can approximate the space between two parallel plates (with a potential difference between them) as having a uniform electric field.
- When electric and magnetic fields interact, forces are exerted. These can be used for electromagnetic applications.
- **HL** Electric potential energy is defined as the total work done when bringing a test charge from infinity to a point in space at a low constant speed.
- **HL** Electric potential is defined as the potential energy per unit mass. The magnitude of potential gradient is equal to the field strength.

413

Practice questions

1. A positive point charge is placed above a metal plate at zero electric potential. Which diagram shows the pattern of electric field lines between the charge and the plate?

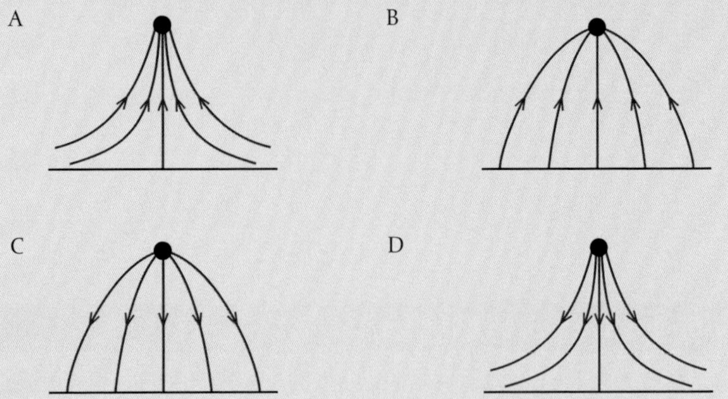

(*Total 1 mark*)

2. The force acting between two point charges is F when the separation of the charges is x. What is the force between the charges when the separation is increased to $3x$?

   A $\dfrac{F}{3}$ B $\dfrac{F}{3x^2}$ C $\dfrac{F}{9}$ D $\dfrac{F}{9x^2}$

   (*Total 1 mark*)

3. Two charges Q_1 and Q_2, each equal to $2\,nC$, are separated by a distance $3\,m$ in a vacuum. What is the electric force on Q_2 and the electric field due to Q_1 at the position of Q_2?

	Electric force on Q_2	Electric field due to Q_1 at the position of Q_2
A	$4 \times 10^{-9}\,N$	$2\,N\,C^{-1}$
B	$4\,N$	$2\,N\,C^{-1}$
C	$4 \times 10^{-9}\,N$	$2 \times 10^{-9}\,N\,C^{-1}$
D	$4\,N$	$2 \times 10^{-9}\,N\,C^{-1}$

(*Total 1 mark*)

4. What is the unit of electrical potential difference expressed in fundamental SI units?

   A $kg\,m\,s^{-1}\,C^{-1}$ C $kg\,m^2\,s^{-3}\,A^{-1}$

   B $kg\,m^2\,s^{-2}\,C^{-1}$ D $kg\,m^2\,s^{-1}\,A$ (*Total 1 mark*)

5. In an experiment, oil droplets of mass m and charge q are dropped into the region between two horizontal parallel plates. The electric field E between the plates can be adjusted. Air resistance is negligible. Which is correct when the droplets fall vertically at constant velocity?

   A $E = 0$ B $E < \dfrac{mg}{q}$ C $E = \dfrac{mg}{q}$ D $E > \dfrac{mg}{q}$

   (*Total 1 mark*)

6. Magnetic field lines are an example of what?

A A discovery that helps us understand magnetism

B A model to aid in visualization

C A pattern in data from experiments

D A theory to explain concepts in magnetism   *(Total 1 mark)*

7. An electron is placed at a distance of 0.40 m from a fixed point charge of −6.0 mC.

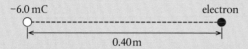

(a) Show that the electric field strength due to the point charge at the position of the electron is $3.4 \times 10^8 \, \text{N C}^{-1}$.   (2)

(b) Calculate the magnitude of the initial acceleration of the electron.   (2)

(c) Describe the subsequent motion of the electron.   (3)

(Total 7 marks)

8. An ionization chamber is a device that can be used to detect charged particles.

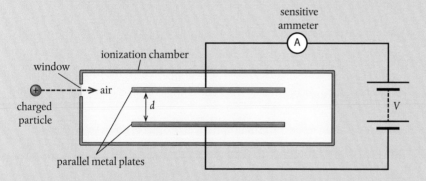

The charged particles enter the chamber through a thin window. They then ionize the air between the parallel metal plates. A high potential difference across the plates creates an electric field that causes the ions to move toward the plates. Charge flows around the circuit and a current is detected by the sensitive ammeter.

(a) Copy the diagram and draw the shape of the electric field between the plates.   (2)

The separation of the plates d is 12 mm and the potential difference V between the plates is 5.2 kV. An ionized air molecule M with charge $+2e$ is produced when a charged particle collides with an air molecule.

(b) Calculate the electric field strength between the plates.   (1)

(c)   **HL**   Determine the change in the electric potential energy of M as it moves from the positive plate to the negative plate.   (3)

(Total 6 marks)

9. (a) If something gets a positive electric charge, then what happens to something else? (1)

    A It becomes equally positively charged.

    B It becomes equally negatively charged.

    C It becomes negatively charged, but not necessarily *equally* negatively charged.

    D It becomes magnetized.

    (b) Explain your answer. (1)

    (Total 2 marks)

10. (a) A bird is sitting on a bare high-voltage line. Will the bird get a shock? (1)

    A Yes   **B** No

    (b) Explain your answer. (1)

    (Total 2 marks)

11. Millikan's oil drop experiment was the first experiment to determine the size of the elementary charge. Charged oil droplets are sprayed into an air-filled chamber and observed through a microscope inserted in the side of the chamber. The droplets fall under gravity with a terminal velocity. A uniform electric field between the top and bottom plates can be used to hold the charged oil drops in a fixed vertical position in the electric field. The potential between the plates is measured. The polarity can be changed in order to pull the droplets back up against gravity. The drag force on the droplets due to air resistance is known as viscous drag. The charge on an oil droplet may change due to random ionization.

    The viscous drag force on a small spherical droplet is given by $F = 6\pi\eta a u$, where a is the radius of the droplet, u is the terminal velocity of the droplet, η is the viscosity of the air ($\eta = 1.82 \times 10^{-5}$ kg m^{-1} s^{-1}). In two successive measurements of an oil droplet, the rise times are 42 s and 78 s. The distance traveled upward at constant velocity is 1.00 cm, the potential difference across the plates is 5000 V and the plate separation is 1.50 cm. The radius of the droplet is $a = 2.76 \times 10^{-6}$ m.

    (a) Calculate the change in the number of electrons in the droplet between the two measurements. (8)

    (b) Two droplets with the same density, radii r_1 and r_2, identical charges Q, and terminal velocities u_1 and u_2 join together. How is the final terminal velocity v of the droplet related to the initial velocities? (5)

    (Total 13 marks)

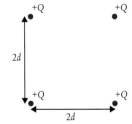

12. **HL** Four identical, positive, point charges of magnitude Q are placed at the vertices of a square of side $2d$ (see figure on the left). What is the electric potential produced at the center of the square by the four charges?

    A 0   **B** $\dfrac{4kQ}{d}$ **C** $\dfrac{\sqrt{2}\,kQ}{d}$ **D** $\dfrac{2\sqrt{2}\,kQ}{d}$

    (Total 1 mark)

13. `HL` The points X and Y are in a uniform electric field of strength E. The distance OX is x and the distance OY is y. What is the magnitude of the change in electric potential between X and Y?

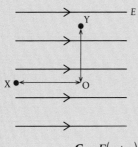

A Ex **B** Ey **C** $E(x + y)$ **D** $E\sqrt{x^2 + y^2}$

(Total 1 mark)

14. `HL` A particle with charge -2.5×10^{-6} C moves from point X to point Y due to a uniform electrostatic field. The diagram shows some equipotential lines of the field.

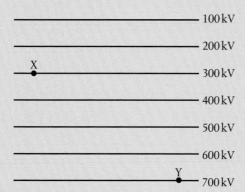

What is correct about the motion of the particle from X to Y and the magnitude of the work done by the field on the particle?

	Motion of the particle from X to Y	Magnitude of the work done by the field on the particle
A	uniform linear	0 J
B	uniform linear	1 J
C	uniform accelerated	0 J
D	uniform accelerated	1 J

(Total 1 mark)

15. **HL** The graph shows the variation of electric field strength E with distance r from a point charge. The shaded area X is the area under the graph between two separations r_1 and r_2 from the charge. What is X?

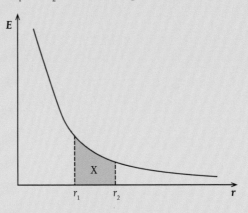

A The electric field average between r_1 and r_2

B The electric potential difference between r_1 and r_2

C The work done in moving a charge from r_1 to r_2

D The work done in moving a charge from r_2 to r_1

(Total 1 mark)

16. **HL** A vertical wall has a uniform positive charge on its surface. This produces a uniform horizontal electric field perpendicular to the wall. A small positively charged ball is suspended in equilibrium from the vertical wall by a thread of negligible mass.

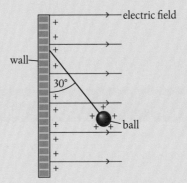

The charge per unit area on the surface of the wall is σ. It can be shown that the electric field strength E due to the charge on the wall is given by the equation:

$$E = \frac{\sigma}{2\varepsilon_0}$$

(a) Demonstrate that the units of the quantities in this equation are consistent. (2)

(b) The thread makes an angle of 30° with the vertical wall. The ball has a mass of 0.025 kg. Determine the horizontal force that acts on the ball. (3)

(c) The charge on the ball is 1.2×10^{-6} C. Determine σ. (2)

The center of the ball, still carrying a charge of 1.2×10^{-6} C, is now placed 0.40 m from a point charge Q. The charge on the ball acts as a point charge at the center of the ball. P is the point on the line joining the charges where the electric field strength is zero. The distance PQ is 0.22 m.

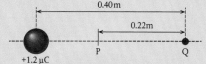

(d) Calculate the charge on Q. State your answer to an appropriate number of significant figures. (2)

(Total 9 marks)

17. **HL** The diagram shows the electric field lines of a positively charged conducting sphere of radius R and charge Q.

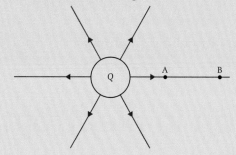

(a) Points A and B are located on the same field line. Explain why the electric potential decreases from A to B. (2)

(b) Copy the axes below and draw the variation of electric potential V with distance r from the center of the sphere. (2)

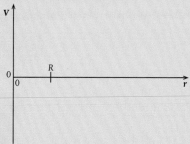

A proton is placed at A and released from rest. The magnitude of the work done by the electric field in moving the proton from A to B is 1.7×10^{-16} J. Point A is 5.0×10^{-2} m from the center of the sphere. Point B is 1.0×10^{-1} m from the center of the sphere.

(c) Calculate the electric potential difference between points A and B. (1)

(d) Determine the charge Q of the sphere. (2)

(e) The concept of potential is also used in the context of gravitational fields. Suggest why scientists developed a common terminology to describe different types of fields. (1)

(Total 8 marks)

D.3

Motion in electromagnetic fields

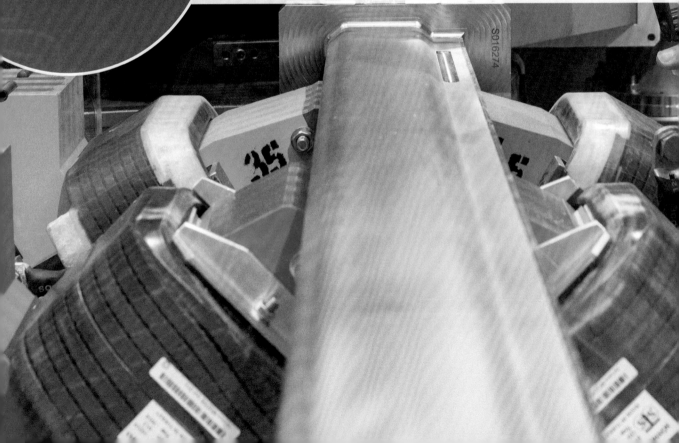

◀ The particle accelerator equipment pictured is located at SESAME (Synchrotron-light for Experimental Science and Applications in the Middle East), which opened in 2017. The facility was the first in the world to be powered entirely by renewable energy. A combination of electric and magnetic fields is used to accelerate the electrons within.

Guiding Questions

How do charged particles move in magnetic fields?

What can be deduced about the nature of a charged particle from observations of it moving in electric and magnetic fields?

If you connect a battery between a hot wire and a metal plate such that the plate is at a higher potential than the wire, current will flow in the circuit. Somehow, charge is able to cross the gap. This is the basis for an electron gun.

If you make a hole in the plate, a faint blue line can be seen coming out of the hole. If the blue line is passed through a uniform electric field, it is deflected into a parabola. This is very similar to the path of a projectile thrown in a uniform magnetic field. If we assume that the blue line is made of particles with a charge of 1.60×10^{-19} C, we can model the situation using the same equation we used for projectile motion (A.1).

Being able to see how an equation in one area of the subject can be used in a different area is a useful skill and is one reason why good physics students are snapped up by employers!

Students should understand:

the motion of a charged particle in a uniform electric field
the motion of a charged particle in a uniform magnetic field
the motion of a charged particle in perpendicularly orientated uniform electric and magnetic fields
the magnitude and direction of the force on a charge moving in a magnetic field as given by $F = qvB\sin\theta$
the magnitude and direction of the force on a current-carrying conductor in a magnetic field as given by $F = BIL\sin\theta$
the force per unit length between parallel wires as given by $\frac{F}{L} = \mu_0 \frac{I_1 I_2}{2\pi r}$ where r is the separation between the two wires.

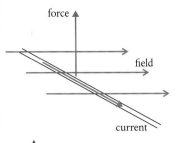

D.3 Figure 1 Force, field and current are at right angles to each other.

D.3 Figure 2 The field into the page can be represented by crosses, and the field out of the page by dots. Think what it would be like looking at an arrow from each end.

D.3 Figure 3 Using Fleming's left-hand rule to find the direction of the force.

What causes circular motion of charged particles in a field? (A.2)

Force on a current-carrying conductor

We have seen that when a small magnet is placed in a magnetic field, each end experiences a force that causes it to turn. If a straight wire is placed in a magnetic field but not parallel to the field, it also experiences a force. However, in the case of a wire, the direction of the force does not cause rotation – the force is in fact perpendicular to the direction of both current and field (Figure 1).

The size of the force depends on the size of current, length of conductor and flux density. We can therefore write that $F \propto BIL$. We can now define the unit of flux density in terms of this force to make the constant of proportionality equal to 1.

The tesla (T)

A flux density of 1 tesla would cause a 1 m long wire carrying a current of 1 A perpendicular to the field to experience a force of 1 N. So if B is measured in T, $F = BIL$.

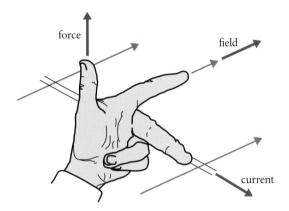

Parallel current-carrying conductors

Due to the difficulty in measuring the amount of charge flowing in a given time, the ampere was not defined in terms of charge until 2019. Before this, the ampere was defined in terms of the force between two parallel conductors.

We know that a current-carrying conductor has an associated magnetic field and that this field exerts a force on permanent magnets. What happens when two current-carrying conductors are placed side by side?

Consider the two wires shown in Figure 4. Each wire carries a current so is creating a magnetic field around it. The wires are next to each other so each wire is in the field of the other. The magnetic field lines produced by wire X are concentric circles that cut wire Y at right angles, as can be seen in the end view. Using Fleming's left-hand rule, we can determine that the direction of force is directed toward X. Likewise, we can show that the force on X is directed toward Y.

Parallel wires

The force F acting per unit length is calculated using Ampère's force law, with the equation for the special case of two straight parallel wires being:

$$\frac{F}{L} = \mu_0 \frac{I_1 I_2}{2\pi r}$$

where L is the total parallel length, μ_0 is the magnetic permeability of a vacuum ($4\pi \times 10^{-7}$ T m A^{-1}), I_1 and I_2 are the currents in the two conductors, and r is the distance by which they are separated.

One ampere is the current that would cause a force of 2×10^{-7} N per meter between two long parallel conductors separated by 1 m in a vacuum.

Exercise

Q1. A straight wire of length 0.5 m carries a current of 2 A in a north–south direction. If the wire is placed in a magnetic field of 20 μT directed vertically downward:

   (a) what is the size of the force on the wire?

   (b) what is the direction of the force on the wire?

Q2. A vertical wire of length 1m carries a current of 0.5 A upward. If the wire is placed in a magnetic field of strength 10 μT directed toward the N geographic pole:

   (a) what is the size of the force on the wire?

   (b) what is the direction of the force on the wire?

Q3. Use Fleming's left-hand rule to find the direction of the force in the following examples below.

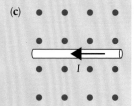

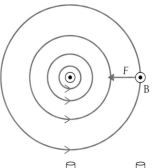

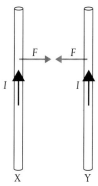

D.3 Figure 4 Two long wires.

423

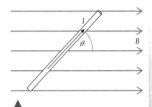

Non-perpendicular fields

If the wire is *not* perpendicular to the field, then to calculate the magnitude of the force, you need to use the component of field that is perpendicular to the wire.

So in the example shown in Figure 5, the component of B perpendicular to the wire = $B \sin \theta$ so the force, $F = B \sin \theta \times IL = BIL \sin \theta$.

The direction of the force can be found by using Fleming's left-hand rule, lining the **F**irst finger with the perpendicular component of the **F**ield. In the case shown here, that would result in a force into the page.

Charges in magnetic fields

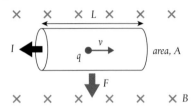

D.3 Figure 6 The force experienced by each electron is in the downward direction. Remember the electrons flow in the opposite direction to conventional current.

From the microscopic model of electrical current in a metal wire, we believe that current is made up of charged particles (electrons) moving through the metal. Each electron experiences a force as it travels through the magnetic field. The sum of all these forces gives the total force on the wire: $F = BIL = BnAveL$ where n is the number of electrons per unit volume, A is the cross-sectional area, v is the net speed (sometimes referred to as **drift velocity**) of the electrons and L is the length of conductor. In turn, $F = BNev$, where N is the number of electrons in this length of wire, so the force on each electron is Bev. If a free charge moves through a magnetic field, then it will also experience a force. The direction of the force is always perpendicular to the direction of motion, and this results in a circular path.

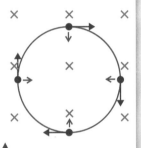

D.3 Figure 7 Wherever you apply Fleming's left hand rule, the force is always toward the center.

The force on each charge q is given by the formula:

$$F = Bqv.$$

The electron gun

In the chapter introduction, we learned about an electron 'gun'. The electron gun shown in Figure 8 is used to produce a stream of electrons that can be projected into a region of electric or magnetic field. The gun is housed in a glass tube, which has had most of the air removed by use of a vacuum pump. The electrons move freely in this low-pressure gas, but collisions with the remaining molecules excite atomic electrons, resulting in the emission of light. This causes a blue glow revealing the path of the electron beam.

How can the orbital radius of a charged particle moving in a field be used to determine the nature of the particle? (A.2)

D.3 Figure 8 An electron gun.

The elections are released from a thin wire heated by an electric current. They enter an electric field created by applying a potential difference between the hot wire and a metal plate. They are accelerated by an electric field toward the positive anode and pass through a hole in the anode plate.

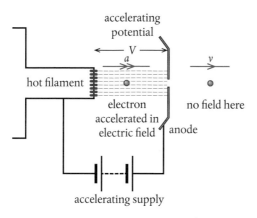

D.3 Figure 9 Electrons are accelerated in the electric field.

As the electron moves within the electric field, its electric potential energy is transferred to kinetic energy:

$$eV = \frac{1}{2}mv^2$$

Once through the hole in the plate, where there is no field, the electrons continue in a straight line with constant velocity.

Motion of electrons in a uniform electric field

If a charge Q is projected perpendicular to a uniform electric field, it will have constant velocity perpendicular to the field lines but acceleration parallel to the field lines. This is just like the motion of a projectile in a uniform gravitational field and results in a parabolic trajectory. We can use the same equations to model the motion, except that acceleration is equal to $\frac{EQ}{m}$ instead of g.

The charged particles traveling around the huge circular rings of the particle accelerator at CERN are kept in a circular path using a magnetic field. These particles move so fast and have such a large mass that the field has to be very strong.

How can conservation of energy be applied to motion in electromagnetic fields? (A.3)

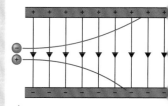

D.3 Figure 10 The paths of positive and negative charges projected into a uniform electric field.

How are the concepts of energy, forces and fields used to determine the size of an atom? (A.3, A.2)

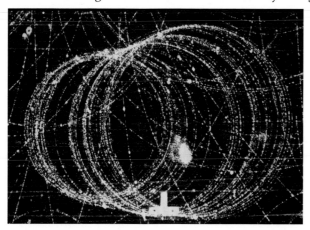

Spiral particle tracks made visible in a cloud chamber.

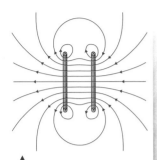

D.3 Figure 11 Uniform magnetic field between Helmholtz coils.

Motion of electrons in a uniform magnetic field

If electrons are passed into a uniform magnetic field with a velocity that is not parallel to the field lines, they will experience a force perpendicular to their direction of motion (and which remains perpendicular even when the direction of the electrons changes). This results in a circular trajectory.

A uniform magnetic field can be created between two large diameter coils (Helmholtz coils).

We have seen how the force on each electron in a current-carrying wire is $F = Bev$. This force, known as the magnetic or Lorentz force, is the sole provider of the centripetal force. We can therefore write:

$$Bev = \frac{mv^2}{r}$$

D.3 Figure 12 Path of electron with magnetic field into the page.

and simplify:

$$Be = \frac{mv}{r}$$

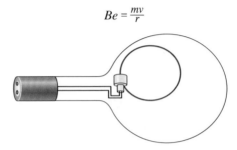

To enable the electrons to travel in a full circle, we can aim the electron gun perpendicular to the field in the spherical tube.

We know that if the accelerating potential difference is V, then the velocity, $v = \sqrt{\frac{2Ve}{m}}$.

By measuring the radius and calculating the magnetic field strength from the dimensions of the coil and current, we can find a value for $\frac{e}{m}$. The electron tube contains a low-pressure gas, which allows the electrons to maintain a constant speed. If they were to travel in a denser gas, they would move in a spiral.

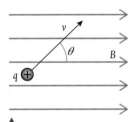

D.3 Figure 13 A positive particle moving in a non-perpendicular magnetic field.

Non-perpendicular fields

If a charged particle moves through a magnetic field at an angle, then you can calculate the magnetic force experienced by it using the component of velocity perpendicular to the magnetic field. In the example shown in Figure 13, this would be $B \sin \theta$, which results in a force, $F = Bqv \sin \theta$.

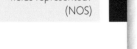

How are the properties of electric and magnetic fields represented? (NOS)

Using Fleming's left-hand rule, this will give a force into the page (note that the charge is positive so it moves in the direction of the current).

So considering the two components of the motion, perpendicular to the field, the particle will travel in a circle, while parallel to the field, the velocity is uniform. The resulting motion is helical as shown in Figure 14.

D.3 Figure 14 Helical path of a charge in a non-perpendicular magnetic field.

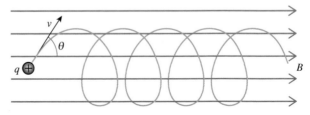

Electrons in perpendicularly orientated electric and magnetic fields

An electron moving at a constant speed in perpendicularly orientated electric and magnetic fields can be made to remain at a constant speed. We know from Newton's laws that forces must be balanced in this case.

If the magnetic field is (say) into the page and the electron moves from left to right, the force exerted will be vertically downward. An electric field orientated vertically can be used to supply an equal and opposite upward force.

Since, in this case, $F_B = F_E$: $\qquad Bev = Ee$

$$Bv = E$$

$$v = \frac{E}{B}$$

In other words, the speed of the electron can be determined from the ratio of the field strengths.

Knowing this speed, with the magnetic field then switched off, the specific charge (charge per unit mass) of the electron can be determined. You might like to research J. J. Thomson's experiment to find out more.

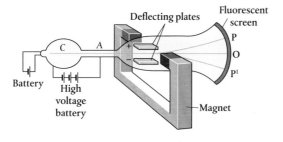

D.3 Figure 15 Arrangement of J. J. Thomson experiment to determine the specific charge of an electron.

Exercise

(Electron charge, $e = 1.6 \times 10^{-19}$ C)

Q4. Calculate the force experienced by an electron traveling through a magnetic field of flux density 5 mT with velocity 500 m s^{-1}.

Q5. An electron is accelerated through a potential difference of 500 V then passed into a region of magnetic field perpendicular to its motion, causing it to travel in a circular path of radius 10 cm. Calculate:

(a) the kinetic energy of the electron in joules

(b) the velocity of the electron

(c) the flux density of the magnetic field.

Q6. A proton (same charge as an electron but positive) passes into a region of magnetic field of flux density 5 mT as shown to the right with a velocity of 100 m s^{-1}. Calculate the force experienced by the proton.

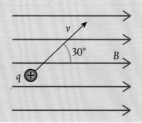

Guiding Questions revisited

How do charged particles move in magnetic fields?

What can be deduced about the nature of a charged particle from observations of it moving in electric and magnetic fields?

In this chapter, we have explored forces and motion due to different combinations of current-carrying conductors, charged particles and electric and magnetic fields to consider how:

- When a charged particle enters a uniform electric field, the force acts in the direction of the field, resulting in parabolic motion.
- When a charged particle enters a uniform magnetic field, the force acts perpendicular to the particle's velocity and the field, resulting in circular motion.
- The force acting on a charged particle in a magnetic field increases with field strength, charge and speed. It is maximized when the velocity is perpendicular to the field.
- The force acting on a current-carrying conductor in a magnetic field increases with field strength, current and conductor length. It is maximized when the orientation of the conductor is perpendicular to the field.
- Two separate current-carrying conductors exert forces on each other, with the force being proportional to the product of the currents and inversely proportional to the separation distance of the conductors.
- Charged particles in electric fields act like particles with mass act in gravitational fields. Magnetic forces can result in circular motion for charged particles moving at right angles to the field, just as particles with mass can orbit other massive bodies with a velocity that is perpendicular to the field lines.

Practice questions

1. An electron enters the region between two charged parallel plates, initially moving parallel to the plates. What can be stated about the electromagnetic force acting on the electron?

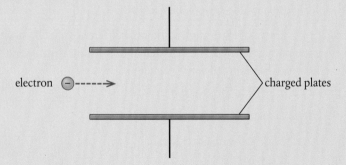

A It causes the electron to decrease its horizontal speed.

B It causes the electron to increase its horizontal speed.

C It is parallel to the field lines and in the opposite direction to them.

D It is perpendicular to the field direction.

(Total 1 mark)

2. A proton of velocity *v* enters a region of electric and magnetic fields. The proton is not deflected. An electron and an alpha particle (helium nucleus comprising two protons and two neutrons) enter the same region with velocity *v*. Which is correct about the paths of the electron and the alpha particle?

	Path of electron	Path of alpha particle
A	deflected	deflected
B	deflected	not deflected
C	not deflected	deflected
D	not deflected	not deflected

(Total 1 mark)

3. A beam of electrons moves between the poles of a magnet. What is the direction in which the electrons will be deflected?

beam of electrons

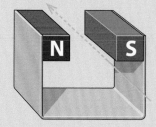

A   Downward   C   Toward the S pole of the magnet

B   Toward the N pole of the magnet   D   Upward

(Total 1 mark)

4. A liquid that contains negative charge carriers is flowing through a square pipe with sides A, B, C and D. A magnetic field acts in the direction shown across the pipe. On which side of the pipe does negative charge accumulate?

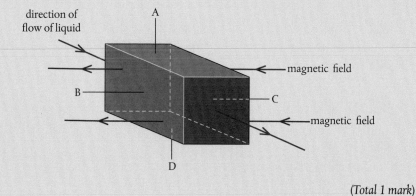

(Total 1 mark)

5. The diagram shows two current-carrying wires, P and Q, that both lie in the plane of the paper. The arrows show the conventional current direction in the wires.

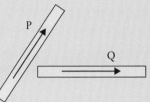

The electromagnetic force on Q is in the same plane as that of the wires. What is the direction of the electromagnetic force acting on Q?

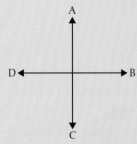

(Total 1 mark)

6. A positively charged particle moves parallel to a wire that carries a current upward. What is the direction of the magnetic force on the particle?

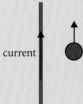

A To the left   **B** To the right   **C** Into the page   **D** Out of the page

(Total 1 mark)

7. Two currents of 3 A and 1 A are established in the same direction through two parallel straight wires, R and S. What is correct about the magnetic forces acting on each wire?

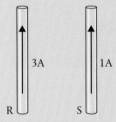

A Both wires exert equal magnitude attractive forces on each other.

B Both wires exert equal magnitude repulsive forces on each other.

C Wire R exerts a larger magnitude attractive force on wire S.

D Wire R exerts a larger magnitude repulsive force on wire S.

(Total 1 mark)

8. A non-uniform electric field, with field lines as shown, exists in a region where there is no gravitational field. X is a point in the electric field. The field lines and X lie in the plane of the paper.

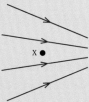

    (a) Outline what is meant by electric field strength. (2)

    (b) An electron is placed at X and released from rest. Copy the diagram and draw the direction of the force acting on the electron due to the field. (1)

    (c) The electron is replaced by a proton that is also released from rest at X. Compare, without calculation, the motion of the electron with the motion of the proton after release. You may assume that no frictional forces act on the electron or the proton. (4)

(Total 7 marks)

9. A proton is moving in a region of uniform magnetic field. The magnetic field is directed into the plane of the paper. The arrow shows the velocity of the proton at one instant and the dotted circle gives the path followed by the proton.

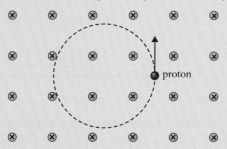

    (a) Explain why the path of the proton is a circle. (2)

The speed of the proton is $2.0 \times 10^6 \text{ m s}^{-1}$ and the magnetic field strength B is 0.35 T.

    (b) Show that the radius of the path is about 6 cm. (2)

    (c) Calculate the time for **one** complete revolution. (2)

    (d) Explain why the kinetic energy of the proton is constant. (2)

(Total 8 marks)

10. An electron moves in circular motion in a uniform magnetic field. The velocity of the electron at point P is $6.8 \times 10^5 \text{ m s}^{-1}$ in the direction shown. The magnitude of the magnetic field is 8.5 T.

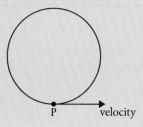

(a) State the direction of the magnetic field. (1)

(b) Calculate, in N, the magnitude of the magnetic force acting on the electron. (1)

(c) Explain why the electron moves at constant speed. (1)

(d) Explain why the electron moves in a circular path. (2)

(Total 5 marks)

11. The diagram shows an arrangement for measuring the force between two parallel sections of the same rigid wire carrying a current as viewed from the front. The supports for the upper section of the wire and the power supply are not shown.

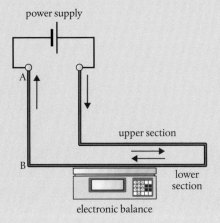

(a) Deduce what happens to the reading on the electronic balance when the current is switched on. (3)

(b) When the current in the wire is 0.20 A, the magnetic field strength at the upper section of wire due to the lower section of wire is 1.3×10^{-4} T. Calculate the magnetic force acting per unit length on the upper section of wire. (1)

(c) Each cubic meter of the wire contains approximately 8.5×10^{28} free electrons. The diameter of the wire is 2.5 mm and the length of wire within the magnetic field is 0.15 m. Using the force per unit length already calculated, deduce the speed of the electrons in the wire when the current is 0.20 A. (4)

(d) The upper section of wire is adjusted to make an angle of 30° with the lower section of wire. Outline how the reading of the balance will change, if at all. (3)

(Total 11 marks)

12. (a) When the electric current in two parallel wires is flowing in the same direction, what will the wires tend to do? (1)

    A Repel each other

    B Attract each other

    C Exert no force on each other

    D Twist at right angles to each other

    E Spin

(b) Explain your answer. (1)

(Total 2 marks)

D.4

HL **Induction**

Microphones, such as those used by podcast creators, convert sound energy into electrical signals using electromagnetic induction. This is the opposite of a loudspeaker.

HL

Guiding Questions

What are the effects of relative motion between a conductor and a magnetic field?

How can the power output of electrical generators be increased?

An electron moving in a magnetic field experiences a force perpendicular to both the field lines and its direction of movement.

If this electron was in a conductor, it would move in the direction of the force. Work is done on electrons in conductors; electrical potential energy is transferred to them. If we have a complete circuit, current will flow.

Moving the conductor changes the area of the magnetic field that is enclosed by the circuit, which leads us to Faraday's law: the induced emf is proportional to the rate of change of flux enclosed by the circuit. It may seem strange that the constant of proportionality is 1. However, we have used a common trick of defining the units so that this is the case (as we previously saw in $F = ma$).

Students should understand:

magnetic flux Φ as given by $\Phi = BA\cos\theta$
a time-changing magnetic flux induces an emf ε as given by Faraday's law of induction $\varepsilon = -N\frac{\Delta\Phi}{\Delta t}$
a uniform magnetic field induces an emf in a straight conductor moving perpendicularly to it as given by $\varepsilon = BvL$
the direction of induced emf is determined by Lenz's law and is a consequence of energy conservation
a uniform magnetic field induces a sinusoidal varying emf in a coil rotating within it
the effect on induced emf caused by changing the frequency of rotation.

Nature of Science

Having discovered that a changing current in one coil causes a current in a second unconnected coil, Faraday could have designed a simple electrical generator without any understanding of how it worked. By performing experiments and changing different variables, Faraday went far beyond his initial discovery to formulate his law of electromagnetic induction.

Conductor moving in a magnetic field

We have considered what happens to free charges moving in a magnetic field, but what happens if these charges are contained in a conductor? Figure 1 shows a conductor of length L moving with velocity v through a perpendicular field of flux density B. We know from our microscopic model of conduction that conductors contain free electrons. As the free electron shown moves downward through the field, it will experience a force. Using Fleming's left-hand rule, we can deduce that the direction of the force is to the left. (Remember, the electron is negative so if it is moving downward the current is upward.)

This force will cause the free electrons to move to the left as shown in Figure 2. We can see that the electrons moving left have caused the lattice atoms on the right to become positive, and there is now a potential difference between the ends of the conductor. The electrons will stop moving when the magnetic force pushing them left is balanced by the electric force pulling them right.

Induced emf

To separate the charges as shown in Figure 2 required work to be done, which means that energy has been transferred to the charges: electrical potential energy. What has happened is rather like what happens in a battery; a potential difference has been created. The amount of work done per unit charge in moving the charges to the ends of the conductor is called the **induced emf**.

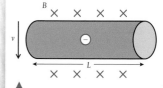

D.4 Figure 1 A conductor moving through a perpendicular field.

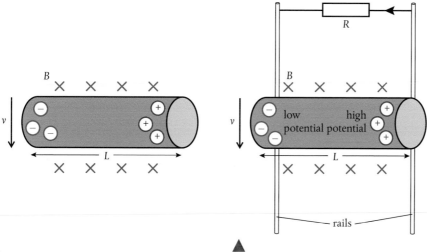

▲
D.4 Figure 2 Charges gather at either end, creating a potential difference.

▲
D.4 Figure 3 Current flows through the stationary circuit.

Induced current

If the conductor is connected to a circuit, then a current will flow from high potential to low potential. Connecting to a circuit is not as easy as it sounds since the circuit must not move through the field with the wire. If it did, then an emf would be induced in the circuit too and no current would flow. Sliding the conductor along static rails as shown in Figure 3 is one way of solving the problem.

As can be seen in Figure 3, the current flows anticlockwise around the circuit so it passes through the moving conductor from left to right. A current-carrying conductor experiences a force when placed in a perpendicular field so the moving conductor will now experience a force.

Using Fleming's left-hand rule, we can determine that the direction of this force is upward. If the conductor is to travel at constant velocity, the forces applied to it must be balanced so there must be an equal force acting downward (Figure 4). This means that to keep the conductor moving with constant velocity, work must be done by the person pushing the conductor. This implies that energy is transferred to the electrical potential energy of the electrons and the electrons in turn transfer energy to the lattice atoms of the resistor, resulting in a rise in temperature.

D.4 Figure 4 If velocity is constant forces must be balanced.

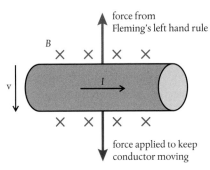

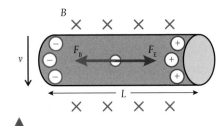

D.4 Figure 5 Magnetic and electric forces act on moving charges.

Calculating induced emf

The maximum potential difference achieved across the conductor is when the magnetic force pushing the electrons left equals the electric force pushing them right. When the forces are balanced, no more electrons will move. Figure 5 shows an electron with balanced forces.

If F_B is the magnetic force and F_E is the electric force, we can say that:

$$F_B = F_E$$

Now we know that if the velocity of the electron is v and the field strength is B, then:

$$F_B = Bev$$

The electric force is due to the electric field E. In this case, the field is uniform so:

potential gradient $= \dfrac{V}{L}$

So:
$$F_E = Ee = \frac{Ve}{L}$$

Equating the forces gives: $\dfrac{Ve}{L} = Bev$

So:
$$V = BLv$$

This is the potential difference across the conductor, which is defined as the work done per unit charge taking a small positive test charge from one side to the other. As current starts to flow in an external circuit, the work done by the pulling force enables charges to move from one end to the other, so the emf (mechanical energy transferred to electrical per unit charge) is the same as this potential difference:

$$\text{induced emf} = BLv$$

In Fleming's right hand rule, the fingers represent the same things as in the left hand rule but it is used to find the direction of induced current if you know the motion of the wire and the field. Try using it on the example in Figure 4.

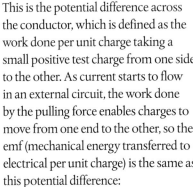

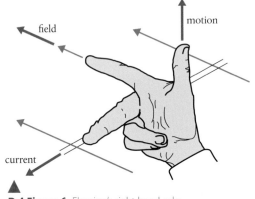

D.4 Figure 6 Fleming's right hand rule.

Q1. A 20 cm long straight wire is traveling at a constant $20\,\mathrm{m\,s^{-1}}$ through a perpendicular magnetic field of flux density $50\,\mu\mathrm{T}$.

   (a) Calculate the emf induced.

   (b) If this wire were connected to a resistance of $2\,\Omega$, how much current would flow?

   (c) How much energy would be transferred to heat in the resistor in 1 s? (power = I^2R)

   (d) How much work would be done by the pulling force in 1 s?

   (e) How far would the wire move in 1 s?

   (f) What force would be applied to the wire?

Faraday's law

From the moving conductor example, we see that the induced emf is dependent on the flux density, speed of movement and length of conductor. These three factors all change the rate at which the conductor cuts through the field lines, so a more convenient way of expressing this is:

The magnitude of the induced emf is equal to the rate of change of flux.

So if in a time Δt the conductor sweeps over an area enclosing flux $\Delta\phi$, then $|\varepsilon| = \dfrac{\Delta\phi}{\Delta t}$.

Rate of change of flux

If flux density is related to the number of field lines per unit area, then flux is related to the number of field lines in a given area. So if, as in Figure 7, area A is perpendicular to a field of flux density B, then the flux enclosed by the area will be BA.

A straight conductor moving perpendicular to a magnetic field will sweep through the field as shown in Figure 8. We can see in this diagram that the circuit encloses an amount of flux BA. As the wire moves, the amount of flux enclosed increases so, according to Faraday's law, emf will be induced. If the conductor moves at velocity v, then it will move a distance v per unit time, so the increase in area per unit time is Lv. Rate of change of flux is therefore BLv.

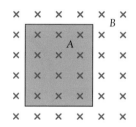

D.4 Figure 7 An area perpendicular to a magnetic field.

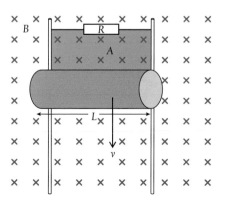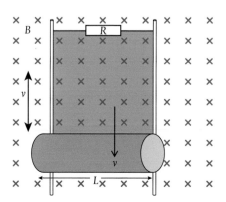

◀ **D.4 Figure 8** The flux enclosed by the circuit increases as the conductor moves forward a distance v in 1 second.

Lenz's law

We noticed that when a current is induced in a moving conductor, the direction of induced current causes the conductor to experience a force that opposes its motion. To keep the conductor moving will therefore require a force to be exerted in the opposite direction. This is a direct consequence of the law of conservation of energy. If it were not true, you would not have to do work to move the conductor, so the energy given to the circuit would come from nowhere. Lenz's law states this fact in a way that is applicable to all examples:

The direction of the induced current is such that it will oppose the change producing it.

So Faraday's law equation can be modified by adding a − sign:

$$\varepsilon = -\frac{\Delta\phi}{\Delta t}$$

Coils in changing magnetic fields

A magnet is moved toward a coil as shown in Figure 9.

Applying Faraday's law

As the magnet approaches the coil, the magnetic field inside the coil increases, and the changing flux enclosed by the coil induces an emf in the coil that causes a current to flow. The size of the emf will be equal to the rate of change of flux enclosed by the coil.

Applying Lenz's law

The direction of induced current will be such that it opposes the change producing it, which in this case is the magnet moving toward the coil. So to oppose this, the current in the coil must induce a magnetic field that pushes the magnet away. This direction is shown in Figure 9.

If the same magnet was pushed into a coil of N turns, then each turn of the coil would enclose the same flux: the flux enclosed would then be BAN. If the flux enclosed changed, each turn would have an equal emf induced in it so the total emf would be given by:

$$\varepsilon = -N\frac{\Delta\phi}{\Delta t}$$

Coil in a changing field

In Figure 10, the magnetic flux enclosed by coil Y is changed by switching the current in coil X on and off.

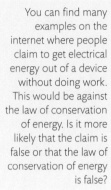

D.4 Figure 9 To oppose the magnet coming into the coil, the coil's magnetic field must push it out. The direction of the current is found using the grip rule.

D.4 Figure 10 Coil X induces a current in coil Y. Use the grip rule to work out the direction of the fields.

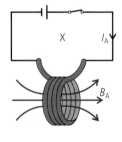
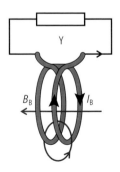

Applying Faraday's law

When the current in X flows, a magnetic field is created that causes the magnetic flux enclosed by Y to increase. This increasing flux induces a current in coil Y.

Applying Lenz's law

The direction of the current in Y must oppose the change producing it, which in this case is the increasing field from X. So to oppose this, the field induced in Y must be in the opposite direction to the field from X, as in the diagram. This is the principle behind the operation of a transformer.

How is the efficiency of electricity generation dependent on the source of energy? (B.5)

> ### Challenge yourself
>
> 1. When a vacuum cleaner is plugged into the mains and switched on, the safety switch cuts off the current to the socket. Why does this happen, and what could be done to stop it happening?

Coil entering and leaving a magnetic field

When a rectangular coil is at rest or moving at a constant speed in a magnetic field, there is no change in magnetic flux and therefore no induced emf. However, an emf is induced when the coil enters and leaves the field.

Since emf is equal to the rate of change of flux:

$$emf = \frac{NB\Delta A}{\Delta t}$$

where N is the number of turns on the coil and A is its cross-sectional area.

For a coil moving at speed v with width w into a perpendicular field:

$$emf = NBwv$$

because the speed is the rate at which the length of area becomes enclosed.

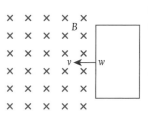

D.4 Figure 11 An emf is induced in a rectangular coil as it enters a magnetic field.

Exercise

Q2. A coil with 50 turns and an area of $2\,cm^2$ encloses a field of flux density $100\,\mu T$ (the field is perpendicular to the plane of the coil).

(a) What is the total flux enclosed?

(b) If the flux density changes to $50\,\mu T$ in $2\,s$, what is the rate of change of flux?

(c) What is the induced emf?

Q3. A rectangular coil, with sides $3\,cm$ and $2\,cm$ and 50 turns, lies flat on a table in a region of magnetic field. The magnetic field is vertical and has flux density $500\,\mu T$.

(a) What is the total flux enclosed by the coil?

(b) If one side of the coil is lifted so that the plane of the coil makes an angle of 30° to the table, what will the new flux enclosed be?

(c) If the coil is lifted in $3\,s$, estimate the emf induced in the coil.

Coil rotating in a uniform magnetic field

D.4 Figure 11 A simple AC generator.

When a generator is connected to a load, the direction of current in the coils opposes the change producing it. To slow down an electric car, the rotating coil of the motor is used as a generator to charge the battery. This provides effective braking and saves energy.

D.4 Figure 12 Looking at the generator from above.

Nature of Science

The invention of the generator made it possible to deliver energy to homes and factories using a network of cables. Generators can be made to deliver either a current that always travels in one direction (DC) or one that changes direction (AC). In the early days of electrification, there was competition between the producers of the different generators who each wanted their system to be adopted. One advantage of AC is that it is relatively simple to step the voltage up and down using a transformer. Supporters of DC saw this as a danger and demonstrated it publicly by electrocuting an elephant called Topsy.

Consider the coil shown in Figure 11. This coil is being made to rotate in a uniform magnetic field by someone turning the handle. The coil is connected to a resistor, but to prevent the wires connected to the coil twisting, they are connected via two slip rings. Resting on each slip ring is a carbon brush, which makes contact with the ring while allowing it to slip past.

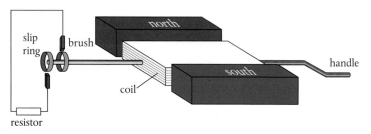

To make the operation easier to understand, a simpler 2D version with only one loop of wire in the coil is shown in Figure 12. As the handle is turned, the wire on the in-between side (AB) moves up through the field. As it cuts the field, a current will be induced. Using Fleming's right-hand rule, we can deduce that the direction of the current is from A to B as shown. The direction of motion of the right-hand side (CD) is opposite so the current is opposite. The result is a clockwise current through the resistor.

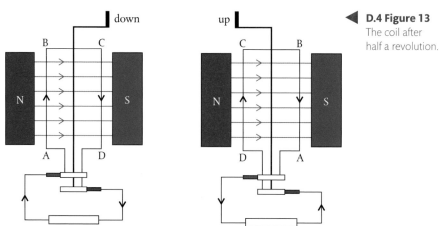

◀ **D.4 Figure 13** The coil after half a revolution.

After turning half a revolution, the coil is in the position shown in Figure 13. Side CD is now moving up through the field. Look carefully at how the slip ring has moved and you will see that, although the current is still clockwise in the coil, it is anticlockwise in the resistor circuit.

The size of the emf induced in a rotating coil

To find the size of the emf, we can use Faraday's law. This states that the induced emf will be equal to the rate of change of flux. The flux enclosed by the coil is related to the angle the coil makes with the field. Figure 14 shows a coil of N turns at time t. At this moment, the normal to the plane of the coil makes an angle θ with the field so flux, $\phi = BA \cos \theta$.

There are N turns so: total flux, $N\phi = BAN \cos \theta$

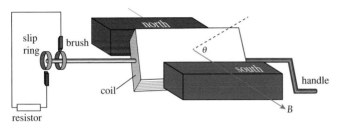

The electrical energy produced by a generator comes from the work done by the person turning the coil. The more current you take from the coil, the harder it is to turn. This follows from Lenz's law: the current in the coil opposes the change producing it. If you do not draw any current, then it is very easy to turn the coil.

D.4 Figure 14 The angle θ is between the rectangular plane of the coil and the magnetic field.

This equation can be represented graphically as in Figure 15.

Note that the starting point is when the angle θ is zero. This is when the coil is in the vertical position.

D.4 Figure 15 Graph of flux vs time showing position of coil.

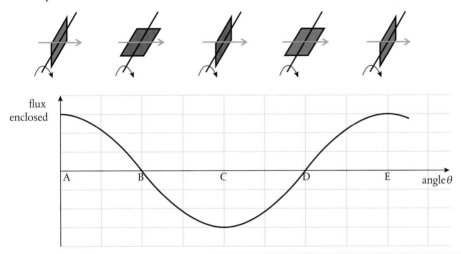

To find the magnitude of the emf, we need the negative value of the rate of change of flux ($\varepsilon = -\frac{\Delta\phi}{\Delta t}$). We can calculate this from the gradient of the graph in Figure 16. Let us consider some specific points.

A   Maximum flux enclosed but the rate of change is zero (zero gradient).

B   No flux enclosed but rate of change of flux is maximum (gradient negative so emf positive).

C   Maximum flux enclosed – rate of change zero.

D   No flux enclosed – rate of change maximum (positive gradient).

E   Back to the start.

D.4 Figure 16 Graph of emf vs time.

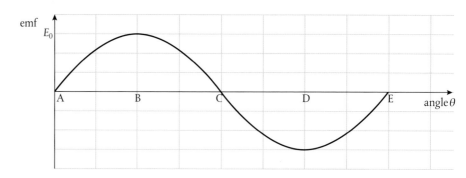

If you have studied differentiation in math, you will understand that Faraday's law can be written:

$$\varepsilon = -\frac{dN\phi}{dt}$$

Then if: $N\phi = BAN\cos\omega t$

$$-\frac{dN\phi}{dt} = BAN\omega\sin\omega t$$

The equation of this line is $\varepsilon = \varepsilon_0 \sin \theta$, where ε_0 is the maximum emf. If the coil took a time t to turn through angle θ, then the angular velocity of the coil, $\omega = \frac{\theta}{t}$.

Rearranging gives: $\theta = \omega t$

So the induced emf in terms of time is given by the equation:

$$\varepsilon = \varepsilon_0 \sin \omega t$$

Effect of increasing angular velocity

D.4 Figure 17 The black line is for the coil rotating at twice the angular speed of the red one.

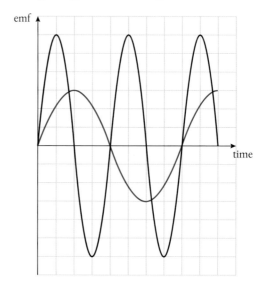

If the speed of rotation is increased, the graph of emf against time will change in two ways, as shown in Figure 17. The time between the peaks will be shorter and the peaks will be higher. This is because if the coil moves faster, then the rate of change of flux will be higher and hence the emf will be greater.

Self-induction

The flow of current in a generator coil is the desired outcome. It is intended that a power station will provide a current to households and industry. However, it is interesting to note that the current itself reduces the effectiveness of the generator because of a process called **self-induction**.

When a changing current flows in the coil, it has an associated changing magnetic field because of electromagnetism. We know from Faraday's law that a coil in a changing magnetic field will have an emf induced. We also know from Lenz's law that the direction of this emf will oppose the change producing it. Sometimes, this self-induced emf is referred to as 'back emf', and the effect is increased with increased current.

With the exception of some small communities, all countries use alternating current for domestic electrical power, but the voltage and frequency is not standardized internationally. Voltages range from 100 V (Japan) to 240 V (Malaysia) and the frequency is either 60 Hz (USA) or 50 Hz (Europe).

Exercise

Q4. A coil similar to the one in Figure 14 has an area of 5 cm² and rotates 50 times a second in a field of flux density 50 mT.

  (a) If the coil has 500 turns, calculate:

  (i) the angular velocity, ω

  (ii) the maximum induced emf

  (b) If the speed is reduced to 25 revolutions per second, what is the new maximum emf?

Q5. Calculate the resistance of a 1000 W light bulb designed to operate at 220 V.

Guiding Questions revisited

What are the effects of relative motion between a conductor and a magnetic field?

How can the power output of electrical generators be increased?

In this chapter, we have considered the physics that lies behind electrical generators, with ideas that include:

- Electromagnetic induction as the process by which an emf is induced through the cutting of magnetic field lines by a conducting wire or coil (which is the opposite of the motor effect).
- Magnetic flux, which is defined as the product of the flux density and the perpendicular cross-sectional area of a loop or coil placed in the field.
- Faraday's law of induction, which states that the emf is proportional to the rate of change of flux.
- Lenz's law, which is a consequence of the conservation of energy and states that the induced current will be in a direction that opposes the change producing it.
- The factors affecting the power output of an AC electrical generator being the angular velocity of the coil, the magnetic field strength, the cross-sectional area of the coil and the number of turns on the coil.

Practice questions

1. **(a)** Define *gravitational field strength*.   (2)

(b) Use the definition of gravitational field strength to deduce that:

$$GM = g_0 R^2$$

where M is the mass of the Earth, R its radius and g_0 is the gravitational field strength at the surface of the Earth. (You may assume that the Earth is a uniform sphere with its mass concentrated at its center.)   (2)

A space shuttle orbits the Earth and a small satellite is launched from the shuttle. The satellite carries a conducting cable connecting the satellite to the shuttle. When the satellite is a distance L from the shuttle, the cable is held straight by motors on the satellite, as shown below.

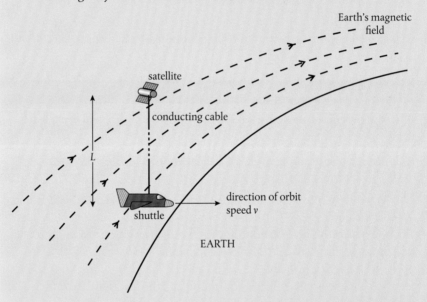

As the shuttle orbits the Earth with speed v, the conducting cable is moving at right angles to the Earth's magnetic field. The magnetic field vector B makes an angle θ to a line perpendicular to the conducting cable as shown below. The velocity vector of the shuttle is directed out of the plane of the paper.

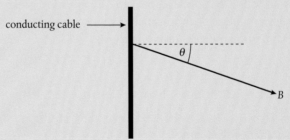

(c) Copy the diagram and draw an arrow to show the direction of the magnetic force on an electron in the conducting cable. Label the arrow *F*. (1)

(d) State an expression for the force *F* on the electron in terms of *B*, *v*, *e* and θ, where *B* is the magnitude of the magnetic field strength and *e* is the electron charge. (1)

(e) Hence, deduce an expression for the emf *E* induced in the conducting wire. (3)

(f) The shuttle is in an orbit that is 300 km above the surface of the Earth. Using the expression:

$$GM = g_0 R^2$$

and given that $R = 6.4 \times 10^6$ m and $g_0 = 10$ N kg^{-1}, deduce that the orbital speed *v* of the satellite is 7.8×10^3 m s^{-1}. (3)

(g) The magnitude of the magnetic field strength is 6.3×10^{-6} T and the angle $\theta = 20°$. Estimate the length *L* of the cable required in order to generate an emf of 1 kV. (2)

(Total 14 marks)

2. A small coil is placed with its plane parallel to a long straight current-carrying wire, as shown below.

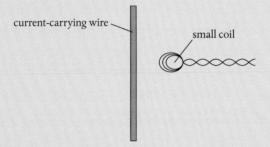

current-carrying wire

small coil

(a) (i) State Faraday's law of electromagnetic induction. (2)

(ii) Use the law to explain why, when the current in the wire changes, an emf is induced in the coil. (1)

445

The current–time graph below shows the variation with time t of the current in the wire.

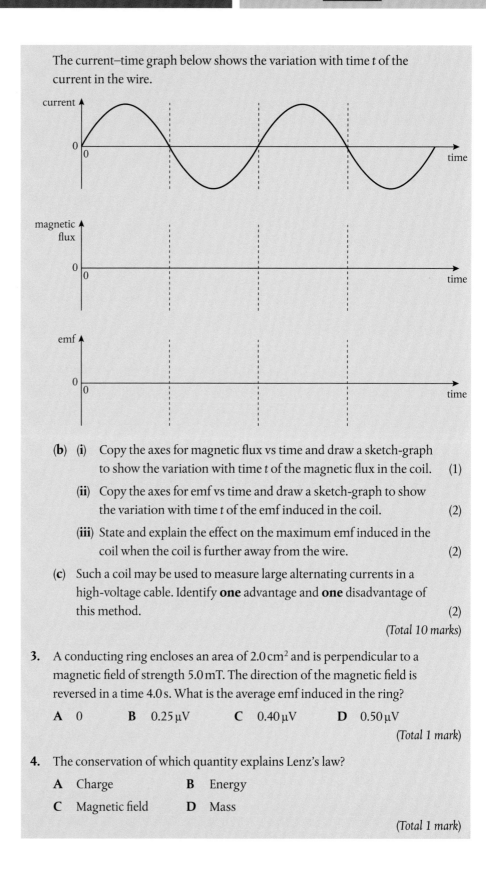

(b) **(i)** Copy the axes for magnetic flux vs time and draw a sketch-graph to show the variation with time t of the magnetic flux in the coil. (1)

(ii) Copy the axes for emf vs time and draw a sketch-graph to show the variation with time t of the emf induced in the coil. (2)

(iii) State and explain the effect on the maximum emf induced in the coil when the coil is further away from the wire. (2)

(c) Such a coil may be used to measure large alternating currents in a high-voltage cable. Identify **one** advantage and **one** disadvantage of this method. (2)

(Total 10 marks)

3. A conducting ring encloses an area of $2.0\,cm^2$ and is perpendicular to a magnetic field of strength $5.0\,mT$. The direction of the magnetic field is reversed in a time $4.0\,s$. What is the average emf induced in the ring?

A 0   **B** $0.25\,\mu V$ **C** $0.40\,\mu V$ **D** $0.50\,\mu V$

(Total 1 mark)

4. The conservation of which quantity explains Lenz's law?

A Charge   **B** Energy

C Magnetic field   **D** Mass

(Total 1 mark)

5. A rectangular coil rotates at a constant angular velocity. At the instant shown, the plane of the coil is at right angles to the line ZZ'. A uniform magnetic field acts in the direction ZZ'. What rotation of the coil about a specified axis will produce the graph of electromotive force (emf) E against time t?

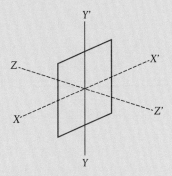 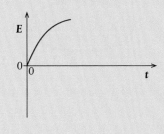

A   Through $\frac{\pi}{2}$ about ZZ'

B   Through π about YY'

C   Through $\frac{\pi}{2}$ about XX'

D   Through π about XX'

(Total 1 mark)

6. The diagram shows a conducting rod of length L being moved in a region of uniform magnetic field B. The field is directed at right angles to the plane of the paper. The rod slides on conducting rails at a constant speed v. A resistor of resistance R connects the rails. What is the power required to move the rod?

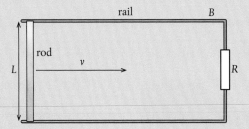

A   Zero

B $\dfrac{vBL}{R}$

C $\dfrac{v^2B^2L^2}{R}$

D $\dfrac{v^2B^2L^2}{R^2}$

(Total 1 mark)

7. A small magnet is dropped from rest above a stationary horizontal conducting ring. The south (S) pole of the magnet is upward.

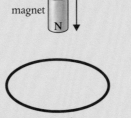
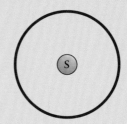

Diagram 1: side view   Diagram 2: view from above

(a) While the magnet is moving toward the ring, state why the magnetic flux in the ring is increasing. (1)

(b) While the magnet is moving toward the ring, copy Diagram 2 and sketch, using an arrow, the direction of the induced current in the ring. (1)

(c) While the magnet is moving toward the ring, deduce the direction of the magnetic force on the magnet. (2)

(Total 4 marks)

8. In an alternating current (AC) generator, a square coil ABCD rotates in a magnetic field. The ends of the coil are connected to slip rings and brushes. The plane of the coil is shown at the instant when it is parallel to the magnetic field. Only one coil is shown for clarity.

The following data are available:

Dimensions of the coil = 8.5 cm × 8.5 cm

Number of turns on the coil = 80

Speed of edge AB = 2.0 m s^{-1}

Uniform magnetic field strength = 0.34 T

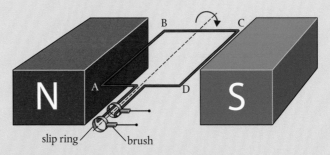

slip ring   brush

(a) Explain, with reference to the diagram, how the rotation of the generator produces an electromotive force (emf) between the brushes. (3)

(b) Calculate, for the position in the diagram, the magnitude of the instantaneous emf generated by a single wire between A and B of the coil. (1)

(c) Hence, calculate the total instantaneous peak emf between the brushes. (1)

(Total 5 marks)

9. **(a)** The picture shows a coil of wire around an iron core. Which statement is true? (1)

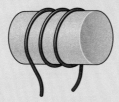

    A   If current is made to flow in the wire, the iron becomes a magnet.

    B   If the iron is a magnet, current is made to flow in the wire.

    C   Both statements A and B are true.

    D   Both statements A and B are false.

(b) Explain your answer. (1)

(Total 2 marks)

10. **(a)** An electric motor and a generator both consist of coils of wire on a rotor that can spin in a magnetic field. The difference is whether electrical energy is the input and mechanical energy the output (a motor) or vice versa (a generator). Current is induced when the rotor is made to spin. When a motor is running, is it also a generator? (1)

    A   Yes, it will send an electrical energy output through the input lines and back to the source.

    B   It would if it was not designed to prevent this problem.

    C   No, the device is either a motor or a generator.

(b) Explain your answer. (1)

(Total 2 marks)

HL end

449

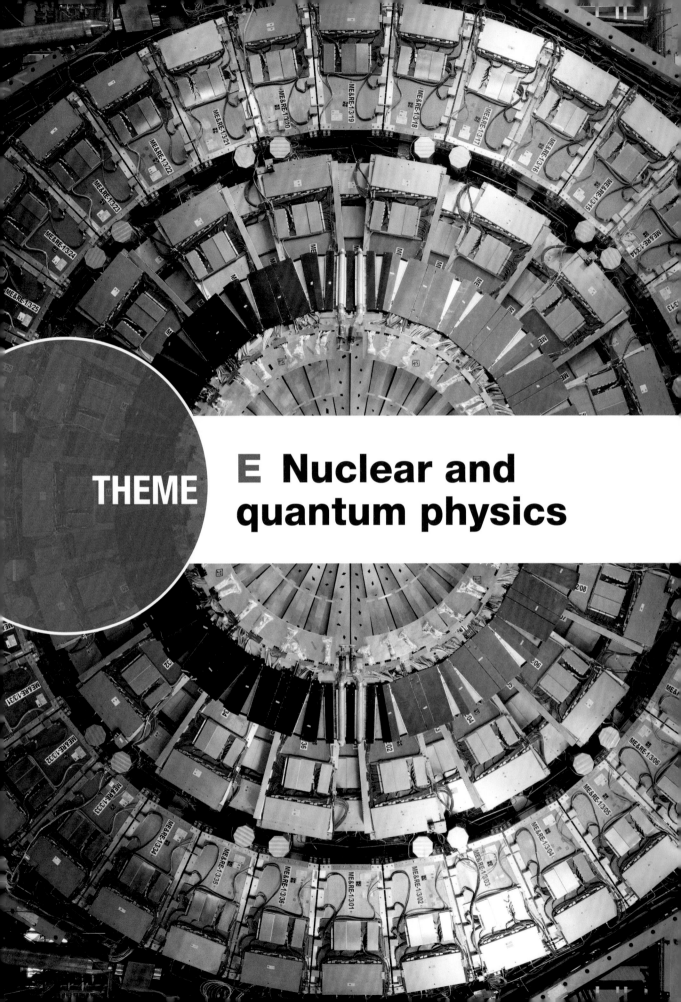

THEME

E Nuclear and quantum physics

E Nuclear and quantum physics

◀ The Standard Model of elementary particles consists of quarks (which make up baryons, such as protons and neutrons, and mesons), leptons (like electrons and neutrinos) and bosons (which transmit forces). The Higgs boson was predicted to exist in the 1960s precisely because of the Standard Model and eventually discovered through ATLAS and CMS collaboration in 2012 at CERN.

Chemical reactions involve electrons. Nuclear reactions involve nucleons. All are found in atoms, the building blocks of matter. The structure of atoms has been investigated for centuries, and the contemporary nuclear model is based on experiments in which only a small proportion of charged particles were found to deflect when fired toward a thin sheet of gold. Experiments also provide the evidence for the energy levels in which electrons reside; the light that is absorbed and released by atoms due to electron energy level transfers has specific wavelengths unique to the element.

Quantum physics is the study of how the smallest particles in the Universe interact. You will find out that all is not as it seems with light; just as there is ample evidence for the wave model of light, there is also experimental evidence that light behaves as discrete particles (or 'quanta'). Unsurprisingly, electrons (and other well-established particles) are just the same! They exhibit wave-like behaviors when manipulated in certain scenarios. Overall, we call these properties wave–particle duality.

Looking at samples of thousands of atoms, the number of physical phenomena only continues to grow – because individual nuclei can undergo decay, fission or fusion in order to become more stable. Radioactive decay is a random and spontaneous process with the format (α, β^-, β^+ or γ) depending on the ratio of neutrons to protons present and in which a constant half-life can be determined for each given isotope. Fission is the route to stability for the heaviest nuclei and fusion the route for the lightest nuclei; binding energy per nucleon increases to a maximum value for iron-56. Chain fission reactions are harnessed using fuel rods, moderators and heat exchangers in nuclear power stations (with control rods and shielding for safety). Fusion is the source of energy in stars, where the density and temperature are high enough for positively-charged nuclei to come together under the strong nuclear force.

Stars can be classified according to their temperature, luminosity, radius and lifecycle stage with the Hertzsprung–Russell diagram being the ultimate for displaying them. You will come to be amazed by the stellar detective work that can be carried out based on just one or two pieces of information.

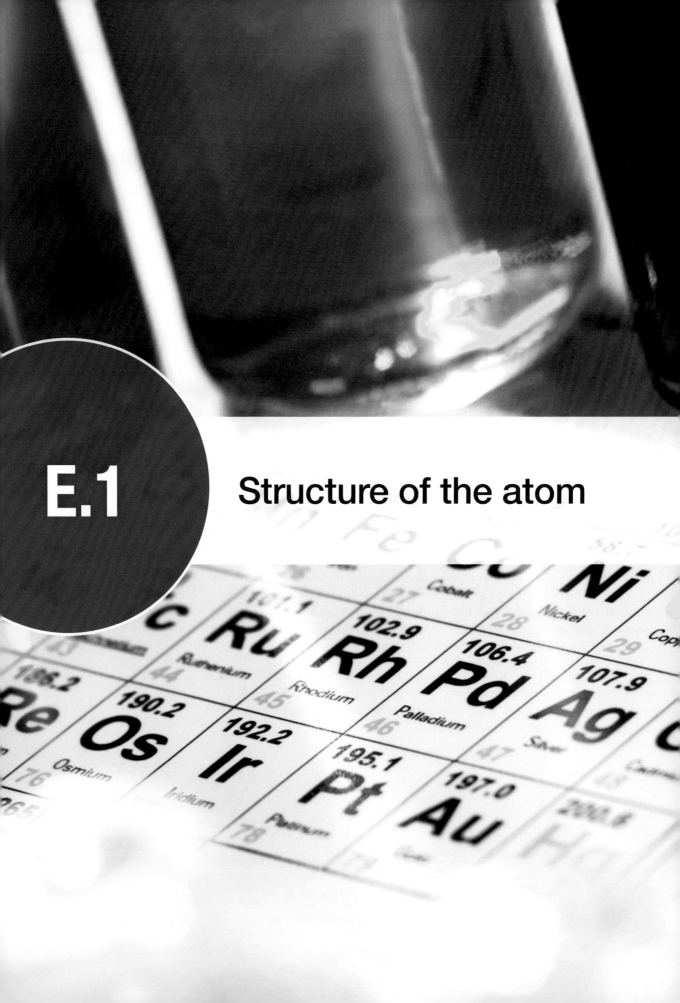

E.1

Structure of the atom

The periodic table is an arrangement of types of atom – the chemical elements. Chemists make use of the reactivity patterns across the periods and down the groups, while physicists pay more attention to the composition of atomic nuclei and the energies of the electrons that surround them.

Guiding Questions

What is the current understanding of the nature of an atom?

What is the role of evidence in the development of the model of the atom?

In what ways are previous models of the atom still valid despite recent advances in understanding?

The development of atomic theory is often used as an example of how the scientific method works, but the process as presented in this chapter is a very selective view of events. This was not the only research being carried out in atomic physics.

Trying to model the atom is like trying to work out what is in a present without opening the packaging.

Imagine you received the object shown in Figure 1. If it was placed in a large box and you rolled it around inside the box, you might think it was a solid cube. However, if you fired pellets at the box, they would pass straight through, except for the occasional pellet that would hit the larger central ball. This would imply that the object was mostly space with a heavy center. The next step might be to blow up the box and attempt to catch the pieces. This would reveal finer details, but only if you were standing in the path of a piece as it moved outward.

E.1 Figure 1 Each of the balls is at the corner of a 'cube'.

Students should understand:

the Geiger–Marsden–Rutherford experiment and the discovery of the nucleus
nuclear notation ^A_ZX where A is the nucleon number, Z is the proton number and X is the chemical symbol
emission and absorption spectra provide evidence for discrete atomic energy levels
photons are emitted and absorbed during atomic transitions
the frequency of the photon released during an atomic translation depends on the difference in energy level as given by $E = hf$
emission and absorption spectra provide information on the chemical composition
HL the relationship between the radius and the nucleon number for a nucleus as given by $R = R_0 A^{\frac{1}{3}}$ and implications for nuclear densities
HL deviations from Rutherford scattering at high energies
HL the distance of closest approach in head-on scattering experiments
HL the discrete energy levels in the Bohr model for hydrogen as given by $E = -\frac{13.6}{n^2} \text{eV}$
HL the existence of quantized energy and orbits arise from the quantization of angular momentum in the Bohr model for hydrogen as given by $mvr = \frac{nh}{2\pi}$.

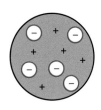

▲
E.1 Figure 2 Thomson's model: a positive pudding with negative 'plums'.

The electron is a fundamental particle with a charge of -1.60×10^{-19} C and a mass of 9.110×10^{-31} kg.

E.1 Figure 3 To see small details, we need to use a small projectile.

The arrangement of charge in the atom

We already know that matter is made up of particles (atoms) and we used this model to explain the thermal properties of matter. We also used the idea that matter contains charges to explain electrical properties. Since matter contains charge and is made of atoms, it seems logical that atoms must contain charge. But how is this charge arranged?

There are many possible ways that charges could be arranged in the atom, but since atoms are not themselves charged, they must contain equal amounts of positive and negative charge. Maybe half the atom is positive and half is negative, or perhaps the atom is made of two smaller particles of opposite charge?

The discovery of the electron by J. J. Thomson in 1897 added a clue that helps to solve the puzzle. The electron is a small negative particle that is responsible for carrying charge when current flows in a conductor. By measuring the charge-to-mass ratio of the electron, Thomson realized that electrons were very small compared to the whole atom. He therefore proposed a possible arrangement for the charges as shown in Figure 2. This model was called the 'plum pudding' model. This model was accepted for some time until, under the direction of Ernest Rutherford, Geiger and Marsden performed an experiment that proved it could not be correct.

Scattering experiments

The problem with trying to find out what is inside an atom is that the atom is far too small to see.

Imagine you have four identical boxes and each contains one of the following: a large steel ball, a glass ball, air or sand. You have to find out what is inside the boxes without opening them. One way of doing this is to fire a pellet at each. Here are the results:

1. Shattering sound → glass ball

2. Bounces back → steel ball

3. Passes straight through → air

4. Does not pass through → sand

Different situations need different projectiles. If, for example, one box contained a large cube, then a projectile smaller than the cube would be fine. If the big cube was made out of smaller cubes, you would need a projectile so small that it could pass between the cubes or one with so much energy that it would knock some of the small cubes out of the box.

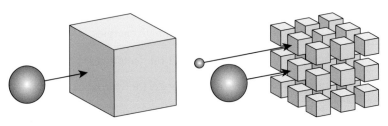

The Rutherford model

Rutherford's idea was to shoot alpha particles at a very thin sheet of gold to see what would happen. In 1909, very little was known about alpha particles – only that they were fast and positive. In accordance with normal scientific practice, Rutherford applied the model of the day to predict the result of the experiment. The current model was that the atom was like a small plum pudding, so a sheet of gold foil would be like a wall of plum puddings, a few puddings thick. Shooting alpha particles at the gold foil would be like firing pellets at a wall of plum puddings. If we think what would happen to the pellets and puddings, it will help us to predict what will happen to the alpha particles.

If you shoot a pellet at a plum pudding, it will pass through and out the other side. If you were to shoot a positive alpha particle at 'plum pudding' atoms, within which negative charges are evenly distributed, the alpha particles, overall, should be undeflected. What actually happened was, most alpha particles passed through without changing direction, but a significant number were deflected and a few even came straight back, as shown in Figure 4. This was so unexpected that Rutherford said: 'It was quite the most incredible event that ever happened to me in my life. It was almost as incredible as if you had fired a 15-inch shell at a piece of tissue paper and it came back and hit you.' We know from our study of collisions that you can only get a ball to bounce off another ball if the second ball is much heavier than the first. This means that there must be something heavy in the atom. The fact that most alpha particles pass through means that there must be a lot of space. If we put these two findings together, we conclude that the atom must have something heavy and small within it. This small, heavy thing is not an electron since they are too light. It must therefore be the positive part of the atom. We call this dense, positive center of the atom the **nucleus**. This would explain why the alpha particles come back, since they are also positive and are repelled from it.

TOK This is a good example of scientific discovery in practice.

ⓘ **HL** Very high-energy alpha particles get so close to the nucleus that they become attracted by the strong nuclear force. This gives a different scatter pattern to that predicted.

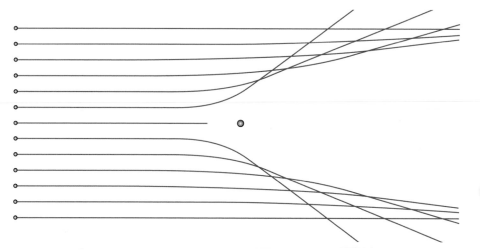

E.1 Figure 4 The paths of alpha particles deflected by a nucleus. At this scale, it looks like most are deflected, but if the nucleus was this size, the next atom would be about 100 m away.

🔗 How have observations led to developments in the model of the atom? (NOS)

Charge and mass

Nature of Science

Nuclear research was undoubtedly hurried along by the strategic race to be the first to build an atom bomb. Many other advancements in science have also been pushed forward due to military, political or financial concerns.

In the 1860s, chemists calculated the relative mass of many elements by measuring how they combine to form compounds. If placed in order of atomic mass, the chemical properties of the elements seemed to periodically repeat themselves. This led to the periodic table that chemistry students will be familiar with. There were, however, some anomalies where the order in terms of chemical properties did not match the order of mass. In 1911, Rutherford's scattering experiments not only revealed the existence of the nucleus but made it possible to calculate the charge of the nucleus. This was found to be the same whole number of positive electron charges for all atoms of the same element. This number is not the same as the mass number, but when the elements were placed in order of this 'charge number', then the anomalies were sorted. To summarize:

The unified mass unit (u) is the unit of atomic mass. 1 u is defined as the mass of $\frac{1}{12}$ of an atom of a carbon-12 atom.

- The mass of all atoms is (approximately) a multiple of the mass of a hydrogen atom, the 'mass number'.
- The charge of all nuclei is a multiple of the charge of a hydrogen nucleus, the 'charge number'.
- The 'charge number' is not the same as the 'mass number'.

Since all nuclei are multiples of hydrogen nuclei, one might imagine that all nuclei are made of hydrogen nuclei. If this was the case, then the charge number would be the same as the mass number, but it is not. For example, helium has a relative atomic mass of 4 but a charge of +2e. There appear to be two extra particles that have approximately the same mass as a hydrogen nucleus but no charge. This particle is the neutron, which was discovered by Chadwick in 1932.

So the nucleus contains two types of particle, shown in Table 1.

E.1 Table 1

	Mass/kg	Mass/u	Charge/C
Proton	1.673×10^{-27}	1.007276	$+1.60 \times 10^{-19}$
Neutron	1.675×10^{-27}	1.008665	0

In past IB papers, these quantities have different names.
A = atomic mass number
Z = atomic number
Note that unknown elements are given the generic chemical symbol X until the proton number can be compared with the periodic table.

Nucleons are the particles of the nucleus (protons and neutrons). A particular combination of nucleons is called a **nuclide**. Each nuclide is defined by three numbers:

- **nucleon number (A)** = number of protons + neutrons (defines the *mass* of the nucleus)
- **proton number (Z)** = number of protons (defines the *charge* of the nucleus)
- **neutron number (N)** = number of neutrons (A − Z).

Isotopes are nuclides with the same proton number but different nucleon numbers.

Periodic Table

Key:
- Atomic number
- **Element**
- Relative atomic mass

1	2	3	4	5	6	7	8	9	10	11	12	13	14	15	16	17	18
1 **H** 1.01																	2 **He** 4.00
3 **Li** 6.94	4 **Be** 9.01											5 **B** 10.81	6 **C** 12.01	7 **N** 14.01	8 **O** 16.00	9 **F** 19.00	10 **Ne** 20.18
11 **Na** 22.99	12 **Mg** 24.31											13 **Al** 26.98	14 **Si** 28.09	15 **P** 30.97	16 **S** 32.07	17 **Cl** 35.45	18 **Ar** 39.95
19 **K** 39.10	20 **Ca** 40.08	21 **Sc** 44.96	22 **Ti** 47.87	23 **V** 50.94	24 **Cr** 52.00	25 **Mn** 54.94	26 **Fe** 55.85	27 **Co** 58.93	28 **Ni** 58.69	29 **Cu** 63.55	30 **Zn** 65.38	31 **Ga** 69.72	32 **Ge** 72.63	33 **As** 74.92	34 **Se** 78.96	35 **Br** 79.90	36 **Kr** 83.80
37 **Rb** 85.47	38 **Sr** 87.62	39 **Y** 88.91	40 **Zr** 91.22	41 **Nb** 92.91	42 **Mo** 95.96	43 **Tc** (98)	44 **Ru** 101.07	45 **Rh** 102.91	46 **Pd** 106.42	47 **Ag** 107.87	48 **Cd** 112.41	49 **In** 114.82	50 **Sn** 118.71	51 **Sb** 121.76	52 **Te** 127.60	53 **I** 126.90	54 **Xe** 131.29
55 **Cs** 132.91	56 **Ba** 137.33	57 **La** † 138.91	72 **Hf** 178.49	73 **Ta** 180.95	74 **W** 183.84	75 **Re** 186.21	76 **Os** 190.23	77 **Ir** 192.22	78 **Pt** 195.08	79 **Au** 196.97	80 **Hg** 200.59	81 **Tl** 204.38	82 **Pb** 207.20	83 **Bi** 208.98	84 **Po** (209)	85 **At** (210)	86 **Rn** (222)
87 **Fr** (223)	88 **Ra** (226)	89 **Ac** ‡ (227)	104 **Rf** (267)	105 **Db** (268)	106 **Sg** (269)	107 **Bh** (270)	108 **Hs** (269)	109 **Mt** (278)	110 **Ds** (281)	111 **Rg** (281)	112 **Cn** (285)	113 **Nh** (286)	114 **Fl** (289)	115 **Mc** (288)	116 **Lv** (293)	117 **Ts** (294)	118 **Og** (294)

† Lanthanides:

58 **Ce** 140.12	59 **Pr** 140.91	60 **Nd** 144.24	61 **Pm** (145)	62 **Sm** 150.36	63 **Eu** 151.96	64 **Gd** 157.25	65 **Tb** 158.93	66 **Dy** 162.50	67 **Ho** 164.93	68 **Er** 167.26	69 **Tm** 168.93	70 **Yb** 173.05	71 **Lu** 174.97

‡ Actinides:

90 **Th** 232.04	91 **Pa** 231.04	92 **U** 238.03	93 **Np** (237)	94 **Pu** (244)	95 **Am** (243)	96 **Cm** (247)	97 **Bk** (247)	98 **Cf** (251)	99 **Es** (252)	100 **Fm** (257)	101 **Md** (258)	102 **No** (259)	103 **Lr** (262)

You might find it interesting to have this Periodic Table to hand as you work through the chapters in this theme.

Example

Lithium-7 has a nucleon number of 7 and a proton number of 3, so it has 3 protons and 4 neutrons. This nuclide can be represented by the symbol $^{7}_{3}\text{Li}$.

$^{6}_{3}\text{Li}$ and $^{8}_{3}\text{Li}$ are isotopes of lithium.

$^{6}_{3}\text{Li}$ $^{7}_{3}\text{Li}$

▲
E.1 Figure 5 Isotopes of lithium.

Exercise

Q1. How many protons and neutrons are there in the following nuclei?

(a) $^{35}_{17}\text{Cl}$

(b) $^{58}_{28}\text{Ni}$

(c) $^{204}_{82}\text{Pb}$

Q2. Calculate the charge in coulombs and mass in kilograms of a $^{54}_{26}\text{Fe}$ nucleus.

Q3. An isotope of uranium (U) has 92 protons and 143 neutrons. Write the nuclear symbol for this isotope.

Q4. Describe the structure of another isotope of uranium, having the symbol $^{238}_{92}\text{U}$.

HL

Size of the nucleus

The size of the nucleus can be determined by conducting a similar experiment to the alpha scattering experiment of Geiger and Marsden. The alpha particles that come straight back off the gold foil must have approached a nucleus head-on, following a path as shown in Figure 6.

E.1 Figure 6 An alpha particle approaches a nucleus head-on.

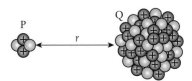

When Geiger and Marsden did their gold foil experiment, they did not have alpha particles with enough energy to get close enough to the gold nucleus to give a very accurate value for its radius. They did however get quite close to aluminum nuclei.

Applying the law of conservation of energy to this problem, we can deduce that at point P, where the alpha particle stops, the original kinetic energy has been transferred to electrical potential energy.

$$\frac{1}{2}mv^2 = k\frac{Qq}{r}$$

where: Q = the charge of the nucleus ($+Ze$)

and: q = the charge of the alpha ($+2e$)

The kinetic energy of the alpha particle can be calculated from the change in the mass of a nucleus when it is emitted (more about this later), so, knowing this, the distance r can be calculated.

To determine the size of the nucleus, faster and faster alpha particles are sent toward the nucleus until they no longer come back. The fastest ones that return have got as close to the nucleus as possible.

This is just an estimate of the nuclear radius, especially since (as is the case for all particles) the position of the particles that make up the nucleus is determined by a probability function. This will make the definition of the edge of the nucleus rather fuzzy.

From the results of experiments like this, we know that the radius of a nucleus is approximately 10^{-15} m. By measuring the radii of different nuclei, it has been found that $R = R_0 A^{\frac{1}{3}}$, in which R_0 is a constant known as the Fermi radius, 1.20×10^{-15} m.

This experimental result reveals that the density of all nuclei is the same. Cubing both sides of the equation shows that R^3 is proportional to A, which means that volume is proportional to mass, while density is constant.

This also tells us that the nuclear force is very short range. If it obeyed an inverse square law, like gravity, the nucleons on the surface would be attracted to all the nucleons inside, which would have the effect of squashing the inside, making heavier nuclei more dense (as happens in stars).

How is the distance of closest approach calculated using conservation of energy? (A.3)

Exercise

Q5. It is found that alpha particles with a kinetic energy of 7.7 MeV bounce back off an aluminum target. If the charge of an aluminum nucleus is 12.1×10^{-18} C, calculate:

(a) the velocity of the alpha particles (they have a mass of 6.7×10^{-27} kg)

(b) the distance of closest approach to the nucleus.

HL end

Electrons

So, the atom consists of a heavy but very small positive nucleus surrounded by negative electrons. But what stops the electrons falling into the nucleus? One idea could be that the atom is like a mini solar system with electrons orbiting the nucleus, similar to how the planets orbit the Sun. The circular motion of the electrons would make it possible for them to accelerate toward the center without getting any closer. One problem with this model is that if an electron were to move in this way, it would create a changing electric and magnetic field, resulting in emission of electromagnetic radiation. This would lead to a loss of energy and the electron would spiral into the nucleus. To gain more insight into the structure of the atom, we need to look in detail at the relationship between light and matter.

The connection between atoms and light

There is a very close connection between matter and light. For example, if we give thermal energy to a metal, it can give out light. Light is an electromagnetic wave so must come from a moving charge. Electrons have charge and are able to move, so it would be reasonable to think that the production of light is something to do with electrons. But what is the mechanism inside the atom that enables this to happen?

Before we can answer that question, we need to consider the nature of light, in particular, light that comes from isolated atoms. We must look at isolated atoms because we need to be sure that the light is coming from single atoms and not the interaction between atoms. A single atom would not produce enough light for us to see, but low-pressure gases have enough atoms far enough apart not to interact.

Atomic spectra

To analyze the light coming from an atom, we need to give the atom energy. This can be done by adding thermal energy or electrical energy. The most convenient method is to apply a high potential to a low-pressure gas contained in a glass tube (a discharge tube). This causes the gas to give out light, and already you will notice (see Figure 7) that different gases give different colors. To see exactly which wavelengths make up these colors, we can split up the light using a prism (or diffraction grating). To measure the wavelengths, we need to know the angle of refraction. This can be measured using a spectrometer.

E.1 Figure 7 Discharge tubes containing bromine, hydrogen and helium.

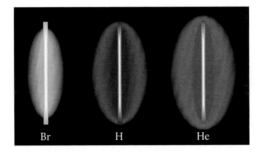

Br   H   He

The hydrogen spectrum

E.1 Figure 8 The line spectrum for hydrogen.

Hydrogen has only one electron, so it is the simplest atom and the one we will consider first. Figure 8 shows the spectrum obtained from a low-pressure discharge tube containing hydrogen. The first thing you notice is that, unlike a usual rainbow, which is continuous, the hydrogen spectrum is made up of thin lines. Light is a form of energy, so whatever the electrons do, they must lose energy when light is emitted. If the color of light is associated with different energies, then, since only certain energies of light are emitted, the electron must only be able to release certain amounts of energy. This would be the case if the electron could only have certain amounts of energy in the first place. We say the energy is **quantized**.

When a metal rod is heated, it first glows red, but as it gets hotter, it will glow white. It seems reasonable to assume that the color of light is related to energy, red being the lowest energy and blue/violet the highest energy.

To help us understand this, we can consider an analogous situation of buying sand. You can buy sand loose or in 50 kg bags, and we say the 50 kg bags are quantized, since the sand comes in certain discrete quantities. So if you buy loose sand, you can get any amount you want, but if you buy quantized sand, you have to have multiples of 50 kg. If we make charts showing all the possible quantities of sand you could buy, then they would be as shown on Figure 9; one has discrete values and the other is continuous.

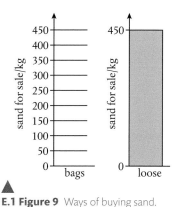

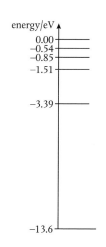

E.1 Figure 9 Ways of buying sand.

◀ **E.1 Figure 10** The electron energy levels of hydrogen.

If the electron in the hydrogen atom can only have discrete energies, then when it changes energy, these changes must also be in discrete amounts. We represent the possible energies on an energy level diagram (Figure 10), which looks rather like the sand diagram.

For this model to fit together, each of the lines in the spectrum must correspond to a different energy change. Light therefore must be quantized and this does not tie in with our classical view of light being a continuous wave that spreads out like ripples in a pond.

The quantum nature of light

Light definitely has wave-like properties; it reflects, refracts, diffracts and interferes. But sometimes light does things that we do not expect a wave to do, and one of these things is the photoelectric effect. The photoelectric effect is an interaction between particles of light (photons) and electrons, in which photons of sufficient energy can cause the emission of electrons from a metal surface after being absorbed by atoms. The energy of a photon is related to the frequency of the electromagnetic radiation. This is the mechanism found in solar panels. Higher Level students will explore the photoelectric effect in detail in E.2.

Quantum explanation of atomic spectra

We can now put our quantum models of the atom and light together to explain the formation of atomic spectra. To summarize what we know so far:

- Atomic electrons can only exist in certain discrete energy levels.
- Light is made up of photons.
- When electrons lose energy, they give out light.
- When light is absorbed by an atom, it gives energy to the electrons.

We can therefore deduce that when an electron changes from a higher energy level to a lower energy level, a photon of light is emitted. Since the electron can only exist in discrete energy levels, there are a limited number of possible changes that can take place. This gives rise to the characteristic line spectra that we have observed. Each element has a different set of lines in its spectrum because each element has different electron energy levels. To make this clear, we can consider a simple atom with electrons in the four energy levels shown in Figure 11.

How can emission spectra be used to calculate the distances to and velocities of celestial bodies? (C.5)

Remember 1 eV is the kinetic energy gained by an electron accelerated through a potential difference of 1 V.
1 eV = 1.60 × 10⁻¹⁹ J

- The average kinetic energy of: an atom in air at 20 °C is about 0.02 eV, a red light photon is 1.75 eV and a blue light photon is 3.1 eV.

- The energy released by one molecule in a chemical reaction is typically 50 eV.

- The energy released by one atom of fuel in a nuclear reaction is 200 MeV.

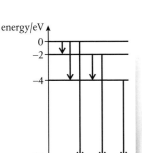

E.1 Figure 11 The four energy levels result in six magnitudes of energy transition.

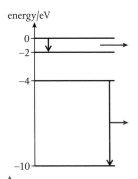

E.1 Figure 12 Blue light photons contain more energy than red light photons.

As you can see in the diagram, there are six possible ways that an electron can change from a higher energy level to a lower energy level. Each transition will give rise to a photon of different energy and hence a different line in the spectrum. To calculate the photon frequency, we use the formula:

$$\text{change in energy, } \Delta E = hf$$

So the bigger energy changes will give lines on the blue end of the spectrum and smaller energy changes will give lines on the red end of the spectrum, as shown in Figure 12.

Example

A change from the −4 eV to the −10 eV energy level will result in a change of 6 eV. This is $6 \times 1.60 \times 10^{-19} = 9.6 \times 10^{-19}$ J.

This will give rise to a photon of frequency given by:

$$\Delta E = hf$$

Rearranging gives: $f = \dfrac{\Delta E}{h} = \dfrac{9.6 \times 10^{-19}}{6.63 \times 10^{-34}} = 1.45 \times 10^{15}$ Hz

This is a wavelength of 207 nm, which is UV.

Ionization

Ionization occurs when the electrons are added to or removed from an atom, leaving a charged atom called an ion. Electrons can be removed if the atom absorbs a high-energy photon or the electron could be 'knocked off' by a fast-moving particle like an alpha particle. These interactions are quite different. When a photon interacts with an atom, it is absorbed. However, alpha particles are positively charged, which means they can attract electrons.

Removal of an electron from the ground state of a hydrogen atom requires 13.6 eV of energy (Figure 13).

Absorption of light

A photon of light can only be absorbed by an atom if it has exactly the right amount of energy to excite an electron from one energy level to another. If light containing all wavelengths (white light) is passed through a gas, then the photons with the right energy to excite electrons will be absorbed. The spectrum of the light emitted will have lines missing. This is called an **absorption spectrum** and is further evidence for the existence of electron energy levels.

The core of the Sun is very hot so the atoms move around with high speed. When the atoms collide with each other, they knock all of the electrons out of their energy levels. This material is called a plasma and because the atoms do not contain electrons, they cannot emit light of visible light frequencies. Instead, they emit gamma radiation, which is absorbed by the outer layer of the star, increasing the temperature.

The outer layer atoms still have electrons in place so give out light of all wavelengths. This light passes through the outermost low-density gas, where certain wavelengths are absorbed by discrete energy changes in the atomic electrons. This leaves dark absorption lines in the spectrum of light from any given star, enabling us to identify the chemical composition (which elements are present).

E.1 Figure 13 Energy levels for hydrogen.

A charge coupled device (CCD) is the light-sensitive part of a digital camera. It contains millions of tiny photodiodes that make up the pixels. When a photon of light is incident on a photodiode, it causes an electron to be released, resulting in a potential difference that is converted to a digital signal.

Use the energy level diagram in Figure 13 to answer the following questions.

Q6. How many possible energy transitions are there in this atom?

Q7. Calculate the maximum energy (in eV) that could be released and the frequency of the photon emitted.

Q8. Calculate the minimum energy that could be released and the frequency of the associated photon.

Q9. How much energy would be required to completely remove an electron that was in the lowest energy level? Calculate the frequency of the photon that would have enough energy to do this.

How can emission and absorption spectra allow for the properties of stars to be deduced? (E.5)

HL

The Bohr model

We can see from the spectrum of hydrogen that the energy of atomic electrons can only have discrete values, but we do not have a model for how the electrons could be arranged around the atom. In 1913, Niels Bohr proposed that if the electrons were in certain specific orbits then they would not emit electromagnetic radiation. The radii of these orbits were defined in terms of the angular momentum of the electron: this is the angular equivalent of linear momentum. If a body, mass m, is traveling in a circle radius r with constant speed v, it will have angular momentum mvr. As with linear momentum, angular momentum is conserved, provided no tangential forces act on the body. According to the Bohr model, an electron will be in a stable orbit if $mvr = \frac{nh}{2\pi}$, where n is a whole number called the **quantum number**.

We can show how this leads to a quantization of energy by considering the orbiting electron in Figure 14:

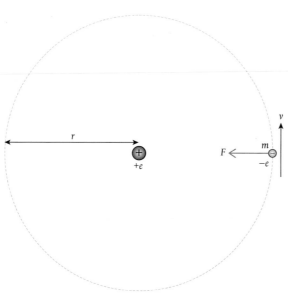

E.1 Figure 14 The Bohr atom.

centripetal force = electrostatic attraction

So:

$$\frac{mv^2}{r} = \frac{ee}{4\pi\varepsilon_0 r^2}$$ equation (1)

Rearranging and multiplying by mr gives: $m^2v^2r^2 = \frac{me^2r}{4\pi\varepsilon_0}$

So, according to Bohr:

$$\frac{me^2r}{4\pi\varepsilon_0} = \left(\frac{nh}{2\pi}\right)^2$$

This implies that:

$$r = \frac{\varepsilon_0 n^2 h^2}{\pi me^2}$$

This gives the radii of all allowed orbits.

Now, the energy of the electron is: $E_k + E_p = \frac{1}{2}mv^2 - \frac{e^2}{4\pi\varepsilon_0 r}$

But from equation (1), $\frac{1}{2}mv^2 = \frac{e^2}{8\pi\varepsilon_0 r}$ so: energy $= \frac{-e^2}{8\pi\varepsilon_0 r}$

Substituting for r gives:

$$E = \frac{-me^4}{8n^2\varepsilon_0^2 h^2} = \frac{-13.6}{n^2} \text{ eV}$$

This predicts that the electron energies will have discrete values that get closer together as the energies increase. This closely matches the energy level diagram that was derived from spectral analysis.

<figure>
energy/eV
0.00
−0.54
−0.85
−1.51

−3.39

−13.6

E.1 Figure 15 The electron energy levels of hydrogen.
</figure>

Exercise

Q10. Calculate the radius of the lowest orbit of an electron in a hydrogen atom based on the Bohr model.

Q11. Use the Bohr model to calculate the frequency of electromagnetic radiation emitted when an electron in a hydrogen atom moves from the second orbit down to the first.

From close observation of spectral lines, we see that they are not all the same intensity. This implies that not all transitions are equally probable. Bohr's model cannot predict this detail and, although it works very nicely for hydrogen, it does not work for any other atom. To create a more accurate model, we would need to look at matter in a different way.

HL end

Under what circumstances does the Bohr model fail? (NOS)

Guiding Questions revisited

What is the current understanding of the nature of an atom?

What is the role of evidence in the development of the model of the atom?

In what ways are previous models of the atom still valid despite recent advances in understanding?

In this chapter, we have explored the experimental evidence relating to the nature of the atom to give the current understanding that:

- Positive protons are located in a dense central nucleus, based on the findings of the Geiger–Marsden–Rutherford experiment.
- Neutral neutrons are also located in the nucleus because the masses and charges of consecutive nuclei do not increase in equal steps.
- **HL** Nuclei have constant density irrespective of the number of nucleons, based on closest-approach measurements to estimate nuclear radii.
- Electrons are located in quantized energy levels, and we can measure the differences between them using the emission spectra of excited atoms.
- **HL** The Bohr model accurately predicts the energy levels for hydrogen.
- Although human understanding of atomic structure has improved in its precision over time because of scientific open-mindedness, some aspects of previous models (e.g. the existence of the electron and the balance of charges) remain.

Practice questions

1. (a) The element helium was first identified from the *absorption spectrum* of the Sun.

    (i) Explain what is meant by the term *absorption spectrum*. (2)

    (ii) Outline how this spectrum may be experimentally observed. (2)

   (b) One of the wavelengths in the absorption spectrum of helium occurs at 588 nm.

    (i) Show that the energy of a photon of wavelength 588 nm is 3.38×10^{-19} J. (2)

    (ii) The diagram represents some of the energy levels of the helium atom. Use the information in the diagram to explain how absorption at 588 nm arises. (3)

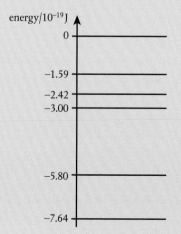

(c)   HL   The Bohr model has been developed to explain the existence of atomic energy levels. It is able to predict the principal wavelengths present in the spectrum of atomic hydrogen. Outline this model.   (3)

(Total 12 marks)

2. (a) The diagram shows the three lowest energy levels of a hydrogen atom as predicted by the Bohr model.

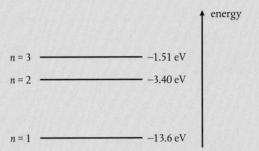

State **two** physical processes by which an electron in the ground state energy level can move to a higher energy level state.   (2)

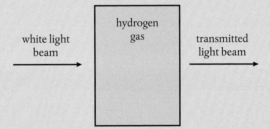

(b) A parallel beam of white light is directed through monatomic hydrogen gas as shown in the diagram. The transmitted light is analyzed.

White light consists of photons that range in wavelength from approximately 400 nm for violet to 700 nm for red light.

(i)   Determine that the energy of photons of light of wavelength 658 nm is about 1.89 eV.   (2)

(ii) The intensity of light of wavelength 658 nm in the direction of the transmitted beam is greatly reduced. Using the energy level diagram in part (**a**), explain this observation.   (3)

(Total 7 marks)

3.   HL   Deviations from Rutherford scattering are detected in experiments carried out at high energies. What can be deduced from these deviations?

A   The impact parameter of the collision

B   The existence of a force different from electrostatic repulsion

C   The size of alpha particles

D   The electric field inside the nucleus

(Total 1 mark)

4. HL The diameter of a nucleus of a particular nuclide X is 12 fm. What is the nucleon number of X?

A 5   **B** 10   **C** 125   **D** 155

(Total 1 mark)

5. HL The diameter of a silver-108 $\left(^{108}_{47}\text{Ag}\right)$ nucleus is approximately three times that of the diameter of a nucleus of which atom?

A $^{4}_{2}\text{He}$ **B** $^{7}_{3}\text{Li}$ **C** $^{11}_{5}\text{B}$ **D** $^{20}_{10}\text{Ne}$

(Total 1 mark)

6. HL Which of the following is true about nuclear density?

A It is constant because the volume of a nucleus is proportional to its nucleon number.

B It is constant because the volume of a nucleus is proportional to its proton number.

C It depends on the nucleon number of the nucleus.

D It depends on the proton number of the nucleus.

(Total 1 mark)

7. Which statement about atomic spectra is **not** true?

A They provide evidence for discrete energy levels in atoms.

B Emission and absorption lines of equal frequency correspond to transitions between the same two energy levels.

C Absorption lines arise when electrons gain energy.

D Emission lines always correspond to the visible part of the electromagnetic spectrum.

(Total 1 mark)

8. A simple model of an atom has three energy levels. The differences between adjacent energy levels are shown below. What are the two smallest frequencies in the emission spectrum of this atom?

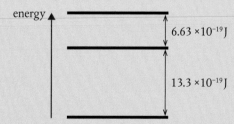

A $0.5 \times 10^{15}\,\text{Hz}$ and $1.0 \times 10^{15}\,\text{Hz}$ **B** $0.5 \times 10^{15}\,\text{Hz}$ and $1.5 \times 10^{15}\,\text{Hz}$

C $1.0 \times 10^{15}\,\text{Hz}$ and $2.0 \times 10^{15}\,\text{Hz}$ **D** $1.0 \times 10^{15}\,\text{Hz}$ and $3.0 \times 10^{15}\,\text{Hz}$

(Total 1 mark)

9. The diagram shows four energy levels for the atoms of a gas. The diagram is drawn to scale. The wavelengths of the photons emitted by the energy transitions between levels are shown. What are the wavelengths of spectral lines, emitted by the gas, in order of decreasing frequency?

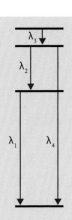

A $\lambda_3, \lambda_2, \lambda_1, \lambda_4$ **C** $\lambda_4, \lambda_3, \lambda_2, \lambda_1$

B $\lambda_4, \lambda_1, \lambda_2, \lambda_3$ **D** $\lambda_4, \lambda_2, \lambda_1, \lambda_3$

(Total 1 mark)

10. **HL** Three possible features of an atomic model are:

I orbital radius

II quantized energy

III quantized angular momentum.

Which of these are features of the Bohr model for hydrogen?

A I and II only   **C** II and III only

B I and III only   **D** I, II and III   *(Total 1 mark)*

11. (a) Outline how the evidence supplied by the Geiger–Marsden–Rutherford experiment supports the nuclear model of the atom.   (4)

(b) Outline why classical physics does not permit a model of an electron orbiting the nucleus.   (3)

(Total 7 marks)

12. The diagram shows the position of the principal lines in the visible spectrum of atomic hydrogen and some of the corresponding energy levels of the hydrogen atom.

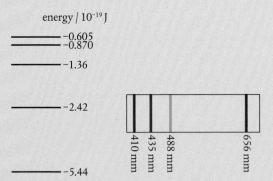

energy / 10^{-19} J

- −0.605
- −0.870
- −1.36
- −2.42
- −5.44

410 mm 435 mm 488 mm 656 mm

(a) Determine the energy of a photon of blue light (435 nm) emitted in the hydrogen spectrum.   (3)

(b) Copy the diagram and identify, with an arrow labeled B, the transition in the hydrogen spectrum that gives rise to the photon with the energy in **(a)**.   (1)

(c) Explain your answer to **(b)**.   (2)

(Total 6 marks)

13. **HL** In a classical model of the singly ionized helium atom, a single electron orbits the nucleus in a circular orbit of radius r.

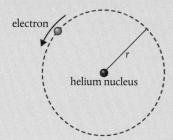

(a) Show that the speed v of the electron with mass m is given by $v = \sqrt{\dfrac{2ke^2}{mr}}$. (1)

(b) Hence, deduce that the total energy of the electron is given by
$E_{\text{TOTAL}} = -\dfrac{ke^2}{r}$. (2)

(c) In this model, the electron loses energy by emitting electromagnetic waves. Describe the predicted effect of this emission on the orbital radius of the electron. (2)

The Bohr model for hydrogen can be applied to the singly ionized helium atom. In this model, the radius r, in m, of the orbit of the electron is given by $r = 2.7 \times 10^{11} \times n^2$, where n is a positive integer.

(d) Show that the de Broglie wavelength λ of the electron in the $n = 3$ state is $\lambda = 5.1 \times 10^{-10}$ m.

The formula for the de Broglie wavelength of a particle is $\lambda = \dfrac{h}{mv}$. (2)

(e) Estimate for $n = 3$, the ratio $\dfrac{\text{circumference of orbit}}{\text{de Broglie wavelength of electron}}$. State your answer to one significant figure. (1)

(Total 8 marks)

14. **HL** Bohr modified the Rutherford model by introducing the condition $mvr = n\dfrac{h}{2\pi}$.

(a) Outline the reason for this modification. (3)

(b) Show that the speed v of an electron in the hydrogen atom is related to the radius r of the orbit by the expression:
$$v = \sqrt{\dfrac{ke^2}{m_e r}}$$
where k is the Coulomb constant. (1)

(c) Deduce that the radius r of the electron's orbit in the ground state of hydrogen is given by the following expression:
$$r = \dfrac{h^2}{4\pi^2 k m_e e^2}$$ (2)

(d) Calculate the electron's orbital radius. (1)

(Total 7 marks)

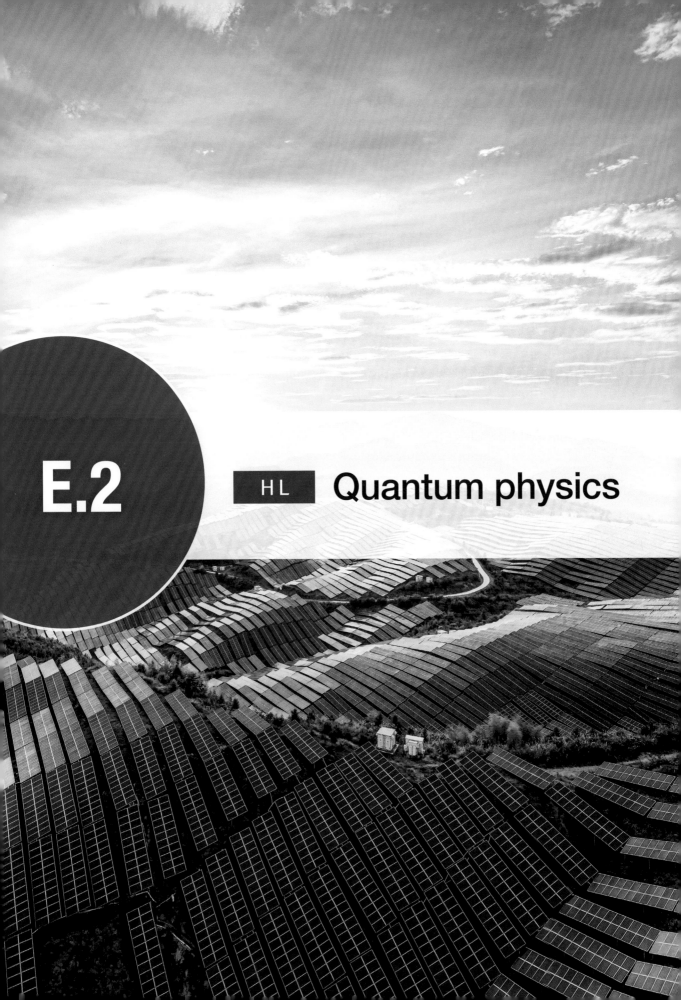

E.2

HL Quantum physics

◀ Non-physicists might mistakenly assume that quantum physics is complicated, abstract and irrelevant to their everyday lives. However, the photoelectric effect is used across the world for electricity generation, such as at this solar power station in Fujian, China. Perhaps there are solar panels on a building near you, where electrons are released due to high-energy particles of light.

HL

Guiding Questions

How can light be used to create an electric current?

What is meant by wave–particle duality?

Wave–particle duality is the way that matter exhibits the properties of both waves and particles. But what are waves and particles?

In the chapters on space, time and motion, we talked about small balls as particles. But the idea of a small ball is not always what we mean by a particle. This is because a small ball has a shape and size – it is a hard little object.

A better way of thinking about a particle is as a small region of space that has some properties assigned to it. These properties tell us how particles interact with each other. A small ball has mass, charge, energy and momentum and so does an electron. If two positively charged small balls are on a collision course, they will repel each other because of their charges. If a small ball is moving at constant velocity, it has kinetic energy and momentum. If it does not interact with anything, these quantities will remain constant. Wherever the ball goes, its properties follow. If it collides with another ball, it can do work on the other ball, transferring energy and momentum. It is the same for electrons, except there are no hard surfaces.

A wave is different. It spreads out, so you cannot transfer all the energy in a wavefront to another wave. If two waves meet, they pass through each other, while balls bounce off each other. When a wave passes through a narrow opening, it spreads out or diffracts. Particles do not do this.

So, when we discover that the energy from a very dim light source arrives on a metal plate rather than spreading out, it makes us think of a particle and not a wave. When an electron diffracts, it is more like a wave than a particle.

Students should understand:

the photoelectric effect as evidence of the particle nature of light
photons of a certain frequency, known as the threshold frequency, are required to release photoelectrons from the metal
Einstein's explanation using the work function and the maximum kinetic energy of the photoelectrons as given by $E_{max} = hf - \Phi$ where Φ is the work function of the metal
diffraction of particles as evidence of the wave nature of matter
matter exhibits wave–particle duality
the de Broglie wavelength for particles as given by $\lambda = \frac{h}{p}$

Compton scattering of light by electrons as additional evidence of the particle nature of light
photons scatter off electrons with increased wavelength
the shift in photon wavelength after scattering off an electron as given by $\lambda_f - \lambda_i = \Delta\lambda = \frac{h}{m_e c}\left(1 - \cos\theta\right)$.

The quantum nature of light

The photoelectric effect

Consider ultraviolet light shining on a negatively charged zinc plate. Will the light cause the charge to leave the plate? To answer this question, we can use the wave model of light, but we cannot see what is happening inside the metal. So to help us visualize this problem, we will use an analogy.

What are the defining features and behaviors of waves (C.3)?

The swimming pool analogy

Imagine a ball floating near the edge of a swimming pool as in Figure 1. If you are at the other side of the swimming pool, could you get the ball out of the pool by sending water waves toward it? To get the ball out of the pool, we need to lift it as high as the edge of the pool. In other words, we must give it enough potential energy to reach this height. We can do this by making the amplitude of the wave high enough to lift the ball. If the amplitude is not high enough, the ball will not be able to leave the pool unless we build a machine (as in Figure 2) that will collect the energy over a period of time. In this case, there will be a time delay before the ball gets out.

E.2 Figure 1 The larger the amplitude, the more likely the ball is to escape the pool.

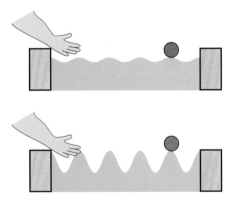

To relate this to the zinc plate, according to this model:

* Electrons will be emitted only if the light source is very bright. (Brightness is related to amplitude of the wave.)

* If the source is dim, we expect no electrons to be emitted. If electrons are emitted, we expect a time delay while the atoms collect energy from the wave.

* If we use lower frequency light, electrons will still be emitted if it is bright enough.

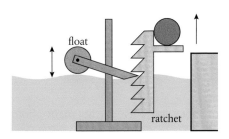

E.2 Figure 2 A ratchet is a model for the time delay in energy transfer via waves. With each subsequent rise and fall of the slot, the ratchet would be elevated higher.

The zinc plate experiment

To find out if electrons are emitted or not, we can put the zinc plate on an electroscope and shine UV light on it as in Figure 3. If electrons are emitted, charge will be lost and the electroscope gold leaf will fall. The results are not entirely as expected:

- The gold leaf does go down, indicating that electrons are emitted from the surface of the zinc plate.

- The gold leaf goes down even if the UV light is very dim. When very dim, the leaf takes longer to go down but there is no time delay before it starts to drop.

- If light of lower frequency (for example, visible light) is used, the leaf does not go down, showing that no electrons are emitted. This is the case even when very intense low-frequency light is used.

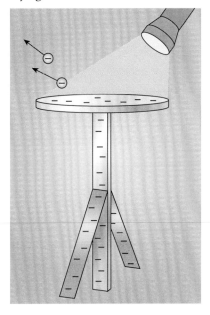

E.2 Figure 3 UV radiation is absorbed and electrons are emitted, causing the gold leaf to fall.

These results can be explained if we consider light to be quantized.

Quantum model of light

In the quantum model of light, light is made up of packets called photons. Each photon is a section of wave with an energy E that is proportional to the wave frequency, f.

$$E = hf$$

where h is Planck's constant (6.63×10^{-34} J s).

How did the explanation of the photoelectric effect lead to the incorrect idea that light was purely a wave? (NOS)

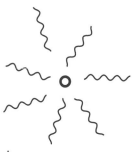

E.2 Figure 4 Light radiating, according to the wave model and the quantum model.

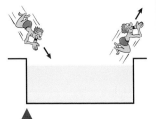

E.2 Figure 5 The quantum swimming pool.

The intensity of light is related to the number of photons, not the amplitude, as in the classical wave model. Using this model, we can explain the photoelectric effect.

- UV light has a high frequency, so when an individual photon of UV light is absorbed by the zinc, it gives enough energy to a zinc electron to enable it to leave the surface.

- When the intensity of light is low, there are not many photons, but each photon has enough energy to enable an electron to leave the zinc.

- Low-frequency light is made of photons with low energy. These do not have enough energy to liberate electrons from the zinc. Intense low-frequency light is simply more low energy photons, none of which have enough energy.

If a swimming pool were like this, then if someone jumped into the pool, the energy they gave to the water would stay together in a packet until it met another swimmer. When this happened, the other swimmer would be ejected from the pool.

To get a deeper understanding of the photoelectric effect, we need more information about the energy of the photoelectrons.

Millikan's photoelectric experiment

Millikan devised an experiment to measure the kinetic energy of photoelectrons. He used an electric field to stop the electrons completing a circuit and used that 'stopping potential' to calculate the kinetic energy. A diagram of the apparatus is shown in Figure 6.

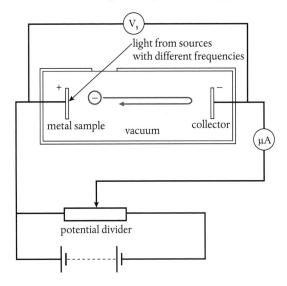

E.2 Figure 6 The stopping potential stops the electrons from reaching the collector.

Light from a source of known frequency passes into the apparatus through a small window. If the photons have enough energy, electrons will be emitted from the metal sample. Some of these electrons travel across the tube to the collector, causing a current to flow in the circuit. This current is measured by the microammeter. The potential divider is adjusted until none of the electrons reach the collector (as in the diagram). We can now use the law of conservation of energy to find the kinetic energy of the fastest electrons, which are *just* stopped by the stopping potential difference provided.

$$\text{loss of } E_k = \text{gain in electrical } E_p$$

$$\frac{1}{2}mv^2 = v_s e$$

So maximum kinetic energy: $E_{max} = v_s e$

The light source is now changed to one with a different frequency and the procedure is repeated.

The graphs in Figure 7 show two aspects of the results.

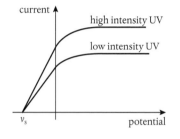
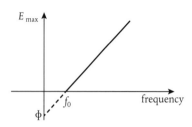

The most important aspect of the first graph in Figure 7 is that, for a given potential, increasing the intensity increases the current but does not change the stopping potential. This is because when the intensity is increased, the light contains more photons (so liberates more electrons) but does not increase the energy (so v_s is the same).

The second graph in Figure 7 shows that the maximum kinetic energy of the electrons is linearly related to the frequency of the photons. Below a certain value, f_0, no photoelectrons are liberated. This is called the **threshold frequency**.

Einstein's photoelectric equation

Einstein explained the photoelectric effect and derived an equation that relates the kinetic energy of the photoelectrons to the frequency of light.

maximum photoelectron kinetic energy = energy of photon − energy needed to get photon out of metal

$$E_{max} = hf - \Phi$$

Φ is called the **work function**. If the photon has only enough energy to get the electron out, then it will have zero kinetic energy, and this is what happens at the threshold frequency, f_0. At this frequency:

$$E_{max} = 0 = hf_0 - \Phi$$

So: $\Phi = hf_0$

We can now rewrite the equation as:

$$E_{max} = hf - hf_0$$

Light intensity is related to the energy per photon and the number of photons incident per unit area in one second. So a red light source will emit more photons per unit area in a given time than a blue one of the same intensity. Increasing the wavelength of light while keeping the intensity constant will result in a higher current.

E.2 Figure 7 Graphs of current vs potential and maximum kinetic energy against frequency. Note that the values of E_{max} less than 0 represent an extrapolation of the relationship for frequencies greater than the threshold frequency.

Exam questions can include many different graphs related to the photoelectric effect. Make sure you look at the axes carefully.

The equation of a straight line is of the form $y = mx + c$, where m is the gradient and c is the y-intercept. You should be able to see how the equation $E_{max} = hf - \Phi$ is the equation of the line in the second graph in Figure 7.

You will find it easier to work in eV for questions 2, 3 and 4.

Can the Bohr model help explain the photoelectric effect? (NOS)

Exercise

Q1. A sample of sodium is illuminated by light of wavelength 422 nm in a photoelectric tube. The potential across the tube is increased to 0.6 V. At this potential, no current flows across the tube.

Calculate:

(a) the maximum kinetic energy of the photoelectrons

(b) the frequency of the incident photons

(c) the work function of sodium

(d) the lowest frequency of light that would cause photoelectric emission in sodium.

Q2. A sample of zinc is illuminated by UV light of wavelength 144 nm. If the work function of zinc is 4.3 eV, calculate:

(a) the photon energy in eV

(b) the maximum kinetic energy of photoelectrons

(c) the stopping potential

(d) the threshold frequency.

Q3. If the zinc in Q2 is illuminated by the light in Q1, will any electrons be emitted?

Q4. The maximum kinetic energy of electrons emitted from a nickel sample is 1.4 eV. If the work function of nickel is 5.0 eV, what frequency of light must have been used?

In thermionic emission, atomic electrons gain enough energy from thermal excitation to leave the surface of the metal.

The wave nature of matter

Rutherford used alpha particles to probe inside the atom. Another particle that can be scattered by matter is the electron. Electrons are released from a metal when it is heated and can be accelerated in a vacuum through a potential difference as shown in Figure 8.

The kinetic energy of the electrons in eV is numerically the same as the accelerating potential difference. If: potential difference = 200 V, E_k = 200 eV

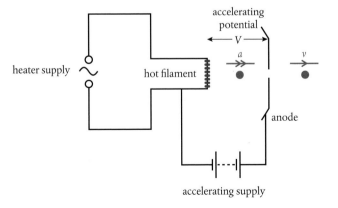

E.2 Figure 8 A simple electron gun.

The filament, made hot by an AC supply, liberates electrons, which are accelerated toward the anode by the accelerating potential difference. Electrons traveling in the direction of the hole in the anode pass through and continue with constant velocity.

We can calculate the speed of the electrons, using the law of conservation of energy.

$$\text{loss of electrical } E_p = \text{gain in } E_k$$

$$Ve = \frac{1}{2}mv^2$$

$$v = \sqrt{\frac{2Ve}{m}}$$

Detecting electrons

You cannot see electrons directly, so you need a detector to find out where they go. When electrons collide with certain atoms, they give the atomic electrons energy (we say the electrons are 'excited'). When the atomic electrons go back down to a lower energy level, they give out light. This is called **phosphorescence** and can be used to see where the electrons land. Zinc sulfide is one such substance. It is used to coat glass screens so that light is emitted where the electrons collide with the screen. This is how the older types of TV screens work.

Electron diffraction

If a beam of electrons is passed through a thin film of graphite, an interesting pattern is observed when they hit a phosphor screen. The pattern looks very much like the diffraction pattern caused when light is diffracted by a small circular aperture. Perhaps the electrons are being diffracted by the atoms in the graphite. Assuming this to be true, we can calculate the wavelength of the wave that has been diffracted.

The de Broglie hypothesis

In 1924, Louis de Broglie proposed that 'all matter has a wave-like nature' and that the equivalent wavelength (known as the 'de Broglie wavelength') could be found using the equation:

$$\lambda = \frac{h}{p}$$

where p is the momentum of the particle.

The original TV screens were called cathode ray tubes. They consisted of an electron gun and a fluorescent screen. The picture was formed by scanning the beam across the screen many times a second. As the beam scanned across, its intensity was varied, resulting in a picture. The problem with this technology is that the electron gun needs to be placed far enough from the screen so that the electrons hit all parts of it. Modern LED and plasma screens have arrays of small lights so they can be made much thinner.

The aurora borealis is caused when charged particles from the Sun excite atoms in the atmosphere, causing light to be emitted.

How can particles diffract (C.3)?

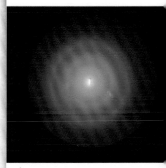

Diffraction of electrons by a graphite film.

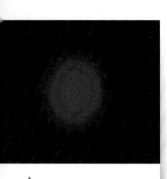

▲
Diffraction of light by a small circular aperture.

Using this equation, we can calculate the wavelength of an electron accelerated through a potential difference of 50 V.

kinetic energy gained = $50\,\text{eV} = 50 \times 1.6 \times 10^{-19} = 8 \times 10^{-18}\,\text{J}$

$E_k = \frac{1}{2}mv^2$ so $v = \sqrt{\left(\frac{2 \times E_k}{m}\right)} = \sqrt{\left(\frac{2 \times 8 \times 10^{-18}}{9.1 \times 10^{-31}}\right)} = 4.2 \times 10^6\,\text{m s}^{-1}$

momentum = $mv = 9.1 \times 10^{-31} \times 4.2 \times 10^6 = 3.8 \times 10^{-24}\,\text{N s}$

So: wavelength, $\lambda = \frac{h}{mv} = 1.7 \times 10^{-10}\,\text{m}$

Given this wavelength, we can use the same equation that was used for the diffraction of light by a diffraction grating to calculate the angle of the first maximum in the diffraction pattern.

$$\sin\theta = \frac{\lambda}{D}, \text{ where } D \text{ is the atomic spacing} \approx 2 \times 10^{-10}\,\text{m}$$

This gives $\theta \approx 60°$, which is consistent with the observable pattern. The first minimum will be at half this angle.

Exercise

Q5. An electron is accelerated by a potential difference of 100 V. Calculate its:

   (a) kinetic energy in eV

   (b) kinetic energy in joules

   (c) de Broglie wavelength.

Q6. Calculate the de Broglie wavelength for a car of mass 1000 kg traveling at $15\,\text{m s}^{-1}$. Why will you never see a car diffract?

What evidence indicates the diffraction of a wave (C.3)?

Probability waves

To understand the relationship between an electron and its wave, let us take a fresh look at the interference of light now that we know light is made up of photons. When light is passed through two narrow slits, it diffracts at each slit and then light from each slit interferes in the region of overlap as shown in Figure 9. This is very much the same as what happens to water waves and sound.

E.2 Figure 9 Two-slit interference of light.

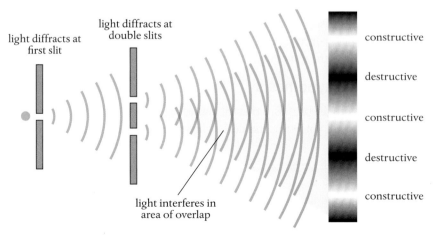

We now know that light is made up of photons so the bright areas in the interference pattern will be areas where there are a lot of photons and in the dark areas there will be few photons. Thus we can say that the probability of the photon landing at a particular spot is given by the intensity of the light, which is proportional to the square of the amplitude of the wave. It is the same with electrons and all other particles. If electrons were passed through the two narrow slits in Figure 9, then a similar pattern would be obtained except that the slits would have to be closer together. The wave does not only give the probability of the particle hitting the screen, it also gives the probability of the particle being at a given position at any time. However, we only really notice the particle properties of the particle when it hits something like the screen.

When electrons are fired from an electron gun, we think of them as moving in straight lines. We can even see this if the region is filled with low-pressure air. How can we model this with a wave? Waves spread, which would mean that the electron could be anywhere and that does not seem to be happening. The way this can be resolved is if the probability wave is thought to be made up of many waves superposed on each other such that they cancel out everywhere except in a small region of space. The probability of the particle being at a given position would then be zero almost everywhere except in the region where the waves add. This is called a **wave packet** (shown in Figure 10).

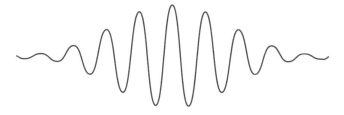

◀ **E.2 Figure 10** A wave packet.

If the wave packet moves forward, then the position where the particle is most likely to be found also moves forward. So the most probable position of an electron moving in a straight line is given by a moving wave packet, which is the resultant of many guide waves, which cancel each other out everywhere except in the line of flight of the electron. If a slit is placed in the electron's path, then guide waves diffract, altering the probable path of the electron.

We only know where the particle is when it hits the screen. If we try to track its position, we run into difficulties.

The Compton effect

You may have thought that a photon is not a real particle like an electron as it is a packet of waves. If it was a particle, surely it would bounce off other particles to conserve momentum. How can a photon, which has no mass, have momentum?

Photons do have momentum and do collide with electrons. When this happens, kinetic energy and momentum are conserved. This was demonstrated by Arthur H. Compton in 1923 and the effect is named after him.

X-rays are scattered by a carbon target. The wavelength of the scattered photons is changed, and so the effect cannot be explained by diffraction. The greater the extent of the scattering, the longer the wavelength becomes. Let us apply conservation of energy and momentum considerations to the collision between a photon and an electron in the carbon atom.

How is a photon scattering off an electron similar to and different from the collision of two solid balls (A.2 and A.3)?

From relativistic mechanics, it can be shown that the energy of a particle is given by the equation:

$$E^2 = m^2 c^4 + p^2 c^2$$

Taking the mass of a photon as zero:

$$E = pc$$

So:

$$p = \frac{E}{c}$$

We know that the energy of a photon is:

$$E = hf = \frac{hc}{\lambda}$$

So:

$$p = \frac{h}{\lambda}$$

A change in momentum will result in a change of wavelength. If we consider a red ball hitting a blue ball, the change in momentum of the red ball will depend on the angle of deflection.

E.2 Figure 11 The change in momentum is related to the angle of deflection.

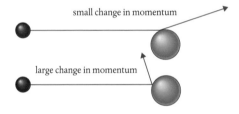

small change in momentum

large change in momentum

If the red ball just hits the edge of the blue ball, its momentum will hardly change but a head-on collision will cause a large change. It is the same with photons. The bigger the angle of deviation, the greater the change in momentum and hence the change in wavelength.

The equation relating the change of wavelength to the angle of deviation is:

$$\lambda_f - \lambda_i = \Delta\lambda = \frac{h}{m_e c}\left(1 - \cos\theta\right)$$

E.2 Figure 12 The loss of momentum causes an increase in wavelength.

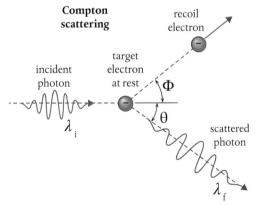

Compton scattering

recoil electron

incident photon

target electron at rest

Φ

λ_i

θ

scattered photon

λ_f

Worked example

Calculate the wavelength of a 0.002 nm photon scattered at 90° by an electron.

Solution

$$\lambda_f - \lambda_i = \Delta\lambda = \frac{h}{m_e c}\left(1 - \cos\theta\right)$$

$$\Delta\lambda = \frac{h}{m_e c}(1 - \cos 90) = \frac{h}{m_e c}$$

$$\Delta\lambda = \frac{6.63 \times 10^{-34}}{9.110 \times 10^{-31} \times 3 \times 10^8} = 2.43 \times 10^{-12}\text{ m} = 0.0024\text{ nm}$$

wavelength of scattered photon = 0.002 + 0.0024 = 0.0044 nm

Exercise

Q7. A 0.002 nm photon is scattered by an electron, resulting in a 0.001 nm change in wavelength. Calculate the scattering angle.

Why is Compton scattering more convincing evidence for the particle nature of light than the photoelectric effect? (NOS)

Guiding Questions revisited

How can light be used to create an electric current?

What is meant by wave–particle duality?

In this chapter, we have weighed up the evidence that suggests that light and electrons have both wave-like and particle-like properties (known as wave–particle duality), which includes:

- The photoelectric effect, in which photons of light with sufficiently high frequency (irrespective of intensity) can provide electrons with enough energy to leave the surface of a metal, resulting in the formation of a current.
- Diffraction of electrons by layers of atoms, with electrons or any other particles having an equivalent de Broglie wavelength that is inversely proportional to their momentum and which can be verified by measuring the angle to the first diffraction maximum or minimum.
- The exchange of momentum between photons and electrons in Compton scattering, with the light increasing in wavelength as its momentum decreases.

Practice questions

1. **(a)** State **one** aspect of the photoelectric effect that **cannot** be explained by the wave model of light. Describe how the photon model provides an explanation for this aspect. (2)

Light is incident on a metal surface in a vacuum. The graph shows the variation of the maximum kinetic energy E_{max} of the electrons emitted from the surface with the frequency f of the incident light.

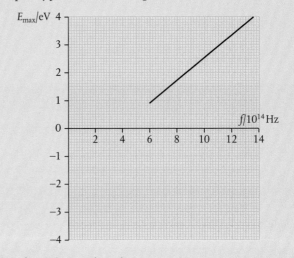

(b) Use data from the graph to determine:

  (i)   the threshold frequency (2)

  (ii) a value of the Planck constant (2)

  (iii) the work function of the surface. (2)

The threshold frequency of a different surface is 8.0×10^{14} Hz.

(c) Copy the graph and draw a line to show the variation with frequency f of the maximum kinetic energy E_{max} of the electrons emitted. (2)

(Total 10 marks)

2. **(a)** Describe the concept of matter waves and state the de Broglie hypothesis. (3)

(b) An electron is accelerated from rest through a potential difference of 850 V. For this electron:

  (i)   calculate the gain in kinetic energy (1)

  (ii) deduce that the final momentum is 1.6×10^{-23} N s (2)

  (iii) determine the associated de Broglie wavelength. (Electron charge $e = 1.6 \times 10^{-19}$ C, Planck constant $h = 6.6 \times 10^{-34}$ J s) (2)

(Total 8 marks)

3. A photon has a wavelength λ. What are the energy and momentum of the photon?

	Energy of photon	Momentum of photon
A	$\dfrac{hc}{\lambda}$	$\dfrac{h}{\lambda}$
B	$\dfrac{hc}{\lambda}$	$\dfrac{\lambda}{h}$
C	$\dfrac{h\lambda}{c}$	$\dfrac{h}{\lambda}$
D	$\dfrac{h\lambda}{c}$	$\dfrac{\lambda}{h}$

(Total 1 mark)

4. A particle has a de Broglie wavelength λ and kinetic energy E. What is the relationship between λ and E?

A $\lambda \propto E^{\frac{1}{2}}$ **B** $\lambda \propto E$ **C** $\lambda \propto E^{-\frac{1}{2}}$ **D** $\lambda \propto E^{-1}$

(Total 1 mark)

5. Which of the following is evidence for the wave nature of the electron?

A Continuous energy spectrum in β^- decay

B Electron diffraction from crystals

C Existence of atomic energy levels

D Existence of nuclear energy levels

(Total 1 mark)

6. Monochromatic electromagnetic radiation is incident on a metal surface. Which is the correct statement for the kinetic energy of the electrons released from the metal?

A It is constant because the photons have a constant energy.

B It is constant because the metal has a constant work function.

C It varies because the electrons are not equally bound to the metal lattice.

D It varies because the work function of the metal is different for different electrons.

(Total 1 mark)

7. Different metal surfaces are investigated in an experiment on the photoelectric effect. A graph of the variation of the maximum kinetic energy of photoelectrons with the frequency of the incident light is drawn for each metal. Which statement is correct?

A All graphs have the same intercept on the frequency axis.

B The work function is the same for all surfaces.

C All graphs have the same slope.

D The threshold frequency is the same for all surfaces.

(Total 1 mark)

8. In a photoelectric effect experiment, a beam of light is incident on a metallic surface W in a vacuum.

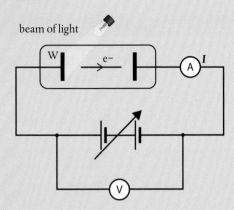

The graph shows how the current *I* varies with the potential difference V when three different beams, X, Y and Z, are incident on W at different times.

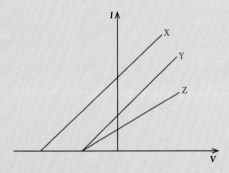

I   X and Y have the same frequency.

II   Y and Z have different intensity.

III   Y and Z have the same frequency.

Which statements are correct?

A   I and II only   **C**   II and III only

B   I and III only   **D**   I, II and III

(Total 1 mark)

9. Monochromatic light is incident on a metal surface and electrons are released. The intensity of the incident light is increased. What changes, if any, occur to the rate of emission of electrons and to the kinetic energy of the emitted electrons?

	Rate of emission of electrons	Kinetic energy of the emitted electrons
A	increase	increase
B	decrease	no change
C	decrease	increase
D	increase	no change

(Total 1 mark)

10. Monochromatic light of very low intensity is incident on a metal surface. The light causes the emission of electrons almost instantaneously. Explain how this observation:

(a) does not support the wave nature of light (2)

(b) does support the photon nature of light. (2)

In an experiment to demonstrate the photoelectric effect, light of wavelength 480 nm is incident on a metal surface.

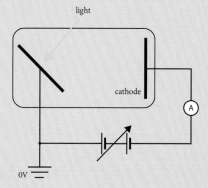

The graph shows the variation of the current I in the ammeter with the potential V of the cathode.

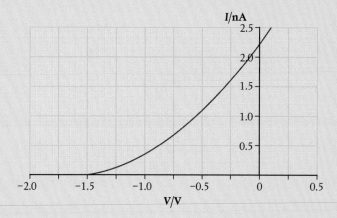

(c) Calculate, in eV, the work function of the metal surface. (3)

The intensity of the light incident on the surface is reduced by half without changing the wavelength.

(d) Copy the graph and draw the variation of the current I with potential V after this change. (2)

(Total 9 marks)

11. The de Broglie wavelength λ of a particle accelerated close to the speed of light is approximately:

$$\lambda \approx \frac{hc}{E}$$

where E is the energy of the particle.
A beam of electrons of energy 4.2×10^8 eV is produced in an accelerator.

(a) Show that the wavelength of an electron in the beam is about 3×10^{-15} m. (1)

The electron beam is used to study the nuclear radius of carbon-12. The beam is directed from the left at a thin sample of carbon-12. A detector is placed at an angle θ relative to the direction of the incident beam.

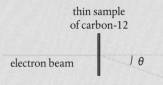

thin sample
of carbon-12

electron beam

θ

detector

The graph shows the variation of the intensity of electrons with θ. There is a minimum of intensity for $\theta = \theta_0$.

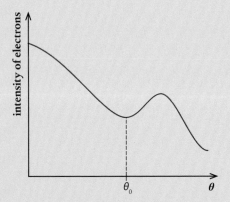

(b) Discuss how the results of the experiment provide evidence for matter waves. (2)

(c) The accepted value of the diameter of the carbon-12 nucleus is 4.94×10^{-15} m. Estimate the angle θ_0 at which the minimum of the intensity is formed. (2)

(d) Outline why electrons with energy of approximately 10^7 eV would be unsuitable for the investigation of nuclear radii. (2)

(e) Experiments with many nuclides suggest that the radius of a nucleus is proportional to $A^{\frac{1}{3}}$, where A is the number of nucleons in the nucleus. Show that the density of a nucleus remains approximately the same for all nuclei. (Hint: Revisit E.1) (2)

(Total 9 marks)

12. (a) Which is smaller? (1)

    A An atom

    B A light wave

    C Both are about the same size

  (b) Suggest an implication of your answer. (1)

(Total 2 marks)

13. The scattering of photons (Compton scattering) can be used to identify the composition of materials by the intensity of the scattered radiation. The scattered radiation is at a different frequency. An incident photon, frequency f, momentum $\frac{hf}{c}$, is scattered by a stationary electron, producing a scattered photon of frequency $f - \Delta f$, where Δf is small compared with f. This photon travels in a direction that makes an angle θ with the direction of the incident photon. The electron, mass m_e, acquires a non-relativistic speed v.

  (a) Draw a labeled vector triangle of the momentum of the particles. (3)

  (b) Write down the equation relating the magnitude of the momentum of the electron to that of the photons. (4)

  (c) Obtain the equation for energy conservation. (2)

  (d) Deduce an equation for Δf. When Δf is much less than f and hf much less than $m_e c^2$, obtain the approximation:

$$\Delta f = \frac{hf^2(1 - \cos\theta)}{m_e c^2}$$

    (7)

  (e) Sketch graphs of:

    (i) Δf against f for constant θ

    (ii) Δf against θ for constant f

    For what angle(s) is Δf greatest? State the value(s) of Δf. (4)

(Total 20 marks)

HL end

487

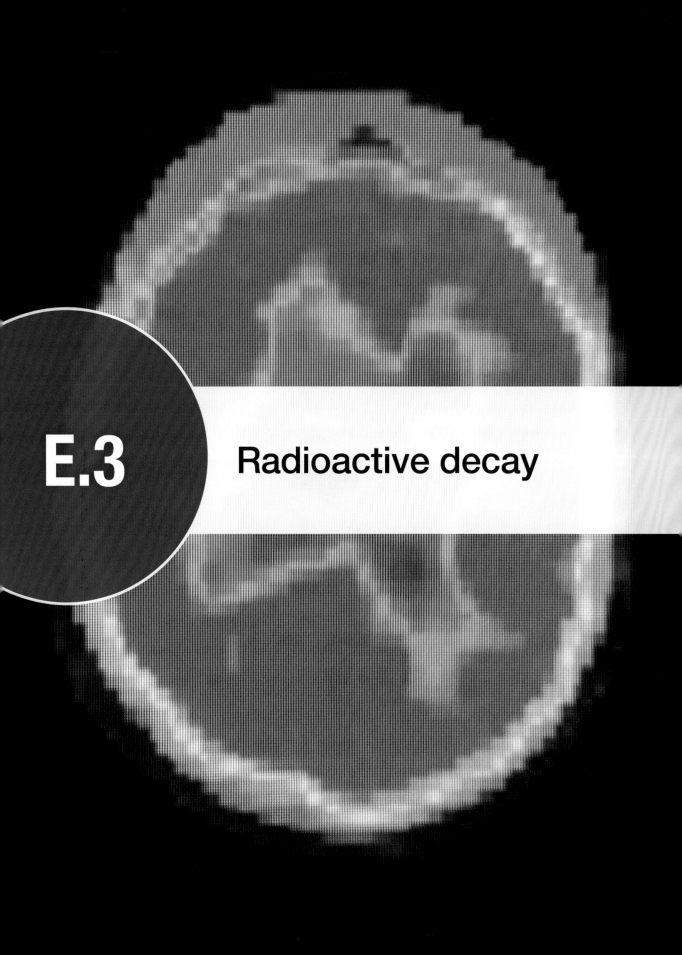

E.3   Radioactive decay

◀ Ionizing radiation from radioactive isotopes can cause harm. However, used in the right way, it also has uses in medical treatments, the home and in industry. This is a PET scan, which detects the gamma radiation released when positrons injected into the body (a tracer) annihilate with electrons already present. PET scans give doctors an understanding of how organs are functioning.

Guiding Questions

Why are some isotopes more stable than others?

In what ways can a nucleus undergo change?

How do large, unstable nuclei become more stable?

How can the random nature of radioactive decay allow for predictions to be made?

A pendulum bob comes to rest at the equilibrium position. Conventional current flows from high potential to low potential. Atomic electrons return to a lower energy level after emitting a photon of light. Thermal energy flows from hot regions to cold regions. A ball placed on a hill rolls down.

All physical systems return to the lowest possible energy state, unless it is made impossible, for example, if the ball on the hill was placed in a bucket.

The same is true for nuclei. If the binding energy per nucleon of a nucleus can increase by changing the composition of neutrons and protons, then this will, if possible, happen. This is why some large nuclei emit ionizing radiation.

How can increasing the binding energy mean that the nucleus is in a lower energy state? Binding energy is the energy released when the nucleus is formed. Increasing binding energy means energy must be released and stability is greater.

Students should understand:

isotopes
nuclear binding energy and mass defect
the variation of the binding energy per nucleon with nucleon number
the mass–energy equivalence as given by $E = mc^2$ in nuclear reactions
the existence of the strong nuclear force, a short-range, attractive force between nucleons
the random and spontaneous nature of radioactive decay
the changes in the state of the nucleus following alpha, beta and gamma radioactive decay
the radioactive decay equations involving $\alpha, \beta^-, \beta^+, \gamma$
the existence of neutrinos ν and antineutrinos $\bar{\nu}$
the penetration and ionizing ability of alpha particles, beta particles and gamma rays
the activity, count rate and half-life in radioactive decay

the changes in activity and count rate during radioactive decay using integral values of half-life
the effect of background radiation on count rate.
HL the evidence for the strong nuclear force
HL the role of the ratio of neutrons to protons for the stability of nuclides
HL the approximate constancy of binding energy curve above a nucleon number of 60
HL the spectrum of alpha and gamma radiations provides evidence for discrete nuclear energy levels
HL the continuous spectrum of beta decay as evidence for the neutrino
HL the decay constant λ and the radioactive decay law as given by $N = N_0 e^{-\lambda t}$
HL the decay constant approximates the probability of decay in unit time only in the limit of sufficiently small λt
HL the activity as the rate of decay as given by $A = \lambda N = \lambda N_0 e^{-\lambda t}$
HL the relationship between half-life and the decay constant as given by $T_{\frac{1}{2}} = \frac{\ln 2}{\lambda}$.

The nuclear force

We have seen that the energy of an atomic electron is in the region of −1 eV. This means that to remove an electron from an atom requires about 1 eV of energy. This does not happen to atoms of air at room temperature but if, for example, air is heated, colliding atoms excite electrons, resulting in the emission of light. The amount of energy required to remove an electron from an atom is directly related to the strength of the force holding it around the nucleus. The fact that it takes a million times more energy to remove a nucleon from the nucleus indicates that whatever force is holding nucleons together must be much stronger than the electromagnetic force holding electrons in position.

Unlike the gravitational and electric forces, the force between two nucleons is *not* inversely proportional to their distance apart. If it was, then all nuclei would be attracted each other. The **strong nuclear force** is in fact very short range, only acting up to distances of about 10^{-15} m: attractive when nucleons are separated but repulsive when closer together.

From data gathered about the mass and radius of nuclei, it is known that all nuclei have approximately the same density, independent of the combination of protons and neutrons.

This implies that the nuclear force is the same for proton–proton, proton–neutron and neutron–neutron interactions. If the force for neutrons was greater than that between protons, you might expect nuclides with large numbers of neutrons to be more dense.

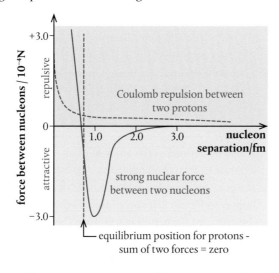

◀ **E.3 Figure 1** The magnitude of the attracting strong nuclear force is greatest between approximately 1 and 3 fm. Beyond this, the electric repulsion between two protons exceeds it.

HL Note that if the strong nuclear force did not exist, nuclei of more than one proton would not exist either. The electric force that repels protons would exceed the gravitational force attracting them.

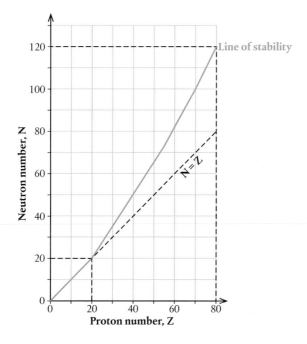

◀ Stable nuclei require an increasing ratio of neutrons to protons as the proton number increases.

The strong nuclear force is so short in its range that it does not act fully across large nuclei. However, the repulsive electric force does. For this reason, nuclei with large numbers of nucleons require a higher ratio of neutrons to protons than small nuclei to ensure stability. Neutrons provide equal amounts of the strong nuclear force and provide additional spacing between protons (reducing the electric force).

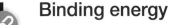

Would a nucleus be able to exist if only gravitational and electric forces were considered?
(D.1, D.2)

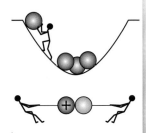

E.3 Figure 2 Separating the nucleons (work done).

The difference between the mass of a nucleus and its constituents is very small, but when multiplied by c^2, gives a lot of energy. It is this energy that is utilized in the fission reactors of today and the fusion reactors of tomorrow.

E.3 Figure 3 Combining nucleons (energy lost).

How does equilibrium within a star compare to stability within the nucleus of an atom?
(E.5)

Binding energy

The binding energy of a nucleus is defined as the amount of energy required to separate a nucleus into its constituent protons and neutrons. Since the nuclear force is very strong, the work done is relatively large, leading to a measurable increase in the mass of the nucleons ($E = mc^2$). This difference in mass, the **mass defect**, can be used to calculate the binding energy of the nucleus. It is helpful to think of nucleons as balls in a hole as in Figure 2. The balls do not have energy in the hole, but to get them out, we would need energy to do work. Similarly, binding energy is not something that the nucleons possess. It is something that we would have to possess if we wanted to separate them.

The binding energy can also be defined as the amount of energy released when the nucleus formed from separate protons and neutrons. Again, the balls in a hole might help us to understand the concept. Imagine that you have three perfectly elastic balls and a perfectly elastic hole as in Figure 3. If you simply throw all the balls in the hole, they will probably just bounce out. To get them to stay in the hole, you need to remove the energy they had before they were dropped in. One way this could happen is if the balls could collide in such a way so as to give one of the balls all of their energy. This ball would fly out of the hole very fast, leaving the other two settled at the bottom. It is the same with nuclei.

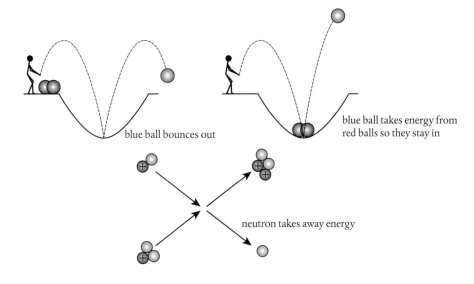

blue ball bounces out

blue ball takes energy from red balls so they stay in

neutron takes away energy

Conversion from u to MeV

The atomic mass unit, u, is defined as $\frac{1}{12}$ of an atom of carbon-12. This is $1.66053878 \times 10^{-27}$ kg.

If we convert this into energy using the formula $E = mc^2$, we get:

$$1.66053878 \times 10^{-27} \times (2.99792458 \times 10^8)^2 = 1.49241783 \times 10^{-10} \, \text{J}$$

To convert this to eV, we divide by the charge of an electron to give:

$$\frac{1.49241783 \times 10^{-10}}{1.60217653 \times 10^{-19}} = 931.494 \, \text{MeV}$$

(The number of significant figures used in the calculation is necessary to get the correct answer.)

This is a very useful conversion factor when dealing with nuclear masses:
$1\ u = 1.661 \times 10^{-27}\ kg = 931.5\ MeV\ c^{-2}$

Calculating binding energy

Tables normally contain atomic masses rather than nuclear masses. This is the mass of a neutral atom so, it also includes Z electrons. Here, we will use atomic masses to calculate mass defect so to make sure the electron mass cancels out, we use the mass of a hydrogen atom instead of the mass of a proton, as shown in Table 1.

The amount of energy required to split iron into its parts can be calculated from Δmc^2, where Δm is the difference in mass between the iron nucleus and the nucleons that it contains.

From Table 1, we can see that the iron nucleus has 26 protons and $(54 - 26) = 28$ neutrons.

mass defect = mass of parts − mass of nucleus
= (mass of 26 protons + mass of 28 neutrons) − mass of iron nucleus

But if we use atomic masses:

$\Delta m = [(26 \times mass\ (hydrogen\ atom) - 26\ m_e) + 28\ m_n] - [mass\ (^{54}Fe\ atom) - 26\ m_e]$

The electrons cancel so:

$\Delta m = [26 \times mass\ (hydrogen\ atom) + 28\ m_n] - [\ mass\ (^{54}Fe\ atom)]$

$\Delta m = 26 \times 1.00782 + 28 \times 1.00866 - 53.9396 = 0.5062$

This is equivalent to: $0.5062 \times 931.5 = 471.5\ MeV$

Binding energy per nucleon

Larger nuclei generally have higher binding energies than smaller nuclei since they have more particles to separate. To compare one nucleus with another, it is better to calculate the **binding energy per nucleon**. This gives an indication of relative stability.

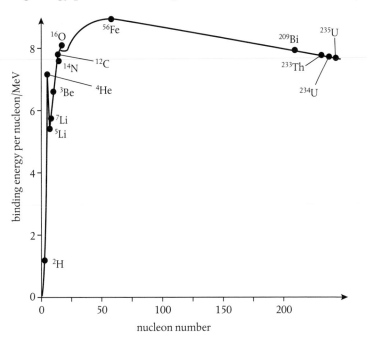

	Atomic mass/u
Hydrogen	1.00782
Neutron	1.00866
Iron ($^{54}_{26}$Fe)	53.9396

▲ **E.3 Table 1**

 You do not need to add the electrons in every time – just use atomic masses plus the mass of hydrogen instead of the proton mass and it will all work out.

E.3 Table 2
▼

	Atomic masses		
Z	Symbol	A	Mass/u
1	H	1	1.0078
1	D	2	2.0141
2	He	3	3.0160
3	Li	6	6.0151
4	Be	9	9.0122
7	N	14	14.0031
17	Cl	35	34.9689
26	Fe	54	53.9396
28	Ni	58	57.9353
36	Kr	78	77.9204
38	Sr	84	83.9134
56	Ba	130	129.9063
82	Pb	204	203.9730
86	Rn	211	210.9906
88	Ra	223	223.0185
92	U	233	233.0396

◀ **E.3 Figure 4** Graph of binding energy per nucleon vs nucleon number.

Figure 4 is a very
important graph, which
is used to explain
almost every aspect of
nuclear reactions from
the energy production
in stars to problems
with nuclear waste.

So for the previous example, the binding energy per nucleon of ^{54}Fe is $\frac{471.5}{54} = 8.7$ MeV/
nucleon. Figure 4 shows the binding energy per nucleon plotted against nucleon
number for a variety of nuclei.

> **HL** Challenge yourself
>
> **1.** Notice that the binding energy per nucleon changes a great deal more for
> nucleon numbers below 60 than above 60. Considering the forces in the nucleus,
> why do you think this is?

From Figure 4, we can see that some nuclei are more stable than others: iron and nickel
are in the middle so they are the most stable nuclei. If small nuclei join to make larger
ones, the binding energy per nucleon increases, resulting in a release of energy, but to
make nuclei bigger than iron, energy would have to be put in. This tells us something
about how the different elements found on Earth must have been formed. Small nuclei
are formed in the center of stars as the matter comes together under gravity, releasing
energy in the form of the light we see. When big nuclei form, energy is absorbed – this
happens toward the end of a star's life.

Remember: 1 u
is equivalent to
931.5 MeV.

> ### Exercise
>
> **Q1.** Find uranium in Table 2.
>
> (a) How many protons and neutrons does uranium have?
>
> (b) Calculate the total mass, in unified mass units, of the protons and
> neutrons that make uranium.
>
> (c) Calculate the difference between the mass of uranium and its constituents
> (the mass defect).
>
> (d) What is the binding energy of uranium in MeV?
>
> (e) What is the binding energy per nucleon of uranium?
>
> **Q2.** Enter the data from Table 2 into a spreadsheet. Add formulae to the
> spreadsheet to calculate the binding energy per nucleon for all the nuclei and
> plot a graph of binding energy per nucleon against nucleon number.

> Challenge yourself
>
> **2.** Estimate the amount of energy in joules required to break 3 g of copper (^{63}Cu)
> into its constituent nucleons.

Radioactive decay

To explain why a ball rolls down a hill, we can say that it is moving to a position of
lower potential energy. In the same way, the combination of protons and neutrons in
a nucleus will change if it results in an increased binding energy. This sounds like it is
the wrong way round but, remember, binding energy is the energy released when a
nucleus is formed. There are three main ways that a nucleus can change:

- **Alpha emission**: Alpha particles are helium nuclei. Emission of a helium nucleus results in a smaller nucleus so, according to the binding energy per nucleon curve, this would only be possible in large nuclei.
- **Changing a neutron to a proton**: This results in the emission of an electron (beta minus).
- **Changing a proton to a neutron**: This results in the emission of a positive electron (positron) (beta plus). This is quite rare.

Notice that all three changes result in the emission of a particle that makes the nucleus more stable. Energy can also be lost by the emission of high-energy electromagnetic radiation (gamma). The amount of energy associated with nuclear reactions is in the order of MeV so these particles are ejected at high speed. Ionizing an atom requires only a few electronvolts, so one of these particles can ionize millions of atoms as they travel through the air. This makes them harmful, but also easy to detect.

Detecting radiation

A Geiger–Müller, or GM, tube is a type of ionization chamber. It contains a low-pressure gas which, when ionized by a passing particle, allows a current to flow between two electrodes as in Figure 5.

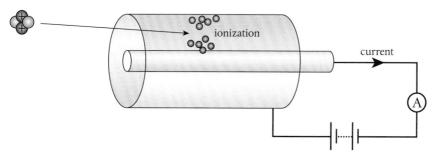

E.3 Figure 5 An ionization chamber.

By detecting this flow of charge, we can count individual particles. However, since all of the particles are ionizing, we cannot tell which type of radiation it is. To do this, we can use their different penetrating powers: most alpha particles, having the highest mass and charge, can be stopped by a sheet of paper; beta particles can pass through paper but are stopped by a thin sheet of aluminum; gamma rays, which are the same as high-energy X-rays but with higher frequency, are the most penetrating and will even pass through lead. For the same reason, the different radiations have varying ranges in air: alpha particles only travel a few centimeters and beta particles travel about 10 times further. Gamma radiation travels furthest but interacts slightly differently, resulting in an inverse square reduction in intensity with increasing distance.

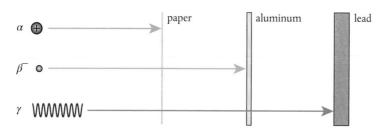

E.3 Figure 6 Absorption of radiation.

An alternative way of detecting radioactive particles is using a cloud chamber. This contains a vapor that turns into droplets of liquid when an ionizing particle passes through. The droplets form a visible line showing the path of the particle, rather like the vapor trail behind an airplane. A bubble chamber is similar but the trail is a line of bubbles.

A bubble chamber photograph from the CERN accelerator. There is a magnetic field directed out of the chamber, causing positive particles to spiral clockwise.

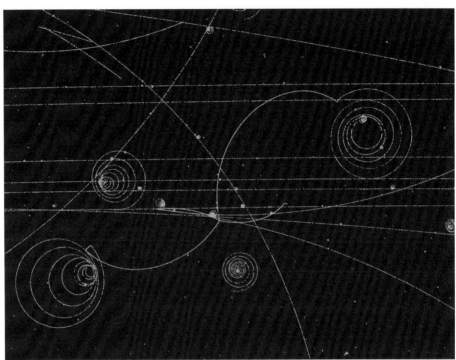

Alpha particles leave the thickest tracks as they are most ionizing, while beta particles leave a thinner track. A single gamma photon is least ionizing and would not leave a track unless absorbed by an atomic electron, which then, if given enough energy, could leave a track as it is ejected from the atom.

E.3 Table 3

Particle	Mass/u	Charge/e	Penetration power
Alpha (α)	4	+2	stopped by paper
Beta minus (β^-)	0.0005	−1	stopped by aluminum
Gamma (γ)	0	0	attenuated by lead (or, less so, by concrete or water) with reduced intensity as thickness increases

Alpha decay

When a nucleus emits an alpha particle, it loses 2 protons and 2 neutrons. The reaction can be represented by a nuclear equation. For example, radium decays into radon when it emits an alpha particle.

$$^{226}_{88}\text{Ra} \rightarrow {}^{222}_{86}\text{Rn} + {}^{4}_{2}\text{He}$$

4 nucleons are emitted, so the nucleon number is reduced by 4.

2 protons are emitted, so the proton number is reduced by 2.

Alpha particles are emitted by large nuclei that gain binding energy by changing into smaller nuclides.

Energy released

When radium changes to radon, the binding energy is increased. This leads to a drop in total mass, this mass having been converted to energy.

mass of radium > (mass of radon + mass of alpha)

energy released = $\{mass_{Ra} - (mass_{Rn} + mass_{alpha})\}c^2$

$mass_{Ra} = 226.0254\,u$

$mass_{Rn} = 222.0176\,u$

$mass_{He} = 4.002602\,u$

change in mass = $0.005198\,u$

This is equivalent to energy of $0.005198 \times 931.5\,MeV$.

energy released = $4.84\,MeV$

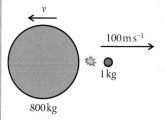
The alpha source americium-241 is used in many smoke detectors. The alpha radiation ionizes the air between two parallel plates, allowing charge to flow in a circuit. When smoke enters the gap between the plates, the current falls. It is this drop in current that triggers the alarm. Even though they contain a radioisotope, smoke detectors do not present a health risk since alpha radiation will not be able to penetrate the casing of the detector.

HL

Nuclear energy levels

When an alpha particle is ejected from a nucleus, it is like an explosion where a small ball flies apart from a big one, as shown in Figure 7. An amount of energy is released during the explosion, which gives the balls the kinetic energy they need to fly apart.

Applying the principle of conservation of momentum:

momentum before = momentum after

$$0 = 100 \times 1 - 800v$$

$$v = 0.125\,m\,s^{-1}$$

So: kinetic energy of the large ball = $\frac{1}{2} \times 800 \times 0.125^2 = 6.25\,J$

kinetic energy of the small ball = $\frac{1}{2} \times 1 \times 100^2 = 5000\,J$

E.3 Figure 7 An explosion analogy.

From this example, we can see that the small particle gets almost all of the energy. The same is true for the alpha decay. So for the example of radium, all alpha particles should have a kinetic energy of $4.84\,MeV$. Some isotopes emit two alpha particles that differ in energy, one taking more of the energy and a second with less energy, for example, a $4.84\,MeV$ alpha particle and a $4.74\,MeV$ alpha particle. Emission of the lower energy alpha particle would leave the nucleus in an unstable excited state. The remaining $0.1\,MeV$ of energy can be lost later through the emission of gamma radiation. This can be explained by considering a situation similar to electrons in an atom; the nucleus has energy levels too.

Are there differences between the photons emitted as a result of atomic transitions and the photons emitted as result of nuclear transitions? (E.1)

HL end

Exercise

Q3. Use the table to calculate the amount of energy released when $^{212}_{84}$Po decays into $^{208}_{82}$Pb.

Nuclide	Atomic mass/u
Polonium (Po) 212	211.988842
Lead (Pb) 208	207.976627

Beta minus (β^-) decay

Beta minus particles are electrons. They are exactly the same as the electrons outside the nucleus but they are formed when a neutron changes to a proton. When this happens, a particle called an antineutrino is also produced.

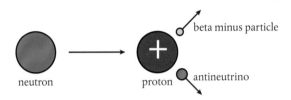

E.3 Figure 8 Beta minus decay.

If a neutron can change into a proton, it would be reasonable to think that a neutron was a proton with an electron stuck onto it. However, electrons cannot be confined to the nucleus. Electrons are not affected by the nuclear force.

Effect on nucleus

When a nucleus emits a beta particle, it loses 1 neutron and gains 1 proton.

Carbon-14 decays into nitrogen-14 when it emits a beta particle.

$$^{14}_{6}C \rightarrow {}^{14}_{7}N + {}^{0}_{-1}e + \bar{\nu}$$

Nuclei with too many neutrons decay by beta minus decay.

Energy released

When carbon-14 decays, the binding energy is increased, resulting in a drop in the total mass.

energy released = $[m_{C\,nucleus} - (m_{N\,nucleus} + m_e)]c^2$

If using atomic masses, then we should subtract the electrons.

energy released = $[(m_{C\,atom} - 6m_e) - (m_{N\,atom} - 7m_e + m_e)]c^2$

We can see that all the electrons cancel out, leaving $[m_{C\,atom} - m_{N\,atom}]c^2$.

$m_{C\,atom} = 14.003241$
$m_{N\,atom} = 14.003074$

So: energy released = 0.0001×931.5 MeV = 0.16 MeV

Living organisms are constantly taking carbon either directly from the atmosphere or by eating plants that are taking it from the atmosphere. In this way, the carbon content of the organism is the same as the atmosphere. The ratio of carbon-12 to carbon-14 is constant in the atmosphere so living organisms have the same ratio as the atmosphere. When organisms die, the carbon is no longer replaced, so as the carbon-14 starts to decay, the ratio of carbon-12 to carbon-14 increases. By measuring this ratio, it is possible to determine the approximate age of dead organic material.

The size of a beta particle is much smaller than the nucleus so we would expect that the beta particle would take all of the energy. However, this is not the case. Beta particles are found to have a range of energies, as shown by the energy spectrum in Figure 9.

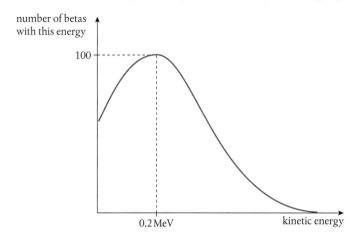

E.3 Figure 9 This graph shows the spread of beta particle energy. It is not really a line graph but a bar chart with thin bars. Notice how many beta particles have zero kinetic energy – this is because they are slowed down by electrostatic attraction to the nucleus.

Nature of Science

The neutrino was first hypothesized to explain how beta decay can take place without violating the law of conservation of energy. The alternative, that energy was not conserved, would have had too many consequences.

When beta radiation was first discovered, the existence of the neutrino was unknown, so physicists were puzzled as to how the beta particle could have a range of energies. It just is not possible if only one particle is ejected from a heavy nucleus. To solve this problem, Wolfgang Pauli suggested that there must be an 'invisible' particle coming out too. The particle was called a **neutrino** ('small neutral one') since it had no charge and negligible mass. Neutrinos are not totally undetectable and were eventually discovered about 26 years later. However, they are very unreactive and they can pass through thousands of kilometers of lead without an interaction.

It was later realized that the extra particle involved in β^- decay is actually the electron antineutrino.

HL end

How did conservation of energy lead to experimental evidence of the neutrino? (NOS)

Exercise

Q4. By calculating the change in binding energy, explain why $^{213}_{84}$Po will not decay by β^- emission into $^{214}_{85}$At.

Q5. Use the information in the table to calculate the maximum possible energy of the β^- radiation emitted when $^{139}_{56}$Ba decays into $^{139}_{57}$La.

Nuclide	Atomic mass /u
Polonium (Po) 213	212.992833
Astatine (At) 213	212.992911
Barium (Ba) 139	138.908835
Lanthanum (La) 139	138.906347

Beta plus (β^+) decay

E.3 Figure 10 Beta plus decay.

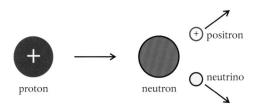

A beta plus particle is a positive electron, or positron. They are emitted from the nucleus when a proton changes to a neutron. When this happens, a neutrino is also produced.

A positron has the same properties as an electron but it is positive (it is the antiparticle of an electron), so beta plus particles have very similar properties to beta minus particles. They penetrate paper, have a range of about 30 centimeters in air, and are weakly ionizing. The beta plus track in a cloud chamber also looks the same as a beta minus track, unless a magnetic field is added. The different charged paths then curve in opposite directions as shown in the photo on page 493.

Effect on nucleus

When a nucleus emits a beta plus particle, it loses 1 proton and gains 1 neutron. An example of a beta plus decay is the decay of sodium into neon.

$$^{22}_{11}\text{Na} \rightarrow {}^{22}_{10}\text{Ne} + {}^{0}_{1}e + \nu$$

In common with negative electrons, positive electrons are not affected by the nuclear force. Once emitted, the positive electrons lose energy as they ionize atoms in the material they pass through. As they slow down, they are able to annihilate with an electron, resulting in gamma radiation.

E.3 Figure 11 Increasing stability through radioactive decay.

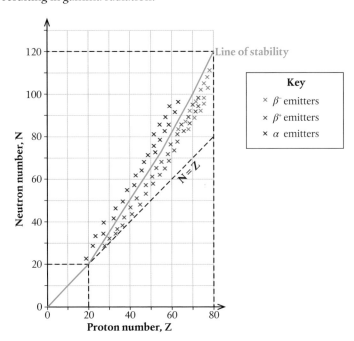

Challenge yourself

3. Show that the energy released when ^{22}Na decays by beta plus decay is 1.82 MeV.

Gamma radiation (γ)

Gamma radiation is electromagnetic radiation, so when it is emitted, there is no change in the particles of the nucleus – they just lose energy.

$$^A_Z X^* \rightarrow {}^A_Z X + {}^0_0 \gamma$$

Each time a nucleus decays, a photon is emitted. As we have seen, the energy released from nuclear reactions is very high, so the energy of each gamma photon is also high. The frequency of a photon is related to its energy by the formula $E = hf$. This means that the frequency of gamma photons is very high. Their high energy also means that if they are absorbed by atomic electrons, they give the electrons enough energy to leave the atom. In other words, they are ionizing, which means they can be detected with a GM tube, photographic paper or a cloud chamber. As they pass easily through human tissue, gamma rays have many medical applications.

HL The gamma radiation emitted from a given nucleus is quantized, which means that isotopes have characteristic 'fingerprints'. The emission spectrum can be compared with reference information to determine the isotope that emitted it. The quantization of gamma radiation, along with there being specific alpha radiation kinetic energies, is further evidence for nuclear energy levels.

Exercise

Q6. Americium-241 emits alpha particles with energies 5.485 MeV and 5.443 MeV. Calculate the frequency of gamma radiation that might also be emitted.

Decay

Radioactive decay is a **random** process, meaning that we cannot predict the moment in time when a nucleus will decay nor which nucleus will be the next to decay. Radioactive decay is also **spontaneous**; the rate of decay is unaffected by external factors (such as temperature). Quantum physics only applies to small-scale systems like the atom and nucleus. However, it might help us to understand some of the concepts if we consider something we can observe more directly, like bursting bubbles. If you blow a bubble with soapy water, it is very difficult to predict when exactly it is going to pop. It might last a few seconds or several minutes, so we can say that the popping of a bubble is fairly random. If we have a lot of bubbles, then lots of them will be popping at any given time. The number popping each second is directly proportional to the number of bubbles.

Soda bubbles.

SKILLS

When a fizzy drink is poured, bubbles form on the top. Normally you do not want too many bubbles, but in this experiment, you do as they are going to be measured as they pop. It would be rather difficult to count the bubbles, but as they burst, the level of the drink goes up so this can be used to measure how many bubbles have burst. You need a drink where the bubbles stay for a long time, like non-alcoholic beer; soda is a bit too fast.

The drink is poured into a measuring cylinder and level of the drink is measured every 5 or 10 seconds until most of the foam has gone. A good way of doing this is to take a video of the foam and measure its height by analyzing the frames.

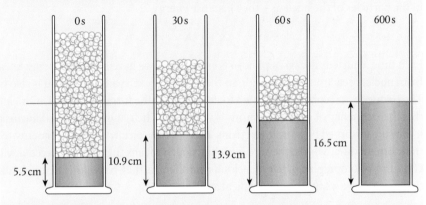

E.3 Figure 11 The height of liquid is measured as the bubbles burst.

Figure 11 shows the drink at four different times, showing how the height of the liquid changes. However, we want to measure the amount of foam. To do this, the drink is left for 10 minutes until all of the bubbles have burst. This is taken to be the zero foam level. If the liquid height is subtracted from this, we get the amount of liquid that was foam.

Plotting the amount of foam vs time, we get a curve like the one shown in Figure 12. The best-fit curve is an exponential decay curve with equation:

$$h = h_0 e^{-\lambda t}$$

SKILLS

A worksheet with full details of how to carry out this experiment is available on this page of the eBook.

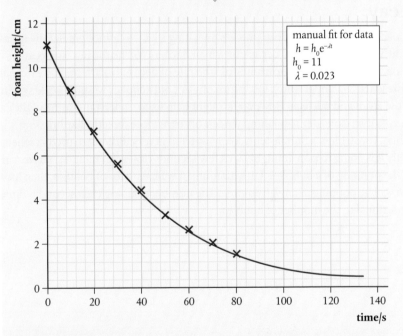

manual fit for data
$h = h_0 e^{-\lambda t}$
$h_0 = 11$
$\lambda = 0.023$

E.3 Figure 12 The exponential decay of foam.

The special thing about an exponential curve is that the gradient is proportional to the y-value. This means that the rate of change of height is proportional to the amount of foam. This is what we expected, since the rate at which the bubbles pop is proportional to the number of bubbles.

Higher Level students should take note of the form of the foam decay curve equation as it is similar to radioactive decay formulae. Standard Level students should simply note the shape of the curve and notice that the foam height halves in equal times, irrespective of the chosen starting height.

Decay constant, λ

The decay constant, λ, is the probability of decay per unit time. It is constant for a given isotope, irrespective of how many nuclei are present or how long the decay has running for. To understand this, we can plot exponential decay curves with a spreadsheet as shown in Figure 13.

Which areas of physics involve exponential change? (NOS)

E.3 Figure 13 Spreadsheet to plot exponential curves.

Challenge yourself

4. The decay constant being equal to the probability of decay requires that either the probability itself or the time interval be small. Why would large probabilities or time intervals invalidate this?

The equation used to calculate $e^{-\lambda t}$ is shown in the formula bar. The $ sign is used so that when the formula is copied into other cells, the cell locations do not change. From these curves, we can see that a higher value of λ gives a faster rate of decay.

HL end

Exponential decay of radioactive isotopes

Higher Level students are required to calculate changes during decay for any time interval using exponential relationships. Standard Level students are required to become familiar with the same vocabulary, but knowledge of the exponential relationships is not required.

Since radioactive decay is a random process, the rate of decay of nuclei is also proportional to the number of nuclei remaining. This results in an exponential decay, the same as with the bubbles. The equation for the decay is:

$$N = N_0 e^{-\lambda t}$$

where N_0 is the number of nuclei present at the start, N is the number after time t and λ is the decay constant. The decay constant can be thought of as the probability of a nucleus decaying in one second: the higher the probability, the faster the decay. Its magnitude depends on how unstable the nucleus is. In general, the more energy released by the decay, the more unstable the nucleus.

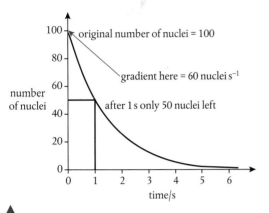

E.3 Figure 14 Decay curve showing half-life.

Half-life

The half-life is the time taken for *half* the nuclei to decay. This gives us some idea of the rate of decay; a *short* half-life implies a *rapid* decay. In Figure 14 the half-life is the time taken for the nuclei to decay to half the original value of 100; this is 1 s. The half-life for a given nuclide to decay is constant, so after a further 1 s, there will only be 25 nuclei left, and so on, until there are so few nuclei that statistical approximations become invalid.

Activity

Each time a nucleus decays, a radioactive particle is emitted so the rate of decay is the same as the rate of emission. This is called the **activity** of the sample and is measured in becquerel (emissions per second).

$$A = \lambda N$$

Since the rate of decay is directly proportional to the number of nuclei, the graph of activity vs time will also be exponential. Figure 15 shows the activity of the sample from Figure 14. Notice that this is simply the gradient of the previous graph.

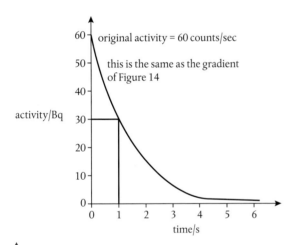

E.3 Figure 15 Activity vs time.

The equation for this line is:

$$A = \lambda N_0 e^{-\lambda t} = A_0 e^{-\lambda t}$$

Count rate

The higher the activity of a source, the higher the rate at which a detector will 'count' ionizing particles. Similar to both the number of undecayed nuclei and the activity, and in the presence of no other sources, count rate also follows an exponential decay relationship:

$$C = C_0 e^{-\lambda t}$$

Background radiation

Radioactivity is a natural phenomenon as there are radioactive isotopes in rocks, building materials, the air, and even our own bodies. This all adds up to give what is known as **background radiation**. If a Geiger–Müller tube is left in a room, then it will detect this radiation. This is normally in the region of 20 counts per minute. When performing experiments on radioactive decay in the laboratory, we can use lead to shield our detector. This will prevent most of the radiation getting to our detector.

However, some will get through, resulting in a systematic error. If the background count is measured separately, the count rate can be corrected accordingly. The source of this radiation is:

- the surrounding rock and soil; some rocks contain radioactive material such as uranium and thorium that decay over a long period of time.
- the air, which may contain radon gas, which is emitted when radium present in some rocks decays by emitting alpha particles. Some houses built into rock particularly rich in radium can have high concentrations of radon gas. People living in these houses are advised to have good ventilation.
- the Sun, which emits fast-moving particles that arrive at the surface of the Earth; these are called cosmic rays.

Relationship between λ and half-life

The half-life, $T_{\frac{1}{2}}$ is the time taken for the number of nuclei to decay to half the original value. So if the original number of nuclei = N_0, then after $T_{\frac{1}{2}}$ seconds, there will be $\frac{N_0}{2}$ nuclei.

HL

If we substitute these values into the exponential decay equation, we get:

$$\frac{N_0}{2} = N_0 e^{-\lambda T_{\frac{1}{2}}}$$

Canceling N_0:

$$\frac{1}{2} = e^{-\lambda T_{\frac{1}{2}}}$$

Taking natural logs of both sides gives: $\quad \ln(\frac{1}{2}) = -\lambda T_{\frac{1}{2}}$

This is the same as: $\quad \ln 2 = \lambda T_{\frac{1}{2}}$

So: $\quad T_{\frac{1}{2}} = \dfrac{0.693}{\lambda}$

So if we know the decay constant, we can find the half-life or vice versa.

HL end

The half-life is also the time taken for the activity or corrected count rate to halve. If the time for decay is a whole number of half-lives, then it is simply a matter of halving the original number, or the activity, the appropriate number of times.

Worked example

Cobalt-60 decays by beta emission and has a half-life of approximately 5 years. If a sample of cobalt-60 emits 40 beta particles per second, how many will the same sample be emitting in 15 years' time?

Solution

After 5 years, the activity will be 20 particles per second.

After another 5 years, it will be 10 particles per second.

Finally, after a further 5 years, it will emit 5 particles per second (5 Bq).

SKILLS

To find the decay constant and hence half-life of short-lived isotopes, the change in activity can be measured over a period of time using a GM tube. The activity of long-lived isotopes takes a long time to change so, in this case, the parent and daughter nuclei are separated.

HL Then, knowing the number of parent nuclei and the activity of the sample, the decay constant can be found from $A = -\lambda N$.

Exercise

Q7. ^{17}N decays into ^{17}O with a half-life of 4 s. How much ^{17}N will remain after 16 s, if you start with 200 g?

Q8. ^{11}Be decays into ^{11}B with a half-life of 14 s. If the ^{11}Be emits 100 particles per second, how many particles per second will it emit after 42 s?

Q9. A sample of dead wood contains $\frac{1}{16}$ of the amount of ^{14}C that it contained when alive. If the half-life of ^{14}C is 6000 years, how old is the sample?

H L

If the time is not a whole number of half-lives, then you have to use the exponential decay equation.

Worked example

Protactinium has a half-life of 70 s. A sample has an activity of 30 Bq. Calculate its activity after 10 minutes.

Solution

We are going to use the equation $A_t = A_0 e^{-\lambda t}$. First we need to find the decay constant $\lambda = \dfrac{0.693}{T_{\frac{1}{2}}} = 9.9 \times 10^{-3}\,\text{s}^{-1}$

The activity after 600 s is now $30 \times e^{-9.9 \times 10^{-3} \times 600} = 0.08\,\text{Bq}$

H L end

Applications of radioactive isotopes

There are two factors to consider when selecting isotopes for real-world use: how penetrating the ionizing particle is and the half-life of any given sample.

Here are some examples:

- Metastable technetium-99 (^{99}Tc) is used as a tracer in the human body to identify anomalies in blood flow or organ function. Its half-life is 6 hours, which is long enough to attribute variations in activity to a diagnosis (rather than decay) but short enough not to cause harm over time. It emits gamma radiation, which is penetrating enough for detection outside the body and not as damaging to DNA (through ionization) as a beta emitter.
- Leaks in underground pipes can be detected by gamma-emitting tracers, but the half-life can be longer than those used in the human body.
- Cobalt-60 (^{60}Co) is used in 'gamma knife' radiotherapy because it emits gamma radiation, which is sufficiently high in energy to kill cancer cells and penetrating enough to reach the tumor. Its half-life is in the order of years, but the treatment is relatively quick. Alternatively, small samples of isotopes can be injected into tumors in radioisotope therapy, which typically emit both beta and gamma radiation and have half-lives in the order of days.

- Factories that produce sheets of material rely on isotopes with large half-lives in the order of years for consistency. Beta emitters can be used in automated control of paper mills, because the count rate is reduced when the paper is too thick.

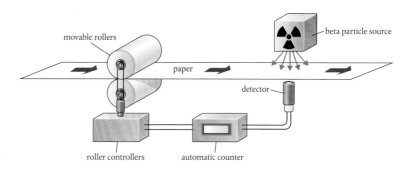

- The half-life of carbon-14 (^{14}C) is roughly 5700 years, which makes radioactive dating of objects that were once alive (and contain atmospheric ratios of carbon-12 and carbon-14) possible. Calculations of age often provide an underestimate, because the atmosphere may have contained more carbon-14 at the time the material was living.
- The americium used in smoke detectors because of its alpha decay (as previously discussed) has a half-life of approximately 400 years, which means the detector will not need to be replaced by a household.

E.3 Figure 16 The GM detector is on the opposite side of the paper to the beta emitter.

Exercise

Q10. A sample has an activity of 40 Bq. If it has a half-life of 5 mins, what will its activity be after 12 mins?

Q11. The activity of a sample of strontium-90 decreases from 20 Bq to 15.7 Bq in 10 years. What is the half-life of strontium 90?

Q12 Cobalt-60 has a half-life of 5.27 years. Calculate:
- (a) the half-life in s
- (b) the decay constant in s^{-1}
- (c) the number of atoms in 1 g of ^{60}Co
- (d) the activity of 1 g of ^{60}Co
- (e) the amount of ^{60}Co in a sample with an activity of 50 Bq.

You do not have to convert everything into seconds. If half-life is in years, then the unit of γ is year^{-1}.

Guiding Questions revisited

Why are some isotopes more stable than others?

In what ways can a nucleus undergo change?

How do large, unstable nuclei become more stable?

How can the random nature of radioactive decay allow for predictions to be made?

In this chapter, we have looked in detail at a specific part of the atom to explain that:
- The stability of a nucleus can be approximated by using binding energy per nucleon.
- **HL** Binding energy is released when nucleons fuse together under the strong nuclear force.

- In order to become more stable (and increase binding energy per nucleon), a nucleus can release ionizing radiation through alpha, beta or gamma decay. Nuclear reactions releasing alpha (α) and beta (β^- or β^+) particles change the composition of the nucleus, whereas gamma (γ) radiation is energy in the form of photons.
- α particles are helium nuclei. These are released from large, unstable nuclei that have a deficit of neutrons.
- β^- particles are electrons that are ejected from a nucleus in which a neutron changes to a proton. β^+ particles are positrons that are ejected from a nucleus in the rare event that a proton changes to a neutron.
- The output of each form of decay has distinctive properties and uses.
- **HL** The spectrum of possible energies provides evidence for the existence of the neutrino and nuclear energy levels.
- We can observe patterns when we have an isotope sample. Although we do not know which nucleus will be the next to decay nor precisely when, we do know that the lower the number of undecayed particles remaining, the lower the number of decays per unit time. The half-life of a given isotope is constant.
- **HL** Number of undecayed nuclei, activity and count rate all vary exponentially with time.

Practice questions

1. (a) Copy and complete the table by placing a tick ($\checkmark$) in the relevant columns to show how an increase in each of the following properties affects the rate of decay of a sample of radioactive material. (2)

Property	Effect on rate of decay		
	Increase	Decrease	Stays the same
Temperature of sample			
Pressure on sample			
Amount of sample			

Radium-226 ($^{226}_{88}$Ra) undergoes natural radioactive decay to disintegrate spontaneously with the emission of an alpha particle (α-particle) to form radon (Rn). The masses of the particles involved in the reaction are:

radium: 226.0254 u

radon: 222.0176 u

α-particle: 4.0026 u

(b) (i) Copy and complete the nuclear reaction equation below for this reaction.

$$^{226}_{88}\text{Ra} \rightarrow \underline{\quad}......+\underline{\quad}\text{Rn}$$ (2)

(ii) Calculate the energy released in the reaction. (3)

(c) The radium nucleus was stationary before the reaction.

(i) Explain, in terms of the momentum of the particles, why the radon nucleus and the α-particle move off in opposite directions after the reaction. (3)

(ii) The speed of the radon nucleus after the reaction is v_R and that of the α-particle is v_α. Show that the ratio $\frac{v_\alpha}{v_R}$ is equal to 55.5. (3)

(**iii**) Using the ratio given in (ii) above, deduce that the kinetic energy of the radon nucleus is much less than the kinetic energy of the α-particle. (3)

(Total 16 marks)

2. **HL** An isotope of helium $^6_2$He decays by emitting a β^- particle.

(**a**) State the name of the other particle that is emitted during this decay. (1)

(**b**) Explain why a sample of $^6_2$He emits β^- particles with a range of energies. (2)

(**c**) The half-life for this decay is 0.82 s. Determine the percentage of a sample of $^6_2$He that remains after a time of 10 s. (3)

(Total 6 marks)

3. **HL** A nucleus of the radioactive isotope potassium-40 decays into a stable nucleus of argon-40.

(**a**) Copy and complete the equation below for the decay of a potassium-40 nucleus.

$$^{40}_{19}\text{K} \longrightarrow {}^{40}_{18}\text{Ar} +$$ (2)

A certain sample of rocks contains 1.2×10^{-6} g of potassium-40 and 7.0×10^{-6} g of trapped argon-40 gas.

(**b**) Assuming that all the argon originated from the decay of potassium-40 and that none has escaped from the rocks, calculate what mass of potassium was present when the rocks were first formed. (1)

The half-life of potassium-40 is 1.3×10^9 years.

(**c**) Determine:

(i) the decay constant of potassium-40 (2)

(ii) the age of the rocks. (2)

(Total 7 marks)

4. Which of the following is the correct definition of the binding energy of a nucleus?

A The product of the binding energy per nucleon and the nucleon number

B The minimum work required to completely separate the nucleons from each other

C The energy that keeps the nucleus together

D The energy released during the emission of an alpha particle

(Total 1 mark)

5. A graph of the variation of average binding energy per nucleon with nucleon number has a maximum. What is indicated by the region around the maximum?

A The position below which radioactive decay cannot occur

B The region in which fission is most likely to occur

C The position where the most stable nuclides are found

D The region in which fusion is most likely to occur *(Total 1 mark)*

6. **HL** In a hydrogen atom, the sum of the masses of a proton and of an electron is larger than the mass of the atom. Which interaction is mainly responsible for this difference?

   A Electromagnetic

   B Strong nuclear

   C Weak nuclear

   D Gravitational   (Total 1 mark)

7. Which of the following lists the particles emitted during radioactive decay in order of increasing ionizing power?

   A γ, β, α

   B β, α, γ

   C α, γ, β

   D α, β, γ (Total 1 mark)

8. Which property of a nuclide does not change as a result of beta decay?

   A Nucleon number

   B Neutron number

   C Proton number

   D Charge   (Total 1 mark)

9. Bismuth-210 $\left(^{210}_{83}\text{Bi}\right)$ is a radioactive isotope that decays as follows:

   $$^{210}_{83}\text{Bi} \xrightarrow{\beta^-} \text{X} \xrightarrow{\alpha} \text{Y}$$

   What are the mass number and proton number of Y?

	Mass number	Proton number
A	206	86
B	206	82
C	210	82
D	214	83

                                                                       (Total 1 mark)

10. Three particles are produced when the nuclide $^{23}_{12}\text{Mg}$ undergoes beta plus (β^+) decay. What are two of these particles?

    A $^{23}_{11}\text{Na}$ and $^{0}_{0}\nu_e$

    B $^{0}_{-1}e$ and $^{0}_{0}\nu_e$

    C $^{23}_{11}\text{Na}$ and $^{0}_{0}\overline{\nu}_e$

    D $^{0}_{1}e$ and $^{0}_{0}\overline{\nu}_e$

                                                                       (Total 1 mark)

11. `HL` Photons of discrete energy are emitted during gamma decay. What is this evidence for?

A Atomic energy levels

B Nuclear energy levels

C Pair annihilation

D Quantum tunneling

(Total 1 mark)

12. `HL` What was a reason to postulate the existence of neutrinos?

A Nuclear energy levels had a continuous spectrum.

B The photon emission spectrum only contained specific wavelengths.

C Some particles were indistinguishable from their antiparticle.

D The energy of emitted beta particles had a continuous spectrum.

(Total 1 mark)

13. `HL` The positions of stable nuclei are plotted by neutron number n and proton number p. The graph indicates a dotted line for which $n = p$. Which graph shows the line of stable nuclides and the shaded region where unstable nuclei emit beta minus (β^-) particles?

A
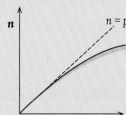

B
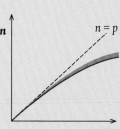

C
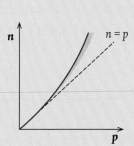

D
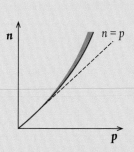

(Total 1 mark)

14. `HL` A pure sample of a radioactive nuclide contains N_0 atoms at time $t = 0$. At time t, there are N atoms of the nuclide remaining in the sample. The half-life of the nuclide is $T_{\frac{1}{2}}$. What is the decay rate of this sample proportional to?

A N **B** $N_0 - N$ **C** t **D** $T_{\frac{1}{2}}$

(Total 1 mark)

15. **HL** The graphs show the variation with time of the activity and the number of remaining nuclei for a sample of a radioactive nuclide. What is the decay constant of the nuclide?

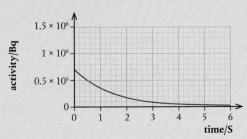

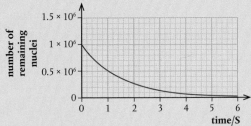

A 0.7 s^{-1}

B 1 s^{-1}

C $\frac{1}{0.7} \text{ s}^{-1}$

D 1.5 s^{-1}

(Total 1 mark)

16. **HL** The graph shows the variation of the natural log of activity, ln (activity), against time for a radioactive nuclide. What is the decay constant, in days^{-1}, of the radioactive nuclide?

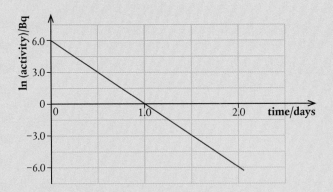

A $\frac{1}{6}$

B $\frac{1}{3}$

C   3

D   6

(Total 1 mark)

17. A radioactive nuclide is known to have a very long half-life. Three quantities known for a pure sample of the nuclide are:

I   the activity of the nuclide

II the number of nuclide atoms

III the mass number of the nuclide.   *(Total 1 mark)*

What quantities are required to determine the half-life of the nuclide?

A   I and II only

B   I and III only

C   II and III only

D   I, II and III   *(Total 1 mark)*

18. The half-life of a radioactive nuclide is 8.0 s. The initial activity of a pure sample of the nuclide is 10 000 Bq. What is the approximate activity of the sample after 4.0 s?

A   2500 Bq

B   5000 Bq

C   7100 Bq

D   7500 Bq   *(Total 1 mark)*

19. What is the definition of the unified atomic mass unit?

A $\frac{1}{12}$ the mass of a neutral atom of carbon-12

B   The mass of a neutral atom of hydrogen-1

C $\frac{1}{12}$ the mass of a nucleus of carbon-12

D   The mass of a nucleus of hydrogen-1   *(Total 1 mark)*

20. What is the energy equivalent to the mass of one proton?

A $9.38 \times (3 \times 10^8)^2 \times 10^6$ J

B $9.38 \times (3 \times 10^8)^2 \times 1.6 \times 10^{-19}$ J

C $\frac{9.38 \times 10^8}{1.6 \times 10^{-19}}$ J

D $9.38 \times 10^8 \times 1.6 \times 10^{-19}$ J   *(Total 1 mark)*

21. The mass defect for deuterium is 4×10^{-30} kg. What is the binding energy of deuterium?

A 4×10^{-7} eV

B 8×10^{-2} eV

C 2×10^6 eV

D 2×10^{12} eV   *(Total 1 mark)*

22. Meteorites contain a small proportion of radioactive aluminum-26 ($^{26}_{13}$Al) in the rock. The amount of $^{26}_{13}$Al is constant while the meteorite is in space due to bombardment with cosmic rays. Aluminum-26 decays into an isotope of magnesium (Mg) by β^+ decay:

$$^{26}_{13}\text{Al} \rightarrow \,^{X}_{Y}\text{Mg} + \beta^+ + Z$$

(a) Identify X, Y and Z in this nuclear decay process. (2)

(b) Explain why the beta particles emitted from the aluminum-26 have a continuous range of energies. (2)

After reaching Earth, the number of radioactive decays per unit time in a meteorite sample begins to diminish with time. The half-life of aluminum-26 is 7.2×10^5 years.

(c) State what is meant by half-life. (1)

(Total 5 marks)

23. A nucleus of phosphorus-32 ($^{32}_{15}$P) decays by beta minus (β^-) decay into a nucleus of sulfur-32 ($^{32}_{16}$S). The binding energy per nucleon of $^{32}_{15}$P is 8.398 MeV and for $^{32}_{16}$S it is 8.450 MeV.

(a) Determine the energy released in this decay. (2)

The graph shows the variation with time t of the activity A of a sample containing phosphorus-32 ($^{32}_{15}$P).

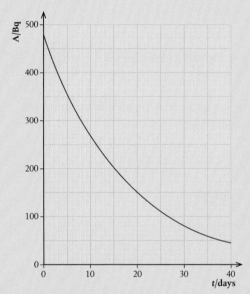

(b) Determine the half-life of $^{32}_{15}$P. (1)

(Total 3 marks)

24. **HL** The first scientists to identify alpha particles by a direct method were Rutherford and Royds. They knew that radium-226 ($^{226}_{86}$Ra) decays by alpha emission to form a nuclide known as radon (Rn).

(a) Write down the nuclear equation for this decay. (2)

At the start of the experiment, Rutherford and Royds put 6.2×10^{-4} mol of pure radium-226 in a small closed cylinder A. Cylinder A is fixed in the center of a larger closed cylinder B.

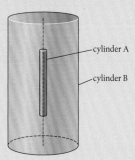

cylinder A

cylinder B

The experiment lasted for 6 days. The decay constant of radium-226 is 1.4×10^{-11} s^{-1}.

(b) Deduce that the activity of the radium-226 is almost constant during the experiment. (2)

(c) Show that about 3×10^{15} alpha particles are emitted by the radium-226 in 6 days. (3)

At the start of the experiment, all the air was removed from cylinder B. The alpha particles combined with electrons as they moved through the wall of cylinder A to form helium gas in cylinder B.

(d) The wall of cylinder A is made from glass. Outline why this glass wall had to be very thin. (1)

(e) The experiment was carried out at a temperature of 18 °C. The volume of cylinder B was 1.3×10^{-5} m^3 and the volume of cylinder A was negligible. Calculate the pressure of the helium gas that was collected in cylinder B over the 6-day period. Helium is a monatomic gas. (3)

(Total 11 marks)

25. (a) On a distant planet, you are going to leave a base station powered by a radioactive energy supply. You have a choice between two supplies of equal mass. Supply 1 uses a radioisotope with a six-month half-life. Supply 2 uses a different radioisotope, which is only half as radioactive as the first radioisotope but it has a one-year half-life. Which power supply will run the base longest? (1)

A Supply 1   **B** Supply 2   **C** Either

(b) Explain your answer. (1)

(Total 2 marks)

26. (a) Out of 1000 babies in a particular country, only 500 are expected to be alive at age 70. Suppose the radioisotope 'Humanitron' has a half-life of 70 years and you start with 1000 babies and 1000 'Humanitron' atoms. What will you find?   (1)

   A   The surviving number of people and atoms will always be approximately equal.

   B   During the first 70 years, the average number of surviving atoms will be larger than the average number of surviving people, but after 70 years, there will always be more surviving people.

   C   During the first 70 years, the average number of surviving people will be larger than the average number of surviving atoms, but after 70 years, there will always be more surviving atoms.

(b) Explain your answer.   (1)

(Total 2 marks)

27. HL   A radioactive substance, with a half-life of T, contains a particular nucleus that has **not** decayed over an observational period of $5T$. What is the probability that it will decay over a further period of:

(i) T (2)

(ii) $3T$ (2)

(Total 4 marks)

28. HL   The *Physical Review* is a physics journal published since 1893. A volume is published twice a year. After 1935, there was an increase in the number of articles published in each volume, making it thicker and thicker, so when it was stacked on the library shelf every six months, the front cover of the journal could be said to be moving along with an ever-increasing velocity. A physicist pointed out that if this continued, then the front cover would eventually exceed the speed of light (but according to relativity theory, there would be no information transmitted). We assume a simple model:
* The number of articles per volume increases exponentially, doubling every six months (the articles are of similar length).
* At the beginning of 1935, the volume was 1 cm thick.
* The velocity of the front cover is the thickness of that volume divided by the number of seconds in six months.

(a) By what factor does the thickness of the volume increase each year?   (1)

(b) What would be the thickness of the volume put on the shelf at the beginning of 1940?   (2)

(c) Write down your answer to (b) to one significant figure and in standard form.   (1)

(d) Using your answer to (c), write down the thickness of each volume for the next three years.   (2)

(e) Determine the year when the front cover of the volume stacked will exceed the velocity of light. You may find it helpful to write your answer to (d) including a term of the form 4^n.   (4)

(Hint: Use logs to base 10.)   *(Total 10 marks)*

E.4

Fission

Lise Meitner (1878–1968) was the second woman to be awarded a doctorate in physics at the University of Vienna and the first to become a professor of physics in Germany. In 1938, she and Otto Hahn discovered nuclear fission when bombarding various isotopes (including uranium) with neutrons, despite fleeing to Sweden that year because of the anti-Jewish Nuremberg Laws.
Unlike her colleague, Meitner was never awarded a Nobel Prize, although the radioactive element meitnerium (created in 1982) is named after her.

Guiding Questions

In which form is energy stored within the nucleus of the atom?

How can the energy released from the nucleus be harnessed?

When a 1 kg block of ice is lifted to a height of 1 m from the ground, approximately 10 J of work is done. Melting 1 kg of ice requires 334 000 J of energy. The reason that the amount of energy required to melt the ice is so much more than to lift it is that the force that holds the molecules together, the electric force, is much greater than the gravitational force.

The force holding the nucleus together is the strong nuclear force, which is much larger than the electric force. Consequently, the energy associated with nuclear reactions is more than that involved in chemical reactions or changes of state. This means that the energy released through fission per kilogram of uranium is much bigger than the energy released through combustion per kilogram of a fuel such as coal. This is an advantage because the cost of transporting the fuel is reduced.

Students should understand:

energy is released in spontaneous and neutron-induced fission
the role of chain reactions in nuclear fission reactions
the role of control rods, moderators, heat exchangers and shielding in a nuclear power plant
the properties of the products of nuclear fission and their management.

Nuclear fission

Nuclear fission is the splitting of large nuclei into two smaller ones. If we look at the graph of binding energy per nucleon vs nucleon number in Figure 1, we can see that the binding energies per nucleon of barium and krypton are higher than that of uranium. So if a uranium nucleus splits into a barium nucleus and a krypton nucleus, energy is released. This energy is transferred to the kinetic energy of the daughter nuclei and any excess neutrons. Uranium-236 nuclei do not tend to remain whole for millions of years without splitting. The fission process is spontaneous, which means it cannot be slowed down or speeded up by external efforts. However, when a uranium-235 nucleus absorbs a neutron, it temporarily turns into uranium-236. This increases the binding energy (because there is another neutron in the nucleus) but reduces the binding energy per nucleon enough to enable the nucleus to split.

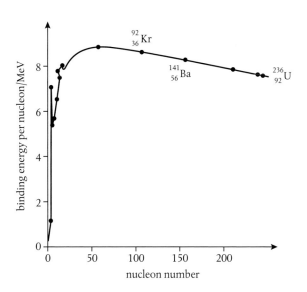

E.4 Figure 1 Binding energy per nucleon vs nucleon number curve revealing fission possibility.

The phrase 'binding energy' sounds as though an increase in binding energy should make the nucleus more stable and not less stable. In fact, the binding energy per nucleon is relatively low when the nucleus is in an excited state and the nucleons shake about. Energy is released when the nucleus becomes more stable and the binding energy per nucleon increases, for example, by the emission of gamma radiation or during fission.

You may have noticed that the numbers do not add up. Uranium has 144 neutrons but the total number of neutrons in barium and krypton is 141. These extra neutrons are expelled from the reaction and can be used in nuclear reactors to induce further fission reactions.

Worked example

Find the energy released if uranium-236 splits into krypton-92 and barium-141.

Solution

binding energy of ^{236}U = 236 × 7.6

$\qquad$ = 1793.6 MeV

binding energy of ^{141}Ba = 141 × 8

$\qquad$ = 1128 MeV

binding energy of ^{92}Kr = 92 × 8.2

$\qquad$ = 754.4 MeV

gain in binding energy = (1128 + 754.4) − 1793.6

$\qquad$ = 88.8 MeV

Since this leads to a release of energy, this process is possible.

Start by multiplying the binding energy per nucleon (from the vertical axis of the graph) by the number of nucleons for each of the nuclei. Then calculate the difference in binding energy between the original nucleus and the daughter nuclei to see whether the binding energy (i.e. the energy released) has increased. Note that any neutrons released are not 'bound' to anything, so there is no binding energy to be considered.

To answer these questions (in which we have not been provided with binding energy information), you must find the difference in mass between the original nucleus and the products. To convert to MeV, simply multiply by 931.5.

Finland produces 30% of its electricity from nuclear power so has a lot of nuclear waste to dispose of. To solve this problem, they are constructing many kilometers of tunnels into the very old and geologically sound rock close to Olkiluoto. The facility, known as Onkalo, will eventually contain 9000 tonnes of spent fuel.

How is binding energy used to determine the rate of energy production in a nuclear power plant (E.3)?

E.4 Figure 2 A chain reaction from nuclear fission.

Exercise

Use the table to answer the following questions.

Q1. If ^{236}U splits into ^{100}Mo and ^{126}Sn, how many neutrons will be produced? Calculate the energy released in this reaction.

Q2. ^{233}U splits into ^{138}Ba and ^{86}Kr plus nine neutrons. Calculate the energy released when this takes place.

Z	Symbol	A	Mass/u
92	U	233	233.039628
92	U	236	236.045563
42	Mo	100	99.907476
50	Sn	126	125.907653
56	Ba	138	137.905233
36	Kr	86	85.910615
0	n	1	1.008664

Nuclear power

When a large nucleus splits into two smaller ones, the total binding energy increases, resulting in the release of about 100 MeV of energy. This is approximately 100 million times more energy per molecule than when coal burns, so any material that does this would be a valuable source of energy.

If a uranium-235 nucleus absorbs a neutron, it will undergo fission, but naturally occurring uranium only contains 0.7% ^{235}U. The remainder is mainly ^{238}U, which absorbs neutrons without undergoing fission. Before naturally occurring uranium can be used as a fuel, the amount of ^{235}U must be increased to about 3%. This is called **enrichment**.

Fuel	Energy density /MJ L^{-1}
Uranium fuel	1 534 000 000
Coal	72.4

E.4 Table 1

The chain reaction

Splitting a ^{235}U nucleus requires some energy because the nucleons are held together by a strong force. This energy can be supplied by adding a neutron to a ^{235}U nucleus. This reduces the stability of the nucleus so it splits in two. As a result, there are too many neutrons for the smaller daughter nuclei and some are released. These neutrons can be captured by more ^{235}U nuclei, and so on, leading to a chain reaction.

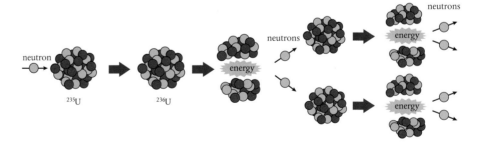

Moderation

The neutrons will only be absorbed if they are traveling slowly – otherwise, they will pass straight through. In terms of kinetic energy, this would be about 1 eV, which is much less than the MeV they possess after being expelled during the fission process. To achieve a chain reaction, we need these neutrons to be absorbed so they need to be slowed down or **moderated**. This is done by introducing some small nuclei between the ^{235}U (Figure 3). Water is sometimes used as a moderator because it contains hydrogen nuclei. The neutrons collide with these nuclei and, because they have a similar mass to the neutrons, they receive energy and the neutrons slow down: a near-elastic collision.

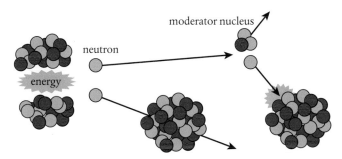

Critical mass

Another critical factor that determines whether a chain reaction can take place is the size of the piece of uranium. If it is too small, then before the neutrons have traveled far enough to be slowed down, they will have left the reacting piece of uranium. The minimum mass required for a chain reaction is called the **critical mass**.

The nuclear power station

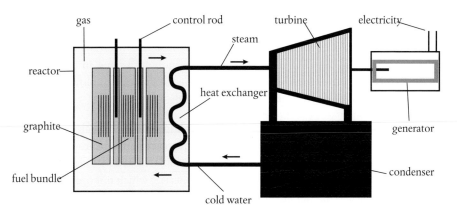

Control in a nuclear reactor

The rate of reaction in a nuclear reactor is limited by the fact that the fuel contains a high proportion of ^{238}U, which absorbs neutrons. Since this cannot easily be altered, it cannot be used to slow the reaction down if it goes too fast. Instead, this is done by introducing control rods of a neutron-absorbing material, such as boron, between the fuel rods (Figure 4).

Using the word *moderation* sounds like the moderator slows down the reaction, but it does not. The moderator slows down the *neutrons*, enabling the chain reaction. If the moderator was removed, the reaction would slow down.

E.4 Figure 3 Only slow-moving neutrons are absorbed.

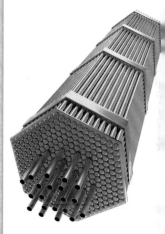

Pellets of nuclear fuel are stacked into tubes, which are bundled together before being put into the reactor.

E.4 Figure 4 An advanced gas-cooled reactor (AGR).

The efficiency of the nuclear reactor is not as high as might be expected. Firstly, the fuel has to be enriched, which takes a lot of energy. Then, it is not possible to get all the energy from the fuel, because when the amount of ^{235}U falls below a certain value, a chain reaction can no longer be sustained.

In which form is energy released as a result of nuclear fission (A.3)?

E.4 Figure 5 Sankey diagram for a nuclear reactor.

A nuclear reactor does not burn fuel so is not dependent on oxygen for it to function. This makes nuclear power particularly useful for powering submarines.

To what extent is there a role for fission in addressing climate change? (NOS)

Different countries have different policies regarding the use of nuclear power. What is the policy where you live?

There are many different designs of nuclear reactor but they all have a nuclear reaction at the core. The energy released when the nuclei split is given to the fission fragments. The temperature of a body is related to the average kinetic energy of the particles, which means that the temperature of the fuel increases. The hot fuel can then be used to boil water and drive a turbine as in the coal-fired power station.

The nuclear reactor is the part that produces heat and contains the fuel rods surrounded by, in this case, a graphite moderator (pale orange in Figure 4). The control rods can be raised and lowered to control the rate of reaction. The nuclear reactor is housed in a pressure vessel in which a gas is circulating (blue). This picks up heat from the fuel rods and transfers it to water in the heat exchanger. This water turns to steam and turns the turbine. The steam cools down and turns back to water in the condenser and is recirculated.

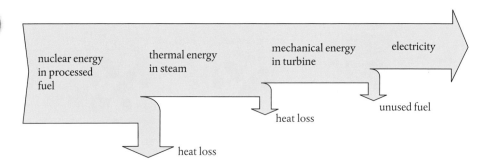

Since there is no burning of fuel involved in the nuclear reactor, no carbon dioxide is produced. This makes it favorable when compared to the use of fossil fuels to produce electricity. However, there are other problems.

Meltdown

If the nuclear reaction is not controlled properly, it can overheat and the fuel rods can melt. This is called **meltdown**. When this happens, the fuel cannot be removed and may cause the pressure vessel to burst, sending radioactive material into the atmosphere. Situations like this have occurred in Chernobyl, Ukraine, in 1986 and in Fukushima, Japan, in 2011. However, it is not possible for a reactor to blow up as an atom bomb, since the fuel is not of a high enough grade.

Meltdown can be caused by a malfunction in the cooling system or a leak in the pressure vessel. It would result in severe damage to the reactor, maybe leading to complete shutdown. Further damage outside the reactor is limited by the containment building, an airtight steel construction covered in concrete, which not only prevents dangerous material leaking out (an effect known as **shielding**), but will withstand a missile attack from the outside. Improved reactor design and construction, coupled with computer monitoring of possible points of weakness, has reduced the possibility of any failure of the structure that might lead to meltdown. However, it is almost impossible to design a reactor that will withstand the force of a major earthquake.

Low level waste

The extraction of uranium from the ground, the process of fuel enrichment and the transfer of heat from the fuel rods all leave some traces of radioactive material that must be carefully disposed of. The amount of radiation given off by this material is not great, but it must be disposed of in places away from human contact for 100–500 years.

Old reactors are another form of low level waste. They cannot simply be knocked down and recycled since most of the parts will have become radioactive. Instead, they must be left untouched for many years before demolition, or they can be encased in concrete.

High level waste

The biggest problem faced by the nuclear power industry is the disposal of spent fuel rods. Some of the product isotopes they contain are radioactive and have a half-life of thousands of years so need to be placed in safe storage for a very long time. Plutonium fuel rods are not considered safe for at least 240 000 years. There have been many suggestions regarding the waste: sending it to the Sun, putting it at the bottom of the sea, burying it in the icecap, or dropping it into a very deep hole. For the moment, most of it is dealt with in one of two ways:

- It is stored under water at the site of the reactor for several years to cool off, then sealed in steel cylinders.

- It is reprocessed to separate the plutonium and any remaining useful uranium from the fission fragments. This results in waste that is high in concentrations of the very radioactive fission fragments, but the half-life of these fragments is much shorter than either uranium or plutonium, so the need for very long-term storage is reduced.

 TOK

There seem to be a lot of problems associated with the production of energy from nuclear fuel. However, these problems are not insoluble. Scientists found a way of producing the energy, so it seems likely that they will be able to find solutions to these problems.

Exercise

Q3. Barium-142 ($^{142}_{56}$Ba) is a possible product of the fission of uranium-236. It decays by β^- decay to lanthanum (La) with a half-life of 11 months.

   (a) Write the equation for the decay of barium.

   (b) Estimate how long will it take for the activity of the barium in a sample of radioactive waste to fall to $\frac{1}{1000}$ of its original value.

Q4. Plutonium-239 splits into zirconium-96 and xenon-136. Use Table 2 to answer the following questions.

   (a) How many neutrons will be emitted?

   (b) Write the nuclear equation for the reaction.

   (c) How much energy is released when the fission takes place?

   (d) What is the mass of 1 mole of plutonium?

   (e) How many atoms are there in 1 kg of plutonium?

   (f) How much energy in eV is released if 1 kg of plutonium undergoes fission?

   (g) Convert the answer to part (f) into joules.

Isotope	Mass/u
^{239}Pu	239.052158
^{96}Zr	95.908275
^{136}Xe	135.907213
Neutron	1.008664

E.4 Table 2

Q5. A sample of nuclear fuel contains 3% ^{235}U. If the energy density of ^{235}U is 9×10^{13} J kg^{-1}, how much energy will 1 kg of fuel release?

Q6. An individual uses around 10 000 kW h of energy in a year.

(a) How many joules is this?

(b) Using the energy density information in Table 2, calculate how much nuclear fuel this amounts to.

Guiding Questions revisited

In which form is energy stored within the nucleus of the atom?

How can the energy released from the nucleus be harnessed?

In this chapter, we have looked at the binding energies within nuclei and the structural components involved in commercial fission reactors to reveal how:

- Energy is released when nucleons come together in the nucleus under the nuclear strong force and we refer to this as the 'binding energy'. Energy input is required to separate the nucleons.
- If the binding energy per nucleon increases during a nuclear fission reaction, energy will be released in the form of the kinetic energy of the products. This can be harnessed by boiling water in a heat exchanger, with the resulting steam used to turn a turbine and power an electrical generator.
- Control rods to absorb excess neutrons and slow the reaction, moderators to slow neutrons, enabling a continued chain reaction, and shielding to prevent the release of ionizing radiation are also required. Because the products of fission are radioactive, preparations must be made for their storage when the fuel rods are spent.

Practice questions

1. (a) (i) Distinguish between *fission* and *radioactive decay*. (4)

A nucleus of uranium-235 ($^{235}_{92}$U) may absorb a neutron and then undergo fission to produce nuclei of strontium-90 ($^{90}_{38}$Sr) and xenon-142 ($^{142}_{54}$Xe), and some neutrons.

The strontium-90 and the xenon-142 nuclei both undergo radioactive decay with the emission of β^- particles.

(ii) Write down the nuclear equation for this fission reaction. (2)

(iii) State the effect, if any, on the nucleon number and on the proton number of a nucleus when the nucleus undergoes β^- decay. (2)

The uranium-235 nucleus is stationary at the time that the fission reaction occurs. In this fission reaction, 198 MeV of energy is released. Of this total energy, 102 MeV and 65 MeV are the kinetic energies of the strontium-90 and xenon-142 nuclei respectively.

(**b**) (**i**) Calculate the momentum of the strontium-90 nucleus. (4)

(**ii**) Explain why the momentum of the strontium-90 nucleus is not exactly equal to the momentum of the xenon-142 nucleus. (2)

In the diagram, the circle represents the position of a uranium-235 nucleus before fission. The momentum of the strontium-90 nucleus after fission is represented by the arrow.

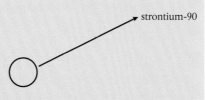

(**iii**) Copy the diagram and draw an arrow to represent the momentum of the xenon-142 nucleus after the fission. (2)

(**c**) In a fission reactor for the generation of electrical energy, 25% of the total energy released in a fission reaction is transferred to electrical energy.

(**i**) Using the data in (b), calculate the electrical energy, in joules, produced as a result of nuclear fission of one nucleus. (2)

(**ii**) The specific heat capacity of water is $4.2 \times 10^3 \, \text{J kg}^{-1} \, \text{K}^{-1}$. Calculate the energy required to raise the temperature of 250 g of water from 20 °C to its boiling point (100 °C). (3)

(**iii**) Using your answer to (c)(i), determine the mass of uranium-235 that must be fissioned in order to supply the amount of energy calculated in (c)(ii). The mass of a uranium-235 atom is $3.9 \times 10^{-25} \, \text{kg}$. (4)

(Total 25 marks)

2. (**a**) When a neutron 'collides' with a nucleus of uranium-235 ($^{235}_{92}\text{U}$), the following reaction can occur:

$$^{235}_{92}\text{U} + ^{1}_{0}\text{n} \rightarrow ^{144}_{56}\text{Ba} + ^{90}_{36}\text{Kr} + 2^{1}_{0}\text{n}$$

(**i**) State the name given to this type of nuclear reaction. (1)

(**ii**) Energy is liberated in this reaction. In what form does this energy appear? (1)

(**b**) Describe how the neutrons produced in this reaction may initiate a chain reaction. (1)

The purpose of a nuclear power station is to produce electrical energy from nuclear energy. The diagram is a schematic representation of the principal components of a nuclear reactor 'pile' used in a certain type of nuclear power station.

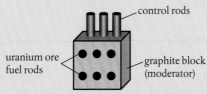

The function of the moderator is to slow down neutrons produced in a reaction such as that described in part (**a**).

(c) **(i)** Explain why it is necessary to slow down the neutrons. (3)

(ii) Explain the function of the control rods. (2)

(d) Describe briefly how the energy produced by the nuclear reactions is extracted from the reactor pile and then transferred to electrical energy. (4)

(Total 12 marks)

3. During the nuclear fission of nucleus X into nucleus Y and nucleus Z, energy is released. The binding energies per nucleon of X, Y and Z are B_X, B_Y and B_Z respectively. What is true about the binding energy per nucleon of X, Y and Z?

A $B_Y > B_X$ and $B_Z > B_X$

B $B_X = B_Y$ and $B_X = B_Z$

C $B_X > B_Y$ and $B_X > B_Z$

D $B_X = B_Y + B_Z$

(Total 1 mark)

4. A neutron collides head-on with a stationary atom in the moderator of a nuclear power station. The kinetic energy of the neutron changes as a result. There is also a change in the probability that this neutron can cause nuclear fission. What are these changes?

	Change in kinetic energy of the neutron	Change in probability of causing nuclear fission
A	increase	increase
B	decrease	increase
C	increase	decrease
D	decrease	decrease

(Total 1 mark)

5. A nuclear reactor contains atoms that are used for moderation and atoms that are used for control. What are the ideal properties of the moderator atoms and the control atoms in terms of neutron absorption?

	Ideal moderator atom	Ideal control atom
A	poor absorber of neutrons	poor absorber of neutrons
B	poor absorber of neutrons	good absorber of neutrons
C	good absorber of neutrons	poor absorber of neutrons
D	good absorber of neutrons	good absorber of neutrons

(Total 1 mark)

6. What is the function of control rods in a nuclear power plant?

A To slow neutrons down

B To regulate fuel supply

C To exchange thermal energy

D To regulate the reaction rate

(Total 1 mark)

7. A nuclear power station contains an alternating current generator. Which energy transfer is performed by the generator?

A Electrical to kinetic

B Kinetic to electrical

C Nuclear to kinetic

D Nuclear to electrical

(Total 1 mark)

8. One possible fission reaction of uranium-235 (U-235) is:

$$^{235}_{92}U + ^{1}_{0}n \rightarrow ^{140}_{54}Xe + ^{94}_{38}Sr + 2^{1}_{0}n$$

Mass of one atom of U-235 = 235 u
Binding energy per nucleon for U-235 = 7.59 MeV
Binding energy per nucleon for Xe-140 = 8.29 MeV
Binding energy per nucleon for Sr-94 = 8.59 MeV

(a) State what is meant by the binding energy of a nucleus. (1)

(b) Outline why quantities such as atomic mass and nuclear binding energy are often expressed in non-SI units. (1)

(c) Show that the energy released in the reaction is about 180 MeV. (1)

A nuclear power station uses U-235 as fuel. Assume that every fission reaction of U-235 gives rise to 180 MeV of energy.

(d) Estimate, in $J\,kg^{-1}$, the specific energy of U-235. (2)

(e) The power station has a useful power output of 1.2 GW and an efficiency of 36%. Determine the mass of U-235 that undergoes fission in one day. (2)

(f) The specific energy of fossil fuel is typically 30 MJ kg^{1}. Suggest **one** advantage of U-235 compared with fossil fuels in a power station. (1)

A sample of waste produced by the reactor contains 1.0 kg of strontium-94 (Sr-94). Sr-94 is radioactive and undergoes beta minus (β^-) decay into a daughter nuclide X. The reaction for this decay is:

$$^{94}_{38}Sr \rightarrow X + \bar{v}_e + e$$

(g) Write down the proton number of nuclide X. (1)

The graph shows the variation with time of the mass of Sr-94 remaining in the sample.

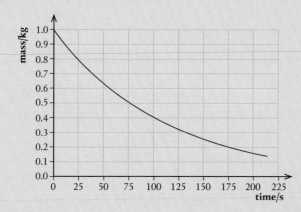

(h) State the half-life of Sr-94. (1)

(i) Calculate the mass of Sr-94 remaining in the sample after 10 minutes. (2)

(Total 12 marks)

E.5

Fusion and stars

◄ This photograph, captured in the infrared region by NASA's James Webb Space Telescope, was published in July 2022. It shows a galaxy cluster, a group of galaxies, as it appeared 4.6 billion years ago across an expanse of space measuring the size of a grain of sand held at arm's length. Each point of light represents one galaxy and each galaxy contains billions (if not hundreds of billions) of stars. Stars are powered by nuclear fusion.

Guiding Questions

How are elements created?

What physical processes lead to the evolution of stars?

Can observations of the present state of the Universe predict the future outcome of the Universe?

Stars are hot spheres of gas formed when particles come together due to the force of gravity. As the particles approach each other, they lose gravitational potential energy and gain kinetic energy. This results in an increase in temperature. The center (or core) of the star becomes very hot and dense, which enables nuclei to fuse. Fusion releases energy, which radiates outward, causing a force that opposes the gravitational attraction. Most of the stars we see are in this balanced **main sequence** state, when hydrogen is fusing.

The energy from the fusion in the core heats the outer layers, causing them to give out visible light. This is the light that we see, and by analyzing it, we can find out the temperature and composition of the star's surface.

The **brightness** of a star as seen on the Earth is determined by the amount of power received per unit area, which is related to the distance between the star and the Earth and how much power the star radiates. The power of a star is called **luminosity**.

The distance to a given star can be found using a variety of methods, but parallax is the simplest. Parallax is the change in position of stars relative to more distant background stars as the Earth moves through space in its orbit around the Sun. Imagine looking at four trees: A is closest, followed by B, and C and D are in the far distance. If you move your head from side to side, the two distant trees, C and D, will not move relative to each other. However, the closer trees will move relative to the distant ones. The closest tree, A, will appear to move more than tree B. Stars 'move' in the same way, so we can use this to measure their distance.

If we measure the distance to a star and its brightness, we can calculate its luminosity. If we know its temperature, we can determine how big it is. The relationship between these quantities can be represented on a Hertzsprung–Russell (HR) diagram. HR diagrams enable us to determine what type of star we are observing. Types of star range from very large red giants to small white dwarfs, and represent different stages of a star's life cycle.

The evolution of a star depends on its mass, but all stars spend the longest part of their lives as stable main sequence stars, which is why there are more main sequence stars than any other type.

Students should understand:

the stability of stars relies on an equilibrium between outward radiation pressure and inward gravitational forces
fusion is a source of energy in stars
the conditions leading to fusion in stars in terms of density and temperature
the effect of stellar mass on the evolution of a star
the main regions of the Hertzsprung–Russell (HR) diagram and how to describe the main properties of stars in these regions
the use of stellar parallax as a method to determine the distance d to celestial bodies as given by $d(\text{parsec}) = \dfrac{1}{p(\text{arc second})}$
how to determine stellar radii.

An artist's impression of the fusion of ^{2}H and ^{3}H to form ^{4}He.

Nuclear fusion

Nuclear fusion is the joining up of two small nuclei to form one big one.

If we look at the binding energy per nucleon vs nucleon number curve (Figure 1), we see that the line initially rises steeply. If you were to add two $^2_1$H nuclei to get one $^4_2$He nucleus, then the He nucleus would have more binding energy per nucleon; the He nucleus is more stable than the two fusing nuclei.

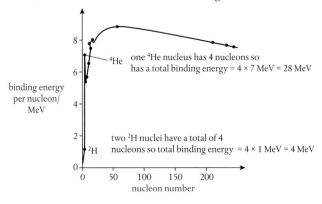

E.5 Figure 1 Binding energy per nucleon vs nucleon number curve, showing fusion possibility.

If we add up the total binding energy for the helium nucleus, it has 24 MeV more binding energy than the two hydrogen nuclei. This means that 24 MeV would be released. This could be by the emission of gamma radiation.

Worked example

Calculate the energy released by the following reaction:

$$\mathrm{^2_1H + ^3_1H \rightarrow ^4_2He + ^1_0n}$$

Solution

If the masses are added, we find that the mass of the original nuclei is greater than the mass of the final ones. This mass has been converted to energy.

mass difference = 0.018883 u

1u is equivalent to 931.5 MeV so energy released = 17.6 MeV

How is fusion like and unlike fission? (E.4)

It would not be possible to conserve both kinetic energy and momentum if two fast-moving nuclei collided and fused together. That is why all the reactions result in two particles not one big one.

Each small nucleus has a positive charge so they will repel each other. To make the nuclei come close enough for the strong force to pull them together, they must be thrown together with very high velocity. For this to take place, the matter must be heated to temperatures as high as the core of the Sun (about 13 million kelvin) or the particles must be thrown together in a particle accelerator.

The fusion reaction produces a lot of energy per unit mass of fuel and much research has been carried out to build a fusion reactor.

Exercise

Q1. Use the data in Table 1 to calculate the change in mass and hence the energy released in the following examples of fusion reactions:

(a) $\mathrm{^2_1H + ^2_1H \rightarrow ^3_2He + ^1_0n}$

(b) $\mathrm{^2_1H + ^2_1H \rightarrow ^3_1H + ^1_1p}$

(c) $\mathrm{^2_1H + ^3_2He \rightarrow ^4_2He + ^1_1p}$

Nuclide	Mass/u
^{1}H	1.007 825
^{2}H	2.014 101
^{3}H	3.016 049
^{3}He	3.016 029
^{4}He	4.002 603
^{1}n	1.008 664

E.5 Table 1

Astronomical distances

Astronomical distances are so large that the meter is not a particularly useful unit. Instead, the **light year** (ly) is often used. This is the distance traveled in a vacuum by light in 1 year:

$$1\ \mathrm{ly} = 9.46 \times 10^{15}\ \mathrm{m}$$

Another useful unit for distances on a solar system scale is the **astronomical unit** (AU). This is the average distance between the Sun and the Earth:

$$1\ \mathrm{AU} = 1.5 \times 10^{11}\ \mathrm{m}$$

One degree can be split up into 60 arc minutes and each arc minute into 60 arc seconds, so there are 3600 arc seconds in one degree.

To measure astronomical distances, astronomers cannot measure lengths directly but instead use the angles subtended by objects as the Earth orbits the Sun. In this case, the **parsec** is a more convenient unit to use since it can be found directly from the angle. The parsec is defined by the triangle in Figure 2. If the angle subtended between two points separated by 1 AU and a distant star is 1 arc second, then the distance to the star is 1 parsec (pc).

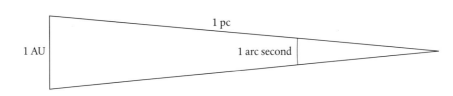

So if the angle is smaller, the distance is larger. The distance in pc = $\dfrac{1}{\text{angle in arc seconds}}$

$$1 \text{ pc} = 3.26 \text{ ly}$$

Although not developed specifically for use with telescopes, astronomers were quick to realize the potential of digital photography and have made many contributions toward the development of this technology.

Getting an idea of the relative size of the different structures in the Universe is very difficult as we cannot draw them all on the same page. However, we can build up step by step, as in Figure 3. For instance, if the size of the Sun was the size of a small insect, then the solar system would be the size of the great pyramid. If the solar system was the size of the insect, the Milky Way would be the size of Mount Everest, etc.

TOK

From an early age, we are taught to appreciate the relative sizes of objects by comparing them to each other. This becomes problematic when trying to comprehend the difference in size between the Sun and the Universe. No matter how small you draw the Sun, you cannot get the Universe on the same page. To manage this, the Sun would have to be smaller than a proton.

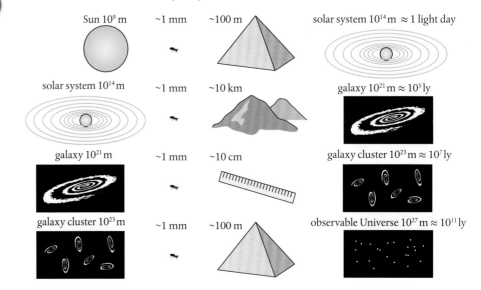

E.5 Figure 3 The relative sizes of different structures in the Universe.

Exercise

Q2. The distance to the nearest star is 4.3×10^{13} km. What is this in light years?

Q3. How long does it take light to travel from the Sun to the Earth?

Q4. How long would it take for a rocket traveling at 30 000 km h⁻¹ to travel to the nearest star from the Earth?

Q5. What is the distance to the nearest star in parsecs? What angle does this star subtend to the Earth when the Earth has moved a distance of 2 AU from one side of the Sun to the other? Verify that the distance in pc agrees with the angle.

Stellar parallax

As mentioned previously, astronomers cannot measure distances directly but use the angle subtended by a star as the Earth moves around the Sun. This technique is known as stellar parallax. To reduce uncertainties, the biggest angle possible should be measured, so the angles are measured when the Earth is on opposite sides of the Sun, which means measuring the position of a star once then again six months later.

Parallax is the way objects move relative to each other as you move past them. If you look straight ahead and move your head to the left, then objects closer to you will move to the right relative to objects further away (assuming you are not looking at a blank wall). Very distant objects do not move at all so can be used as a reference direction when measuring the angles. Consider the simplified version in Figure 4. To find the distance to the red star, the telescope is lined up at position A with the distant blue star. The telescope is then rotated to the red star and the angle measured. Six months later, the blue star is still in the same position but, due to parallax, the red star has moved relative to it. The angle between the stars is measured again. The distance is now:

$$d \text{ (parsec)} = \frac{1}{p \text{ (arc second)}}$$

For distant stars, the angle can be a fraction of an arc second. This would be very difficult to measure by rotating the telescope for each star so photographs are used to measure the angles as in Figure 5. This can be done by calibrating the photograph by rotating the telescope through a known angle, which will cause all the stars in the photograph to move to one side. The distance moved is proportional to the angle, so the angle subtended by the stars six months later can be found by measuring how far they move compared to the distant stars that do not move. Note that negatives are used to make the background transparent. This means that the photographs can be placed on top of each other, making the measurements easier.

This method is limited by the smallest angle that can be measured. This is around 0.01 arc seconds for a terrestrial telescope (one on the Earth) and 0.001 arc seconds for a space telescope such as the Hubble Space Telescope. This is equivalent to a distance of 1 kpc which does not even extend beyond our galaxy. There are other methods for more distant stars but we will not consider them here.

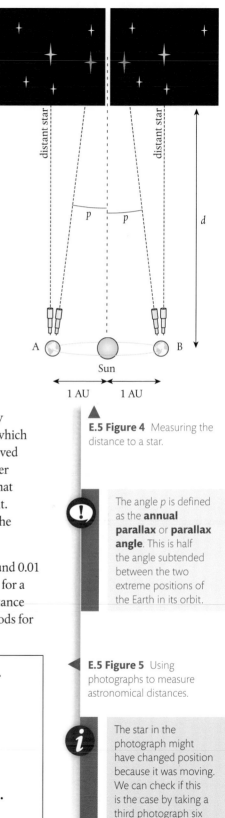

E.5 Figure 4 Measuring the distance to a star.

The angle p is defined as the **annual parallax** or **parallax angle**. This is half the angle subtended between the two extreme positions of the Earth in its orbit.

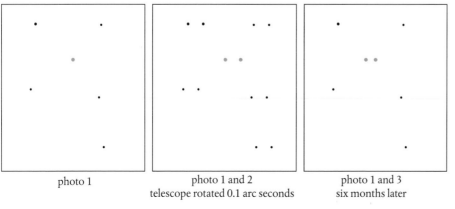

| photo 1 | photo 1 and 2 telescope rotated 0.1 arc seconds | photo 1 and 3 six months later |

E.5 Figure 5 Using photographs to measure astronomical distances.

The star in the photograph might have changed position because it was moving. We can check if this is the case by taking a third photograph six months later to see if it is back in the original place.

In the twentieth century, people were employed to study photographs of stars looking for stars, that moved or varied in brightness. Today, the Internet can be used to distribute images all around the world, involving thousands of amateurs in the hunt for interesting features. This is an example of citizen science.

Exercise

Q6. Using the images in Figure 5, measure the distance from the Earth to the blue star.

Q7. Calculate the parallax angle for a star on the other side of our galaxy (a distance of 10^{21} m). Would the movement of this star be visible using the telescope in Figure 5?

EM radiation from stars

All that we know about stars has been deduced by measuring the radiation they emit. This radiation has two important pieces of information: **intensity** (related to brightness) and **wavelength** (related to the color). Knowing the distance to a star, we can use this information to calculate the star's temperature, radius and the amount of energy it radiates per second.

Brightness, b

If you look at the stars at night, you will notice that some stars are brighter than others. You may also think that some look bigger than others, but this is not the case. All stars, except the Sun, are so far away that they appear as points of light. The only objects that have size (and do not appear as point sources) are the planets, the Moon and the occasional comet. The effect is caused by poor focus, movements of the air and your brain telling you that brighter must be bigger. If you look at a photograph of the stars, the brighter ones still look bigger. This is due to the way the camera works but gives us a useful method for measuring the relative brightness of stars and is the way brightness is indicated on a star map. In the early days of astronomy, the stars were put in order of brightness from 1 to 6, 1 being the brightest and 6 the least bright visible with the unaided eye. This number is called the **magnitude** and is still used today.

Exercise

Q8. Given that Rigel is a magnitude 0 star, use the photograph to estimate the magnitudes of the other stars labeled in the star chart.

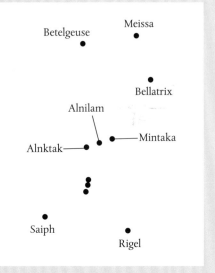

With today's technology, it is possible to measure the brightness of a star directly using a digital photograph, the potential difference across each pixel being directly related to the number of photons absorbed. Brightness is the amount of power per unit area perpendicular to the direction of the radiation.

The **brightness** of a star is the brightness measured from the Earth. This depends on how much power the star is emitting and how far away it is. The unit of brightness is $W\,m^{-2}$.

Luminosity, L

The **luminosity** of a star is the total amount of energy emitted per unit time (power). The unit of luminosity is the watt.

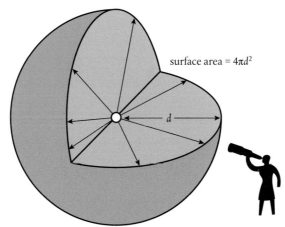

surface area $= 4\pi d^2$

Some of the brightest night sky objects are actually two very close stars.

◄ **E.5 Figure 6** As the light travels away from the star, the energy is spread over a bigger area.

In what ways has technology helped to collect data from observations of distant stars? (NOS)

The energy is radiated equally in all directions, so at a distance d, it will be spread out over the surface of a sphere of surface area $4\pi d^2$. The brightness is the power per unit area so the brightness b at distance d will be given by the equation:

$$b = \frac{L}{4\pi d^2}$$

The Sun has a luminosity, $L_\odot = 3.84 \times 10^{26}\,W$. The luminosity of other stars is normally quoted as a multiple of this value $L_\odot$.

Exercise

Q9. The luminosity of the Sun is $3.839 \times 10^{26}\,W$ and its distance from the Earth is $1.5 \times 10^{11}\,m$. Calculate its

(a) brightness

(b) brightness at a distance of 10 pc.

Q10. Sirius, the brightest star, has a luminosity 25 times greater than the Sun and is 8.61 light years from the Earth. Calculate:

(a) its brightness

(b) its brightness at a distance of 10 pc.

Q11. If the luminosity of a star is $5.0 \times 10^{31}\,W$ and its brightness is $1.4 \times 10^{-9}\,W\,m^{-2}$, calculate its distance from the Earth in light years.

Stellar spectra

Stars are almost perfect radiators. This means the intensity distribution of electromagnetic radiation they emit is the same as the characteristic pattern of the black-body spectrum as shown in Figure 7.

E.5 Figure 7 The intensity distribution for bodies of different temperature. The spectrum indicates where the visible region lies on the scale.

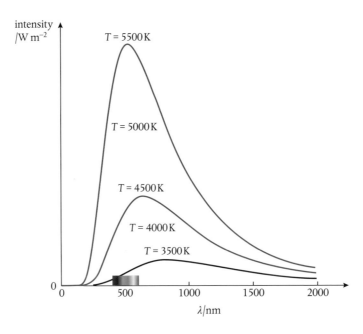

We can see from this set of curves that increasing the temperature reduces the average wavelength but increases the area under the graph. This means that the total power radiated per unit area has increased. The relationship between power per unit area and temperature in kelvin is given by the **Stefan–Boltzmann law**:

$$\frac{\text{power}}{\text{area}} = \sigma T^4$$

where $\sigma = 5.6 \times 10^{-8} \, \text{W m}^{-2} \text{K}^{-4}$ (the Stefan–Boltzmann constant).

The total power radiated (luminosity) by a star of surface area A is therefore:

$$L = A\sigma T^4$$

The relationship between the peak wavelength and the temperature in kelvin is given by the **Wien displacement law**:

$$\lambda_{\text{max}} = \frac{2.9 \times 10^{-3} \, \text{m K}}{T}$$

The unit m K of the constant is *meter* kelvin.

If we plot the spectrum for a star, we can calculate its temperature from the peak wavelength. This would mean measuring the intensity of light at many different wavelengths. Luckily, there is a shortcut. Since we know the shape of the curve is the same as a black-body spectrum, we only need a few points to be able to determine which of the different temperature curves represents its spectrum. This is done by using three filters: ultraviolet, blue and green (called UBV; the V stands for *visual* but the color is *green*).

How can the understanding of black-body radiation help determine the properties of stars? (B.1)

Worked example

The maximum in the black-body spectrum of the light emitted from the Sun is at 480 nm. Calculate the temperature of the Sun and the power emitted per square meter.

Solution

Using Wien's law:

$$\lambda_{max} = \frac{2.90 \times 10^{-3}}{T}$$

$$T = \frac{2.9 \times 10^{-3}}{\lambda_{max}} = \frac{2.9 \times 10^{-3}}{480 \times 10^{-9}} = 6000\,K$$

Now using the Stefan-Boltzmann law:

$$\text{power per unit area} = 5.6 \times 10^{-8} \times (6000)^4 = 7.3 \times 10^7\,W\,m^{-2}$$

If the radius of the Sun is 7.0×10^8 m, what is the luminosity?

$$\text{surface area of the Sun} = 4\pi r^2 = 6.2 \times 10^{18}\,m^2$$

$$\text{total power radiated} = 6.2 \times 10^{18} \times 7.3 \times 10^7 = 4.5 \times 10^{26}\,W$$

Exercise

Q12. The star Betelgeuse has a radius of 3.1×10^{11} m and a surface temperature of 2800 K. Find its luminosity.

Q13. The intensity peak in a star's spectrum occurs at 400 nm. Calculate:

    (a) its surface temperature

    (b) the power radiated per square meter.

Absorption lines

As the black-body radiation passes through the outer layers of a star, some of it is absorbed by the gases found there. This leads to dark **absorption lines** in the otherwise continuous spectrum. These lines are unique for each element and can be used to determine the chemical composition of the outer layers. The spectrum of light from the Sun includes the spectral lines of some 67 different elements. Studies show that most stars have similar composition: 72% hydrogen, 25% helium and 3% other elements.

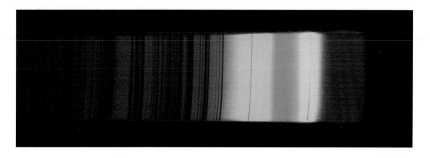

The absorption lines give information about the composition of the outer layers of gas surrounding a star. However, the layers of a star are continually mixing so the outer layers have the same composition as the rest of the star.

The spectrum of the light from the Sun showing the absorption lines for many elements.

How do emission and absorption spectra provide information about observations of the cosmos? (E.1)

Colors of stars

We have seen that the spectrum of a star is related to its surface temperature and chemical composition. This also determines its color. If the peak is at the blue end (high temperature so short wavelength), it will be blue, and if at the red end (low temperature so large wavelength), then it will be red.

The spectra and star color for different stars, starting with hot at the top and ending with cool at the bottom.

Note that you will not be asked about spectral classifications in the exam but you might see them if you look up stars in a database.

Nature of Science

By plotting the position of thousands of stars on a luminosity vs temperature graph, it is possible to see patterns that reveal the way stars are thought to evolve. Without this visual aid, it would be very difficult to see any pattern in the data.

Hertzsprung–Russell (HR) diagrams

A Hertzsprung–Russell diagram is a graph on which the temperature of a star is plotted against its luminosity as shown in Figure 8.

When interpreting this diagram, you need to look closely at the axes. The y-axis is luminosity, which is logarithmic. The x-axis is surface temperature, which is non-linear and goes from hottest on the left to coldest on the right. A star at the top right-hand corner is cold but luminous. This means (based on the Stefan–Boltzmann law) that it is a big star. At the other extreme, bottom left, the stars are hot but not luminous so must be small. This means that we can easily deduce the size of a star from its position on the diagram. The diagonal lines on the diagram indicate stars of equal radius, so if we know the temperature and luminosity of a star, we can plot it on the HR diagram and determine its radius.

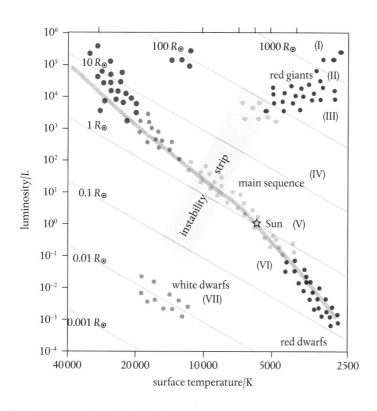

 E.5 Figure 8 HR diagram showing the position of the Sun.

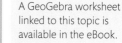

 A GeoGebra worksheet linked to this topic is available in the eBook.

When all the stars are plotted on the diagram, we see some interesting trends. First, they are not uniformly distributed but seem to be arranged in groups.

Main sequence

90% of the stars are in a diagonal band, called the main sequence. This band includes the Sun. The **main sequence** ranges from large hot blue stars on the left to small cool red stars on the right. Like the Sun, all main sequence stars have a core that is undergoing fusion from hydrogen to helium. This radiates energy, causing a pressure that prevents the star from collapsing under the force of gravity. These stable stars will remain at the same point on the diagram for a long time. That is why most stars are main sequence.

Red giants

A cool star that gives out a lot of energy must be very big, so these are called giants. The coolest M class stars are called **red giants** due to their color. The luminosity of a giant is about 100 times greater than the Sun. If they are the same temperature as the Sun, they must have an area 100 times bigger, therefore a radius 10 times bigger. If their temperature is lower, they can be even larger.

Supergiants

A **supergiant** is a very big cool star. With luminosities 10^6 times greater than the Sun, they have radii up to 1000 times that of the Sun. These are very rare stars but one is very easy to spot. Betelgeuse is the right shoulder of Orion and you can see it in the photo in exercise question 8.

HR diagrams have been helpful in the classification of stars by finding patterns in their properties. What other areas of physics use classification to help our understanding? (NOS)

White dwarfs

A **white dwarf** is a small hot star, hotter than the Sun but only the size of the Earth. They have a low luminosity so it is not possible to see them without a telescope.

Variable stars

A variable star has a changing luminosity, so its position on the HR diagram is not constant. This is due to a change in the size of the star. As it gets bigger, its luminosity increases. This variation is sometimes cyclic as in a **Cepheid variable**. These stars appear in the instability strip on the HR diagram.

The birth of a star

The life cycle of a star takes billions of years so we are never going to see the whole cycle from birth to death. However, by measuring the light emitted from stars, we have discovered that stars differ in mass, temperature, radius and composition, leading to the classifications we have plotted on the HR diagram. Trying to deduce the life cycle of stars from this is a bit like an alien trying to make some sense of the human race from one photograph of a crowd. The alien would notice that all the humans were basically the same, and although most are the same size, there are some very small ones and some old wrinkly ones. Maybe the alien would work out that these are not three types of creature but are different stages in the life of the same thing. The alien might also deduce that, since there are more upright large ones, this is the longest part of the human's life cycle, but would be unlikely to work out where they came from or that in the end they died. Applying the same logic to stars, we deduce that the different types of star are different stages in a life cycle. To complete the picture, we can use what we know about the way matter interacts on Earth to work out how this happens.

Stars start their life inside **giant molecular clouds** called nebulae, which are swirling clouds of gas and dust left over after the formation of a galaxy, made up of mainly hydrogen but also larger elements and molecules. There are several thousands of these clouds in our galaxy, such as the Horsehead Nebula in the photo.

The Horsehead Nebula: the gas around the cloud appears pink as hydrogen is excited by UV radiation emitted from new stars inside the cloud. However, these are color composites, so someone used a computer program to assign color. Color on astronomical images is arbitrary, although astronomers try to follow guidelines. Ultraviolet and beyond is purple-ish while infrared and beyond is usually red, etc.

The temperature of a giant molecular cloud is only about 10 K, which is why molecules are able to be present. The clouds are held together by gravity but they are kept from collapsing by the pressure of the molecules moving about in random motion. If, however, the gas is compressed by the shock wave from an exploding star or the collision between two clouds, the gravity overcomes the thermal pressure and the cloud begins to collapse.

When hit by a shockwave, areas of the cloud compress and collapse, forming many stars of different sizes. They are difficult to see as they are inside the cloud. However, as they collapse, they get hot, emitting IR radiation that can be detected with an IR telescope. Some of the bigger stars get very hot and cause strong winds to blow through the cloud, compressing more of the dust to create even more stars. Eventually, all the dust is used up.

Gravity keeps on collapsing the new star (**protostar**) until the center becomes so dense and hot that hydrogen nuclei start to fuse to make helium. This reaction releases energy due to the fact that the mass of the products is less than the mass of the original hydrogen nuclei. This mass is converted to energy.

Once the core starts to undergo nuclear fusion, the outward radiation pressure counteracts the inward gravitation, preventing further collapse. The star is now a stable main sequence star. The position of the star on the main sequence depends upon its mass. Figure 9 shows the changes from protostar to main sequence represented on an HR diagram.

Note that it appears that a red giant has turned into a main sequence star, but it has not. The large protostar was in the same region of the diagram but had very different properties to a red giant.

Main sequence

Once on the main sequence, the star is stable and will remain that way until most of the hydrogen is used up. The amount of time that a star is a main sequence star depends on how much hydrogen the star contains and the rate at which it is used up. You may think that a bigger star will last longer since they have more fuel but that is not the case. More massive stars are more luminous so use up their fuel more quickly.

After the main sequence

A star will stay on the main sequence until it uses up almost all of the hydrogen in the core. This is only about 10% of the total amount of hydrogen in the star so the mass of the star does not change a great deal. This means that its position on the HR diagram stays almost the same as at its point of entry. As the hydrogen fuses to helium, the heavier helium sinks to the center of the core, which is the densest hottest part of the star. Hydrogen fusion continues outside this central core until the pressure and temperature are no longer great enough and the fusion slows. It is the pressure caused by the fusion that stops the star from collapsing, so when the rate of fusion gets less, the core starts to collapse, resulting in an increase in core temperature. This heats the outer layers of the star, causing them to expand and changing the main sequence star into a red giant as represented on the HR diagram in Figure 10.

What happens next depends on the mass of the star. We will start by considering roughly Sun-sized stars.

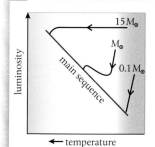

E.5 Figure 9 The HR diagram for three stars as they turn from protostar to main sequence. Note how the luminosity of the big ones stays constant. This is because they are getting smaller but hotter. A Sun-sized star also shrinks but its outside stays cool until the inside gets so hot that it heats the outer layers. The core of a small star never gets that hot, so it gets less and less bright as it contracts.

How can gas laws be used to model stars? (NOS)

A star smaller than about 0.25 $M_\odot$ will never get to the point where fusion starts. It would simply cool down into a lump of matter called a **brown dwarf**.

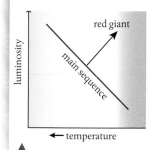

E.5 Figure 10 HR diagram representing the change from main sequence to red giant.

Sun-sized stars

As the core is compressed by the gravitational attraction of the surrounding matter, it reaches a point when the electrons cannot get any closer. After this point is reached, the core cannot get any smaller but continues to get hotter. When the core temperature exceeds 10^8 K, helium can fuse to form beryllium:

$$^4_2\text{He} + {}^4_2\text{He} \rightarrow {}^8_4\text{Be} + \gamma$$

Beryllium then fuses with more helium to form carbon:

$$^8_4\text{Be} + {}^4_2\text{He} \rightarrow {}^{12}_6\text{C} + \gamma$$

which in turn fuses with more helium to form oxygen:

$$^{12}_6\text{C} + {}^4_2\text{He} \rightarrow {}^{16}_8\text{O} + \gamma$$

During this stage in the star's life, the outer layers are very far from the central core so the force of gravity holding them together is not very strong. Any increased activity in the core can cause these outer layers to blow away. This happens over a period of time, leaving the core surrounded by the remains of the outer layers. The core is no longer producing energy so contracts until the electrons prevent it getting any smaller. It is now called a **white dwarf**. The whole process is represented on the HR diagram in Figure 11.

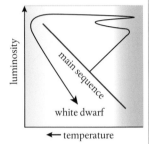

E.5 Figure 11
The evolutionary path of a Sun-sized star.

The Helix Nebula. This object is a planetary nebula, a dying star ejecting its dusty outer layers. The image was obtained by combining infrared (yellow, green and red) and ultraviolet (UV in blue) data from NASA's Spitzer Space Telescope and Galaxy Evolution Explorer (GALEX). The ejected layers are glowing due to the intense UV radiation from the collapsed stellar core, a white dwarf (not visible at this scale).

Large stars

The pressure generated in the core by stars with masses over $4\,M_\odot$ is enough to enable the carbon and oxygen to fuse into larger elements such as neon and magnesium. For example:

$$^{12}_6\text{C} + {}^{12}_6\text{C} \rightarrow {}^{20}_{10}\text{Ne} + {}^4_2\text{He}$$

$$^{12}_6\text{C} + {}^{12}_6\text{C} \rightarrow {}^{24}_{12}\text{Mg} + \gamma$$

As this happens, the heavier elements sink to the center of the core. For larger stars with masses over $8\,M_\odot$, this process continues until iron is produced. If you remember the binding energy per nucleon vs nucleon number curve, you will know that iron is at the top of the peak, so fusing iron with other elements to produce larger nuclei will not liberate energy. This makes iron the end of the road as far as energy production in stars is concerned. However, as the core runs out of nuclear fuel, it collapses, resulting in an increase in temperature. This allows iron to fuse, resulting in the absorption of energy. This reduces the outward pressure preventing the core from collapsing so the core collapses, causing electrons to combine with protons to form neutrons.

$$p^+ + e^- \rightarrow n + \nu_e$$

This continues until the core contains only neutrons. This collapse takes only about 0.25 s so leaves a gap, between the core and the outer layers. The outer layers fall into the gap, resulting in a rapid rise in temperature, which causes a huge explosion that blows away everything except the core. This is called a **type II supernova** and what remains of the core is a **neutron star** or **black hole**.

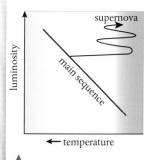

E.5 Figure 12
The evolutionary path of a large star.

Guiding Questions revisited

How are elements created?

What physical processes lead to the evolution of stars?

Can observations of the present state of the Universe predict the future outcome of the Universe?

In this chapter, we have considered fusion, a nuclear process that releases energy when small nuclei come together to form larger nuclei, and which powers stars:

- Elements are defined by the number of protons in the nucleus. The simplest element is hydrogen (one proton). If two hydrogen isotopes fuse, then helium (two protons) is produced. Fusion continues to release energy (due to increasing binding energy per nucleon) up to iron, above which heavier elements are formed only in stellar explosions.

- A star's evolution depends on its mass. The higher the mass, the shorter the main sequence because the increased rate of fusion is enough to outweigh the increased fuel supply. The main sequence concludes when the radiation pressure is no longer sufficient to balance the inward force of gravity. Following collapse, the star's increased temperature causes the outer layers to expand. Lighter stars move from the main sequence to become red giants and, eventually, white dwarfs. Heavier stars form supergiants followed by neutron stars or black holes.

- Parallax observations of nearby stars enables us to determine how far away they are. Analysis of their spectra enables us to determine their luminosity and temperature. When the distance and luminosity of stars is known, the properties of other stars can be determined through comparison. By scaling up measurements of spectra to whole galaxies, we can deduce that the Universe is expanding because the galaxies furthest from our own are moving fastest away from our galaxy and apart from one another.

- Stars of known luminosity and temperature can be plotted on a Hertzsprung–Russell diagram to determine their radii.

Practice questions

1. (a) The helium in the Sun is produced as a result of a nuclear reaction. Explain whether this reaction is burning, fission or fusion. (2)

   At a later stage in the development of the Sun, other nuclear reactions are expected to take place. One such overall reaction is given below:

   $$_2^4He + {}_2^4He + {}_2^4He \rightarrow C + \gamma + \gamma$$

   (b) (i) Identify the proton number and the nucleon number of the isotope of carbon C that has been formed. (2)

   (ii) Use the information below to calculate the energy released in the reaction.

   Atomic mass of helium = $6.648\,325 \times 10^{-27}$ kg

   Atomic mass of carbon = $1.993\,200\,0 \times 10^{-26}$ kg (3)

   (Total 7 marks)

2. (a) State what is meant by a *fusion reaction*. (3)

   (b) Explain why the temperature and pressure of the gases in the Sun's core must both be very high for the Sun to produce its radiant energy. (5)

   (Total 8 marks)

3. (a) (i) Define *nucleon*. (1)

   (ii) Define *nuclear binding energy of a nucleus*. (1)

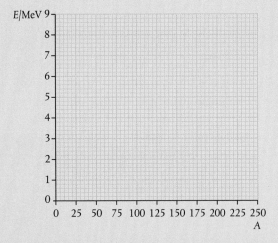

   The axes on the graph show values of nucleon number A (horizontal axis) and average binding energy per nucleon E (vertical axis). (Binding energy is taken to be a positive quantity.)

   (b) Copy the graph and mark on the y-axis (E/MeV) the approximate position of:

   (i) the isotope $_{26}^{56}Fe$ (label this F) (1)

   (ii) the isotope $_1^2H$ (label this H) (1)

   (iii) the isotope $_{92}^{238}U$ (label this U). (1)

(c) Draw a graph to show the variation with nucleon number A of the average binding energy per nucleon E. (2)

(d) Use the following data to deduce that the binding energy per nucleon of the isotope $^3_2$He is 2.2 MeV.

Nuclear mass of $^3_2$He = 3.01603 u
Mass of proton = 1.00728 u
Mass of neutron = 1.00867 u (3)

In the nuclear reaction $^2_1$H + $^2_1$H $\rightarrow$ $^3_2$He + $^1_0$n, energy is released.

(e) (i) State the name of this type of reaction. (1)

(ii) Use your graph in (c) to explain why energy is released in this reaction. (2)

(Total 13 marks)

4. A graph of the variation of average binding energy per nucleon with nucleon number has a maximum. What is indicated by the region around the maximum?

A   The position below which radioactive decay cannot occur

B   The region in which fission is most likely to occur

C   The position where the most stable nuclides are found

D   The region in which fusion is most likely to occur

(Total 1 mark)

5. What gives the total change in nuclear mass and the change in nuclear binding energy as a result of a nuclear fusion reaction?

	Nuclear mass	Nuclear binding energy
A	decreases	decreases
B	decreases	increases
C	increases	decreases
D	increases	increases

(Total 1 mark)

6. Two photographs of the night sky are taken, one six months after the other. When the photographs are compared, one star appears to have shifted from position A to position B, relative to the other stars.

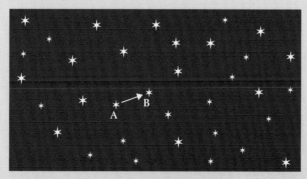

(a) Outline why the star appears to have shifted from position A to position B. (1)

The observed angular displacement of the star is θ and the diameter of the Earth's orbit is d. The distance from the Earth to the star is D.

(b) Draw a diagram showing d, D and θ.   (1)

(c) Explain the relationship between d, D and θ.   (2)

(d) One consistent set of units for D and θ are parsecs and arc seconds. State **one** other consistent set of units for this pair of quantities.   (1)

(Total 5 marks)

7. Mintaka is one of the stars in the constellation Orion.

(a) The parallax angle of Mintaka measured from Earth is 3.64×10^{-3} arc seconds. Calculate, in parsecs, the approximate distance of Mintaka from Earth.   (1)

(b) State why there is a maximum distance that astronomers can measure using stellar parallax.   (1)

(Total 2 marks)

8. **(a)** Describe **one** key characteristic of a nebula.   (1)

(b) Beta Centauri is a star in the southern skies with a parallax angle of 8.32×10^{-3} arc seconds. Calculate, in meters, the distance of this star from Earth.   (2)

(c) Outline why astrophysicists use non-SI units for the measurement of astronomical distance.   (1)

(Total 4 marks)

9. A Hertzsprung–Russell (HR) diagram is shown below.

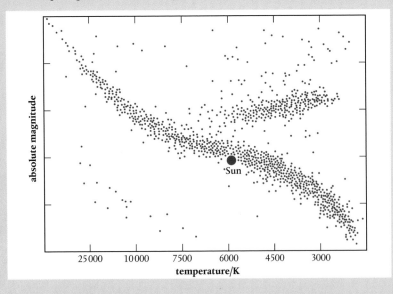

The following data are given for the Sun and a star Vega:

Luminosity of the Sun = 3.85×10^{26} W

Luminosity of Vega = 1.54×10^{28} W

Surface temperature of the Sun = 5800 K

Surface temperature of Vega = 9600 K

(a) Determine, using the data, the radius of Vega in terms of solar radii. (3)

(b) Outline how observers on Earth can determine experimentally the temperature of a distant star. (3)

(Total 6 marks)

10. The Hertzsprung–Russell (HR) diagram shows several star types. The luminosity of the Sun is $L_\odot$.

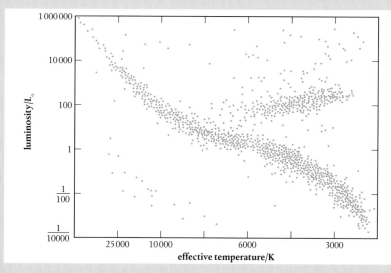

(a) Identify, on the HR diagram, the position of the Sun. Copy the HR diagram and label the position S. (1)

(b) Suggest the conditions that will cause the Sun to become a red giant. (3)

(c) During its evolution, the Sun is likely to be a red giant of surface temperature 3000 K and luminosity $10^4 L_\odot$. Later, it is likely to be a white dwarf of surface temperature 10 000 K and luminosity $10^{-4} L_\odot$. Calculate the $\dfrac{\text{radius of the Sun as a white dwarf}}{\text{radius of the Sun as a red giant}}$. (2)

(Total 6 marks)

11. (a) Main sequence stars are in equilibrium under the action of forces. Outline how this equilibrium is achieved. (2)

The following data apply to the star Gacrux:

$$\begin{aligned}
\text{Radius} &= 58.5 \times 10^9 \text{ m} \\
\text{Temperature} &= 3600 \text{ K} \\
\text{Distance} &= 88 \text{ ly}
\end{aligned}$$

(b) The luminosity of the Sun $L_\odot$ is 3.85×10^{26} W. Determine the luminosity of Gacrux relative to the Sun. (3)

(c) The distance to Gacrux can be determined using stellar parallax. Outline why this method is not suitable for all stars. (1)

A Hertzsprung–Russell (HR) diagram is shown below.

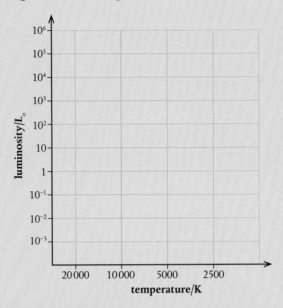

Copy the HR diagram.

(**d**) Draw the main sequence.   (1)

(**e**) Plot the position, using the letter G, of Gacrux.   (1)

(**f**) Discuss, with reference to its change in mass, the evolution of a star
from the main sequence until its final stable phase.   (3)

(Total 11 marks)

12. The diagram shows the structure of a typical main sequence star.

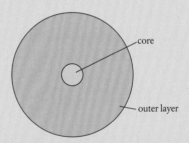

(**a**) State the most abundant element in the core and the most abundant
element in the outer layer.   (2)

The Hertzsprung–Russell (HR) diagram shows two main sequence stars, X
and Y, and includes lines of constant radius. *R* is the radius of the Sun. Star X is
likely to evolve into a neutron star.

(Total 2 marks)

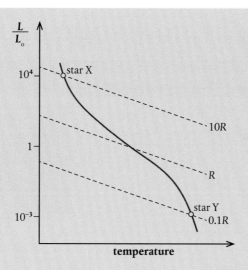

(b) Copy the HR diagram and draw a line to indicate the evolutionary path of star X. (1)

(c) The radius of a typical neutron star is 20 km and its surface temperature is 10^6 K. Determine the luminosity of this neutron star. (2)

(d) Determine the region of the electromagnetic spectrum in which the neutron star emits most of its energy. (2)

(Total 5 marks)

13. (a) The natural uranium in the Earth was probably formed by the fusion of iron nuclei inside ancient stars. Which statement about the nuclear fusion is correct? (1)

    A It cooled the star.

    B It heated the star.

    C It could have done either.

(b) Explain your answer. (1)

(Total 2 marks)

Theory of Knowledge in physics

'The task is not to see what has never been seen before, but to think what has never been thought before about what you see every day.'

Erwin Schrödinger

The knowledge framework

In the Theory of Knowledge (TOK) course, you will be asked to analyze and discuss the different areas of knowledge. One area of knowledge is the natural sciences, which physics is an example of. What makes the natural sciences different from other areas such as the arts, mathematics, history or the human sciences? In this chapter, we will look at elements of the knowledge framework in physics, so that you can make comparisons with your other subjects in the TOK Essay:

- What **methods and tools** are involved?
- What is the **scope** of physics?
- What links are there to **perspectives** and culture?
- What impact does **ethics** have on knowledge production?

What are the ethical responsibilities of a particle physicist?

'It is not the result of scientific research that ennobles humans and enriches their nature, but the struggle to understand while performing creative and open-minded intellectual work.'

Albert Einstein

As you read this chapter, think carefully about what ideas might be tested in physics and which might tested in TOK.

For example, physics questions might include:
1. What is the range of a projectile, assuming air resistance is not present?
2. What is the ratio of the gravitational and electric forces acting between two electrons?
3. What are the fundamental units in physics?
4. What is Brownian motion and how does this relate to our understanding of matter?
5. What are the components of a nuclear fission reactor?

Knowledge questions relating to these physics questions might include:
1. How can a model be useful even if it has limitations?
2. What are the unanswered questions in this area of knowledge?
3. What is the significance of key historical developments within this theme?
4. What constitutes 'good evidence'?
5. What responsibilities rest on the knower?

How do physicists test models of the atom?

The scientific method

The scientific method is the way that scientists work to invent new theories and to discover new laws, and it is also the way that you will have been working in the practical aspect of the course. There are actually many variations to this process and many exceptions, where new theories have come about without following any strict procedure. However, to make things simple, we will consider just one four-step version of the scientific method.

What brings a sunset into the scope of physics?

1 Observation

Physics is all about making models to help us understand the Universe. Before we can make a model, we must observe what is happening. In physics, the observations are often of the following form:

How does one thing affect another?

2 Hypothesis

Having made an observation, the next step is to use your knowledge to develop an idea of what is causing the event you have observed. What factors cause this thing to happen and what factors are not involved? Having made a hypothesis, is it possible to predict the outcome of a change in one of the variables?

3 Experiment

The experiment is designed to test the hypothesis. It is important to change only the quantity that you think is responsible for the event. You must keep the other variables constant.

If the experiment does not confirm the hypothesis, then you must go back to the observations and think of a new hypothesis (this is often helped by the outcome of the experiment). If the experiment supports the hypothesis, then you can go on to the next step.

4 Theory

If the experiment supports the hypothesis, you can make a theory that relates the variables involved.

'The secret of genius is to carry the spirit of childhood into maturity.'

T. H. Huxley

▲ Franklin's lightning experiment. If Franklin had had a full understanding of electricity, he would not have done this. He was lucky that the kite did not get struck by lightning, and in fact many people died repeating this experiment.

🔒 **A theory is a set of related statements that can be used to make predictions and explain observations.**

▲ Galileo made observations using a telescope.

Experiments are conducted to test hypotheses.

Example – the simple pendulum

'There are children playing in the street who could solve some of my top problems in physics, because they have modes of sensory perception that I lost long ago.'

Robert Oppenheimer

Observation

A student watches a simple pendulum swinging and wonders what factors affect the frequency of the swing.

Hypothesis

The student had not studied the motion of the pendulum previously, but having studied other mechanics concepts, came up with an idea related to the mass of the bob. They thought that since the bigger mass had more weight, then the force pulling it down would be greater, causing it to swing faster. So the hypothesis was that the frequency of the bob was proportional to the mass of the bob.

Experiment

An experiment was carried out measuring the frequency of bobs with different masses. The length of the string, the height of release and all other variables were kept constant. The result showed that there was no change in the frequency. The hypothesis was therefore incorrect.

Back to observations

On observing the pendulum further, the student noticed that if its length were increased, it appeared to swing more slowly. This led to a second hypothesis and the process continued.

The scope of physics

For a theory to be accepted, it must be possible to think of a way that it can be proved wrong. For example, Newton's gravitational theory would be proved wrong if an object with mass was seen to be repelled from the Earth.

However, the theory that the Earth is inhabited by invisible creatures with eyes on each finger is not falsifiable since you cannot see the creatures to tell if they have eyes on their fingers or not.

'Just a theory'

If someone says 'special relativity is just a theory', what do they mean?

The use of the word *theory* in the English language can cause some problems for scientists. The word is sometimes used to mean that something is not based on fact. For instance, you could say that you have a theory as to why your friend was annoyed with you last night. In physics, a theory is based on strong experimental evidence.

Occam's razor

A razor is a strange name for a principle. Its name arises because it states that a theory should not contain any unnecessary assumptions – it should be reduced down to its bare essentials.

This is the same as the KIS principle: 'keep it simple'.

For example, a theory for gravitational force could be that there is a force between all masses that is proportional to the product of their mass and is caused by invisible creatures with very long arms. The last bit about the invisible creatures is unnecessary so can be cut out of the theory (using Occam's razor).

Discoveries in physics

Discoveries are not always made by following a rigorous scientific method – sometimes luck plays a part. Serendipity is the act of finding something when you were looking for something else. For example, you could be looking for your car keys but find your sunglasses. There are some famous examples of this in physics.

Hans Christian Oersted discovered the relationship between electricity and magnetism when he noticed the needle of a compass moving during a lecture on electric current.

Arno A. Penzias and Robert Woodrow Wilson discovered radiation left over from the Big Bang, while measuring the microwave radiation from the Milky Way. They at first thought their big discovery was just annoying interference.

These serendipitous discoveries were all made by people who had enough knowledge to know that they had found something interesting. If they were not expert physicists, would they have realized that they had discovered something new?

Standing on the shoulders of giants

When you do practical work for the IB Diploma, is the principle behind what you are doing the same as that used by physicists working in research departments of universities around the world?

What you are doing is using the knowledge learned in class to develop your hypothesis. You should find out that, if you apply your knowledge correctly, your experiment will support your hypothesis. At the cutting edge of science, scientists are developing new theories, so the experiment is used to test the theory, not to test if they have applied accepted theory correctly. To find a new way of relating quantities requires imagination, but what you are being asked to do in your physics lab is to use accepted theory and not to use imagination. As a student being trained to apply strict physical laws, how prepared do you feel for a career in physics where imagination is key?

▲ Hans Christian Oersted experimenting with magnets and current after a chance observation.

Scientific perspectives and culture

A paradigm is a set of rules that make up a theory that is accepted by the scientific community.

Having completed this course, you will have accepted certain paradigms. We see and interpret the world using paradigms and theories. Newtonian mechanics is a paradigm – we apply Newton's laws of motion to balls, electrons, planes and cars. The theory works well and is accepted by the scientific community. The way we treat almost everything as a particle is another paradigm. This paradigm is so much a part of the way that we think that it is almost impossible to think of matter not being made of particles. How could you have a gas that was continuous? Before anyone thought of matter being made of particles, this would not have been a problem, but now it is. To change your way of thinking requires a big change in your previously accepted perspective, and this is called a paradigm shift. Throughout the development of physics, there have been many paradigm shifts.

The solar system

In 1543, when Copernicus suggested that the Sun was the center of the solar system, it went against a theory that had been accepted for over a thousand years. Furthermore, it not only went against scientific theory but it went against common sense. How can we be going around the Sun when we are quite obviously standing still? It required a totally new way of thinking to accept this new idea. At the time, the evidence was not strong enough to be convincing and the old paradigm remained. It was not until Galileo provided more evidence and Newton developed an explanation later that the shift took place.

Time and relativity

Before Einstein, it was accepted that time is the same everywhere, the length of a body is the same as measured by everyone, and the mass of a body is constant. Einstein showed that time, length and mass all depend on the relative velocity of two observers. After this discovery, it was not possible to carry on as if nothing had happened. What was required was a new way of thinking and a new set of laws. However, when a paradigm shift takes place, the old laws do not suddenly become obsolete; they just obtain limits. Newtonian mechanics is still fine when relative velocities are much less than the speed of light, and that is why it is still included in physics courses such as this.

The next paradigm shift

Will there be another paradigm shift? Can physics advance without one? Because there have been paradigm shifts in the past, does that mean that there has to be another one to advance physics further? In 1900, Lord Kelvin famously said 'There is nothing new to be discovered in physics now, all that remains is more and more precise measurement.' He was certainly wrong. Does that mean that if someone said the same thing today they would also be wrong? One problem is that, as students go through the process of education, they can get entrenched in the ways of thinking of their teachers, so the leap in imagination to make that new paradigm shift becomes bigger and bigger as time goes on.

Ethics and physics

Ethics is the study of right and wrong. It is sometimes not easy to decide when a course of action is right or wrong, and in these cases, it is useful to have a moral code or set of guidelines to refer to. In physics, there are two areas where ethical considerations are important:

1 The way physicists work in relation to other physicists; for example, they should not copy each other's work or make up data.

2 The way their actions affect society; for example, physicists should not work on projects that will endanger human life.

Whether a particular piece of research is ethical or not can be difficult to determine, especially when you do not know what the results of the experiment might lead to.

- Should Rutherford have performed his experiments in nuclear physics, since the discovery of the nucleus led to the invention of the atom bomb?
- Can it be ethical to work in the weapons industry?
- Who should decide whether a piece of research is carried out: physicists or governments?
- If you left your body to science, would it be OK if it were used to test car seat belts? How about if it were used to test how far different types of bullets penetrate flesh?
- Is it ethical to spend billions to carry out an experiment to test someone's hypothesis?

The TOK Essay is a response to one of six titles, which are published a few months before the submission date. It is worth two-thirds of your TOK grade.

Some advice to bear in mind:

1. Answer the question and use its words throughout your essay.
2. Construct your essay as a coherent argument, with every paragraph and section providing evidence in your favor. It is great if you can (briefly) acknowledge and address possible counterarguments, but there are not many words to spare.
3. Use an example that is typical of the area of knowledge you are discussing. If you were writing about the natural sciences, you might refer to an experiment in which the scientific method was used (such as the gas laws) or an explanation of the available evidence (such as the nuclear model of the atom). Having a deep understanding of your example and relating it to your title is better than choosing a title and then searching for the 'perfect' evidence.
4. When summarizing a particular area of knowledge, try putting a finger over the name of the area. If the sentence could be true for another area, then aim to make your writing more specific.

These are two titles that students were required to link to the natural sciences in 2022. Which other area might you choose to compare and contrast with?

- Is there solid justification for regarding knowledge in the natural sciences more highly than another area of knowledge? (May 22)
- Within an area of knowledge, is it more important to have credibility or power? (Nov 22)

The following titles (from assessments prior to 2022) did not prescribe particular areas of knowledge. As a challenge, consider how Boyle's law could link to these:

- 'There is a sharp line between describing something and offering an explanation of it.' To what extent do you agree with this claim? (May 20)
- 'Too much of our knowledge revolves around ourselves, as if we are the most important thing in the Universe' (adapted from Carlo Rovelli). Why might this be problematic? (Nov 20)
- Within areas of knowledge, how can we differentiate between change and progress? (May 21)
- If all knowledge is provisional, when can we have confidence in what we claim to know? (Nov 21)

This textbook is full of physics content, any aspect of which could potentially help you to respond to your TOK Essay title, if you choose to refer to the natural sciences. Remember to communicate regularly with your TOK teacher, submitting a plan and a draft ahead of your interactions with them.

Themes

As well as areas of knowledge, in your TOK course, you will also study themes. Themes are related to all areas, perhaps because they have influenced the formulation of new ideas or been affected by the sharing of knowledge. You are likely to select one theme for your TOK Exhibition.

Themes include language, technology, politics, religion, indigenous societies and the knower.

Language

Can a battleship float in a bucket of water? To solve this problem, you could use Archimedes' principle that tells you that a body will float if it displaces its own weight of fluid. You might therefore conclude that the ship cannot float in a bucket of water, because it has to displace (move out of the way) its own weight of water, and there is not enough water in the bucket. However, this is a misinterpretation of the theory.

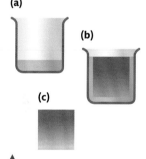

(a) Bucket with a bit of water. **(b)** Large object floating in bucket. The bucket is now almost full because the object is taking the place of the water. The object displaces the water. **(c)** Archimedes says that the weight of fluid displaced equals the weight of the object. So this amount of water will have the same weight as the object.

How much water do you need to float this huge oil tanker?

555

What Archimedes meant is that, when a boat is floating, if you filled the space in the water taken up by the boat with water, then it would be the same weight as the boat. This still might not make sense, which is why physicists use so many diagrams.

This demonstrates how important it is to understand language.

In physics, we use language in a very precise way. Every time a quantity is named, it is given a specific definition. For example, *velocity* means one thing and one thing only: the rate of change of displacement. In normal use of language, words can mean more than one thing. Ambiguous terminology is often used in jokes, poems and literature but not in physics.

Sometimes, the other meanings of a word can lead to confusion. *Potential energy* sounds like a body could possibly have energy, but it actually means the energy a body has due to its position. *Electron spin* sounds like the electron is a little ball spinning. A spinning charge would indeed have the properties exhibited by the electron; however, these properties do not arise for this reason.

To get around this problem, physicists sometimes use words that cannot be confused with other meanings. Who would think that a charm quark was actually charming, for instance?

Technology

The relationships between technology and physics cannot be ignored.

In reading this book, you are benefiting from the printing press for information sharing and from the Internet for facilitating collaboration between the two authors. During your course, you become skilled in using analog and digital apparatus for making measurements and gained awareness of how modern equipment can reduce uncertainties. Your smartphone can be used as a data logger with an appropriate app or as a video camera to record an experiment. How else has technology facilitated progress in physics?

Memory foam for shock absorption was developed by NASA for astronaut seating and can now be found in mattresses and sports equipment. The (lack of) penetrating power of alpha particles means that radioactive americium-241 nuclei can be found in household smoke detectors worldwide. What other real-world applications have emerged from physics?

Politics

The pursuit of knowledge in physics generally requires funding, because of the costs of facilities, equipment and salaries of team members. The allocation of funding for science comes down to political decisions – whether at the level of national governments, international organizations or research institutions. These decisions, in turn, typically rest on a panel's impression of a research proposal, which includes the aims of the research, a project plan, the experience of the researcher(s) and the anticipated impact from the work. Another political consideration is how leaders represent scientific findings.

What degree of importance does science have in the manifestos of political organizations where you live? What political changes are taking place in your country? How might these affect the work of scientists? What areas of physics do you perceive as deserving highest priority? What difference would you or your community like to make?

Religion

Take care before speaking only of contradictions if you refer to religion and science in your exhibition.

It can be tempting to assume that religion has been responsible for holding back scientific progress over the centuries, perhaps because there are conflicting views among certain groups on the age of the Universe and the emergence of different species of organisms.

There are also differences in how religious organizations and scientists claim to know something – based on faith or based on experimental evidence.

However, religion and physics have many connections. 'Light' is an example of a topic that has been of interest to religious writers and physicists. Religious architecture has required the use of physics in its design, and structural engineers still study the arches, domes, flying buttresses and towers of churches, mosques, stupas, synagogues and temples in their training. Physics was once known as natural philosophy (with the University of Cambridge not appointing a chair in physics until the late 19th century); philosophy offers an examination of religious traditions.

Indigenous societies

Knowers belong to communities. These communities might include indigenous societies. An indigenous society is a culturally distinct ethnic group, whose members are directly descended from the earliest known inhabitants of a particular geographic region and who, to some extent, maintain the language and culture of those original peoples.

Take a moment to reflect on the themes in your physics course. Which of them could benefit from an understanding of indigenous societies? Which of them did indigenous peoples (implicitly) understand?

The kinematics of tools for hunting, thermal energy transfer considerations in shelter, starlight as a mechanism for navigation, the use of tides in transportation, and curiosities about the elemental make-up of the physical world are some possibilities. Physicists have also played a part in impacting indigenous societies, for example, through the harmful effects of nuclear testing, or ignoring them, for example, in climate science innovation.

The knower

As a knower, you will possess knowledge, beliefs and opinions. There is an abundance of knowledge claims in physics. However, the emphasis on evidence and proof makes belief less relevant.

If you see a painting, read a book or look at the news, you will probably formulate some opinion about it. In many subjects that you study, you are actively encouraged to develop opinions and discuss them in class. For example, you could think a painting is beautiful or you could think that it is horrible. Either way is fine, because it is your opinion and you can have whatever opinion you like when it comes to such things.

Can you have an opinion in physics? Is it OK to say that in your opinion Newton was wrong when he said that force was proportional to rate of change momentum and that you think that they are independent? In physics, opinions do not count for much, although they can sometimes be the beginning of the formulation of a testable hypothesis. The final element of the TOK course is the concepts: interpretation, justification, culture, power, perspective, values, truth, certainty, objectivity, evidence, explanation and responsibility. Any one of these concepts could be linked to any element of physics. Some examples follow in this section.

The TOK Exhibition is typically related to one of the themes. You will link three objects under your chosen prompt. You can locate the prompts in the TOK Guide, but many successful students opt to choose three objects that they find interesting before selecting a prompt.

If you visited the Science Museum in London in June 2022, you might have been interested in a sample of rock from the Moon, an invitation to a party from Stephen Hawking to time travelers (that was circulated after the party took place), and the film *Planet Science* projected onto a sphere, which showed representations of planetary surfaces. Notice that these objects are linked to a particular location and time; none of them were created for the TOK Exhibition. Of course, your objects may have nothing to do with physics.

> Next, you might think about the themes. The knower is a theme that can be related to all objects. Perhaps religion comes to mind, because of the differences in how religious communities and scientific communities view the world. Or maybe you opt for technology, because scientific endeavor and technological developments are often intertwined.
>
> Finally, you turn to the prompts. Those relating to evidence, ownership, explanations and certainty might stand out, but all could be used. If you were to opt for 'Why do we seek knowledge?', you might discuss the fascination that humans have had for the night sky, the quest for evidence relating to abstract scientific concepts and the interest that effective communication can engender. Or, in terms of 'Are some things unknowable?', you could raise the limitations of sampling during fieldwork, the importance of falsification in science and the developments of imaging relating to wavelengths and resolution.
>
> The exhibition is to be enjoyed and is an opportunity for you to test your ideas, as it is the 950-word written commentary that is marked. Your task is to convince your reader that you understand how TOK manifests in the world around us. The exhibition is worth one-third of your TOK grade.

'All truths are easy to understand once they are discovered; the point is to discover them.'

Galileo Galilei

▲ Galaxy cluster MS0735.6+7421. The red part of this image is radio, the blue is X-ray and the yellow is visible light. You could never see this.

▲ Can you see a hidden face in this picture? Apparently those with 'physical brains' take a long time. See below for a hint.

Hint: Look between 3 o'clock and 5 o'clock.

Concepts

The knowledge framework allows you to draw comparisons between areas of knowledge. The themes provide lenses for considering the two-way relationships between knowledge and knowers. Finally, we have the 12 concepts, which can crop up in TOK at any time.

Interpretation, justification and culture

In physics, we use the term *law* quite a lot, for example, Newton's laws of motion, the law of conservation of energy and Ohm's law. The laws are generalized descriptive **interpretations** of observations that are used to solve problems and make predictions. If we want to know what height a ball will reach when thrown upward, then we can use the law of conservation of energy to find the answer. When you use a law to solve a problem, you have a solid foundation for **justifying** your solution. If someone were to disagree with your solution, then they are disagreeing not only with you, but with the law (assuming you applied it correctly). Laws sometimes give easy answers to difficult problems. If someone comes to you with a design of a machine that is 100% efficient, you do not need to study the details, because you can simply apply the second law of thermodynamics and say it will not work.

Some of the laws in physics are called universal laws, for example, Newton's universal law of gravitation. A universal law applies to the whole Universe, but it is possibly naïve (or arrogant) to think that we can write laws that apply to the whole Universe, when we can only make measurements from one very small part of it. Today, scientists are more modest in their claims and accept that there are probably parts of the Universe that do not behave in the same way as things in our solar system; the **culture** has changed.

Power, perspective and values

Physics is based on observation and observations are made with our senses. This was certainly true hundreds of years ago but today, although the information finally arrives into our brain via our senses, the observation itself is often done via some instrument. Copernicus experienced issues related to **power**. He had difficulty convincing anyone about his theory that the planets orbited the Sun, because he did not have any convincing observations. He had predicted that Venus would have phases like the Moon, but could not observe this. Galileo used the telescope to observe that Venus did indeed have phases like the Moon. At first, this was not accepted, since people did not trust the telescope and wanted to 'see it with their own eyes'.

A camera operates on the same physical principle as the human eye. Visible light is reflected from an object and focused by a lens onto a screen. It is reasonable to think that a picture is a good record of what we see. Digital technology offers new **perspectives**. It is now possible to recreate pictures from light that we cannot see. Is this seeing? Can we say that we have seen a distant galaxy when we look at a picture constructed from radio waves? Can we say that we have seen the face of a flea when the picture was constructed from the diffraction pattern of electrons?

During your IB physics course, you will have been asked to make observations and devise research questions. Is this easier when you have studied the topic already or when it is something totally new? When you already know what you are supposed to be looking for, it is often easier to get started. However, if you have no preconceived ideas, you might have more chance of spotting something new. These considerations relate to your **values**.

Truth, certainty and objectivity

One of the problems with studying physics is that we all live in the physical world and have all seen how bodies interact. We know that if you drop something, it falls to Earth, and if you push someone on a swing, they will move back and forth. These **true** observations give us a feeling for what is going to happen in other instances. We call this feeling intuition, the ability to sense or know what is going to happen without reasoning. In physics, we create models to help give a reason for what is happening. This all works fine until intuition gives us a different answer to the laws of physics or until we acknowledge the lack of **certainty** in our model. Here is an example.

Consider a metal bar floating in space, where the gravitational field strength is zero. If you apply two forces to the bar as shown in the diagram, what happens? Intuition will probably tell you that the bar will rotate about point A. This is because if you do this yourself, that is what will happen. However, that answer is wrong. Let us now apply Newton's laws of motion to the problem.

Newton's first law states that a body will remain at rest or with uniform motion in a straight line unless acted upon by an unbalanced force. The forces in this example are balanced, so the center of mass of the bar (B) will not move. We can see however that the turning effect (torque) of these forces is not balanced, so the rod will turn. If the rod turns but point B does not, then the rod must turn about B. And that is what happens.

The reason for the difference in these two predictions is that this rod is not in a gravitational field. When we try to do this with a rod on Earth, there are other forces acting. Intuition was wrong and physics was right. The laws of physics can tell you what happens even if you cannot do it or see it yourself. They offer **objectivity**.

Evidence, explanation and responsibility

One of the strengths of modern scientific practice is that every new discovery goes through a rigorous process of peer review. Before a theory is published, it is sent to other scientists working in the same field so that the **evidence** and the **explanations** that follow can be checked. They give feedback to the research team before the theory is published. In this way, mistakes can be spotted and problems ironed out. It also provides the opportunity for other groups working in the same field to think of experiments that could be conducted to prove the theory wrong. One **responsibility** of scientists is to continually look for ways to prove theories wrong so that, when a theory is accepted by the scientific community, one can be sure that it has been rigorously tested.

The face of a flea. Can you really say that this is what it looks like since you can never see it directly?

What happens when the bar experiences these forces?

'The most exciting phrase to hear in science, the one that heralds new discoveries, is not 'Eureka!' (I found it!) but 'That's funny…'

Isaac Asimov

'There are many hypotheses in science which are wrong. That's perfectly alright; they're the aperture to finding out what's right.'

Carl Sagan

Internal Assessment

Internal Assessment in physics consists of one 10-hour scientific investigation of up to 3000 words, which is worth 20% of the final assessment. The complexity of work should be in line with the other practical work done during the course and should therefore have the same level of analysis and a detailed evaluation of results. There are a range of tasks that could be completed for the Internal Assessment.

A hands-on laboratory investigation

A hands-on laboratory investigation is similar to the practical work that you will be doing throughout the course as outlined in the different chapters of this book. However, simply repeating a standard laboratory experiment like measuring the acceleration due to gravity using a range of drop distances and times will *not* be enough to earn full marks. You need to be a bit more inventive to meet every criterion. Maybe you could extend the investigation by measuring the acceleration of different-sized balls to investigate the effect of air resistance, or turn the investigation upside down and measure the upward acceleration of a helium balloon.

Research question

The investigation should start with a research question such as:

> 'What is the relationship between the radius of a ball and its terminal velocity when falling freely through air?'

Research questions in physics are usually of the form 'What is the relationship between x and y?' Try to keep it simple by only considering one research question in your investigation. If you really want to consider more than one aspect, then split the report into two sections.

Variables

The variables should then be clearly stated. For example:

Independent variable (this is the variable you are going to change)

- Radius of the ball.

Dependent variable (this is the variable that changes because of the independent variable you changed)

- Terminal velocity of the ball.

Controlled variables (these are things that could be changed but you are not going to change them)

- Mass and material of ball.
- Air temperature, density and humidity.

Hypothesis

A hypothesis is your prediction of the outcome of your experiment. This should not be a wild guess. Try to use the physics you have learned in class to make a mathematical model of the situation that you are investigating. It does not have to give the correct

The research question might come before or after an introduction. When introducing your work, you might like to include the topic of your investigation, how it links to your course, your source of inspiration, what specific aims you have and what real-world relevance there is. You could also provide some signposts for the reader about what to expect from your report as a whole, for example your methodology and findings. It is a misconception that you need to present your research as though it has been a source of fascination from birth.

A literature review is not required in the physics Internal Assessment, but you should cite any sources that you have used in formulating your hypothesis.

answer but you should apply what you know correctly. Although not essential, having a mathematical model will make the analysis of the data easier. It may also be possible to find the relationship from your data by analyzing the graph.

Method

The method should contain all the details of how you did the experiment including:

- how the independent variable was changed
- how the dependent variable was measured
- how the controlled variables were kept constant.

It is always important to collect sufficient data. This means a wide range of many different values for the independent variable and many repeats of each measurement of the dependent variable. This might be limited by the materials that you have but try to get as much data as possible. If you have different ways of testing your hypothesis, these will strengthen your conclusion.

The majority of reports for practical investigations will also include a list of apparatus, a diagram and a risk assessment. Some will also need details about the context in which the work is situated or ethical considerations.

Data collection

- Data is always organized in a table, which should include units and uncertainties in the headers.
- If your data is coming from video analysis or a 'data logger' graph, then include a sample in the report.
- Take time to think through the procedure you will follow with regard to uncertainties and significant figures.
- Do not forget to comment on how you arrived at the uncertainties in your table.
- Include any relevant observations.

Data processing

Processing of data will first involve finding the mean of your repeated measurements. If you have predicted the relationship between the variables, then you might be able to process the data so that you can plot a straight-line graph. This makes the interpretation easier since you will only have the gradient and intercept to explain.

It is usual to use a spreadsheet to process tables of data but make sure that you explain the calculation performed. Remember that software is only as good as the instructions you provide.

Graphing

As in the other experiments performed throughout the course:
- The graph should be correctly labeled including units.
- The equation of the line should be clearly displayed including the gradient and intercept.
- There should be error bars on the points.
- The uncertainty in the gradient should be found either by plotting steepest and least steep lines or by using the inbuilt function of the software used.

These questions might help you to look deeper still:
What intervals and range did you select for the independent variable?
What resolution of measuring instrument did you choose for the dependent variable?
What constant values did you select for the controlled variables?

As well as your method, a data table and a graph, you should also include some text. This might include a consideration of how you went from data collection to a conclusion, a qualitative description of what the data in your table indicates and a discussion of what graph(s) you plotted and why.

If you cannot linearize the relationship between your variables, then you can plot a curve. However, you should have some idea of which curve to plot. Just because the line touches the points does not mean that you have found the relationship; a polynomial best fit will appear to match almost anything.

Conclusion

The conclusion should be based on the results of your experiment, so look at your graph of results and interpret what you see. This is much easier if you have a linear relationship since you only have to explain the relevance of the gradient and the intercept. Make sure you answer your research question.

If there was some known value that could be calculated from the gradient or intercept, then quote it here (with uncertainties) and compare it to the accepted/expected value.

Try to justify any statements made here by referring to the spread of data along the best fit line and how close the intercept is to the origin (or expected value).

If there is an outlier, you could try leaving it out and plotting the graph again but do not delete outliers to make the line fit. Be honest about your data and what analytical processes you have tried.

Evaluation

When writing an evaluation, think like a detective. You must have evidence for any statements you make. The idea is to explain any deviation of your results from what you expected in terms of what you did in the experiment. You should not say that there were big random errors in the results if the data points lie almost perfectly on the best fit line. Do not blame friction if there is no evidence in your results that friction was a problem.

If your results are not what you expected, do not assume that this is because your experimental technique was in some way flawed. It might be that your original hypothesis was incorrect. Maybe you forgot to include something in your derivation of the formula. You will often find that the theory learned in class does not apply exactly to practical situations.

If your gradient is bigger than expected, try to explain what could have made it bigger. If the intercept is negative, then make sure your explanation would cause a negative intercept.

In the case of outliers, go back to your original data table to see if you can see what caused the point to lie outside the best fit line.

- Do your error bars reflect the spread of data?
- Did you control all of the controlled variables or were there things you forgot to consider?

Discussion

Here, you can discuss possible ways of improving the experimental method or elaborate on deviations from the theory that you might have highlighted in the evaluation. Any improvements must relate to problems that were mentioned in the evaluation. There is no point in repeating measurements to reduce random errors if the points were lying almost exactly on the best fit line.

Explain fully any modifications you might make to the apparatus. Diagrams are often useful here.

If possible, do a small experiment to support observations that you made in the evaluation. Alternatively, it might be better to try a simulation to support your evaluation. A simulation will show you what result the theory predicts so might help to highlight any deviations from the expected.

Spreadsheet analysis and modeling

An investigation does not necessarily have to have any 'actual hands-on' measurements. It could be based entirely on a mathematical model that you develop using a spreadsheet. There are several examples in this book where this has been done.

Modeling waves

A spreadsheet can be set up to model simple harmonic motion. A wave is made up of a series of points oscillating with slightly different phase. By incorporating this into your spreadsheet, it is possible to model a wave, but you should work out the details for yourself. Once you have modeled one wave, you can add two waves together to investigate superposition, standing waves and beats. To animate waves in Excel®, a slider can be added that changes time as it is moved.

The greenhouse effect

Very sophisticated spreadsheets are used to model the greenhouse effect but you can make your own. However, it is not enough to simply copy the formulae from this book; you need to develop your own way of doing it. How might you introduce the effect of the concentration of greenhouse gases, for example?

Radioactive decay

Modeling the exponential decay of a radioactive isotope would be fairly straightforward using a spreadsheet but how about modeling a decay chain where isotope A decays into isotope B, which decays into C, etc.

If you undertake a spreadsheet investigation, you need to be sure that you show how this is connected to the real world. Combining a spreadsheet model with a hands-on practical would be a good way to do this.

Databases

A database is a computer program that enables you to store related data, making it relatively easy to find the connections between different parts. Schools often use databases to connect student information to class lists, rooms, grades and reports. There are many scientific databases that can be used as the basis of an investigation. For example, databases containing the physical properties of materials such as Engineering Toolbox are used by engineers to look up data on building materials and elements. Astronomical data and images are stored in databases such as SIMBAD, which includes a star map. This is not particularly easy to use so you will need to read the user guide and have some idea of what it is you want to find out. A lot of the databases you will find on the Internet are for use by professionals so it might be difficult to understand what the data represents. If you do not understand it, then do not use it!

Simulations

There are many excellent simulations on the Internet but they are programed so that they follow the theoretical models. There would be no point in investigating whether a gas represented by a simulation obeys the gas laws because it is programed to do just that. What would be more interesting would be to build your own simulation using GeoGebra®, Interactive Physics®, Algodoo® or something similar. In GeoGebra®, you can plot graphs from equations so you could set up sliders to vary the pressure, volume and temperature of a gas and plot P–V curves. Using Algodoo® or Interactive Physics®, you could model a gas by putting a large number of perfectly elastic balls into a container. Try compressing the gas and see what happens. By building your own simulations, you have much more control of the underlying mathematical model so will have a better chance of making your own personal input. If you do use a ready-made simulation, make sure that it has many controls. Search 'ripple tank applet' and you will find an example that has many possibilities for some interesting investigations.

Assessment objectives

You will find an example of an Internal Assessment with hints and teacher notes on this page of the eBook.

The objectives assessed in the scientific investigation are:

1. To demonstrate knowledge of:
    a. terminology, facts and concepts
    b. skills, techniques and methodologies.

2. To understand and apply knowledge of:
    a. terminology and concepts
    b. skills, techniques and methodologies.

3. To analyze, evaluate and synthesize:
    a. experimental procedures
    b. primary and secondary data
    c. trends, patterns and predictions.

4. To demonstrate the application of skills necessary to carry out insightful and ethical investigations.

Assessment process

The scientific investigation is internally assessed then externally moderated. This means that your teacher will mark the reports and then a sample from your cohort will be sent to the IB and checked by an expert for how the criteria are applied. If the IB expert does not agree with the teacher's grading, the marks of the whole class will be adjusted, which will bring your school into alignment with the global standard.

Assessment criteria

There are four assessment criteria: research design, data analysis, conclusion and evaluation. Each criterion is marked out of 6, giving a possible total of 24. The investigation as a whole is worth 20% of the final grade. It is worth putting some effort into this as it can make a difference of up to 2 points in your final grade out of 7.

Research design

This criterion assesses the extent to which the student effectively communicates the methodology (purpose and practice) used to address the research question.

Marks	Level descriptor
0	The report does not reach the standard described by the descriptors below.
1–2	<ul><li>The research question is stated without context.</li><li>Methodological considerations associated with collecting data relevant to the research question are stated.</li><li>The description of the methodology for collecting or selecting data lacks the detail to allow for the investigation to be reproduced.</li></ul>
3–4	<ul><li>The research question is outlined within a broad context.</li><li>Methodological considerations associated with collecting relevant and sufficient data to answer the research question are described.</li><li>The description of the methodology for collecting or selecting data allows for the investigation to be reproduced with few ambiguities or omissions.</li></ul>
5–6	<ul><li>The research question is described within a specific and appropriate context.</li><li>Methodological considerations associated with collecting relevant and sufficient data to answer the research question are explained.</li><li>The description of the methodology for collecting or selecting data allows for the investigation to be reproduced.</li></ul>

A good introduction will help to introduce the phenomenon under investigation and put the research question into context. The introduction should not be a story about why you are interested in doing this investigation, but you can say how you came up with the idea and, if relevant, its significance. For example, if you are investigating the efficiency of a wind turbine related to the number of blades, you could say how this is important in the quest for renewable sources of energy; do not pretend that this has fascinated you ever since you got a pinwheel at the age of six on a family holiday to the seaside .

The research question should ideally be of the form: *What is the relationship between the efficiency of a wind turbine and the number of blades?* Note that more open questions are not recommended, such as: *What factors affect the efficiency of a wind turbine?*

You should define any terms that are not obvious (such as *efficiency of a wind turbine*). If you are going to measure another variable (such as *emf generated*), then it is simpler to use this in your research question. If there is a variable that must be kept constant, mention it in the research question: *What is the relationship between the emf generated by a wind turbine and the number of blades for the same wind speed?*

The method should be clearly described. Include all the details (such as *how you made sure the blades were all the same length*), including how you kept controlled variables constant: *The wind speed was kept constant by using the same fan at the same distance.* Use any diagrams and photographs that would be needed by an IB physics peer to understand your set-up. If you use a ruler to measure a length, it should be mentioned, including any details such as its range, resolution and how *it was stuck to the bench to prevent movement*. Mention any safety issues and how you addressed them but do not take this to the extreme; no one has ever been injured by a ruler falling off the table!

You will be graded on whether you collected sufficient data, so make sure you articulate how you achieved this. Five different values of the independent variable are the absolute minimum, so aim to use more.

Data analysis

This criterion assesses the extent to which the student's report provides evidence that the student has recorded, processed and presented the data in ways that are relevant to the research question.

Marks	Level descriptor
0	The report does not reach a standard described by the descriptors below.
1–2	• The recording and processing of the data is communicated but is neither clear nor precise. • The recording and processing of data shows limited evidence of the consideration of uncertainties. • Some processing of data relevant to addressing the research question is carried out but with major omissions, inaccuracies or inconsistencies.
3–4	• The communication of the recording and processing of the data is either clear or precise. • The recording and processing of data shows evidence of a consideration of uncertainties but with some significant omissions or inaccuracies. • The processing of data relevant to addressing the research question is carried out but with some significant omissions, inaccuracies or inconsistencies.
5–6	• The communication of the recording and processing of the data is both clear and precise. • The recording and processing of data shows evidence of an appropriate consideration of uncertainties. • The processing of data relevant to addressing the research question is carried out appropriately and accurately.

Data should be recorded and organized in tables that are likely to be copied from your spreadsheets. If you are patient in learning how to use a spreadsheet properly, then you will be rewarded because of the time saving in calculating uncertainties for tens or even hundreds of rows.

Tables should contain independent and dependent variables. The headers should be correctly formatted and include units and uncertainties. Use $\frac{(\max - \min)}{2}$ to estimate the uncertainty in a range of values, checking that this is larger than the measurement uncertainty from the instrument used. Uncertainties should be given to 1 significant figure, which in turn dictates the number of decimal places in the data (to match the uncertainty).

Process data according to your theoretical model. If you think displacement is proportional to the square of time, then either plot s vs t and draw a parabola or plot s vs t^2 and draw a straight line. As it is typically easiest to interpret a straight line ($y = mx + c$), always linearize if you can. Any curve drawn should be based on your theoretical model; do not just plot the one that fits best. You can plot custom curves using LoggerPro®.

Plot error bars for each point. If finding a gradient, plot steepest and least steep lines to find its uncertainty. This will also enable you to find the uncertainty in the intercept(s).

Conclusion

This criterion assesses the extent to which the student successfully answers their research question with regard to their analysis and the accepted scientific context.

Marks	Level descriptor
0	The report does not reach a standard described by the descriptors below.
1-2	• A conclusion is stated that is relevant to the research question but is not supported by the analysis presented. • The conclusion makes superficial comparison to the accepted scientific context.
3-4	• A conclusion is described that is relevant to the research question but is not fully consistent with the analysis presented. • A conclusion is described that makes some relevant comparison to the accepted scientific context.
5-6	• A conclusion is justified that is relevant to the research question and fully consistent with the analysis presented. • A conclusion is justified through relevant comparison to the accepted scientific context.

The reason for the experiment is to test a model based on some accepted theory: the scientific context. In this example, imagine that you have applied the law of conservation of energy to derive an equation for the height of an arrow projected vertically from a bow.

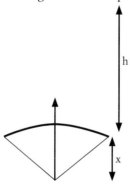

Initial elastic potential energy is equal to final gravitational potential energy:

$$\tfrac{1}{2}kx^2 = mgh$$
$$h = \frac{k}{2mg}x^2$$

According to the scientific context, a graph of h vs x^2 should be a straight line passing through the origin. The gradient should be $\frac{k}{2mg}$. By comparing your graph's equation with what is expected, you are comparing with the accepted scientific context. There are several possible scenarios:

1. Your data is scattered and you cannot get a straight line through your data. Your conclusion might be of the form: 'Due to the large uncertainties, it cannot be concluded that h is proportional to x^2.'

2. The graph is linear, but the gradient does not give $\frac{k}{2mg}$ within the uncertainties of the experiment. Your conclusion might be of the form: 'The graph indicates that the experiment was carried out precisely with small uncertainties. Although the relationship is supported by the data, the value obtained for the gradient suggests that the assumptions made concerning the bow and arrow were incorrect.'

3. The graph is linear and passes close to the origin and the gradient is within the uncertainties equal to $\frac{k}{2mg}$. Your conclusion might be of the form: 'The data supports the theoretical model showing that the assumptions made in deriving the equation (that energy was conserved and the bow obeys Hooke's law) were reasonable.'

4. The graph is a perfect straight line passing through the origin and the gradient is equal to $\frac{k}{2mg}$. This will never happen! If you ever get a perfect straight line, it will be because you have plotted a calculated value of the independent variable against the independent variable by mistake. Remember that you must always quote any value derived from your gradient to being within a range of uncertainty.

Evaluation

This criterion assesses the extent to which the student's report provides evidence of evaluation of the investigation methodology and has suggested improvements.

Marks	Level descriptor
0	The report does not reach a standard described by the descriptors below.
1–2	• The report states generic methodological weaknesses or limitations. • Realistic improvements to the investigation are stated.
3–4	• The report describes specific methodological weaknesses or limitations. • Realistic improvements to the investigation that are relevant to the identified weaknesses or limitations, are described.
5–6	• The report explains the relative impact of specific methodological weaknesses or limitations. • Realistic improvements to the investigation, that are relevant to the identified weaknesses or limitations, are explained.

Anyone can explain what is supposed to happen but only someone with a good understanding of physics can explain what actually happens. In your evaluation, you will need to explain why your data does not exactly agree with your theoretical model. Stating that uncertainties could be reduced by using 'better apparatus' is not sufficient.

If there are only small error bars, then your uncertainties are small so you should not spend time explaining how to reduce the uncertainties. Any problems are likely to be in the model and its assumptions, not the experimental method. If you do have large uncertainties, you should describe in detail how you would minimize them and explain what effect this would have on your data.

This is where you act like a detective. Everything you say should be based on evidence and all the evidence is in your graph. Why is your gradient too small? Why does your line not pass through the origin? Here you could use Algodoo® to help show the effect of changing different variables. It shows promise to say that air resistance would have resulted in the height being lower than expected, but it is better to run the experiment with and without air resistance to show the effect it has on your data.

It might be worth implementing the changes and running a quick version of your experiment again. Do not spend too long on this or it will turn into an Extended Essay and eat into the time you have to learn and consolidate the course content.

External Assessment

Assessment objectives

Physics in the IB Diploma is assessed mostly through two examination papers, which are taken at the end of the course. The assessment objectives tested in the external assessments are:

1. To demonstrate knowledge of terminology, facts, concepts, skills, techniques and methodologies. Question command terms for this objective might include 'draw' and 'state'.

2. To understand and apply knowledge of terminology, concepts, skills, techniques and methodologies. Question command terms for this objective might include 'annotate', 'calculate', 'describe', 'estimate' and 'outline'.

3. To analyze, evaluate and synthesize experimental procedures, primary and secondary data, trends, patterns and predictions. Question command terms for this objective might include 'analyze', 'determine', 'discuss', 'explain', 'predict', 'show', 'sketch' and 'suggest'.

There is an equal weighting in the marks between assessment objectives 1 and 2 combined (50%) and assessment objective 3 (50%).

What to expect

At Standard Level:

- Paper 1 consists of two booklets taken together, which have a total of 45 marks, and is worth 36% of the course. Its duration is 1 hour and 30 minutes. The first booklet, Paper 1A (25 marks), is made up of 25 multiple-choice questions. The second booklet, Paper 1B (20 marks), is made up of questions that will require you to interpret and process data.

- Paper 2 consists of one booklet, which has a total of 55 marks, and is worth 44% of the course. Its duration is 1 hour and 30 minutes. It is made up of short-answer and extended-response questions.

- Only Standard Level material is assessed: all of A.1, A.2, A.3, B.1, B.2, B.3, B.5, C.2, C.4, D.3, E.4 and E.5 and some of C.1, C.3, C.5, D.1, D.2, E.1 and E.3.

At Higher Level:

- Paper 1 consists of two booklets taken together, which have a total of 60 marks, and is worth 36% of the course. Its duration is 2 hours. The first booklet, Paper 1A (40 marks), is made up of 40 multiple-choice questions. The second booklet, Paper 1B (20 marks), is made up of questions that will require you to interpret and process data.

- Paper 2 consists of one booklet, which has a total of 90 marks, and is worth 44% of the course. Its duration is 2 hours and 30 minutes. It is made up of short-answer and extended-response questions.

- Standard Level and additional Higher Level materials are assessed throughout: all of A.1 to E.5.

Irrespective of the level and paper, you will always have access to 5 minutes of reading time, a permissible calculator and a clean copy of the data booklet. Unlike some subjects, you should answer every question as there are no options. You can expect a handful of marks to relate to the nature of science. All IB Diploma physics students receive the remaining 20% of their marks from internal assessment.

Preparing for the examinations

It is likely that, having studied every topic, you will feel more at ease with the course as a whole because of the connections and recurring ideas. You should revise the relevant content by reading, watching videos, using animations and simulations, note-taking, asking questions, discussing with peers and other interactive activities. It is important that you interrogate what you are learning and that you 'tick off' the understandings, Guiding Questions and Linking Questions from the subject guide as you go. You can expect the proportions of marks awarded for each topic to be roughly in line with the recommended course learning hours:

Syllabus content	SL	HL
A. Space, time and motion	25%	23%
B. The particulate nature of matter	22%	18%
C. Wave behavior	15%	16%
D. Fields	17%	21%
E. Nuclear and quantum physics	21%	22%

Once you have memorized the key facts, it is time to practice. (Re)attempt every question in this book and ask your teacher for access to past papers, being sure to put yourself under closed-book, timed conditions where possible. Spend as much time marking your work as you do attempting the problems so that you can appreciate the mindset of your examiner. Use your corrections as the basis for your next round of revision.

During the examinations

Ensure that you read every word and follow every instruction. They all serve a purpose. It is sensible to underline important terms in the question. You should make full use of your calculator and data booklet, and remember to draw diagrams, list variables, structure your calculations and check the units and significant figures. If part (b) of a question requires an answer to part (a), check to see if part (a) provided the response through being 'show that' in format, or make a guess at part (a). Error carried forward marks are applied if your working is clear.

Write your answers within the required spaces. Take care to assign your time appropriately, as later questions may be no more difficult than earlier questions and having a few minutes to spare at the end could enable you to scrape around for extra marks. There is no negative marking and you should attempt all questions (especially multiple-choice) even when you are not sure.

And remember, you might be able to eliminate multiple-choice answers using dimensional analysis or by looking at magnitudes (or directions). You will not always need to perform a calculation when the possible responses are numerical.

After the examinations

You will have to be patient during the wait for results. When you receive them, reflect on how far you have come and the ways in which your study of physics has influenced your future plans. It is possible to request remarks of all externally assessed components and even to re-sit them if things have not gone to plan, but bear in mind that marks can stay the same or go down, and that you might not be as motivated a year later. Your physics teacher and IB coordinator may have some words of advice.

Extended Essay

The Extended Essay is a 4000 word piece of independent research on an IB topic of your choice. Tackling an Extended Essay in physics can be a daunting prospect, but a physics teacher will supervise your research and be on hand to give guidance and help solve any practical problems that you might come across. Your supervisor will also give you an excerpt from the Extended Essay guide, giving information on how to construct the essay, with some specific recommendations for physics.

Choosing physics

Making physics the subject of your Extended Essay is an exciting prospect. You will have the opportunity to conduct an experiment, construct a model, engage with simulations or analyze databases on a topic that you have an interest in. Physics at university can be similar to physics at school, with new material to understand and problems to solve, and you will impress prospective tutors by taking on independent research when you can.

You need to be studying IB Diploma physics to undertake an Extended Essay in physics. Otherwise, it will be tricky to ensure that you are tackling something beyond the scope of the course and of appropriate difficulty. Physics should not be seen as more challenging for an Extended Essay than other subjects. Success comes from having a focused research question, sticking to a structure that resembles research at university level and understanding the criteria.

Choosing a topic

If you have chosen a good topic, then writing an Extended Essay in physics can be quite straightforward. Your supervisor will help you but here are some additional guidelines:

- Do not be too ambitious. Simple ideas often lead to the best essays. Students commonly do not believe that they can write 4000 words on something as simple as a ball of modeling clay being dropped on the floor, but then end up struggling to keep to the word count when they opt for something more complex.
- Make sure your topic is about physics. Avoid anything that overlaps with chemistry or biology and keep well away from metaphysics or bad science.
- Although the essay does not have to be something that has never been done before, it must not be something lifted straight from the syllabus.
- Avoid a purely theoretical essay unless you have specialist knowledge. The essay must include some personal input. This is very difficult if you write about an advanced topic like black holes or superstrings.
- It is best if you can do any experiments you require in your school laboratory under supervision. If you do the experiments at home, keep in contact with your supervisor so your research stays on the right track.
- Choose a topic that interests you as it will be easier to keep motivated.
- Sports offer a wide range of interesting research questions but sometimes it is very difficult to perform experiments. Roberto Carlos' famous free kick is a fascinating topic for an Extended Essay, but not even he would be able to do it every time, let alone with different amounts of spin. If you are keen to do this sort of research, try to think how you can simplify the situation so it can be done in the laboratory and not on the football pitch.
- You must not do anything dangerous or unethical.

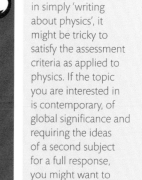

If you have an interest in simply 'writing about physics', it might be tricky to satisfy the assessment criteria as applied to physics. If the topic you are interested in is contemporary, of global significance and requiring the ideas of a second subject for a full response, you might want to consider writing your Extended Essay in world studies instead of physics. Read the world studies chapter in the Extended Essay guide to compare the advice and requirements.

The research question

Once a topic has been decided on, you will have to think of a specific research question. This normally involves some experimental trials and book research. The title of the essay often suggests an inquiry that could be answered in many ways. The research question focuses on the way that you are going to conduct the inquiry. It is important that as you write the essay you refer back to the original topic, title and research question and do not get lost in the details of your experimental method.

From initial ideas to focused research questions

The examples below show how an initial idea can be developed into a focused research question.

Does the depth of a swimming pool affect the maximum speed achieved by a swimmer?

Rather than trying to measure the speed of swimmers in different depth pools, experiments were performed in the physics lab by pulling a floating ball across a ripple tank. This led to the research question 'What is the relationship between the depth of water and the drag experienced by a body moving across the surface?'

Why is it not possible to charge a balloon that is not blown up?

This topic led to the research question 'What is the relationship between the electron affinity of rubber and the amount that the rubber is stretched?' To perform the experiment, a machine was built that could rub different samples of stretched rubber in the same way.

Why does my motorbike lean to the left when I turn the handle bars to the right?

Rather than experimenting on a motorbike, experiments were performed in the lab with a simple gyroscope. The research question was 'How is the rate of precession of a spinning wheel related to the applied torque?'

Performing the practical work

Most Extended Essays will involve some practical work and you should start this as early as possible. If it does not work or you find you do not have the right equipment, you might want to change your research question. You do not have to spend hours and hours on the experiment (although some students do). The whole essay is only supposed to take 40 hours, so keep things in perspective. Make sure the experiments are relevant to the research question and that you consider possible sources of error, as you would in any other piece of practical work. If you get stuck, ask your supervisor for help. Your supervisor cannot do the essay for you but can help you solve problems.

Research

Remember that you are doing research, not a piece of Internal Assessment. This means that you should find out what other people have done and compare their findings with your own. This might be difficult if you have chosen a particularly new topic but most things have been done before. You can try the Internet to hunt for scientific journals, which are often the best source of good information.

Giovanni Braghieri (IB physics student and EE writer) riding his motorbike.

Physics Education and *School Science Review* contain articles written by school science teachers, which might provide examples of how to bridge the gap between the physics experiments in your course and work that can be conducted in a school laboratory.

Writing the essay

Once you have done some research and conducted your experiment(s), you are ready to write the essay. Remember you are trying to answer a research question, so get straight to the point. There is no need to tell a story, such as how this has been your greatest interest since you were a small child. Make a plan of how you want your essay to be structured; the thread running through it should be the research question so do not lose sight of this. Here is a plan for the balloon essay mentioned above:

- Introduction to the topic and research question – how the electron affinity of rubber is connected to the charging of a balloon.
- The theory of charging a balloon and electron affinity.
- Hypothesis based on the theory.
- How you are going to test the hypothesis.
- Details of the experimental technique.
- Results of the experiment.
- Interpretation of results, including evaluation of the method.
- Conclusion – how the results support the hypothesis and the findings of others.
- Why a balloon that is inflated cannot be charged.

What can go wrong

In the real world, things are rarely as simple as they first appear and you might find that your data does not support your original hypothesis. This can be disappointing but should not ruin your chances of writing a good essay. First make sure that you have not made any mistakes in your initial assumptions or analysis of data, then try to think about why the experiment does not match the theory and write this in the conclusion. Do not pretend that it does match if it does not.

Extended Essay assessment

The Extended Essay is marked by experienced physics teachers against five criteria. It is important that you understand the criteria, since if your essay does not satisfy them, it will not score well, irrespective of how much effort you have put in. You can read the full criteria in the official IB Extended Essay guide. Here is some advice on how to get the most from your Extended Essay.

Criterion A: Focus and method: topic, title, research question and methodology (6 marks)

The essay title should be a statement saying what the essay is about, for example, 'A practical investigation into the effect of pool depth on swimming times'. The essay title is fairly self-explanatory. However, the title is not the research question. The research question must be posed as a question that is introduced and put into context in the introduction to the essay. You may have come up with the idea that the drag on a swimmer's body is dependent on the depth of the pool, and that this in turn might affect swimming times. The research question is then: 'What is the relationship between the depth of a swimming pool and the drag experienced by a body moving across the surface?'

It is important that you include consistent and contradictory data collected and give a sense in your report of how your ideas evolved over the course of the project. However, your report need not be chronological. It is possible to be academically honest and write a coherent sequence.

The whole essay should be an attempt to answer the research question, which, once it has been formulated in the introduction, should be followed with some background information to put it into context. You must then apply physical theory to develop a hypothesis. The theory you use does not have to be beyond the syllabus, but its application should be an extension of it. The experimental method should be appropriate given the constraints of a school lab, with attempts made to control all other variables, apart from those mentioned in the research question, and careful measurement of independent and dependent variables. The conclusion should explain how the results answer the research question, or if they do not, you should include an explanation as to why. At the end of the essay, there is room for a discussion, which can bring in other angles. Remember that the examiner will only mark the first 4000 words of the essay.

Criterion B: Knowledge and understanding: context and terminology (6 marks)

When scientists do research, they first find out what has been done already. This is known as 'book research', and you are expected to do the same for your Extended Essay. At university, the main source of information is likely to be scientific papers and journals, but you are likely to find appropriate information in textbooks and on websites. Make sure that websites are published by a reliable source, such as a university, and note the last date on which you accessed the website to verify its contents. Bookmark the pages you have used so that you can add appropriate citations and a bibliography when you come to write the essay. You might like to consider using references management software, such as Zotero®. Do not copy and paste text into your draft unless you intend to cite this quote correctly.

Similar research may have been done by someone else before you. If you find that someone has already derived an equation, for example, for the relationship between pool depth and drag, then it is acceptable to use their equation as long as you quote the source.

This is a physics essay so you must use the language of physics. Make sure you understand what all the words mean and use them in the right context. If you write about something you do not understand, your examiner will be able to tell. There will probably be some mathematics in the essay so make sure you get it right, including the units.

Criterion C: Critical thinking: research, analysis, discussion and evaluation (12 marks)

Critical thinking does not mean that you have to question every piece of knowledge. Instead of wasting time justifying the law of conservation of energy, instead be careful when quoting the results of other experimenters – what did they do to justify their results? Keep an open mind when analyzing and drawing conclusions from your results. Do not try to justify results that clearly do not support your hypothesis. Cast a critical eye over your method and try to work out what its weaknesses were. However, bear in mind that the theory you developed could be wrong, not your method.

Graphs are very useful when testing relationships between two quantities. If the graph is straight, then the relationship is linear, a quadratic relationship gives a parabola, and so on. Simply drawing the closest fitting curve to the data is meaningless unless you have some idea of the relationship you are testing. For example, if your theory predicts that the relationship is sinusoidal, then it would not be useful to plot an automated parabola, even if it was a perfect fit. If you can find the equation of the line, then you will also find the values of any constants. If possible, use known values to test the validity of the graph.

When evaluating your work, think about what specific terms can be used to assess the quality of research in physics. Accuracy, validity, precision, repeatability and reproducibility may get a mention. You should discuss the limitations as well as the strengths.

Criterion D: Presentation: structure and layout (4 marks)

The general structure of the essay should follow a logical progression from introduction to theory, hypothesis, method, results, conclusion, evaluation and discussion. Each of these sections can include subsections with individual subtitles.

Equations should be numbered, axes should be labeled, and all measurements should include uncertainties as well as units. You do not have to include all of your raw data, just enough to make sense. Make sure labels on diagrams and graphs are readable. Learn how to use your word processor, graphics, spreadsheet and graph plotting programs properly. If you use an image from the Internet, make sure you reference it. A title page, contents page, page numbers and a bibliography are all essential.

Criterion E: Engagement: process and research (6 marks)

During the process of writing your essay, you will meet with your supervisor for about four hours in total to reflect on the process. You will be asked to make a written reflection on three occasions: at the start when formulating a working idea and research question, in the middle when data collection is nearing completion and at the end when the essay is finished. The nature of your reflections, together with the final essay, will be used to assess your engagement with the writing process. This means that leaving everything to the night before the final deadline will not score well. The sort of thing you might reflect upon includes any difficulties you faced in finding information, sources you found particularly useful, things you learned along the way and skills you needed to develop, such as using a particular piece of software.

Set yourself realistic deadlines and try to stick to them. Work with your supervisor and respond to the advice that you receive. To ensure timely feedback, keep to the schedule you have agreed upon. Writing an Extended Essay is an opportunity to do some in-depth research into a subject you are interested in, so enjoy the process.